Integrated weed management for sustainable agriculture

It is widely recognised that agriculture is a significant contributor to global warming and climate change. Agriculture needs to reduce its environmental impact and adapt to current climate change whilst still feeding a growing population, i.e. become more 'climate-smart'. Burleigh Dodds Science Publishing is playing its part in achieving this by bringing together key research on making the production of the world's most important crops and livestock products more sustainable.

Based on extensive research, our publications specifically target the challenge of climate-smart agriculture. In this way we are using 'smart publishing' to help achieve climate-smart agriculture.

Burleigh Dodds Science Publishing is an independent and innovative publisher delivering high quality customer-focused agricultural science content in both print and online formats for the academic and research communities. Our aim is to build a foundation of knowledge on which researchers can build to meet the challenge of climate-smart agriculture.

For more information about Burleigh Dodds Science Publishing simply call us on +44 (0) 1223 839365, email info@bdspublishing.com or alternatively please visit our website at www.bdspublishing.com.

Related titles:

Improving organic crop cultivation
Print (ISBN 978-1-78676-184-2); Online (ISBN 978-1-78676-186-6, 978-1-78676-187-3)

Managing soil health for sustainable agriculture Volume 1: Fundamentals
Print (ISBN 978-1-78676-188-0); Online (ISBN 978-1-78676-190-3, 978-1-78676-191-0)

Managing soil health for sustainable agriculture Volume 2: Monitoring and management
Print (ISBN 978-1-78676-192-7); Online (ISBN 978-1-78676-194-1, 978-1-78676-195-8)

Water management for sustainable agriculture
Print (ISBN 978-1-78676-176-7); Online (ISBN 978-1-78676-178-1, 978-1-78676-179-8)

Chapters are available individually from our online bookshop: https://shop.bdspublishing.com

climate·SMART·publishing
IN AGRICULTURAL SCIENCE

BURLEIGH DODDS SERIES IN AGRICULTURAL SCIENCE

NUMBER 42

Integrated weed management for sustainable agriculture

Edited by Emeritus Professor Robert Zimdahl, Colorado State University, USA

burleigh dodds
SCIENCE PUBLISHING

Published by Burleigh Dodds Science Publishing Limited
82 High Street, Sawston, Cambridge CB22 3HJ, UK
www.bdspublishing.com

Burleigh Dodds Science Publishing, 1518 Walnut Street, Suite 900, Philadelphia, PA 19102-3406, USA

First published 2018 by Burleigh Dodds Science Publishing Limited
© Burleigh Dodds Science Publishing, 2018, except the following: Chapters 9 and 17 were prepared by U.S. Department of Agriculture employees as part of their official duties and are therefore in the public domain. Chapter 18 remains the copyright of the author. All rights reserved.

Library of Congress Control Number: 2017956484

British Library Cataloguing in Publication Data
A catalogue record for this book is available from the British Library

ISBN 978-1-78676-164-4 (print)
ISBN 978-1-78676-166-8 (online)
ISBN 978-1-78676-167-5 (online)
ISSN 2059-6936 (print)
ISSN 2059-6944 (online)

Typeset by Deanta Global Publishing Services, Chennai, India
Printed by Lightning Source

Contents

Series list

Preface

Weeds have received a lot of attention, but they have never been respected or understood well. The fact that many people earn a living and serve society by working to control and manage them is often greeted with amusement if not outright laughter. Even scientific colleagues who work in other esoteric disciplines find it hard to believe that another group of scientists could be concerned exclusively with what is perceived to be as mundane and ordinary as weeds.

No agricultural enterprise or part of our environment is immune to the detrimental effects of weeds. They have interfered with human endeavors for a long time. In much of the world weeds are controlled by hand or with a hoe. A person with a hoe may be as close as we can come to a universal symbol for the farmer, even though most farmers in developed countries no longer weed with, or even use, hoes. For many, the hoe and the weeding done with it, symbolize the practice of agriculture. The battle to control weeds, done by people with hoes, is the farmer's primary task in much of the world.

Weed science is vegetation management — the employment of many techniques to manage plant populations in an area. This includes dandelions in turf, poisonous plants on rangeland, and weeds in soybeans. Weed scientists attempt to modify the environment against natural evolutionary trends.

Weed science is not a panacea for the world's agricultural problems. The problems are too complex for any simple solution and students should be suspicious of those who propose simple solutions to complex problems. In fact, the hope should be not to solve but to diminish, not to cure but to alleviate. The work of the weed scientist is fundamental to solving problems of production agriculture in our world. Weeds have achieved respect among farmers who deal with them every year in each crop. Weeds and weed scientists have achieved respect and credibility in academia and the business community. The world's weed scientists are and will continue to be in the forefront of efforts to feed the world's people.

In weed science movements occur, directions change, and progress results. Some movements are called bandwagons with words and phrases that define and identify them. Each movement makes its contribution to the parade of ideas and contributes to the general cacophony of competing ideas, which, one hopes will yield a harmonious new paradigm. Some ideas assume a position at the head of the line. Integrated weed management (IWM) has assumed a position of centrality and leadership. The concept of integrated pest management (IPM), particularly in entomology, can be traced to the late 1800s when ecology was recognized as an essential foundation for scientific plant protection. If integrated IWM systems are to succeed, changes in weed management systems will be required. The direction and scope of changes will determine their success.

There is a risk that integrated weed management may only be another bandwagon, but it makes so much sense that it is likely to endure. It is not perfect but it is better than anything else we have. The evidence in this book is sufficient to demonstrate that weed management systems for crops are incomplete. They are developing, but research gaps exist. Present weed management is dependent on herbicides. Integrated weed management systems that are sustainable over time will emphasize putting components of weed control together. Even casual observation of the presently dominant capital, energy, and chemical dependent agriculture demands answers to questions about the system's sustainability. In view of the rapid appearance of herbicide resistant weeds, continued dependence on herbicides, and emphasis on short-term solutions, does not seem sustainable. The system that now dominates US and other industrialized, developed country agriculture emphasizes production in monocrop plant communities. Many argue that this kind of agriculture cannot achieve what all regard as a proper goal: sustainability. Weed science has been focused on control technology rather than understanding why and how weeds compete so well. Integrated weed management is widely promoted by weed scientists, but it is not widely practiced.

It is likely that successful integrated weed management systems will have to be developed within the opportunities and constraints of agricultural industrialization. Industrialization, is a process whereby agricultural production is structured under the pressure of increasing levels of capital and technology that allow management systems to integrate each step in the economic process to maximize efficiency of capital, labor, and technology. An inevitable question is whether this process is compatible with, and capable of achieving a sustainable agriculture. Some believe the answer is yes, others believe that the industrializing forces of consumer desires and demands, prescription agricultural products, molecular biology, and the changing nature of farming combine to make industrialization inevitable. Many argue that modern industrial agriculture built on and dependent on scientific knowledge is the only way to feed the 9 billion people expected by 2050 and the possible 11 billion by 2100. They argue that if the dominance of energy, chemical and capital dependent agriculture is not drastically changed, the 9 billion will not be fed, because the system is not sustainable. The result of the debate will affect weed science.

Integrated weed management should not limit its focus to weed control. To be successful, the focus must be the total vegetation complex or better, habitat management rather than weed control in a year in a crop. Perhaps it is most correct to say that industrialization should, although it may not, change the scale of concern. Sustainable integrated weed management systems should extend concern to environmental quality and future generations. These are large scale concerns. Small scale concerns such as how to control weeds in a crop in a year have dominated but future agricultural systems will require major changes. Environmental concerns demand large scale thought. Small scale thought suffices for individual concerns. Large thoughts are needed for large systems. Everything needs to be integrated to have a complete crop management system. It won't be easy to do. It is necessary.

Introduction

Since settled agriculture began weeds have been and remain a significant obstacle to increased yields and to feeding a growing world population. Without effective weed management productive, profitable agriculture is not possible. Weed control has relied heavily on the use of herbicides which account for the majority of pesticide use in countries such as the US. However, herbicides suffer from a number of disadvantages, including environmental effects, effects on other species, residues in food and the environment. Weed scientists are aware of and dealing with these concerns as well as herbicide resistance among ever more weed species and the effects of invasive species.

These problems are being addressed by the development of integrated weed management (IWM) systems which include herbicides as part of a broader array of cultural, mechanical and biological methods of control. The chapters in this book review research on IWM directed toward developing sustainable methods of weed management. The volume summarises the latest research on the principles and methods of IWM. Chapters also assess the current challenges facing the use of herbicides, and provide a detailed review of the range of cultural, physical and biological methods of control available for IWM.

Part 1 Weeds

The focus of the first part of the volume is on the ecology of weeds. Chapter 1 addresses the relationship between weed ecology and the population dynamics of weeds, exploring the reasons for abundance of weeds and the effect of weed distribution on overall populations. The chapter examines the 'target transitions' approach to weed control, a technique based on targeting weeds at key life stages. It also examines the place of weeds within on-farm ecosystem communities and agroecosystems, and includes a detailed case study on efforts to mitigate the invasive potential of a bioenergy crop species.

Chapter 2 builds on the themes of Chapter 1 by focussing on weed-plant interactions. Crops or desired plant species co-occur with undesired species which are then classed as weeds. This human-imposed classification is based on the perception that there is an interaction that results in some negative effect of the weed on the crop or desired species. Chapter 2 offers an evolutionary perspective on crop-weed interactions and examines the nature of shared resource pools between desired crops and weeds. The chapter addresses the effects of direct competition between weeds and crops for resources, the indirect effects of competition and the spatial and temporal dynamics of crop-weed interaction.

Complementing the themes of Chapter 2, Chapter 3 concentrates on the nature and effects of invasive weed species. An invasive weed exhibits a tendency to spread rapidly and occupy new niches. The chapter describes ten examples of situations in which invasive weeds directly affect agriculture. The chapter also examines the indirect effects of invasive weeds, and discusses how climate change and globalization interact to promote invasions. The chapter explores the potential contribution of IWM to managing and controlling weed invasions, describing the invasion process and its economic effects on agricultural commodities.

Part 2 IWM principles

The focus of the second part of the volume is on IWM principles, including surveillance, risk assessment and planning an IWM programme. The focus of Chapter 4 is on key issues and challenges in the field of IWM. In order to intensify agricultural productivity while at the same time enhancing ecosystem services, it is necessary to evaluate carefully how current weed management technologies are deployed, including herbicides and herbicide resistant crops. Herbicide chemistries and herbicide resistant crops have provided excellent technologies that have resulted in significant changes to the way weeds can be controlled. Chapter 4 highlights several key components that must form the basis for an effective IWM strategy, including tillage, the importance of understanding weed emergence relative to the crop, critical periods for weed control, crop morphology, row width, nutrient management and crop rotation.

Chapter 5 complements Chapter 4 by discussing ethical issues in integrated weed management. Without an appropriate ethical framework, research runs the risk of pursuing too narrow a focus and thus the wrong goals. As the chapter points out, agriculturalists must see agriculture in its many forms — productive, scientific, environmental, economic, social, political, and moral. It is not sufficient to justify all management activities on the basis of increased production. Other criteria, many with a clear moral foundation, should be included.

Chapter 6 develops the themes of Chapter 5 by examining surveillance and monitoring of weed populations. To implement IWM more effectively, it is necessary to determine the temporal and spatial distribution of weed populations in a field. Weed species tend to be patchy and this influences the ability to calculate average weed densities when conducting a survey. The chapter reviews current and evolving practices for scouting and mapping weed populations both during and across growing seasons. It considers the use of scouts on the ground, UAVs with cameras flying over the fields, and advanced software and computer-based tools to detect, identify, and record weed species. The chapter also discusses the use of regional and global scales to understand changes in the occurrence of herbicide-resistant or invasive weed populations, and includes case studies on how research has been used to improve practice.

Part 3 Using herbicides in IWM

The theme of the third part of the volume is on the role of herbicides in IWM. The focus of Chapter 7 is the challenge of site-specific weed management. Weeds vary in species and density across fields, but uniform management is typical. Chapter 7 reviews the definition and underpinnings of site-specific weed management, and discusses how information about weed spatial and temporal variability can be used to determine if weed management strategies should be varied by location. Building on Chapter 6, the chapter considers how data about weed distribution can be collected using satellites, aerial platforms, and unmanned aerial vehicles (UAVs), and then verified by scouting. The chapter reviews the advantages of site-specific weed management, as well as the major factors which stand in the way of its adoption.

Chapter 8 complements the themes of the preceding chapter by concentrating on the assessment of herbicides and minimisation of their environmental effects. Herbicides are widely used to control weeds but they can have other effects on the environment. Herbicides can move from the site of application through spray drift, volatilization from surfaces, surface runoff or leaching to groundwater. Whether and how far a herbicide will move depends on the physical and chemical properties of the herbicide, the style of application, environmental conditions at the time of and after application, site topography, how tightly the herbicide is bound to soil components, and how quickly the herbicide is degraded. Environmental effects of herbicides include damage to sensitive plants in the environment, damage to aquatic organisms and alterations in microbial populations. Chapter 8 examines the sources and fate of herbicides in the environment, the different types of environmental effects herbicides may have, and the challenge of managing the environmental effects of herbicides.

Chapter 9 switches the focus from herbicides themselves to address trends in the development of herbicide-resistant weeds. Since the mid-1940s, herbicides have been the most cost effective and efficient method of weed control for agronomic crops. Today, herbicide-resistant weeds, in combination with a decline in industry discovery programs and a cessation in discovery of new herbicide sites of action, threaten the continued utility of herbicides. Weeds have evolved resistance to 160 different herbicide active ingredients (23 of the 26 known herbicide sites of action) in 86 crops and in 66 countries. Chapter 9 reviews the various kinds of herbicide-resistance, and then considers resistant weeds by site of action, crop, region and weed family. It considers the available strategies for managing herbicide-resistant weeds, but concludes that although herbicides are likely to remain the backbone of agronomic weed control for the next 30 years, their utility will steadily decline, and we need to begin working on new weed control technologies that will eventually replace herbicides.

Part 4 Cultural and physical methods for weed control

Chapter 10 reviews the development and use of crops resistant to herbicides such as glyphosate, glufosinate, imidazolinone (IMI) and sulfonylurea, as well as the development of multiple herbicide-resistant (HR), stacked-trait crops. Prudent use of HR crops potentially diversifies weed control by enabling use of herbicide tank mixtures, herbicide rotations, or sequential herbicide programs. Instead, as the chapter points out, the simplicity and convenience of glyphosate-based cropping systems using glyphosate-resistant (GR) crops has been over-exploited, with growers often relying on glyphosate only for weed control in GR corn, soybean, and cotton, for example. Over-reliance on HR crop technology over the past two decades, has led to rapid evolution of HR weeds because of massive selection pressure. As the chapter points out, HR crop technology alone cannot provide total weed control. HR crops must be integrated with other weed control tactics. It is best regarded as supplementary to other weed control methods that increase the diversity of weed control tactics. This highlights the need for IWM, a holistic approach that integrates different methods of weed control to manage weeds and maintain crop yields. Integration of HR crop technology with cultural, mechanical, chemical and biological tactics is critical in the management of herbicide resistance.

Chapter 11 develops the theme of non-herbicide-based weed control by examining cultural techniques for managing weeds. Widespread problems with herbicide-resistant weeds, environmental contamination by herbicides, and soil degradation due to excessive cultivation have led to an increasing need for a wide array of cultural techniques to reduce weed population densities, biomass production, and competition against crops. Chapter 11 reviews cultural techniques whose efficacy has been demonstrated in particular farming systems. These include increasing crop population density; increasing crop spatial uniformity; altering planting date; transplanting; the choice of highly competitive and allelopathic cultivars; mulching; and soil fertility and moisture management. The chapter shows how, when used in particular combinations, the cumulative effects of cultural tactics may be substantial and can lessen the burden of crop protection placed on chemical and mechanical controls.

Chapter 12 addresses another non-herbicide based method of weed control, the use of crop rotations and cover crops to manage weeds. Crop rotation has been known for many years as an effective strategy for controlling weeds because it has a disruptive effect on weed populations. Cover crops are important additions to cash crop rotations because they suppress weeds during rotational periods when crops are absent and provide ecosystem services that enhance soil quality and fertility. The chapter describes current research on crop phenological diversity and management disturbance diversity, before suggesting new analytical frameworks for assessing the multifunctional properties and discussing the overall sustainability of cover crops and crop rotations.

Chapter 13 moves the focus from the efficacy of crop rotations against weeds to developments in physical weed control, examining the effects of tillage on weed populations and offering an overview of the methods of physical weed control. The chapter examines the tools for physical weed control and the effect of soil conditions on the effectiveness of these approaches. Addressing in particular the issue of weed-crop selectivity, the chapter examines some of the fundamental problems associated with cultivation and the challenges of weed control. It looks, for example, at how to achieve effective combinations of intra-row weeding tools such as torsion, finger and tine weeders. As the chapter shows, recent advances in GPS and camera-based guidance systems permit increasingly precise, close-to-the-row tool adjustment, even for slow-growing, direct seeded crops.

Continuing the theme of physical methods of weed control, Chapter 14 homes in on techniques of flame weeding. Flaming as a vegetation control method began in the mid-1800s. It is based on utilizing heat for plant control, and has the potential to be used effectively for at least six agronomic crops (field corn, sweet corn, popcorn, sorghum, soybean, and sunflower) when conducted properly at the most tolerant crop growth stage. There has been increasing interest in integrating flame weeding with conventional cropping systems, and in locations where herbicide use is undesirable, such as in cities, parks, and other urban areas. The chapter reviews flame weeding requirements, the mechanism by which the technique reduces weeds, and the potential uses of flame weeding. The chapter also consider its advantages and disadvantages, including its potential environmental effects.

Shifting the focus of the volume to the effect of soil on weed control, Chapter 15 examines the potential of soil solarisation as a sustainable method for weed management. Solar heating of soils involves heating moist and mulched soil (with a transparent polyethylene film) for several weeks. The advantages of the technique include its nonchemical nature and its effective use in a wide range of agricultural areas worldwide. The chapter reviews the use of solarization in sustainable weed management, covering its mode of action, its effects on weeds, soil nutrients and pesticides, and the benefits and limitations of this strategy. The plastic mulching technology required for solarisation is also discussed, along with the significance of the technique for integrated pest management.

Chapter 16 continues the theme of non-chemical techniques of weed control by focussing on weed management in organic crop cultivation. Managing weeds in organic production systems is critical to the economic success of organic farmers, as well as long-term ecological sustainability. Problems with weeds are a major reason why organic operations fail, or never get started. The chapter provides an overview of the range of tools and tactics that can be used to contend with weeds in organic systems and describes the integration of several tools and tactics. The chapter presents several organic farmer case studies to illustrate different types of weed management plans, and looks ahead to future trends in scientific research that will help organic farmers manage weeds while conserving and building soil resources.

Part 5 Biological methods for weed control

The fifth and final part of the volume surveys the available biological techniques for weed control. Chapter 17 examines the use of allelopathy and competitive crop cultivars for weed suppression in cereal crops. Due to the rise of herbicide resistance, diverse weed management tools are required to ensure sustainable weed control. The chapter focuses on competitive cereal crops and cultural strategies for weed management, including the use of weed-suppressive cultivars, post-harvest crop residues and cover crops for managing the weed seed bank and eventual weed suppression. It also addresses factors influencing the effect of allelopathy on weeds, including soil and environmental conditions, which limit or intensify the efficacy of allelochemicals. The chapter reviews the response of some weeds to secondary metabolites released by living cereal crops and/or crop residues (selectivity). The chapter recommends future research areas, aiming to address the knowledge gap regarding the fate of these compounds in the environment and their role in important physiological processes in both plants and microbes in the soil rhizosphere. Case studies are provided on the production of benzoxazinoids in cereal crops and the use of competitive cereal cultivars as a tool in integrated weed management.

Following on from the themes of Chapter 17, Chapter 18 offers an overview of bio-herbicides. Chemical control methods for weeds are widespread, but there are many invasive species for which these are not economically feasible. In addition, there are social, economic and political drivers towards reducing the overall use of pesticides. The chapter considers bioherbicides as an alternative method of weed control. It reviews the use of products based on natural compounds derived from plants or microbes, the classical approach to microbial bioherbicide application, and the use of an inundative approach which applies an endemic pathogen applied in much greater quantity than would be found naturally. The chapter discusses the ways in which bioherbicides can be integrated into weed management programs and the institutional changes needed for biological control adoption.

Complementing Chapter 18, Chapter 19 focusses on the use of microorganisms in integrated weed management. Biological control of weeds by fungal pathogens, bacteria and viruses has been studied for more than three decades, with the aim of suppressing or reducing the weed population below an ecological or economic threshold. The chapter describes the role of biopesticides in weed control, historical accomplishments in biological weed control and recently registered pathogens. The chapter discusses new discoveries currently under development, target weed control, and the role of screening and fermentation technologies, as well as looking ahead to future developments in this area.

Continuing the theme of microorganisms in more detail, Chapter 20 specifically explores the use of bacteria in integrated weed management. Annual grass weeds are increasing as a dominant weed species in the western United States, Canada and Mexico. Downy brome, one of the most widespread, invasive annual grass weeds, negatively affects cereal yields, reduces forage quality in grazing lands, degrades rangelands, and increases the fire frequency of western lands. Based on case studies, the chapter reviews how naturally-occurring bacteria can be screened to find those that suppress downy brome but do not harm native plants and crops. The chapter describes how one such bacterial strain, *Pseudomonas fluorescens* strain ACK55, was identified as able to reduce downy brome root formation, root growth, and tiller initiation. The chapter discusses long-term field trials in the western US, in which application of the bacteria resulted in almost complete suppression of downy brome for three to five years after one application, when desirable plants were present.

The final chapter in the volume, Chapter 21, moves discussion from microorganisms to the use of insects in integrated weed management. Seed predation by insects is a potentially promising approach to the regulation of weeds that could offset herbicide use as part of integrated weed management. Using the example of carabid beetles, as the most intensively studied grouping of insect weed seed predators, the chapter describes the current state of knowledge in this subject area and highlights future research trends. The chapter examines the interaction between weeds and predator communities and assesses how fields and landscapes can be managed to enhance weed seed predation. The chapter looks at the level of weed regulation that can realistically be expected from this approach, and provides a detailed case study from the UK.

Part 1

Weeds

Weed ecology and population dynamics

Adam S. Davis, USDA-ARS, USA

'All happy families resemble each other, but each unhappy family is unhappy in its own way...'

L. Tolstoy, *Anna Karenina*

1 Introduction

Clean production fields look remarkably homogeneous, as anyone who remembers the halcyon days of herbicide-resistant crops in the late 1990s can attest. A scouting trip through the northern US Corn Belt during that time would have revealed row after uniform row of maize and soya beans with barely a weed in sight. Twenty years later, the view is much changed. Completely weed-free grain fields are the exception, not the norm. Driving a transect east from central Illinois, through Indiana and into Ohio this past summer, I saw weeds in nearly every field I passed, and these infestations were true to Tolstoy's allegory in their unending variety. Some fields held many weeds of a single species spread throughout the field in a diffuse, consistent pattern. Other fields showed tight patches of multiple weed species. Novel weeds invaded from the field margins to others. Understanding the sources of diversity of weeds in agricultural fields is central to developing successful long-term weed management approaches. In an era where herbicide resistance has increasingly become common, this understanding

http://dx.doi.org/10.19103/AS.2017.0025.01

is as critical for 'conventional' growers with herbicide-dependent weed management programmes as it is for growers who, for a variety of reasons, have relied less or not at all on herbicides.

Ecology offers a robust scientific framework for understanding, and acting upon, the variety and complexity of agricultural weeds (Liebman et al. 2001). Weed scientists and managers are, in effect, applied ecologists even though they may not recognize it. Agricultural weed problems are a function of local conditions, including the abiotic and biotic dimensions of the growth environment. Using weed ecology to unravel the causes of a particular problem, manage it and prevent its reoccurrence involves scientific knowledge at three nested levels of hierarchical organization (Booth et al. 2003): agroecosystems, communities and populations (Fig. 1). At each level of organization, there are emergent properties that result from the characteristics and processes of the level below. For example, *carrying capacity* is an agroecosystem-level emergent property that results from resource partitioning at the community level interacting with the specific environmental context (spatiotemporal configuration, resource availability and disturbance regime) of a particular agroecosystem. These emergent properties form ecological benchmarks for designing and assessing weed management practices.

The goal of this chapter is to provide an overview of weed ecology as it relates to informed opportunities for prevention and management of weed problems in agricultural systems. These management opportunities and approaches are the primary subject of the remaining chapters of this book. In three sections describing the major levels of hierarchical scale in ecology, I will elucidate how scale-relevant properties can be used to inform the design of weed-suppressive cropping systems and to guide integrated weed management (IWM). Ecophysiological processes will be woven into each section, rather than treating them as an independent level of organization. A case study demonstrating how such ecological concepts may be applied will provide readers with additional sources of information for researching these topics in greater depth with the aim of using them in applied weed management systems.

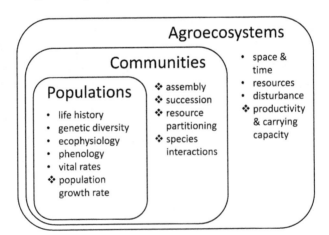

Figure 1 Populations, communities and agroecosystems form nested levels of ecological organization. Diamonds indicate key emergent properties at each level of organization, forming critical management targets.

2 Populations: abundance

Populations consist of individuals of a given species living together in a local setting. Population ecology addresses patterns of species abundance and distribution and the processes that result in these patterns. It also, from the perspective of weed managers, offers the most direct link between weed management tactics and outcomes. This section will concentrate on species abundance, while Section 3 focuses on distribution. Section 4 will then present a population dynamics framework for identifying weed life stages that are particularly influential points of control in the weed life cycle.

Population dynamics, or *demography*, is a branch of ecology concerned with changes in the abundance of populations over time (Cousens and Mortimer 1995). The basic organizational unit of demography is the *life cycle* (Mohler 2001): how a single cohort of organisms grows, survives and reproduces. Plant life cycles are divided into three basic types, or *life histories* (Fig. 2): annual (life cycle is completed in a single year; reproduction is from seed only), biennial (life cycle takes two years to complete; the first year's growth produces a non-flowering rosette, which produces a flowering adult in the second year of growth; reproduction is from seed only) and perennial (rosettes and adult plants may both survive more than one year; reproduction may be sexual or vegetative). Life history and mating schedule have many variations (Harper 1977; Silvertown and Charlesworth 2001). This chapter focuses on the three mentioned above.

For any population to persist, three major processes must be completed: (1) new individuals must enter the population (*recruitment*), (2) individuals must reach maturity (*survival*) and (3) mature individuals must create offspring (*reproduction*) (Cousens and Mortimer 1995). These *vital rates* act together to determine the *population growth rate* (λ), an important emergent property that provides a management benchmark at the population level of organization. The vital rates of a weed species in a particular time and

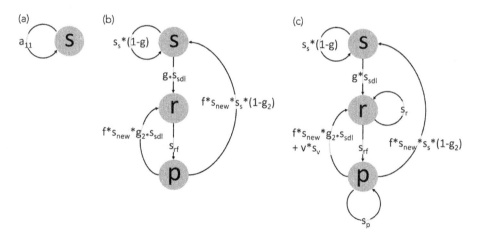

Figure 2 Loop diagrams representing the life stage transitions of weeds with (a) annual, (b) biennial and (c) perennial life histories. Explanation of abbreviations: $a_{11} = s_s*(1-g) + s_s*g*s_{sdl}*f*s_{new}$; f = fecundity; g = germination rate of seeds in soil seedbank; g_2 = germination rate of newly produced seeds; s_{new} = survival of newly produced seeds; s_p = survival of adult plant; s_{rf} = rosette survival; s_s = annual seed survival; s_{sdl} = seedling survival to reproductive maturity; s_v = survival of vegetative propagules; v = production of vegetative propagules. Figure re-drawn from Davis (2006), and used with permission of Allen Press.

place are fundamentally dependent on the interaction of its physiological characteristics, the degree of genetic diversity in the population and the disturbance and resource regime of the growth environment (Lindquist et al. 1995; Forcella et al. 1997; Freckleton and Watkinson 1998; Davis et al. 2005; Averill et al. 2011; Werle et al. 2014; Schwartz et al. 2016).

The process of recruitment depends on the presence of seeds or vegetative propagules in the soil. This subsection focuses on recruitment dynamics from soil seedbanks because of the abundance of scientific information on seeds and the paucity of data on the dynamics of vegetative propagules. One of the common characteristics of weeds is that their newly shed seeds often have some level of primary dormancy (Baskin and Baskin 2006). Seed dormancy links germination to environmental conditions that promote successful seedling establishment. This bet-hedging mechanism enables weed populations to sample environments through time to maximize fitness. The mechanisms of dormancy are complex, involving variation in numerous seed traits at the level of species and genotype, resulting in several distinct dormancy types (Baskin and Baskin 2001). Understanding which weed species has a given dormancy mechanism, and how this interacts with the local abiotic and management environment, is one of the keys to long-term depletion of weed seedbanks.

A new synthesis of the scientific literature on soil seedbanks (Long et al. 2015) aims to explain interspecific and spatiotemporal variation in seed persistence through what the authors call the *resistance-exposure* model of seedbank dynamics. Seeds may exit the seedbank through recruitment, decay, fatal germination, predation and ageing. Within a given seedbank population, buried seeds exhibit a range of resistance to exiting the soil seedbank, on the basis of both species-specific seed traits and abiotic features of the burial site. Weed seeds possess a wide range of physical, biological and chemical defences against predation and decay (Dalling et al. 2011; Davis et al. 2016) that are responsible for much of the observed variation in seedbank persistence in experimental populations. The precise location of seeds in the soil seedbank matters greatly, however, in determining the type and intensity of environmental exposure. Seeds at the soil surface are much more likely to be consumed by granivores than seeds that are buried (Westerman et al. 2006; Westerman et al. 2009); however, successful seedling recruitment (the process whereby seeds germinate and give rise to seedlings that become part of the above-ground plant population) declines rapidly with burial depth (Benvenuti et al. 2001), creating a trade-off between seed protection and recruitment.

Over time, for seeds that remain viable, site-specific variation in exposure to soil moisture, heat, gases and light determines the degree of seedbank depletion through germination by determining seed dormancy status (Benech-Arnold et al. 2000; Long et al. 2015). Just as seed dormancy type and depth are strongly species specific, the breaking of seed dormancy is a complex ecophysiological process that is grounded in seed physiology and seed traits of individual weed species (Baskin and Baskin 2001). For example, spherical seeds or seeds with thick seed coats are more likely to require greater exposure to dormancy-breaking environmental signals than oblong seeds or seeds with thinner seed coats (Gardarin and Colbach 2015). Successful recruitment depends upon newly germinated seedlings being able to reach the soil surface and transition to autotrophy before their seedling reserves run out, resulting in a strong positive relationship between seed size and maximum recruitment depth (Benvenuti et al. 2001). Therefore, to avoid inappropriate germination, seeds must be able to physiologically detect the burial depth at which they reside in the soil. At fine spatial scales, this is accomplished through sensing the degree of fluctuation in soil aqueous, gaseous, light and thermal environment (Buhler 1997; Benech-Arnold et al. 2000; Dekker and Hargrove 2002; Leon et al. 2006; Gardarin and Colbach 2015). Such responses can be exploited in a management tactic known as 'stale seedbed' formation to elicit seedling emergence flushes in response to soil disturbance (Schutte et al. 2014). At regional scales, the phenology of weed seedling emergence may follow clear latitudinal gradients linked to clines in soil temperature and moisture (Forcella et al. 1997), but there is also substantial site-specific variation that departs from these gradients in response to local differences/variation in soil abiotic conditions (Davis et al. 2013; Clay et al. 2014).

In arable systems, weed management efficacy makes the largest contribution to variation in weed seedling survival to reproductive maturity, growth and reproduction (Davis et al. 2004; Freckleton et al. 2008). Elucidating the mechanisms by which this happens comprises the core content of the rest of the chapters in this book. In the absence of control measures, weed seedling survival is largely dependent on local population densities, with negative feedbacks to seedling survival due to intra- and interspecific competition the main cause of seedling mortality (Harrison 1990; Fausey and Renner 1997; Harrison et al. 2001; Davis and Williams 2007). Although per capita fecundity also follows negative density-dependent limitations, population-level reproductive output tends to remain fairly stable across plant population densities (Lindquist et al. 1995; Bensch et al. 2003). Spatiotemporal variation in local abiotic environments forms a secondary filter on seedling survival and fecundity at field to regional scales (Wortman et al. 2012).

3 Populations: weed distribution

Weed managers must pay attention to both the spatial and temporal distributions of weed populations to develop species-appropriate management strategies. Space and time in agroecosystems can be divided into discrete scales (space: patch, field, farm, region, continent; time: weed life stage, season, growing season or year, decade), but can also be measured continuously within or across levels of scale. Understanding which processes are relevant to weed distribution and abundance at a particular scale will aid determination of what management techniques are most likely to reduce weed populations at that scale (Cousens and Mortimer 1995; Booth et al. 2003; Cousens et al. 2008).

Weeds from the three major life history classes are present all over the world (Holm et al. 1997). The weeds that dominate in a field are determined by the schedule of disturbances and resource availability, as well as the crop species, within the agroecosystem. Agronomic operations create soil disturbances, nutrient pulses and an undisturbed growth period tailored to maximize the growth of a primary crop. The presently dominant weeds are thought to have evolved in environments with regular soil disturbance and nutrient flushes, such as riparian areas and desert washes (Sauer 1957; Baker 1974), rendering them pre-adapted to colonization and exploitation of the disturbances and nutrient pulses associated with particular agricultural systems. Successful weeds in any given crop phase will, therefore, be those whose life history and *phenology* (the timing of developmental stages) are best suited to exploiting *safe sites* for recruitment, survival (Harper et al. 1965; Eriksson and Ehrlen 1992) and reproduction created by the crop's growth environment. Thus, summer annual weed species (those that germinate in the spring and produce new seed in late summer or early fall) predominate in the maize–soya bean production systems of the upper Midwest US (Werle et al. 2014), winter annual weeds tend to be associated with winter annual small grain systems of northern Europe (Storkey et al. 2015) and perennial weeds tend to become dominant in cropping systems with longer intervals between soil disturbance events (Bond and Turner 2006a;b). At the level of a patch within a field, dominant weed species will depend not only on the field-scale context but also on microsite properties related to soil heterogeneity and landscape position (Mortensen and Dieleman 1998; Dieleman et al. 2000). For perennial weeds, such as Canada thistle (*Cirsium arvense*) and field bindweed (*Convolvulus arvensis*), the above environmental factors can influence patch success, but the arrival of vegetative propagules is critical in patch formation (Håkansson 1982).

Movement of weed propagules among soil depths within a patch, patches within a field, fields within a farm, farms within a region, regions within continents and among continents all depend upon scale-specific *dispersal* processes (Mohler 2001; Benvenuti 2007; Cousens et al. 2008). At each level of the spatiotemporal scale, different types of biotic and abiotic dispersal vectors are relevant (Fig. 3). At the very fine spatial scale of movement between soil layers or within a patch (10^0 to 10^1 m), dispersal tends to take place at brief timescales, on the order of 10^0 to 10^4 s (minutes to days). Dispersal at this scale is mediated by abiotic factors such as gravity (Vibrans 1999), secondary dispersal of fallen seeds into soil cracks and aggregates by rain (Pareja et al. 1985; Reuss et al. 2001), biotic processes such as ballistic seed release (Garrison et al. 2000) and caching by soil biota (Pemberton and Irving 1990; Regnier et al. 2008), and management activities such as tillage (Mohler et al. 2006; Schemer et al. 2016). Attempts at physical weed control through soil disturbance

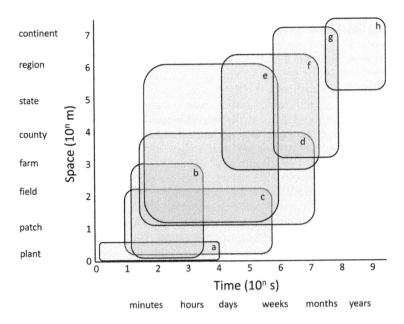

Figure 3 Dispersal vectors operating at different levels of spatial and temporal scale, including a) gravity; b) tillage, mechanical harvest; c) animals; d) water; e) wind; f) contaminated equipment, feed and seed; g) global trade; h) climate change. Figure based on data in Mohler (2001), Benvenuti (2007) and Cousens et al. (2008).

can be counterproductive for many perennial species, since the soil movement further spreads the vegetative propagules throughout the field. At local spatial scales, from field to farm level (10^2 to 10^3 m), processes such as mechanical harvest (Ballare et al. 1987; Woolcock and Cousens 2000; Humston et al. 2005; Walsh et al. 2013), contaminated tillage equipment, wind and overland flow of water act at timescales from hours to weeks (Bryson and Carter 2004; Dauer et al. 2006; Huang et al. 2015). At intermediate spatial scales, from county to region (10^4 to 10^6 m), dispersal is effected at timescales of weeks to months by human-mediated processes such as transport of contaminated agricultural inputs and products (Grundy et al. 1998; Norsworthy et al. 2009) and equipment (Taylor et al. 2012), abiotic processes such as atmospheric mixing and settling (Sosnoskie et al. 2009) and river transport (Wadsworth et al. 2000; Khudamrongsawat et al. 2004), and biotic processes such as waterfowl migration (Farmer et al. 2016). Riparian corridors dominated by rhizomatous perennials are evidence of how effective floods are at dislodging and transporting large rhizome mats to new locations downstream. Finally, at continental to global scales, dispersal is accomplished by forces such as ocean currents (Harries and Clement 2014) and global trade and transport (Chapman et al. 2016; Early et al. 2016).

Weed distributions result from the combined influence of dispersal and suitability of local conditions for persistence of a species. Global change in climate, land use and connectivity are driving ongoing shifts in weed distributions (Pattison and Mack 2008; Storkey et al. 2014). As local conditions change, variation in adaptability among weed species is shaping differential range expansion (Clements et al. 2004; Clements and DiTommaso 2012). Bioclimatic envelope models can help predict the geographic limits of the distributions of weedy and invasive species (Barney and Ditomaso 2008), but they should be interpreted within the context of scale-dependent processes and characteristics of the receiving agroecosystem (Pearson and Dawson 2003). Range expansion of a weed species is not, by itself, cause for concern. Weed managers should focus instead on the *damage niche*, the extent of a species' range in which it can cause economic crop yield loss (McDonald et al. 2009). Both empirical and modelling approaches will be helpful in characterizing the changing damage niche of weeds undergoing range expansion (Stratonovitch et al. 2012; Storkey et al. 2014; Davis et al. 2015; Storkey et al. 2015).

4 Target transitions: a quantitative approach to targeting weed life stages

The annual transitions between weed life stages (seeds, rosettes and adult plants; shown as solid arrows in Fig. 2) that govern recruitment, survival and reproduction may be subdivided into *lower-level demographic rates* (Cousens and Mortimer 1995; Caswell 2001). The empirical estimation of lower-level demographic rates (shown as lower-case letters accompanying the annual transitions in Fig. 2) through field studies of weed populations (Heggenstaller and Liebman 2006) supports two important activities:

- construction of population dynamics simulation models to quantify the fine-scale consequences of management and environmental variation for population abundance over time (Holst et al. 2006) and
- identification of high-priority management interventions for vulnerable life stages, or *target transitions*, for a weed in a particular agroecosystem (Mohler 2001).

The target transition concept (McEvoy and Coombs 1999) is based on perturbation analyses (Caswell 2001) of population dynamics simulation models. Prospective perturbation analyses ask 'what if' questions: How will population growth rate (λ) respond to changes in each lower-level demographic rate, in turn, as other rates are held constant? Retrospective perturbation analyses as 'what happened' questions based on empirical studies: How did variation in various lower-level demographic rates contribute to observed treatment effects on λ? A target transition is a lower-level demographic rate that, when varied, has a large effect on λ. Target transitions may be identified either through prospective or through retrospective perturbation analyses.

When simulation models are parameterized from field data, they can provide an objective means of weighing potential management effects of management interventions for a specific lower-level demographic rate for weeds in a given environment. The target transitions selected through such an analysis can vary widely across life histories and among environments. Figure 4 shows the elasticity of λ to changes in lower-level demographic rates among weeds varying in life history: two annual plants, a biennial and a perennial. In each panel of Fig. 4, the uppermost lines denote the demographic rate with the greatest elasticity values; these are the target transitions. Note that the x-axis

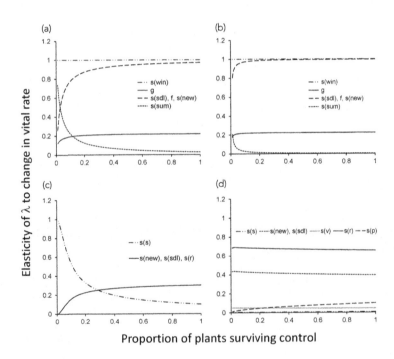

Proportion of plants surviving control

Figure 4 The elasticity of weed population growth rate (λ) to variation in different lower level demographic rates for (a) a summer annual weed with low fecundity, (b) a summer annual weed with very high fecundity, (c) a biennial weed and (d) a perennial weed. Explanation of abbreviations: s(win) = overwinter survival of dormant seeds; s(sum) = survival of dormant seeds during the growing season; g = germination rate; s(sdl) = seedling survival to reproductive maturity; f = fecundity; s(new) = survival of newly produced seeds; s(r) = rosette survival; s(v) = survival of vegetative propagules; s(p) = survival of adult plant. Figure adapted from Davis (2006), and used with permission of Allen Press.

represents a changing environmental context: variation in seedling survival owing to some form of management or environmental variation. In some cases (e.g. panel 4c), the target transition changes with environmental context. For annual plants, all individuals within a population must pass through the seed stage in any given year (Fig. 4a); for this reason, the survival of seeds in the soil seedbank is the most important target transition for annual weeds (Davis 2006; Tidemann et al. 2016). Only recently have IWM technologies, such as the Harrington Seed Destructor, become available, which have a substantial effect on this critical transition (Walsh et al. 2012; Walsh et al. 2013; Norsworthy et al. 2016). For biennials, seeds lose their importance as target transitions; instead, rosette survival becomes more important for management (Davis et al. 2006). In perennials, survival of rosettes is once again the most important management target, followed by survival of new seeds and seedlings (Donald 1994; Heimann and Cussans 1996; Blumenthal and Jordan 2001).

Simulation models may combine demographic and dispersal information to make projections about the change in weed population spatial structure and abundance over time in various landscapes and spatial scales. At the patch to field scale, such models have quantified the role of management in mediating weed population expansion (Ballare et al. 1987; Gonzalez-Andujar and Perry 1995; Woolcock and Cousens 2000; Humston et al. 2005). At larger spatial scales, spatial population dynamics simulation models help to identify the contribution of management and environmental variation at the landscape scale and beyond to the spatial structure of weed populations (Buckley et al. 2005; Skarpaas et al. 2005; Dauer et al. 2007; Jongejans et al. 2007; Skarpaas and Shea 2007; Jongejans et al. 2008; Dauer et al. 2009).

5 Communities in arable systems

When two or more species inhabit the same area within an ecosystem, the association comprises a *community*. Understanding the origins and maintenance of species diversity in ecosystems is at the heart of the discipline of community ecology. Within agroecosystems, communities include crop species and non-crop biota (Fig. 5), including weeds, field margin vegetation, above-ground fauna and soil organisms. Assembly theory offers a conceptual framework for scientific investigations of how communities form. Under this theory, weed communities are assembled from starting materials (species pools) interacting within a specific environmental context or filter (agroecosystem), resulting in context-dependent outcomes (Booth and Swanton 2002). Many agricultural practices act as filters on community assembly (Ryan et al. 2010), including crop diversity (Smith and Gross 2007), cover crops (Smith et al. 2015) and soil disturbance through tillage (Smith 2006).

Communities are not static entities. Once assembled, they continue to change over time in response to species interactions with each other and with their changing environment in a process called *succession*. Many of the dominant agricultural production systems throughout the world are maintained in an enforced state of primary succession, in which repeated soil disturbance and high levels of external inputs are used to create consistent conditions that favour the growth of annual crops in communities with low species diversity (Smith 2015). The empty niches created through this approach to agricultural production are both large and predictable, inviting their colonization by one or two dominant weed

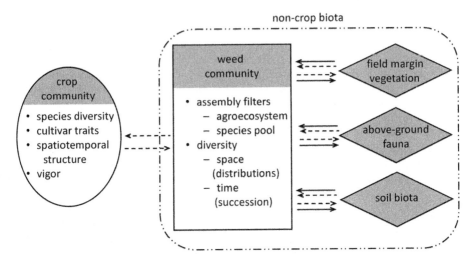

Figure 5 Components of agricultural communities, showing interactions between crop and non-crop biota. Dashed arrows indicate negative interactions, whereas solid arrows show positive interactions.

species (Pollnac et al. 2009; Schutte et al. 2010a, Okada et al. 2013; Ward et al. 2013; Beckie et al. 2014; Evans et al. 2016). As those weed species become increasingly entrenched in time- and management-invariant cropping systems, repeated selection through uniform weed management leads to reduced efficacy of the primary control tactics, as has been seen in the global rise of herbicide-resistant weeds in systems with continuous monocrops (Powles and Yu 2010; Shaner 2014).

In contrast to the stringent, and increasingly untenable, attempts to completely exclude weeds from spatiotemporally homogeneous agricultural systems through one or two heavy-handed tactics, managing weed communities with the goal of increasing species diversity, and therefore reducing the dominance of any one species, may not only improve overall weed management, but also reduce effects on non-target species (Smith 2015). Such an approach will require increasing species and management diversity in cropping systems in both time and space and incorporating the insights and tactics of many disciplines (Liebman and Gallandt 1997; Ward et al. 2014; Jordan and Davis 2015; Jordan et al. 2016; Liebman et al. 2016). Conservation of biodiversity in agroecosystems and adjoining non-arable ecosystems is a matter of increasing urgency, due to the growing use of herbicides prone to drift and volatilization, and therefore non-target plant mortality, in response to herbicide resistance (Egan and Mortensen 2012; Mortensen et al. 2012; Egan et al. 2014). In-field and field margin weed community diversity may not only serve as an indicator of management intensity and potential non-target effects, but also provide valuable ecosystem services to non-crop biota in the form of food, shelter and pollinator resources (Storkey and Westbury 2007; Storkey et al. 2013; Meziere et al. 2015).

Species interactions within agricultural communities can take many forms (Fig. 5). The direction and magnitude of the interactions depends on the species and their environment. In-field plant communities, involving crops and weeds, are generally engaged in *competitive interactions*, either through interference (precluding access to a resource, as in shading and light competition) or through extractive competition (simultaneous use of a resource, such as soil inorganic nutrient supply) (Zimdahl 2004). The competitive effect of weeds on crops, manifested as crop yield loss, varies in response to numerous factors, including but not limited to weed characteristics associated with competitive ability (Bensch et al. 2003; Williams et al. 2007), crop cultivar traits associated with tolerance of weed competition (Williams et al. 2008b), weed population density (Zimdahl 2004), duration of competition (Blackshaw 1993; VanAcker 1993; Blackshaw et al. 1999; Knezevic et al. 2002), source and timing of soil nutrient availability (Davis and Liebman 2001; Blackshaw and Brandt 2008), and light quality in the early growth environment (Rajcan et al. 2004; Page et al. 2010; Page et al. 2012). The *resource pool diversity hypothesis* proposes that increased functional diversity in agroecosystems may mitigate weed–crop competition by facilitating resource partitioning among crops and weeds to over competitive overlap in resource use (Smith et al. 2010). Competitive interactions between weeds and crops may also be mediated through their interactions with other components of the agricultural community, including soil biota (Vatovec et al. 2005; Regnier et al. 2008; Schutte et al. 2010b; Lou et al. 2014; Lou et al. 2016) and above-ground fauna (Westerman et al. 2003; Menalled et al. 2006; Westerman et al. 2011). Understanding how to manage these relationships to facilitate coexistence in agroecosystems is explored in later chapters of this book.

6 Agroecosystems

Ecosystems are comprised of communities of organisms in their abiotic and biotic context. Agroecosystems form a special case, in which the context results from two types of conditions (Fig. 6): those that are beyond a grower's control (local environment) and those that are influenced by grower management decisions (system design). Proper IWM must begin with the agroecosystem. A key consideration is how system design may either work

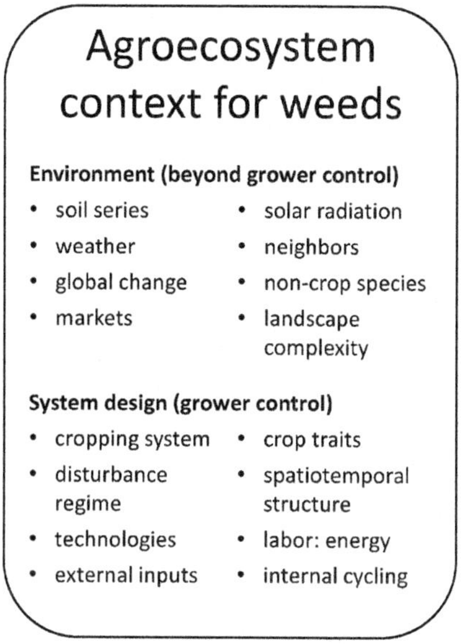

Agroecosystem context for weeds

Environment (beyond grower control)
- soil series
- weather
- global change
- markets
- solar radiation
- neighbors
- non-crop species
- landscape complexity

System design (grower control)
- cropping system
- disturbance regime
- technologies
- external inputs
- crop traits
- spatiotemporal structure
- labor: energy
- internal cycling

Figure 6 The agroecosystem context for weed growth and management is formed by the nexus of abiotic, biotic and management factors.

with or overcome features of the local environment to shape the abiotic and biotic context of crop and weed growth in ways that favour crops and suppress weeds. That is to say, truly effective weed management must begin with an analysis of how the ways by which we practise agriculture influence weeds and in fact may create the weed problem.

There are many dimensions of agroecosystem design, including crop spatial and temporal diversity, crop cultivar traits, crop spacing and population, integration of livestock, timing and intensity of soil disturbance, reliance upon external synthetic inputs relative to internal processes, the degree to which energy substitutes for human labour, spatial extent and level of technological advancement (Liebman et al. 2001; Smith 2015). Each requires growers to consider how environmental characteristics shape the agroecosystem in ways that have consequences for weed management.

Local soil conditions, resulting from variation in parent material and paedogenic processes, form one of the primary constraints on agroecosystem design (Weil and Brady 2016). Soil textural class largely dictates how the manner, timing and intensity of soil disturbance translates into secondary soil structure. This, in turn, shapes local tillage practices, from coarse soils that may be tilled at almost any point during the growing season with no negative consequences, to soils high in shrink-swell clays (e.g. montmorillonite) that rely upon freeze-thaw cycles to disperse clods formed by tillage, to heavy soils best managed under continuous no till. The timing of soil disturbance interacts with the dormancy mechanisms of species present within weed seedbanks to influence seedling recruitment schedules (Schutte et al. 2014), while the depth and type of tillage influences the depth structure of soil seedbanks (Cousens and Moss 1990; Mohler et al. 2006) and recruitment from depth (Benvenuti et al. 2001).

The degree of cropping system diversity has generally declined over past several decades, as production systems have become streamlined to produce fungible commodities, and crop and livestock production have become physically disengaged (Naylor et al. 2005). Separate production of crops and livestock reduces internal cycling of nutrients while reinforcing the demand for steady supply of sole-cropped grain concentrates. Crop production systems optimized for this particular supply chain are reliant upon low crop diversity, large external inputs of nutrients and prophylactic herbicide applications. These features, while spectacularly productive in the short run, have become a source of concern due to non-target species effects, agrichemical water pollution and widespread herbicide resistance in the dominant weeds (Davis et al. 2012). There are legitimate questions about the sustainability of the dominant system. Indeed, the key features of the business-as-usual agroecosystem have created especially challenging weed problems.

Lack of crop diversity, especially in phenology and life history, reinforces the recruitment and fecundity schedules of dominant weed populations and creates optimal conditions for their perpetuation (Mohler 2001). In contrast, there is abundant evidence that crop sequences with temporal diversity in phenology and cultivar traits reduce weed growth and competition with crops (Schreiber 1992; Jordan 1993; Williams et al. 2008b; Williams et al. 2008a; So et al. 2009a; So et al. 2009b; Andrew et al. 2015). Predictable, highly concentrated inorganic nutrient pulses favour the ability of weeds to rapidly and efficiently obtain soil nutrients (Blackshaw et al. 2001; Bonifas et al. 2005; Bonifas and Lindquist 2006), with predictable species-specific effects on weed growth and crop competition effects (Blackshaw and Brandt 2008). Integration of crop, soil and weed management (Liebman and Davis 2000) with crop-centric fertilization strategies (DiTomaso 1995; Rasmussen 2002) and diversification of nutrient pools in time, space and source (Davis and Liebman 2001; Smith et al. 2010) can help reduce weed stimulation through soil fertility management. Increasing diversity of crop and weed management practices, both within and across rotation phases, increases weed management efficacy and reduces the ability of weeds to evolve in response to a particular weed control tactic (Melander et al. 2005; Evans et al. 2016). When these different vulnerabilities are accounted for in agroecosystem design, the result is weed-suppressive cropping systems that balance weed management with productivity, profitability and environmental health (Davis et al. 2012).

7 Case study: mitigating the invasive potential of a bioenergy crop species

Although much of weed science concerns itself with non-crop plant species that invade, and become problematic in, cropping systems, there are also cases where the main crop has the potential to become a weed problem in its own right. This case study presents a body of research related to quantifying and mitigating the potential of a bioenergy crop species to escape production and become naturalized in surrounding non-arable systems. The study includes work focused on the population and ecosystem.

An ongoing need for energy alternatives to fossil fuels, because of geopolitical and environmental concerns, has created growing demand for the development and production of biomass crops as feedstocks for bioenergy supply chains (Hill et al. 2009). The search for highly productive, cost-effective bioenergy crops has highlighted large-statured herbaceous perennial plant species as a prime source of renewable bioenergy. Many of these species are not only prodigious biomass producers, but share numerous traits with known plant invaders (Raghu et al. 2006). In the Upper Midwest of the United States, much attention has focused on warm season grasses in the Eurasian genus *Miscanthus* as ideal biofeedstocks because of their high biomass production and low external input requirements (Heaton et al. 2008). This genus has known invaders among its ranks (Fig. 7) and has generated uncertainty about potential invasions arising from its bioenergy use (Barney and Ditomaso 2008; Davis et al. 2010). This uncertainty indicated the need for further testing, prior to widespread deployment, to quantify the invasive potential of *Miscanthus* species used in bioenergy production and to develop guidelines for mitigating any such risks (Davis et al. 2010).

Two candidates in the *Miscanthus* genus are *Miscanthus sinensis* Anderss. and *Miscanthus x giganteus* Greef et Deu ex Hodkinson et Renvoize. *M. sinensis*, a medium-sized bunchgrass with fertile windborne seeds, was introduced to the United States in the mid-1800s for ornamental purposes, and has produced naturalized populations (Fig. 7) mostly located in the Southeast United States (Quinn et al. 2010). *M. x giganteus* is an exceptionally large-statured bunchgrass that largely consists of a seed-sterile triploid genotype thought to have arisen from the hybridization of *M. sinensis* and *M. sacchariflorus* (Linde-Laursen 1993; Lafferty and Lelley 1994). Subsequent commercial attempts to re-create *M. x giganteus* through recombinant DNA and tissue culture methods have resulted in a seed-fertile genotype (Barney et al. 2016). Both *M. sinensis* and *M. x giganteus* produce between 10^4 and 10^6 light caryopses plant^{-1}, each caryopsis bearing a coma to facilitate wind dispersal for distances of up to 0.1 km (Quinn et al. 2011). These species may also spread locally through growth and fragmentation of deep rhizome networks (Matlaga et al. 2012; West et al. 2014a). A final dispersal mode is through detached green stems, the internodes of which can remain viable propagules during water transport (Mann

Figure 7 Large-statured warm season perennial grasses in the genus *Miscanthus* are being considered for bioenergy use. *Miscanthus* species produce both wind-borne caryopses (a) and spreading rhizomes (b), growing to form dense monospecific clumps (c). Some species in this genus, such as *Miscanthus sinensis*, have escaped cultivation, forming naturalized populations in a variety of environments, including this roadside in North Carolina (d). Photo credits: a & c (A. Davis); b (S. Post), d (B. Baldwin).

et al. 2013). Mortality of newly established seedlings is high in the first year, but survival of juvenile and mature plants remains high thereafter for several years (Matlaga et al. 2012).

Demographic simulation models, parameterized with empirical data on vital rates and dispersal kernels collected across environmental gradients (Quinn et al. 2011; Matlaga et al. 2012; West et al. 2014b), indicate target transitions related to both sexual and vegetative spread for both *M. sinensis* and *M. x giganteus*. Perturbation analyses of vital rates related to sexual reproduction (Fig. 8) indicate that both population growth rate and the rate of spread are related to seed fertility and seedling recruitment levels in these species (Matlaga and Davis 2013). Analysis of vital rates related to vegetative reproduction indicate that rhizome fragmentation and dispersal can drive spread from field margins to adjoining riparian areas (West et al. 2014a). Establishment risk of escaped propagules is greatest in high-light, low-moisture receptor environments, making successional habitats adjoining *Miscanthus* production fields likely receptor habitats (West et al. 2014b).

These analyses have guided the development of management guidelines for reducing invasive risk of *Miscanthus* production for bioenergy purposes (Barney et al. 2016). First, demographic analyses show that a low-invasion potential breeding ideotype will have low seed fertility, no seed dormancy and low seedbank persistence (Matlaga and Davis 2013). Second, at a landscape scale, invasive spread of *Miscanthus* from production areas is likely to be minimized if field margins of at least 10 m are maintained weed-free, and *Miscanthus* plantations are located within a larger annual field crop production matrix subject to stringent weed control (Muthukrishnan et al. 2015; Pitman et al. 2015). Finally, recognizing that both *M. sinensis* and *M. x giganteus* have non-zero invasion risks and that the likelihood that they will be grown over large production areas creates the probability

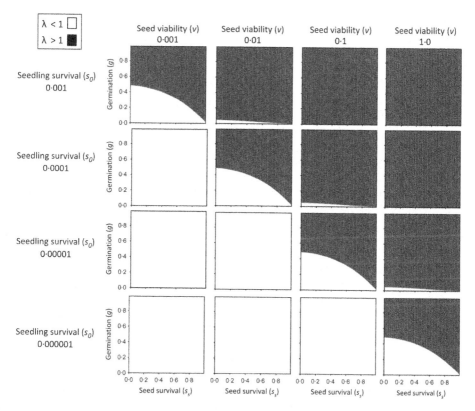

Figure 8 The population growth rate of a fertile cultivar of *Miscanthus x giganteus* is influenced by contributions from its vital rates, including viability of newly produced seed, seed germination rate, seedling survival, and seed survival in the soil seedbank. Populations are growing (λ >1) in shaded regions of the figure. Figure reproduced from Matlaga et al. (2013) with permission from the British Ecological Society.

that some invasive populations will form as a result, it is important to develop legal frameworks to assign responsibility for monitoring and eradicating naturalized populations in advance (Barney et al. 2016).

8 Summary and future trends

Truly effective IWM systems will be built upon ecological principles. For the reasons outlined above, there are no one-size-fits-all prescriptions for IWM. Rather, for the same reason that different weed communities may assemble from the same starting species pool under differing environmental conditions, the most appropriate, effective approaches to IWM are likely to be specific to individual agroecosystems. The route that an individual grower takes to IWM depends on their willingness to redesign the agroecosystem for inherent weed suppressiveness (DeDecker et al. 2014). If agroecosystem redesign is an option, then IWM can be built into the system. If redesign is not an option, then IWM will necessarily be limited to reactive approaches at the community and population levels. Reactive approaches may be therapeutic, but not curative.

Future research into the ecological basis of IWM should follow the lead of research groups that have developed new theories that frame testable hypotheses in weed ecology and management (Booth and Swanton 2002; Page et al. 2010; Smith et al. 2010; Long et al. 2015; Smith 2015). This is an area in weed science that has been neglected for much of its history. Development of new theories over the past decade will help focus the next generation of scholars in productive directions.

Using ecological knowledge to identify effective new technologies is another welcome development of the past decade. Given the extent and severity of herbicide resistance, new thinking and technologies are essential. The development of techniques, such as the Harrington Seed Destructor and windrow burning, to destroy seeds of annual weeds during crop harvest has filled an important management gap that will help integrate physical control into IWM systems at large production scales (Walsh et al. 2013).

Finally, the multi-scale foci of ecological science can help to organize systems-level thinking and transdisciplinarity necessary to build the research and adoption networks that develop future IWM systems that manage weeds in ways that disrupt selection rather than drive it in predictable ways (Ward et al. 2014; Jordan et al. 2016; Liebman et al. 2016).

9 Where to look for further information

There are several foundational texts that will help the interested reader explore the material presented here in greater depth. For a comprehensive presentation of the basic principles in weed ecology, read *Weed ecology in natural and agricultural systems* by Booth et al. (2003). These principles are developed further and put into a strong management context in *Ecological management of agricultural weeds* (Liebman et al. 2001). For greater detail on seedbank ecology, with material relevant to weed populations, see *The ecology of seeds*, by Fenner and Thompson (2005). Cousens et al. (2008) provide a state-of-the-art treatment of the dispersal of seeds and other plant propagules in *Dispersal in plants*. Weed population ecology is described in depth in *Dynamics of weed populations*, by Cousens and Mortimer (Cousens and Mortimer 1995). Matrix population modelling methods, such as those used in the results shown in Fig. 4 and 8, are presented in great detail and clarity in *Matrix population models* (Caswell 2001), with code fragments and exercises.

10 References

Andrew, I. K. S., Storkey, J. and Sparkes, D. L. (2015) A review of the potential for competitive cereal cultivars as a tool in integrated weed management. *Weed Research*, 55, 239–48.

Averill, K. M., DiTommaso, A., Mohler, C. L. and Milbrath, L. R. (2011) Survival, Growth, and Fecundity of the Invasive Swallowworts (Vincetoxicum rossicum and V. nigrum) in New York State. *Invasive Plant Science and Management*, 4, 198–206.

Baker, H. G. (1974) The evolution of weeds. *Annual Review of Ecology and Systematics*, 5, 1–24.

Ballare, C. L., Scopel, A. L., Ghersa, C. M. and Sanchez, R. A. (1987) The Population Ecology of Datura-Ferox in Soybean Crops – a Simulation Approach Incorporating Seed Dispersal. *Agriculture Ecosystems and Environment*, 19, 177–88.

Barney, J. N., Davis, A. S., Porter, R. D. and Simberloff, D. (2016) *A Life-Cycle Approach to Low-Invasion Potential Bioenergy Production.* CAST Commentary QTA 2016-1. Ames, IA.

Barney, J. N. and Ditomaso, J. M. (2008) Nonnative species and bioenergy: Are we cultivating the next invader? *BioScience*, 58, 64–70.

Baskin, C. C. and Baskin, J. M. (2001) *Seeds: Ecology, Biogeography, and Evolution of Dormancy and Germination.* Academic Press, San Diego, CA. p. 666.

Baskin, C. C. and Baskin, J. M. (2006) The natural history of soil seed banks of arable land. *Weed Science*, 54, 549–57.

Beckie, H. J., Sikkema, P. H., Soltani, N., Blackshaw, R. E. and Johnson, E. N. (2014) Environmental Impact of Glyphosate-Resistant Weeds in Canada. *Weed Science*, 62, 385–92.

Benech-Arnold, R. L., Sanchez, R. A., Forcella, F., Kruck, B. C. and Ghersa, C. M. (2000) Environmental control of dormancy in weed seed banks in soil. *Field Crops Research*, 67, 105–22.

Bensch, C. N., Horak, M. J. and Peterson, D. (2003) Interference of redroot pigweed (*Amaranthus retroflexus*), Palmer amaranth (*A. palmeri*), and common waterhemp (*A. rudis*) in soybean. *Weed Science*, 51, 37–43.

Benvenuti, S. (2007) Weed seed movement and dispersal strategies in the agricultural environment. *Weed Biology and Management*, 7, 141–57.

Benvenuti, S., Macchia, M. and Miele, S. (2001) Quantitative analysis of emergence of seedlings from buried weed seeds with increasing soil depth. *Weed Science*, 49, 528–35.

Blackshaw, R. E. (1993) Downy Brome (*Bromus tectorum*) density and relative time of emergence affects interference in winter wheat (*Triticum aestivum*). *Weed Science*, 41, 551–6.

Blackshaw, R. E. and Brandt, R. N. (2008) Nitrogen fertilizer rate effects on weed competitiveness is species dependent. *Weed Science*, 56, 743–7.

Blackshaw, R. E., Brandt, R. N., Janzen, H. H., Entz, T., Grant, C. A. and Derksen, D. A. (2001) Differential response of weed species to added nitrogen. *Weed Science*, 51, 532–9.

Blackshaw, R. E., Muendel, H. H. and Saindon, G. (1999) Canopy architecture, row spacing and plant density effects on yield of dry bean (Phaseolus vulgaris) in the absence and presence of hairy nightshade (Solanum sarrachoides). *Canadian Journal of Plant Science*, 79, 663–9.

Blumenthal, D. and Jordan, N. (2001) Weeds in field margins: a spatially explicit simulation analysis of Canada thistle population dynamics. *Weed Science*, 49, 509–19.

Bond, W. and Turner, R. (2006a) The biology and non-chemical control of Creeping thistle (*Cirsium arvense*). pp. 16. HDRA.

Bond, W. and Turner, R. J. (2006b) The biology and non-chemical control of common couch (*Elytrigia repens* (L.) Nevski). pp. 1–8. HDRA.

Bonifas, K. D. and Lindquist, J. L. (2006) Predicting biomass partitioning to root versus shoot in corn and velvetleaf (*Abutilon theophrasti*). *Weed Science*, 54, 133–7.

Bonifas, K. D., Walters, D. T., Cassman, K. G. and Lindquist, J. L. (2005) Nitrogen supply affects root:shoot ratio in corn and velvetleaf (*Abutilon theophrasti*). *Weed Science*, 53, 670–5.

Booth, B. D., Murphy, S. D. and Swanton, C. J. (2003) *Weed Ecology in Natural and Agricultural Systems.* CABI Publishing, Cambridge, MA. p. 303.

Booth, B. D. and Swanton, C. J. (2002) Assembly theory applied to weed communities. *Weed Science*, 50, 2–13.

Bryson, C. T. and Carter, R. (2004) Biology of pathways for invasive weeds. *Weed Technology*, 18, 1216–20.

Buckley, Y. M., Brockerhoff, E., Langer, L., Ledgard, N., North, H. and Rees, M. (2005) Slowing down a pine invasion despite uncertainty in demography and dispersal. *Journal of Applied Ecology*, 42, 1020–30.

Buhler, D. D. (1997) Effects of tillage and light environment on emergence of 13 annual weeds. *Weed Technology*, 11, 496–501.

Caswell, H. (2001) *Matrix population models: construction, analysis and interpretation*, 2 edn. Sinauer, Sunderland, MA. p. 722.

Chapman, D. S., Makra, L., Albertini, R., Bonini, M., Paldy, A., Rodinkova, V., Sikoparija, B., Weryszko-Chmielewska, E. and Bullock, J. M. (2016) Modelling the introduction and spread of non-native species: international trade and climate change drive ragweed invasion. *Global Change Biology*, 22, 3067–79.

Clay, S. A., Davis, A., Dille, A., Lindquist, J., Ramirez, A. H. M., Sprague, C., Reicks, G. and Forcella, F. (2014) Common sunflower seedling emergence across the US Midwest. *Weed Science*, 62, 63–70.

Clements, D. R. and DiTommaso, A. (2012) Predicting weed invasion in Canada under climate change: Evaluating evolutionary potential. *Canadian Journal of Plant Science*, 92, 1013–20.

Clements, D. R., DiTommaso, A., Jordan, N., Booth, B. D., Cardina, J., Doohan, D., Mohler, C. L., Murphy, S. D. and Swanton, C. J. (2004) Adaptability of plants invading north American cropland. *Agriculture Ecosystems & Environment*, 104, 379–98.

Cousens, R., Dytham, C. and Law, R. (2008) *Dispersal in Plants: A Population Perspective*. Oxford University Press, Oxford, Great Britain. p. 221.

Cousens, R. and Mortimer, M. (1995) *Dynamics of weed populations*. Cambridge University Press, Cambridge, England. p. 332.

Cousens, R. and Moss, S. R. (1990) A Model of the Effects of Cultivation on the Vertical-Distribution of Weed Seeds within the Soil. *Weed Research*, 30, 61–70.

Dalling, J., Davis, A. S., Schutte, B. and Arnold, A. E. (2011) A framework for a seed defence theory: integrating predation, dormancy and the soil microbial community. *Journal of Ecology*, 99, 89–95.

Dauer, J. T., Luschei, E. C. and Mortensen, D. A. (2009) Effects of landscape composition on spread of an herbicide-resistant weed. *Landscape Ecology*, 24, 735–47.

Dauer, J. T., Mortensen, D. A. and Humston, R. (2006) Controlled experiments to predict horseweed (Conyza canadensis) dispersal distances. *Weed Science*, 54, 484–9.

Dauer, J. T., Mortensen, D. A. and Vangessel, M. J. (2007) Temporal and spatial dynamics of long-distance Conyza canadensis seed dispersal. *Journal of Applied Ecology*, 44, 105–14.

Davis, A. S. (2006) When does it make sense to target the weed seed bank? *Weed Science*, 54, 558–65.

Davis, A. S., Cardina, J., Forcella, F., Johnson, G. A., Kegode, G., Lindquist, J. L., Luschei, E. C., Renner, K. A., Sprague, C. L. and Williams II, M. M. (2005) Environmental factors affecting seed persistence of 13 annual weeds across the U.S. corn belt *Weed Science*, 53, 860–8.

Davis, A. S., Clay, S., Cardina, J., Dille, A., Forcella, F., Lindquist, J. and Sprague, C. (2013) Seed burial physical environment explains departures from regional hydrothermal model of giant ragweed (Ambrosia trifida) seedling emergence in US Midwest. *Weed Science*, 61, 415–21.

Davis, A. S., Cousens, R. D., Hill, J., Mack, R. N., Simberloff, D. and Raghu, S. (2010) Screening bioenergy feedstock crops to mitigate invasion risk. *Frontiers in Ecology and the Environment*, 8, 533–9.

Davis, A. S., Dixon, P. M. and Liebman, M. (2004) Using matrix models to determine cropping system effects on annual weed demography. *Ecological Applications*, 14, 655–68.

Davis, A. S., Fu, X., Schutte, B., Berhow, M. and Dalling, J. (2016) Interspecific variation in persistence of buried weed seeds follows trade-offs among physiological, chemical and physical seed defenses. *Ecology and Evolution*, DOI: 10.1002/ece3.2415.

Davis, A. S., Hill, J. D., Johanns, A. M., Chase, C. A. and Liebman, M. (2012) Increasing cropping system diversity balances productivity, profitability and environmental health. *PLOS ONE*, 7, e47149.

Davis, A. S., Landis, D. A., Nuzzo, V., Blossey, B., Hinz, H. and Gerber, E. (2006) Demographic models inform selection of biocontrol agents for garlic mustard (Alliaria petiolata). *Ecological Applications*, 16, 2399–410.

Davis, A. S. and Liebman, M. (2001) Nitrogen source influences wild mustard growth and competitive effect on sweet corn. *Weed Science*, 49, 558–66.

Davis, A. S., Schutte, B. J., Hager, A. G. and Young, B. G. (2015) Palmer Amaranth (Amaranthus palmeri) Damage Niche in Illinois Soybean Is Seed Limited. *Weed Science*, 63, 658–68.

Davis, A. S. and Williams, M. M., II (2007) Variation in wild proso millet (Panicum miliaceum) fecundity in sweet corn has residual effects in snap bean. *Weed Science*, 55, 502–7.

DeDecker, J. J., Masiunas, J. B., Davis, A. S. and Flint, C. G. (2014) Weed Management Practice Selection Among Midwest U.S. Organic Growers. *Weed Science*, 62, 520–31.

Dekker, J. and Hargrove, M. (2002) Weedy adaptation in *Setaria* spp. V. Effects of gaseous environment on giant foxtail (*Setaria faberii*) (Poaceae) seed germination. *American Journal of Botany*, 89, 410–16.

Dieleman, J. A., Mortensen, D. A., Buhler, D. D., Cambardella, C. A. and Moorman, T. B. (2000) Identifying association among site properties and weed species abundance. I. Multivariate analysis. *Weed Science*, 48, 567–75.

DiTomaso, J. M. (1995) Approaches for improving crop competitiveness through the manipulation of fertilization strategies. *Weed Science*, 43, 491–7.

Donald, W. W. (1994) The biology of Canada thistle (*Cirsium arvense*). *Reviews in Weed Science*, 6, 77–101.

Early, R., Bradley, B. A., Dukes, J. S., Lawler, J. J., Olden, J. D., Blumenthal, D. M., Gonzalez, P., Grosholz, E. D., Ibanez, I., Miller, L. P., Sorte, C. J. B. and Tatem, A. J. (2016) Global threats from invasive alien species in the twenty-first century and national response capacities. *Nature Communications*, 7.

Egan, J. F., Bohnenblust, E., Goslee, S., Mortensen, D. and Tooker, J. (2014) Herbicide drift can affect plant and arthropod communities. *Agriculture Ecosystems and Environment*, 185, 77–87.

Egan, J. F. and Mortensen, D. A. (2012) A comparison of land-sharing and land-sparing strategies for plant richness conservation in agricultural landscapes. *Ecological Applications*, 22, 459–71.

Eriksson, O. and Ehrlen, J. (1992) Seed and microsite limitation of recruitment in plant populations. *Oecologia*, 91, 360–4.

Evans, J., Tranel, P. J., Hager, A. G., Schutte, B., Wu, C., Chatham, L. A. and Davis, A. S. (2016) Managing the evolution of herbicide resistance. *Pest Management Science*, 72, 74–80.

Farmer, J. A., Bish, M. D., Long, A., Biggs, M. and Bradley, K. W. (2016) Next day air: waterfowl and weed seed distribution. *WSSA Abstracts*, 56, 28.

Fausey, J. C. and Renner, K. A. (1997) Germination, emergence, and growth of giant foxtail (*Setaria faberi*) and fall panicum (*Panicum dichotomiflorum*). *Weed Science*, 45, 423–5.

Fenner, M. and Thompson, K. (2005) *The Ecology of Seeds*. Cambridge University Press, Cambridge, England. p. 250.

Forcella, F., Wilson, R. G., Dekker, J., Kremer, R. J., Cardina, J., Anderson, R. L., Alm, D., Renner, K. A., Harvey, G. and Clay, S. (1997) Weed seed bank emergence across the corn belt. *Weed Science*, 45, 67–76.

Freckleton, R. P., Sutherland, W. J., Watkinson, A. R. and Stephens, P. A. (2008) Modelling the effects of management on population dynamics: some lessons from annual weeds. *Journal of Applied Ecology*, doi: 10.1111/j.1365-2664.2008.01469.x.

Freckleton, R. P. and Watkinson, A. R. (1998) Predicting the determinants of weed abundance: a model for the population dynamics of *Chenopodium album* in sugar beet. *Journal of Applied Ecology*, 35, 904–20.

Gardarin, A. and Colbach, N. (2015) How much of seed dormancy in weeds can be related to seed traits? *Weed Research*, 55, 14–25.

Garrison, W. J., Miller, G. L. and Raspet, R. (2000) Ballistic seed projection in two herbaceous species. *American Journal of Botany*, 87, 1257–64.

Gonzalez-Andujar, J. L. and Perry, J. N. (1995) Models for the herbicidal control of the seed bank of *Avena sterilis*: the effects of spatial and temporal heterogenity and of dispersal. *Journal of Applied Ecology*, 32, 578–87.

Grundy, A. C., Green, J. M. and Lennartson, M. (1998) The effect of temperature on the viability of weed seeds in compost. *Compost Science and Utilization*, 6, 26–33.

Håkansson, S. (1982) Multiplication, growth and persistence of perennial weeds. *Biology and ecology of weeds* (eds W. Holzner and M. Numata), pp. 123–35. Dr. W. Junk, The Hague, Netherlands.

Harper, J. L. (1977) *Population Biology of Plants*. Academic Press, San Diego. pp. 652–64.

Harper, J. L., Williams, J. T. and Sagar, G. R. (1965) The behaviour of seeds in soil: I. The heterogeneity of soil surfaces and its role in determining the establishment of plants from seed. *Journal of Ecology*, 53, 253–76.

Harries, H. C. and Clement, C. R. (2014) Long-distance dispersal of the coconut palm by migration within the coral atoll ecosystem. *Annals of Botany*, 113, 565–70.

Harrison, S. K. (1990) Interference and seed production by common lambsquarters (*Chenopodium album*) in soybeans (*Glycine max*). *Weed Science*, 38, 113–18.

Harrison, S. K., Regnier, E. E., Schmoll, J. T. and Webb, J. E. (2001) Competition and fecundity of giant ragweed in corn. *Weed Science*, 49, 224–9.

Heaton, E. A., Dohleman, F. G. and Long, S. P. (2008) Meeting US biofuel goals with less land: the potential of *Miscanthus*. *Global Change Biology*, 14, 2000–14.

Heggenstaller, A. H. and Liebman, M. (2006) Demography of *Abutilon theophrasti* and *Setaria faberi* in three crop rotation systems. *Weed Research*, 46, 138–51.

Heimann, B. and Cussans, G. W. (1996) The importance of seeds and sexual reproduction in the population biology of *Cirsium arvense* – A literature review. *Weed Research*, 36, 493–503.

Hill, J., Polasky, S., Nelson, E., Tilman, D., Huo, H., Ludwig, L., Neumann, J., Zheng, H. C. and Bonta, D. (2009) Climate change and health costs of air emissions from biofuels and gasoline. *Proceedings of the National Academy of Sciences of the United States of America*, 106, 2077–82.

Holm, L., Doll, J., Holm, E., Pancho, J. and Herberger, J. (1997) *World weeds: natural histories and distribution*. John Wiley and Sons, Inc., New York, NY. pp. 1129.

Holst, N., Rasmussen, I. A. and Bastiaans, L. (2006) Field weed population dynamics: a review of model approaches and applications. *Weed Research*, 46, 1–14.

Huang, H. Y., Ye, R. J., Qi, M. L., Li, X. Z., Miller, D. R., Stewart, C. N., DuBois, D. W. and Wang, J. M. (2015) Wind-mediated horseweed (Conyza canadensis) gene flow: pollen emission, dispersion, and deposition. *Ecology and Evolution*, 5, 2646–58.

Humston, R., Mortensen, D. A. and Bjornstad, O. N. (2005) Anthropogenic forcing on the spatial dynamics of an agricultural weed: the case of the common sunflower. *Journal of Applied Ecology*, 42, 863–72.

Jongejans, E., Pedatella, N. M., Shea, K., Skarpaas, O. and Auhl, R. (2007) Seed release by invasive thistles: the impact of plant and environmental factors. *Proceedings of the Royal Society B-Biological Sciences*, 274, 2457–64.

Jongejans, E., Shea, K., Skarpaas, O., Kelly, D., Sheppard, A. W. and Woodburn, T. L. (2008) Dispersal and demography contributions to population spread of Carduus nutans in its native and invaded ranges. *Journal of Ecology*, 96, 687–97.

Jordan, N. (1993) Prospects for weed control through crop interference. *Ecological Applications*, 3, 84–91.

Jordan, N. R. and Davis, A. S. (2015) Middle way strategies for sustainable intensification of agriculture. *BioScience*, 65, 513–19.

Jordan, N. R., Schut, M., Graham, S., Barney, J., Child, D., Christensen, S., Cousens, R., Davis, A. S., Eisenberg, H., Fernandez-Quintanilla, C., Harrison, L., Harsch, M., Heijting, S., Liebman, M., Loddo, D., Mirsky, S., Riemens, M., Neve, P., Peltzer, D., Renton, M., Recasens, J. and Sonderskov, M. (2016) Transdisciplinary weed research: new leverage on challenging weed problems? *Weed Research*, 56, 345–58.

Khudamrongsawat, J., Tayyar, R. and Holt, J. S. (2004) Genetic diversity of giant reed (*Arundo donax*) in the Santa Ana River, California. *Weed Science*, 52, 395–405.

Knezevic, S. Z., Evans, S. P., Blankenship, E. E., Van Acker, R. C. and Lindquist, J. L. (2002) Critical period for weed control: the concept and data analysis. *Weed Science*, 50, 773–86.

Lafferty, J. and Lelley, T. (1994) Cytogenetic Studies of Different Miscanthus Species with Potential for Agricultural Use. *Plant Breeding*, 113, 246–9.

Leon, R. G., Bassam, D. C. and Owen, M. D. K. (2006) Germination and proteome analyses reveal intraspecific variation in seed dormancy regulation in common waterhemp (*Amaranthus tuberculatus*). *Weed Science*, 54, 305–15.

Liebman, M., Baraibar, B., Buckley, Y., Childs, D., Christensen, S., Cousens, R., Eizenberg, H., Heijting, S., Loddo, D., Merotto, A., Renton, M. and Riemens, M. (2016) Ecologically sustainable weed management: How do we get from proof-of-concept to adoption? *Ecological Applications*, 26, 1352–69.

Liebman, M. and Davis, A. S. (2000) Integration of soil, crop and weed management in low-external-input farming systems. *Weed Research*, 40, 27–47.

Liebman, M. and Gallandt, E. R. (1997) Many little hammers: ecological approaches for management of crop-weed interactions. *Ecology in agriculture* (edited by Jackson, L. E.), pp. 291–343. Academic Press, San Diego.

Liebman, M., Mohler, C. L. and Staver, C. P. (2001) *Ecological Management of Agricultural Weeds*. Cambridge University Press, Cambridge, England. p. 532.

Linde-Laursen, I. B. (1993) Cytogenetic analysis of *Miscanthus* 'Giganteus', an interspecific hybrid. *Hereditas*, 119, 297–300.

Lindquist, J. L., Maxwell, B. D., Buhler, D. D. and Gonsolus, J. L. (1995) Velvetleaf (*Abutilon theophrasti*) recruitment, survival, seed production, and interference in soybean (*Glycine max*). *Weed Science*, 43, 226–32.

Long, R. L., Gorecki, M. J., Renton, M., Scott, J. K., Colville, L., Goggin, D. E., Commander, L. E., Westcott, D. A., Cherry, H. and Finch-Savage, W. E. (2015) The ecophysiology of seed persistence: a mechanistic view of the journey to germination or demise. *Biological Reviews Cambridge Philosophical Society*, 90, 31–59.

Lou, Y., Clay, S. A., Davis, A. S., Dille, A., Felix, J., Ramirez, A. H. M., Sprague, C. L. and Yannarell, A. C. (2014) An Affinity-Effect Relationship for Microbial Communities in Plant-Soil Feedback Loops. *Microbial Ecology*, 67, 866–76.

Lou, Y., Davis, A. S. and Yannarell, A. C. (2016) Interactions between allelochemicals and the microbial community affect weed suppression following cover crop residue incorporation into soil. *Plant and Soil*, 399, 357–71.

Mann, J. J., Kyser, G. B., Barney, J. N. and DiTomaso, J. M. (2013) Assessment of aboveground and belowground vegetative fragments as propagules in the bioenergy crops *Arundo donax* and *Miscanthus x giganteus*. *Bioenergy Research*, 6, 688–98.

Matlaga, D., Schutte, B. J. and Davis, A. S. (2012) Age-dependent demographic rates of the bioenergy crop *Miscanthus x giganteus* in Illinois. *Invasive Plant Science and Management*, 5, 238–48.

Matlaga, D. P. and Davis, A. S. (2013) Minimizing invasive potential of *Miscanthus × giganteus* grown for bioenergy: identifying demographic thresholds for population growth and spread. *Journal of Applied Ecology*, 50, 479–87.

McDonald, A., Riha, S., DiTommaso, A. and DeGaetano, A. (2009) Climate change and the geography of weed damage: Analysis of US maize systems suggests the potential for significant range transformations. *Agriculture Ecosystems and Environment*, 130, 131–40.

McEvoy, P. B. and Coombs, E. M. (1999) Biological control of plant invaders: regional patterns, field experiments and structured population models. *Ecological Applications*, 9, 387–401.

Melander, B., Rasmussen, I. A. and Barberi, P. (2005) Integrating physical and cultural methods of weed control - Examples from European research. *Weed Science*, 53, 369–81.

Menalled, F. D., Liebman, M. and Renner, K. (2006) The ecology of weed seed predation in herbaceous crop systems. *Handbook of Sustainable Weed Management* (eds Singh, H. P., Batish, D. R. and Kohli, R. K.), pp. 297–327. Haworth Press, Binghamton, NY.

Meziere, D., Colbach, N., Dessaint, F. and Granger, S. (2015) Which cropping systems to reconcile weed-related biodiversity and crop production in arable crops? An approach with simulation-based indicators. *European Journal of Agronomy*, 68, 22–37.

Mohler, C. L. (2001) Weed life history: identifying vulnerabilities. *Ecological Management of Agricultural Weeds* (eds M. Liebman, C. L. Mohler and C. P. Staver), pp. 40–98. Cambridge University Press, Cambridge, UK.

Mohler, C. L., Frisch, J. C. and McCulloch, C. E. (2006) Vertical movement of weed seed surrogates by tillage implements and natural processes. *Soil and Tillage Research*, 86, 110–22.

Mortensen, D. A. and Dieleman, J. A. (1998) Why weed patches persist: dynamics of edges and density. *Precision weed management in crops and pastures* (eds Medd, R. W. and Pratley, J. E.), pp. 14–19. CRC for Weed Management Systems, Adelaide, Wagga Wagga.

Mortensen, D. A., Egan, J. F., Maxwell, B. D., Ryan, M. R. and Smith, R. G. (2012) Navigating a critical juncture for sustainable weed management. *BioScience*, 62, 75–84.

Muthukrishnan, R., West, N. M., Davis, A. S., Jordan, N. R. and Forester, J. D. (2015) Evaluating the role of landscape in the spread of invasive species: The case of the biomass crop Miscanthus x giganteus. *Ecological Modelling*, 317, 6–15.

Naylor, R., Steinfeid, H., Falcon, W., Galloways, J., Smil, V., Bradford, E., Alder, J. and Mooney, H. (2005) Losing the links between livestock and land. *Science*, 310, 1621–2.

Norsworthy, J. K., Korres, N. E., Walsh, M. J. and Powles, S. B. (2016) Integrating Herbicide Programs with Harvest Weed Seed Control and Other Fall Management Practices for the Control of Glyphosate-Resistant Palmer Amaranth (Amaranthus palmeri). *Weed Science*, 64, 540–50.

Norsworthy, J. K., Smith, K. L., Steckel, L. E. and Koger, C. H. (2009) Weed Seed Contamination of Cotton Gin Trash. *Weed Technology*, 23, 574–80.

Okada, M., Hanson, B. D., Hembree, K. J., Peng, Y. H., Shrestha, A., Stewart, C. N., Wright, S. D. and Jasieniuk, M. (2013) Evolution and spread of glyphosate resistance in *Conyza canadensis* in California. *Evolutionary Applications*, 6, 761–77.

Page, E. R., Cerrudo, D., Westra, P., Loux, M., Smith, K., Foresman, C., Wright, H. and Swanton, C. J. (2012) Why Early Season Weed Control Is Important in Maize. *Weed Science*, 60, 423–30.

Page, E. R., Tollenaar, M., Lee, E. A., Lukens, L. and Swanton, C. J. (2010) Shade avoidance: an integral component of crop-weed competition. *Weed Research*, 50, 281–8.

Pareja, M. R., Staniforth, D. W. and Pareja, G. P. (1985) Distribution of weed seed among soil structural units. *Weed Science*, 33, 182–9.

Pattison, R. R. and Mack, R. N. (2008) Potential distribution of the invasive tree *Triadica sebifera* (Euphorbiaceae) in the United States: evaluating CLIMEX predictions with field trials. *Global Change Biology*, 14, 813–26.

Pearson, R. G. and Dawson, T. P. (2003) Predicting the impacts of climate change on the distribution of species: are bioclimate envelope models useful? *Global Ecology and Biogeography*, 12, 361–71.

Pemberton, R. W. and Irving, D. W. (1990) Elaiosomes on weed seeds and the potential for myrmecochory in naturalized plants. *Weed Science*, 38, 615–19.

Pitman, S. E., Muthukrishnan, R., West, N. M., Jordan, N. R., Davis, A. S. and Forester, J. D. (2015) Mitigating the potential for invasive spread of the exotic biofuel crop, *Miscanthus x giganteus*. *Biological Invasions*, 17, 3247–61.

Pollnac, F. W., Maxwell, B. D. and Menalled, F. D. (2009) Using Species-Area Curves to Examine Weed Communities in Organic and Conventional Spring Wheat Systems. *Weed Science*, 57, 241–7.

Powles, S. B. and Yu, Q. (2010) Evolution in action: plants resistant to herbicides. *Annual Review of Plant Biology*, 61, 317–47.

Quinn, L. D., Allen, D. J. and Stewart, J. R. (2010) Invasiveness potential of *Miscanthus sinensis*: implications for bioenergy production in the United States. *Global Change Biology*.

Quinn, L. D., Matlaga, D. P., Davis, A. S. and Stewart, R. (2011) Evaluating the influence of wind speed on caryopsis dispersal of *Miscanthus sinensis* and *Miscanthus x. giganteus*. *Invasive Plant Science and Management*, 4, 142–50.

Raghu, S., Anderson, R. C., Daehler, C. C., Davis, A. S., Wiedenmann, R. N., Simberloff, D. and Mack, R. N. (2006) Adding biofuels to the invasive species fire? *Science*, 313, 1742.

Rajcan, I., Chandler, K. J. and Swanton, C. J. (2004) Red-far red ratio of reflected light: a hypothesis of why early-season weed control is important in corn. *Weed Science*, 52, 774–8.

Rasmussen, K. (2002) Influence of liquid manure application method on weed control in spring cereals. *Weed Research*, 42, 287–98.

Regnier, E., Harrison, S. K., Liu, J., Schmoll, J. T., Edwards, C. A., Arancon, N. and Holloman, C. (2008) Impact of an exotic earthworm on seed dispersal of an indigenous US weed. *Journal of Applied Ecology*, 45, 1621–9.

Reuss, S. A., Buhler, D. D. and Gonsolus, J. L. (2001) Weed seed associations with soil aggregates: distribution and viability in a silt loam soil. *Applied Soil Ecology*, 16, 209–17.

Ryan, M. R., Smith, R. G., Mirsky, S. B., Mortensen, D. A. and Seidel, R. (2010) Management Filters and Species Traits: Weed Community Assembly in Long-Term Organic and Conventional Systems. *Weed Science*, 58, 265–77.

Sauer, J. (1957) Recent migration and evolution of the dioecious Amaranths. *Evolution*, 11, 11–31.

Schemer, A., Melander, B. and Kudsk, P. (2016) Vertical distribution and composition of weed seeds within the plough layer after eleven years of contrasting crop rotation and tillage schemes. *Soil and Tillage Research*, 161, 135–42.

Schreiber, M. M. (1992) Influence of tillage, crop rotation, and weed management on giant foxtail (*Setaria faberi*) population dynamics and corn yield. *Weed Science*, 40, 645–53.

Schutte, B. J., Hager, A. G. and Davis, A. S. (2010a) Respray Requests on Custom-Applied, Glyphosate-Resistant Soybeans in Illinois: How Many and Why. *Weed Technology*, 24, 590–8.

Schutte, B. J., Liu, J. Y., Davis, A. S., Harrison, S. K. and Regnier, E. E. (2010b) Environmental factors that influence the association of an earthworm (*Lumbricus terrestris* L.) and an annual weed (*Ambrosia trifida* L.) in no-till agricultural fields across the eastern US Corn Belt. *Agriculture Ecosystems and Environment*, 138, 197–205.

Schutte, B. J., Tomasek, B. J., Davis, A. S., Andersson, L., Benoit, D. L., Cirujeda, A., Dekker, J., Forcella, F., Gonzalez-Andujar, J. L., Graziani, F., Murdoch, A. J., Neve, P., Rasmussen, I. A., Sera, B., Salonen, J., Tei, F., Torresen, K. S. and Urbano, J. M. (2014) An investigation to enhance understanding of the stimulation of weed seedling emergence by soil disturbance. *Weed Research*, 54, 1–12.

Schwartz, L. M., Norsworthy, J. K., Young, B. G., Bradley, K. W., Kruger, G. R., Davis, V. M., Steckel, L. E. and Walsh, M. J. (2016) Tall Waterhemp (Amaranthus tuberculatus) and Palmer amaranth (Amaranthus palmeri) Seed Production and Retention at Soybean Maturity. *Weed Technology*, 30, 284–90.

Shaner, D. L. (2014) Lessons Learned From the History of Herbicide Resistance. *Weed Science*, 62, 427–31.

Silvertown, J. and Charlesworth, D. (2001) *Introduction to plant population biology*, 4th edn. Wiley-Blackwell, Oxford, UK. p. 360.

Skarpaas, O. and Shea, K. (2007) Dispersal patterns, dispersal mechanisms, and invasion wave speeds for invasive thistles. *American Naturalist*, 170, 421–30.

Skarpaas, O., Shea, K. and Bullock, J. M. (2005) Optimizing dispersal study design by Monte Carlo simulation. *Journal of Applied Ecology*, 42, 731–9.

Smith, R. G. (2006) Timing of tillage is an important filter on the assembly of weed communities. *Weed Science*, 54, 705–12.

Smith, R. G. (2015) A succession-energy framework for reducing non-target impacts of annual crop production. *Agricultural Systems*, 133, 14–21.

Smith, R. G., Atwood, L. W., Pollnac, F. W. and Warren, N. D. (2015) Cover-Crop Species as Distinct Biotic Filters in Weed Community Assembly. *Weed Science*, 63, 282–95.

Smith, R. G. and Gross, K. L. (2007) Assembly of weed communities along a crop diversity gradient. *Journal of Applied Ecology*, 44, 1046–56.

Smith, R. G., Mortensen, D. A. and Ryan, M. R. (2010) A new hypothesis for the functional role of diversity in mediating resource pools and weed-crop competition in agroecosystems. *Weed Research*, 50, 185.

So, Y. F., Williams, M. M. and Pataky, J. K. (2009a) Wild-Proso Millet Differentially Affects Canopy Architecture and Yield Components of 25 Sweet Corn Hybrids. *HortScience*, 44, 408–12.

So, Y. F., Williams, M. M., Pataky, J. K. and Davis, A. S. (2009b) Principal Canopy Factors of Sweet Corn and Relationships to Competitive Ability with Wild-Proso Millet (Panicum miliaceum). *Weed Science*, 57, 296–303.

Sosnoskie, L. M., Webster, T. M., Dales, D., Rains, G. C., Grey, T. L. and Culpepper, A. S. (2009) Pollen grain size, density, and settling velocity for Palmer amaranth (*Amaranthus palmeri*). *Weed Science*, 57, 404–9.

Storkey, J., Brooks, D., Haughton, A., Hawes, C., Smith, B. M. and Holland, J. M. (2013) Using functional traits to quantify the value of plant communities to invertebrate ecosystem service providers in arable landscapes. *Journal of Ecology*, 101, 38–46.

Storkey, J., Holst, N., Bojer, O. Q., Bigongiali, F., Bocci, G., Colbach, N., Dorner, Z., Riemens, M. M., Sartorato, I., Sonderskov, M. and Verschwele, A. (2015) Combining a weed traits database with a population dynamics model predicts shifts in weed communities. *Weed Research*, 55, 206–18.

Storkey, J., Stratonovitch, P., Chapman, D. S., Vidotto, F. and Semenov, M. A. (2014) A Process-Based Approach to Predicting the Effect of Climate Change on the Distribution of an Invasive Allergenic Plant in Europe. *PLOS ONE*, 9.

Storkey, J. and Westbury, D. B. (2007) Managing arable weeds for biodiversity. *Pest Management Science*, 63, 517–23.

Stratonovitch, P., Storkey, J. and Semenov, A. A. (2012) A process-based approach to modelling impacts of climate change on the damage niche of an agricultural weed. *Global Change Biology*, 18, 2071–80.

Taylor, K., Brummer, T., Taper, M. L., Wing, A. and Rew, L. J. (2012) Human-mediated long-distance dispersal: an empirical evaluation of seed dispersal by vehicles. *Diversity and Distributions*, 18, 942–51.

Tidemann, B. D., Hall, L. M., Harker, K. N. and Alexander, B. C. S. (2016) Identifying Critical Control Points in the Wild Oat (Avena fatua) Life Cycle and the Potential Effects of Harvest Weed-Seed Control. *Weed Science*, 64, 463–73.

VanAcker, R. C. (1993) The critical period of weed control in soybean (*Glycine max* (L.) Merr.). *Weed Science*, 41, 194–200.

Vatovec, C., Jordan, N. and Huerd, S. (2005) Responsiveness of certain agronomic weed species to arbuscular mycorrhizal fungi. *Renewable Agriculture and Food Systems*, 20, 181–9.

Vibrans, H. (1999) Epianthropochory in Mexican weed communities. *American Journal of Botany*, 86, 476–81.

Wadsworth, R. A., Collingham, Y. C., Willis, S. G., Huntley, B. and Hulme, P. E. (2000) Simulating the spread and management of alien riparian weeds: are they out of control? *Journal of Applied Ecology*, 37, 28–38.

Walsh, M., Newman, P. and Powles, S. (2013) Targeting weed seeds in-crop: a new weed control paradigm for global agriculture. *Weed Technology*, 27, 431–6.

Walsh, M. J., Harrington, R. B. and Powles, S. B. (2012) Harrington Seed Destructor: A new nonchemical weed control tool for global grain crops. *Crop Science*, 52, 1343–7.

Ward, S. M., Cousens, R. D., Bagavathiannan, M. V., Barney, J. N., Beckie, H. J., Busi, R., Davis, A. S., Dukes, J. S., Forcella, F., Freckleton, R. P., Gallandt, E. R., Hall, L. M., Jasieniuk, M., Lawton-Rauh, A., Lehnhoff, E. A., Liebman, M., Maxwell, B. D., Mesgaran, M. B., Murray, J. V., Neve, P., Nunez, M. A., Pauchard, A., Queenborough, S. A. and Webber, B. L. (2014) Agricultural weed research: a critique and two proposals. *Weed Science*, 62, 672–8.

Ward, S. M., Webster, T. M. and Steckel, L. E. (2013) Palmer Amaranth (*Amaranthus palmeri*): a review. *Weed Technology*, 27, 12–27.

Weil, R. R. and Brady, N. C. (2016) *The nature and properties of soils*, 15th edn. Pearson, Columbus, OH. p. 1071.

Werle, R., Sandell, L. D., Buhler, D. D., Hartzler, R. G. and Lindquist, J. L. (2014) Predicting Emergence of 23 Summer Annual Weed Species. *Weed Science*, 62, 267–79.

West, N. M., Matlaga, D. P. and Davis, A. S. (2014a) Managing spread from rhizome fragments is key to reducing invasiveness in *Miscanthus* × *giganteus*. *Invasive Plant Science and Management*, 7, 517–25.

West, N. M., Matlaga, D. P. and Davis, A. S. (2014b) Quantifying targets to manage invasion risk: light gradients dominate the early regeneration niche of naturalized and pre-commercial *Miscanthus* populations. *Biological Invasions*, 16, 1991–2001.

Westerman, P. R., Dixon, P. M. and Liebman, M. (2009) Burial rates of surrogate seeds in arable fields. *Weed Research*, 49, 142–52.

Westerman, P. R., Liebman, M., Heggenstaller, A. H. and Forcella, F. (2006) Integrating measurements of seed availability and removal to estimate weed seed losses due to predation. *Weed Science*, 54, 566–74.

Westerman, P. R., Luijendijk, C. D., Wevers, J. D. A. and Van der Werf, W. (2011) Weed seed predation in a phenologically late crop. *Weed Research*, 51, 157–64.

Westerman, P. R., Wes, J. S., Kropff, M. J. and Van der Werf, W. (2003) Annual losses of weed seeds due to predation in organic cereal fields. *Journal of Applied Ecology*, 40, 824–36.

Williams, M. M., Boydston, R. A. and Davis, A. S. (2007) Wild proso millet (*Panicum miliaceum*) suppressive ability among three sweet corn hybrids. *Weed Science*, 55, 245–51.

Williams, M. M., Boydston, R. A. and Davis, A. S. (2008a) Crop competitive ability contributes to herbicide performance in sweet corn. *Weed Research*, 48, 58–67.

Williams, M. M., Boydston, R. A. and Davis, A. S. (2008b) Differential tolerance in sweet corn to wild-proso millet (Panicum miliaceum) interference. *Weed Science*, 56, 91–6.

Woolcock, J. L. and Cousens, R. (2000) A mathematical analysis of factors affecting the fate of spread of patches of annual weeds in an arable field. *Weed Science*, 48, 27–34.

Wortman, S. E., Davis, A. S., Schutte, B. J., Lindquist, J. L., Cardina, J., Felix, J., Sprague, C. L., Dille, J. A., Ramirez, A. H. M., Reicks, G. and Clay, S. A. (2012) Local conditions, not spatial gradients, drive demographic variation of *Ambrosia trifida* and *Helianthus annuus* across northern US maize belt. *Weed Science*, 60, 440–50.

Zimdahl, R. L. (2004) *Weed-crop Competition: A Review*, 2nd edn. Blackwell Publishing, Ames. p. 220.

Chapter 2

Weed-plant interactions

Bruce Maxwell, Montana State University, USA

1 Introduction

Crop or desired plant species co-occur with undesired species and thereby the co-occurring species become weeds. This human-imposed quality is based on the perception that there is an interaction that results in some negative effect of the weed on the crop or desired species. At odds with this perception is an evolutionary perspective that would expect co-occurring species to select traits to minimize the interaction in a shared and limited resource pool as well as selection for traits to outcompete co-occurring plants. So which traits are more likely to be selected remains a question: Those that minimize interaction or those that intensify the interaction through competition (Grime 2006). It is quite possible that some co-occurring species may not be interacting in competition for resources, or competition may be minimal or indirect so that control of the undesired species is not required. Clearly, it becomes important to gain an understanding of the potential interaction among desired and undesired species and use that knowledge to instruct a management decision. All too often the decision to manage is reduced to selecting the best tool (usually herbicide) to get rid of the weed driven by the precautionary principle. The precautionary principle applied here is that when in doubt about the interaction between the weed and the crop one should control the weed. Unfortunately, controlling weeds can be expensive and can have environmental side effects. So there are trade-offs in the decision to control weeds and thus knowledge about the interaction between desired and undesired species should be gained to make the best decision about weed management (Swanton et al. 2015; Maxwell and Luschei 2004).

http://dx.doi.org/10.19103/AS.2017.0025.02

2 Crop-weed interactions: an evolutionary perspective

If plants are under selection to minimize interaction through a resource pool, one would expect that species that had co-occurred for many generations would have differentiated their respective *niche* requirements (Zuppinger-Dingley et al. 2014). Narrowly defined a species *niche* is a multidimensional space from which a plant gathers resources for growth and reproduction to sustain a population. Plants contribute the most pollen and seed thereby contribute the most traits to the population following natural selection. Since reproductive output is proportional to resources available to a plant, the plants with the most resources make the greatest contribution of traits. So resource capture by individual plants is not just driven by selection for niche differentiation but also intraspecific competition between individuals of the same species which tends to select for traits that allow plants to optimize resource consumption and use efficiency (Kraft et al. 2015). Interspecific competition between plants (between crops and weeds) may select for more niche differentiation traits.

The complexity of understanding and ultimately predicting the outcome of competitive interactions among individual plants, much less among species, becomes difficult given the different selective constraints or histories (Keddy 1989; Lavorel and Garnier 2002). In addition, one does not need to spend much time observing a multispecies plant community to see that natural selection must vary significantly across space and time so the interplay among the selective processes for niche differentiation, resource consumption, resource use efficiency and other competition-related traits is inconsistent (Rees et al. 2001; Dudley 1996; Kraft et al. 2015). What gives advantage to one set of traits in one place, or at one time, may not pay off for an individual at another point in space or time (e.g. Ellner 1985; Zuppinger-Dingley et al. 2014). Perennial plant species that must endure many conditions over time are challenged differently than annual species (Kunstler et al. 2016). Selection in one generation on one individual could be driven by one set of factors and another individual of the same species even in the same population could be driven by a different set of factors so if they interbreed what traits could we expect to emerge in a population? In fact, there is some consistency of selective forces and thus part of understanding the interaction among plants of different or even the same species is understanding the consistency of the environment that they have been selected upon (Kraft et al. 2015). Rotating crop species is a well-known strategy for weed management because it narrows the spectrum of species that can tolerate the temporal changes in the environment shared with the crop.

The agroecosystem environment is made up of abiotic and biotic components and these are heavily influenced by disturbance and human land management (inputs, etc.). In most crop systems soil and plant management has the goal to make the environment consistent for the crop which makes selection on the co-occurring weed species in any given generation relatively consistent. However, in crop rotations where the crop species with different life cycles, morphologies, physiologies and phenologies can change, selection is heterogeneous and thus successful traits to allow coexistence with the crops would be expected to decrease with increased crop or community diversity (Milbau and Nijs 2004). Alternatively, diverse mixtures of grasses can be selected for greater resource use efficiency traits than the same species grown in monocultures over eight generations (Zuppinger-Dingley et al. 2014). These results suggest that selection in species mixtures may therefore increase species coexistence and allow increased mixture yields in agriculture. These results suggest that crops should be selected under mixture conditions including weeds, cover crops and companion crops.

As atmospheric CO_2 concentrations and subsequent climate changes have occurred, yet another temporal variation is introduced and the successful subset of available weed species to any given site may be even more narrowed and the interaction with the crop fundamentally changed (Blumenthal et al. 2014). Rees et al. (2001) evaluated multiple plant traits and their trade-offs in structuring annual plant communities based on empirical evidence from long-term studies. They identified that seed size, rapid exploitation and differential influence of seed herbivory often determine the importance of competition or colonization as the driving selective forces determining the relative success of different species in annual plant communities. Their observations lead to the *competition-colonization trade-off* hypothesis where small-seeded species are predicted to be good colonists, but poor competitors and large seeded species are good competitors for resources but poor colonists. Although there are exceptions to this hypothesis, consideration of the potential for predicting competitive outcomes between weeds and crops must start with initial plant (seed) size as a critical trait. For example, for two plants with identical abilities to compete for resources that germinate and emerge at the same time, the one that started from a larger seed will get larger than the smaller seeded one. Kraft et al. (2015) quantified average fitness and stabilizing niche differences between 102 plant species pairs and related the differences to 11 functional traits. Competitive fitness traits were correlated with exclusion of one of the species in a pair, but coexistence of the species pairs was not necessarily correlated with traits that they identified as niche differentiating. Coexistence of species pairs could only be associated with combinations of traits, representing differentiation in multiple dimensions. Thus, even in the relatively simple selective arena of agricultural systems there is yet to be a clear emergence of prevalence of competitive or niche differentiating traits in weeds. It is clear, however, that site-specific history could determine trait prevalence but may be not intensity of competition for resources.

3 The nature of shared resource pools

Plants begin growth as an individual genet by drawing on resources from the seed, producing a radicle followed by mainly haustorial cotyledons. True leaves are formed dedicated to photosynthesis and the plant begins drawing on resources from the surrounding environment. Resources shared among plants in the soil include water and nutrients and, above ground, light, physical space and CO_2. If we turn our focus to competition for nutrients and water in the soil profile the multidimensional niche from which an individual plant draws resources is tangible and sharing the same soil volume of root systems from different plants, or among ramets of the same plant, suggests the potential for competition for resources. It turns out that plants may physically overlap, but they may be phenologically offset in time so there is minimal simultaneous growth and draw on the resource pool. For example, some weed species mostly grow after crop harvest and/or crop senescence (e.g. *Kochia scoparia* and *Salsola kali* in dry land small grain systems) which minimizes their competition with the crop (niche differentiation). It is also possible for plants to have mutualistic relationships between species that by close association in space may improve their ability to germinate, grow and ultimately

Table 1 Types of interactions between plants of two different species where the sign + or – indicates positive and negative interactions or 0 means no interaction (adapted from Burkholder 1952)

Types of interaction between species A and B	Species A	Species B
No interaction	0	0
Competition	–	–
Mutualism	+	+
Commensalism	+	0
Amensalism	0	–
Parasitism	+	–

reproduce. There is terminology for the different relationships that can occur between two species sharing the same niche (Table 1).

The main reason to identify the alternatives to competition as a process among weeds and crops or other desired species is that we may identify positive relationships and thus management would not require weed removal. In addition, the process of natural selection acting on weeds that are inherently disadvantaged in most crop systems would be to select traits that allow avoidance of competition with the crop as much as those traits selected to be more competitive with the crop. Thus, there may be specific environmental conditions where the weed does not affect the crop or the threshold for significant economic effect is higher than that for the biological effect (Cousens 1987). The environment influencing the weed-crop interaction is likely to vary on small spatial scales (m²) and from growing season to growing season (Maxwell and Luschei 2004). Weed densities recorded across

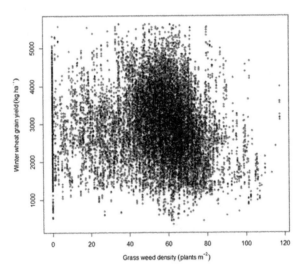

Figure 1 The relationship between dryland winter wheat grain yield and monocot weed density demonstrating variation indicative of variables other than weed abundance determining crop yield.

a wheat production field with site-specific yield information demonstrates the variation in weed impact across the field (Fig. 1). Clearly, factors other than weed density were playing a role in determining crop yield.

4 Direct competition for resources

Direct effects of interactions between plants through the resource pool are based on the study of resource uptake and allocation. Benjamin and Parks (2007) described an ecophysiological model for predicting monoculture and mixed species interactions to predict the effect of weeds on crops. Their 'Conductance' model provided a simple yet mechanistic method of simulating growth and competition for light based on space occupied by crown zones of plants in a range of competitive environments. Berger et al. (2013) similarly presented a spatially explicit, individual-based plant growth model (COMPETE) utilizing components of tested and validated mechanistic biophysical models (Maestra, PNM, LeachN and Gecros). The COMPETE model captures the major dynamics of the growth of a population of competing plants (i.e. formation of canopy hierarchies) and also simulates changes in plant transpiration as a function of plant size and competitive environment. Biophysical process models are gaining popularity particularly to predict plant species interaction changes that may occur with climate change in CO_2-enriched atmosphere. The current utility of these models at management decision spatial scales or special scales that would capture crop–weed interactions is challenged by a large number of parameters requiring measurement on expensive instruments (Deen et al. 2003).

Alternatively, competition for resources among plants may be captured if one assumes that the rate of resource uptake by individual plants is proportional to plant size, resources are finite but renewable by some rate, and plant size is asymptotic (Tilman 1988). For example, the following equations can be used to simulate growth of individual plants in a population over time when they are competing for resource in a shared pool.

$$RS_t = \frac{\delta RS}{\delta t} \tag{1}$$

$$RUN_t = \sum_{i=1}^{N} \left[v_s Si + \left(1 - \frac{q_s d_i}{1 + q_s d_i} \right) \right] \tag{2}$$

$$RAG_t = RAG_{t-1} + RS_t - RUN_t \tag{3}$$

$$GRi_t = \frac{GR_{max} RAG_t}{RAG_t + k_s} \tag{4}$$

$$Si_t = Si_{t-1} + GRi_t \tag{5}$$

where RS is the resource supply rate to an individual target plant's niche (neighbourhood); RUN is the sum of resources used by neighbouring plants over a specified time period (e.g. 1 week) and is a function of the sizes of N neighbour plants (S_i) at the previous time (starting with seed size) and their distances (d_i) from the target plant as well as species-specific resource use efficiency (v_s) and competition intensity (q_s) parameters. *RAG* is the resources available for growth in the neighbourhood for the next time increment based on what resources remain after use in the previous time (RAG_{t-1}) step, the resource supply rate (*RS*) and the *RUN*. *GRi* is the growth rate of the target plant and is a function of the maximum growth rate (GR_{max}), *RAG* and a species-specific growth efficiency parameter (k_s). Then the size of each plant at the end of each time step is Si_t calculated using Equations 1–5 in sequence. Assigning an *x* and *y* coordinate to each plant in a community allows calculation of each individual plant size over time and a mechanistic, yet empirically based, way to determine individual plant performance in a community where resources are shared (Milbau et al. 2007; Weiner 1982). The performance of plants in a multispecies community can also be explicitly considered with species-specific parameter values. In addition, by varying *RS* over time and space one can examine the range of environmental effects on competitive outcomes. This individual-based model can be parameterized by tracking plant sizes through a growing season and then estimating the resource dynamics without actually measuring the resource pool. More importantly it allows an understanding

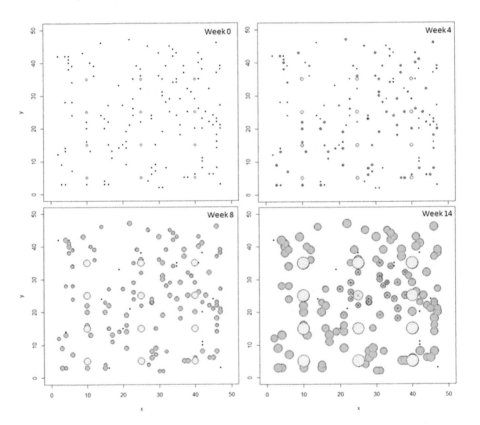

Figure 2 Plant maps showing time steps over growing season when at week 0 plants are seed (black dots) and crop (yellow dots). As plants grow circles enlarge depending on resources available. Weeds are green and crop plants are yellow. Red x indicates plants that died because they did not grow for three consecutive weeks due to lack of resources.

of the complicated dynamics among individual plants drawing on a shared resource pool and thus a theoretical basis for the quantification of impacts of weeds on crops (McGlade 1999). This model provides a visualization of the competition process (Fig. 2). What it does not account for is indirect competition where the same 3-dimensional space is shared by two species but they do not necessarily interact through the same resource pool.

The individual plant approach to characterizing population response avoids the assumptions of population-based theory (Watkinson 1996; Cousens and Mortimer 1995), averaging individual plant responses and thereby ignoring individual variation that could be critical to weed management decisions. For example, the standard for characterizing weed-crop competition outcomes was the use of average plant biomass in a population as the response variable to intra- and interspecific densities (Shinozaki and Kira 1956; Spitters 1983; Firbank and Watkinson 1985; Radosevich 1987; Roush et al. 1989). Cousens (1991) suggested the more direct quantification of competition between crops and weeds with the hyperbolic equation where weed density (N) became a direct predictor of crop yield (Y):

$$Y = Y_{wf}\left(1 - \frac{iN}{1 + iN/a}\right)$$

where Y_{wf} is the weed-free crop yield (assumed to be the maximum possible yield) and i and a are shape parameters. This equation was modified to include emergence time of the crop relative to the weed (Cousens et al. 1987) and incorporated into simulation models to demonstrate a range of alternative weed management practices (Maxwell and O'Donovan 2007). The above models have been useful for understanding first principles of weed–crop interactions, but often require significant effort to estimate the parameters with small plot experiments and the parameter variation is high over crop growing regions and from year to year limiting the application of these responses (Jasieniuk et al. 1999).

The aforementioned approaches facilitated easy quantification of the effect of the weed on the crop and more importantly the effect of the crop on the weed to determine the future ramifications of leaving weeds if not managed. However, these approaches gloss over some aspects of the interaction between crop and weeds that the more individual-based models could pick up. Further, these glossed-over aspects could reveal opportunities for more integrated weed management. For example, the role of seed size of crop or weeds, emergence time of the crop relative to the weeds or crop plant spatial arrangement could all be manipulated in the model to explore ways to reduce the impact of weeds on a crop.

A number of studies have elucidated potentially important interactions among weeds, crops and cover crops that could represent effective ecologically based weed management. However, these empirical studies without a theoretical base often conclude with the requirement for more knowledge about what drives the crop-weed interaction in order to recommend general outcomes. A mounting number of studies have been conducted to evaluate how cover crops can influence weed population dynamics by maximizing competition with weeds (Mirsky et al. 2013; Den Hollander et al. 2007) and, when intercropped, minimize competition with the cash crop while maximizing competition with the weed (Benaragama et al. 2016; Hartwig and Amon 2002). In addition, some cover crops with legume species change the soil nitrogen dynamics favouring the crop or some of the weed species with little knowledge to base generalized weed management recommendations. An interesting study was conducted to evaluate how the relative proportions of cereal rye (Secale cereale L.) and hairy vetch (Vicia villosa Roth) sown in mixtures as a cover crop influenced winter annual weed suppression and vetch nitrogen fixation (Hayden et al. 2014). Increasing the vetch in the cover mix increased nitrogen accumulation and improved weed suppression, but the mechanism for the successful result was not clear. Amini et al. (2014) provided a generalizable result drawing on competitive indices like crop leaf area index (LAI) (Zhao et al. 2006; Hansen et al. 2008) and growth rate to rank different crop varieties competitive potential with a single weed species (Steinmaus and Norris 2002; Kropff and Lotz 1992). If multiple weeds occur in the system the interactions become experimentally intractable. Thus, the understanding of the more complex weed community dynamics and effects on crops may require simulation models based on empirical parameterization of first principle relationships to build an understanding of the breadth of interactions and subsequent outcomes.

5 Indirect effects of competition

A good example of indirect competition for resources that looks like competition would be where three plants share nearly the same space but one has root morphology that places most of its roots deep in the soil profile while the co-occurring species have mid-profile or shallow root systems (Fig. 3). The shallow root system species captures small

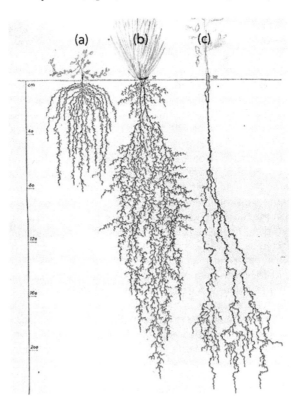

Figure 3 Three co-occurring species with different morphologies enabling them to occupy the same area but minimizing niche overlap.

volume rain event moisture and has adapted this strategy for climates where growing season precipitation comes in small pulses. The other species commits to deeply stored soil moisture that may get stored outside of the growing seasons. Under dry conditions with pulse rain events the shallow-rooted species usurp all of the moisture and are able to increase biomass whereas the deep-rooted species slows its growth as it depletes the deep soil moisture reserve. If our measurement of competition is based on relative growth of these two species one may conclude that the shallow-rooted species outcompetes the deep-rooted species for resources even though the two species are not drawing on the same resource pool (direct competition) and have adapted morphologies to effectively share the same 2-dimensional space. In a dry growing season following a wet winter, the deep-rooted species may gain more biomass and appear as the superior competitor for resources. The weed ecologist may observe these contrasting results in the field in different years and thus design the definitive experiment to determine which species is the superior competitor for resources. He/she plants the two species together in a shallow pot and determines that one species has a higher relative growth rate than the other and definitively concludes that the deep-rooted species is more efficient at drawing on the resource pool. This is direct competition because each species is competing directly with the other for resources in the same space. The results of the experiment, however, may be irrelevant to the field condition where these two species co-occur but never draw on the same resource pool because their root morphologies use different niche space.

Another example of an indirect interaction that could easily be interpreted as direct competition for resources in a shared pool is where one species like a cover crop draws down the resource pool for weed species which in turn releases the crop from competition. Wells et al. (2013) showed that rye (*Secale cereale*) used as a cover crop in soybeans could pull down the soil nitrogen to a level that redroot pigweed (*Amaranthus retroflexus*) could not tolerate but the soybeans could thrive making it look like the competition was directly between the soybeans and pigweed.

In some cases the opposite of competition (mutualism or commensalism) has been shown where co-occurring crops and weeds resulted in enhanced resource availability, creating a direct facilitation (Brooker et al. 2016). Phosphorous availability can be enhanced on acidic soils by the secretion of organic acids and phosphatases by one species (Hinsinger et al. 2011; Li, Zhang and Zhang 2013) or N transfer from nitrogen-fixing legumes to co-occurring crops or weeds (Laberge et al. 2011; Bedousac et al. 2015). Increased availability of water in the upper layers of the soil profile through hydraulic lift is shown to increase soil water for co-occurring plants (Prieto et al. 2012). Niche complementarity can result as facilitation from herbivore protection or pollinator attraction (Brooker et al. 2015). Indirect facilitation through the soil microbial community has also been documented (Van der Heijden et al. 1998; Bennett, Daniell and White 2013). A meta-analysis indicated that arbuscular mycorrhizal fungi can directly suppress weeds that are weak hosts for the fungi, but also can indirectly suppress some strong host weeds in the presence of strong host crops that ultimately obtain the advantage (Li et al. 2016). These facilitative processes can differentially provide advantage to the weeds or crop in any given system at any particular time, so their interpretation must be made with caution and careful understanding of the drivers in variation before applying what appears to be a definitive outcome or assumption about the outcome of the interaction (e.g. Fig. 1 where detection of the negative effect of weeds is hard to distinguish). Direct competition for resources between weeds and crops can be occurring, but may be masked by niche differentiation, facilitative or mutualistic processes.

6 Spatial and temporal dynamics

Most crops have been selected under conditions of plentiful resources, whereas weeds were more likely required to survive and produce offspring in a wide range of growing conditions. Thus one might expect that under a typical crop production scenario the crop might be better physiologically equipped to rapidly use resources than the co-occurring weed species. It has been widely shown that crops can offer significant competition to weeds in many cases (Andrew et al. 2015; Rasmussen 2004). Planting strategies such as narrow row spacing and early planting that maximize competition often provide advantage to the crop over co-occurring weeds and those advantages can translate across generations (Evers and Bastiaans 2016; Fahad et al. 2015; Kolb and Gallandt 2013; Maxwell and O'Donovan 2007). Planting depth and/or planting time can also increase crop competition with weeds by maximizing the likelihood of the crops capturing soil resources ahead of the weeds.

Increasing crop density by decreasing the distance between rows and/or decreasing within-row plant spacing is an optimization problem that maximizes the effect of the crop on the weed without affecting the marketable crop product. Intraspecific competition can be just as effective on the crop as interspecific competition. Generally, rectangularity near 1.0, the ratio between inter-row and intra-row distance between plants (Willey and Heath 1969), has been identified as optimum. However, research that specifically addresses the light interception by the crop canopy resulting in weed shading is also important and may reveal different ideal crop plant arrangements to maximize competition for light (Maddonni et al. 2001).

Competition for light and the influence of light quality may be interpreted as competition. The mere presence of another species can influence the reflected light quality which can have negative effects on the growth of a neighbour plant (Gundel et al. 2014; Balare an Casal 2000). It has been shown that light quality can change plant morphologies to avoid (niche differentiate) competition for light. Liu et al. (2009) proposed that changes in the red to far red ratio of light acts as an early signal of pending competition by a physiologically triggered shade avoidance response that additionally had a physiological cost-limiting plant growth. Traits that provide advantages in a competitive environment can have trade-offs, and sorting among the different mechanisms that determine weed effects on crops is no easy task and experimentally requires special methods (Swanton et al. 2015).

7 Conclusion

There is great need for understanding the potential for weed impact on crops. Plant competition for resources in a shared environment is an important aspect of understanding effects. Thousands of dollars are often spent on a single production field to control weeds and sometimes with significant potential for non-target environmental impact. Unfortunately most weed management is conducted under the precautionary principle, because there is great uncertainty about the extent of weed effects on current or future crops if not managed. Even with weeds that have been extensively studied and their crop effects quantified, translation of study results to field scale has rarely been accomplished. Annual variability and site-to-site variability have challenged the ability to make prescriptive recommendations. It has become clear that fundamental processes including competition for resources among plants and interpreting ecological processes for management recommendations requires an understanding of how to scale those processes appropriately. This remains a great challenge for Weed Scientists.

8 Where to look for further information

Future research into the interaction between weeds and desired plant species would best be focused at the individual spatial scale where each plant neighbourhood is characterized to specifically capture the potential resource interactions (Damgaard and Weiner 2017). Stratified random selection of target individual plants in a plant community followed by neighbour plant quantification by height, canopy cover, distance from target and angular dispersion in the neighbourhood are effective ways to discern the spectrum of plant interactions in a community (Bussler et al. 1995). Neighbourhood approaches to study competition for resources are often considered too detailed to be useful for management. However, a frequency distribution of properly selected neighbourhood conditions in a plant community can provide a mechanism to scale up inference space meaningful to management. Useful publications and papers cited within the chapter are listed at:

http://plen.ku.dk/english/employees/?pure=en%2Fpersons%2Fjacob-weiner(dd3c08ff-333b-4591-9968-3957845a6fa1)%2Fpublications.html.

9 References

Amini, R., Alizadeh, H. and Yousefi, A. (2014). Interference between red kidneybean (*Phaseolus vulgaris* L.) cultivars and redroot pigweed (*Amaranthus retroflexus* L.). *European Journal of Agronomy*, 60, 13–21.

Andrew, I. K. S., Storkey, J. and Sparkes, D. L. (2015). A review of the potential for competitive cereal cultivars as a tool in integrated weed management. *Weed Research*, 55(3), 239–48.

Ballare, C. and Casal, J. J. (2000) Light signals received by crop and weed plants. *Field Crops Research*, 67, 149–60.

Bedousac, L., Journet, E. P., Hauggaard-Nielsen, H., Naudian, C., Corre-Hellou, G., Jensen, E. S., Prieur, L. and Justes, E. (2015) Ecological principles underlying the increase of productivity achieved by cereal-legume intercrops in organic farming. A review. *Agronomy for Sustainable Development*, 35, 911–35. Doi 10.1007/s13593-014-0277-7.

Benaragama, D., Shirtliffe, S. J., Johnson, E. N., Duddu, H. S. N. and Syrovy, L. D. (2016). Does yield loss due to weed competition differ between organic and conventional cropping systems? *Weed Research*, 56(4), 274–83.

Benjamin, L. R. and Park, S. E. (2007). The conductance model of plant growth and competition in monoculture and species mixtures: A review. *Weed Research*, 47(4), 284–98.

Bennett, A. E., Daniell, T. J. and White, P. J. (2013) Benefits of breeding crops for yield response to soil organisms. In: *Molecular Microbial Ecology of the Rhizosphere*, Volume 1 (F. J. de Bruijn and N. J. Hoboken (Eds)), pp. 17–27. Wiley, Hoboken, USA.

Berger, A. G., McDonald, A. J. and Riha, S. J. (2013). Simulating Root Development and Soil Resource Acquisition in Dynamic Models of Crop–Weed Competition. In: Timlin, D. and Ahuja, L. R. (Eds.), *Enhancing Understanding and Quantification of Soil–Root Growth Interactions*, American Society of Agronomy, pp. 229–44.

Blumenthal, D. M., Kray, J. A., Ziska, L. H. and Dukes, J. S. (2014). Climate change, plant traits and invasion in natural and agricultural ecosystems. In: Ziska, L. H. and Dukes, J. S. (Eds.), *Invasive Species and Global Climate Change*, 4, CABI, p. 62.

Brooker, R. W., Bennett, A. E., Cong, W. F., Daniell, T. J., George, T. S., Hallett, P. D., Hawes, C., Iannetta, P. P., Jones, H. G., Karley, A. J. and Li, L. (2015). Improving intercropping: a synthesis of research in agronomy, plant physiology and ecology. *New Phytologist*, 206(1), 107–17.

Brooker, R. W., Karley, A. J., Newton, A. C., Pakeman, R. J. and Schöb, C. (2016). Facilitation and sustainable agriculture: A mechanistic approach to reconciling crop production and conservation. *Functional Ecology*, 30(1), 98–107.

Burkholder, P. R. (1952). Cooperation and conflict among primitive organisms. *American Science* 40, 601–31.

Bussler, B. H., Maxwell, B. D. and Puettmann, K. J. (1995). Using plant volume to quantify interference in corn (*Zea mays*) neighborhoods. *Weed Science*, 43, 586–94.

Cousens, R. (1991). Aspects of the design and interpretation of competition (interference) experiments. *Weed Technology*, 5(3), 664–73.

Cousens, R. and Mortimer, M. (1995). *Dynamics of Weed Populations*. Cambridge University Press, Cambridge, UK.

Cousens, R. (1987). Theory and reality of weed control thresholds. *Plant Protection Quarterly*, 2, 13–20.

Cousens, R., Brain, P., O'Donovan, J. T. and O'Sullivan, P. A. (1987). The use of biologically realistic equations to describe the effects of weed density and relative time of emergence on crop yield. *Weed Science*, 720–5.

Damgaard, C. and Weiner, J. (2017). It's about time: a critique of macroecological inferences concerning plant competition. *Trends in Ecology & Evolution*, 32(2), 86–7.

Deen, W., Cousens, R., Warringa, J., Bastiaans, L., Carberry, P., Rebel, K., Riha, S., Murphy, C., Benjamin, L. R., Cloughley, C., Cussans, J., Forcella, F., Hunt, T., Jamieson, P., Lindquist, J. and Wangs, E. (2003). An evaluation of four crop: Weed competition models using a common data set. *Weed Research*, 43(2), 116–29.

Den Hollander, N. G., Bastiaans, L. and Kropff, M. J. (2007). Clover as a cover crop for weed suppression in an intercropping design: II. Competitive ability of several clover species. *European journal of Agronomy*, 26(2), 104–12.

Dudley, S. A. (1996). Differing selection on plant physiological traits in response to environmental water availability: A test of adaptive hypotheses. *Evolution*, 92–102.

Ellner, S. (1985). ESS germination strategies in randomly varying environments. II. Reciprocal yield-law models. *Theoretical Population Biology*, 28(1), 80–116.

Evers, J. B. and Bastiaans, L. (2016). Quantifying the effect of crop spatial arrangement on weed suppression using functional-structural plant modelling. *Journal of Plant Research*, 129, 339–51.

Fahad, S., Hussain, S., Chauhan, B. S., Saud, S., Wu, C., Hassan, S., Tanveer, M., Jan, A. and Huang, J. 2015. Weed growth and crop yield loss in wheat as influenced by row spacing and weed emergence times. *Crop Protection*, 71, 101–8.

Firbank, L. G. and Watkinson, A. R. (1985). On the analysis of competition within two-species mixtures of plants. *Journal of Applied Ecology*, 503–17.

Grime, J. P. (2006). Trait convergence and trait divergence in herbaceous plant communities: Mechanisms and consequences. *Journal of Vegetation Science*, 17(2), 255–60.

Gundel, P. E., Pierik, R., Mommer, L. and Ballaré, C. L. (2014). Competing neighbors: Light perception and root function. *Oecologia*, 176(1), 1–10.

Hansen, P. K., Kristensen, K. and Willas, J. (2008). A weed suppressive index for spring barley (Hordeum vulgare) varieties. *Weed research*, 48(3), 225–36.

Hartwig, N. L. and Ammon, H. U. (2002). Cover crops and living mulches. *Weed science*, 50(6), 688–99.

Hayden, Z. D., Ngouajio, M. and Brainard, D. C. (2014). Rye–vetch mixture proportion tradeoffs: Cover crop productivity, nitrogen accumulation, and weed suppression. *Agronomy Journal*, 106(3), 904–14.

Hinsinger, P., Betencourt, E., Bernard, L., Brauman, A., Plassard, C., Shen, J., Tang, X. and Zhang, F. (2011) P for two, sharing a resource: Soil phosphorus acquisition in the rhizosphere of intercropped species. *Plant Physiology*, 156, 1078–86.

Jasieniuk, M., Maxwell, B. D., Anderson, R. L., Evans, J. O., Lyons, D. J., Miller, S. D., Morishita, D. W., Ogg Jr., A. G., Seefeldt, S., Stahlman, P. W., Northam, F. E., Westra, P., Kebede, Z. and Wicks, G. A. 1999. Site-to-site and year-to-year variation in *Triticum aestivum - Aegilops cylindrica* interference relationships. *Weed Science*, 47, 529–37.

Keddy, P. A. (1989) *Competition*. Chapman & Hall, London, UK. (2nd Ed. 2001 Kluwer Academic Publishers, Dordrecht, NL.)

Kolb, L. N. and E. R. Gallandt. (2013) Modelling population dynamics of *Sinapis arvensis* in organically grown spring wheat production systems. *Weed Research*, 53, 201–12.

Kraft, N. J., Godoy, O., and Levine, J. M. (2015). Plant functional traits and the multidimensional nature of species coexistence. *Proceedings of the National Academy of Sciences*, 112(3), 797–802.

Kropff, M. J. and Lotz, L. A. P. (1992). Systems approaches to quantify crop-weed interactions and their application in weed management. *Agricultural Systems*, 40(1–3), 265–82.

Kunstler, G., Falster, D., Coomes, D. A., Hui, F., Kooyman, R. M., Laughlin, D. C., Poorter, L., Vanderwel, M., Vieilledent, G., Wright, S. J., Aiba, M., Baraloto, C., Caspersen, J., Cornelissen, J. H., Gourlet-Fleury, S., Hanewinkel, M., Herault, B., Kattge, J., Kurokawa, H., Onoda, Y., Peñuelas, J., Poorter, H., Uriarte, M., Richardson, S., Ruiz-Benito, P., Sun, I. F., Ståhl, G., Swenson, N. G., Thompson, J., Westerlund, B., Wirth, C., Zavala, M. A., Zeng, H., Zimmerman, J. K., Zimmermann, N. E. and Westoby, M. 2016. Plant functional traits have globally consistent effects on competition. *Nature*, 529(7585), 204–7.

Laberge, G., Haussmann, B. I.G., Ambus, P. and Høgh-Jensen, H. (2011) Cowpea N rhizodeposition and its below-ground transfer to a co-existing and to a subsequent millet crop on a sandy soil of the Sudano-Sahelian eco-zone. *Plant and Soil*, 340, 369–82.

Lavorel, S. and Garnier, E. (2002). Predicting changes in community composition and ecosystem functioning from plant traits: Revisiting the Holy Grail. *Functional Ecology*, 16(5), 545–56.

Li, L., Zhang, L.-Z. and Zhang, F.-Z. (2013) Crop mixtures and the mechanisms of overyielding. In: *Encyclopedia of Biodiversity*, 2nd Edition, Volume 2 (S. A. Levin(Ed.)), pp. 382–95. Academic Press, Waltham, Massachusetts, USA.

Li, M., Jordan, N. R., Koide, R. T., Yannarell, A. C. and Davis, A. S. (2016). Meta-analysis of crop and weed growth responses to arbuscular mycorrhizal fungi: Implications for integrated weed management. *Weed Science*, 64(4), 642–52.

Liu, J. G., Mahoney, K. J., Sikkema, P. H. and Swanton, C. J. (2009). The importance of light quality in crop–weed competition. *Weed Research*, 49(2), 217–24.

Maddonni, G. A., Chelle, M., Drouet, J. L. and Andrieu, B. (2001). Light interception of contrasting azimuth canopies under square and rectangular plant spatial distributions: Simulations and crop measurements. *Field Crops Research*, 70(1), 1–13.

Maxwell, B. D. and Luschei, E. (2004). The ecology of crop-weed interactions: towards a more complete model of weed communities in agroecosystems. *Journal of Crop Improvement*, 11(1–2), 137–51.

Maxwell, B. D. and O'Donovan, J. T. (2007). Understanding weed–crop interactions to manage weed problems. In: *Non-Chemical Weed Management: Principles, Concepts and Technology* (M. K. Upadhyaya and R. E. Blackshaw (Eds), pp. 17–33. CAB International, Oxfordshire, UK.

McGlade, J. (1999). Individual-based models in ecology. In: *Advanced Ecological Theory: Principles and Applications* (J. McGlade (Ed.)). Oxford, UK: Blackwell Science, Inc.

Milbau, A., Reheul, D., De Cauwer, B. and Nijs, I. (2007). Factors determining plant–neighbour interactions on different spatial scales in young species-rich grassland communities. *Ecological Research*, 22(2), 242–7.

Milbau, A. and Nijs, I. (2004). The Role of Species Traits (Invasiveness) and Ecosystem Characteristics (Invasibility) in Grassland Invasions: A Framework 1. *Weed Technology*, 18(sp1), 1301–4.

Mirsky, S. B., Ryan, M. R., Teasdale, J. R., Curran, W. S., Reberg-Horton, C. S., Spargo, J. T. and Moyer, J. W. (2013). Overcoming weed management challenges in cover crop-based organic rotational no-till soybean production in the Eastern United States. *Weed Technology*, 27(1), 193–203.

Prieto, I., Armas, C. and Pugnaire, F. I. (2012) Water release through plant roots: New insights into its consequences at the plant and ecosystem level. *New Phytologist*, 193, 830–41.

Radosevich, S. R. (1987). Methods to study interactions among crops and weeds. *Weed Technology*, 1, 190–8.

Rasmussen, I. A. (2004). The effect of sowing date, stale seedbed, row width and mechanical weed control on weeds and yields of organic winter wheat. *Weed Reserach*, 44, 12–20.

Rees, M., Condit, R., Crawley, M., Pacala, S. W. and Tilman, D. (2001). Long-term studies of vegetation dynamics. *Science*, 293, 650–5.

Roush, M. L., Radosevich, S. R., Wagner, R. G., Maxwell, B. D. and Petersen, T. D. (1989). A comparison of methods for measuring effects of density and proportion in plant competition experiments. *Weed Science*, 37(2), 268–75.

Shinozaki, K. and Kira, T. 1956. Intraspecific competition among higher plants. VII. Logistic theory of the C-D Effect. *Journal of the Institute of Polytechnics, Osaka City University*, 7, 35–72.

Spitters, C. J. T. (1983). An alternative approach to the analysis of mixed cropping experiments. I. Estimation of competition effects. *Netherlands Journal of Agricultural Science*, 31, 1–11.

Steinmaus, S. J. and Norris, R. F. (2002). Growth analysis and canopy architecture of velvetleaf grown under light conditions representative of irrigated Mediterranean-type agroecosystems. *Weed Science*, 50(1), 42–53.

Swanton, C. J., Nkoa, R. and Blackshaw, R. E. (2015). Experimental methods for crop–weed competition studies. *Weed Science*, 63(sp1), 2–11.

Tilman, D. 1988. *Plant Strategies and the Dynamics and Structure of Plant Communities*. Princeton University Press, Princeton, NJ.

Van der Heijden, M. G. A., Klironomos, J. N., Ursic, M., Moutoglis, P., Streitwolf-Engel, R. and Boller, T., Wiemken, A. and Sanders, I. R. (1998) Mycorrhizal fungal diversity determines plant biodiversity, ecosystem variability and productivity. *Nature*, 396, 69–72.

Watkinson, A. R. (1996) Plant population dynamics. In: *Plant Ecology*, 2nd Edition (M. J. Crawley(Ed.)), pp. 359–400. Blackwell Publishing Ltd., Oxford, UK. doi:10.1002/9781444313642.ch12.

Weiner, J. (1982) A neighbourhood model of annual-plant interference. *Ecology*, 63, 1237–41.

Wells, M. S., Reberg-Horton, S. C., Smith, A. N. and Grossman, J. M. (2013). The reduction of plant-available nitrogen by cover crop mulches and subsequent effects on soybean performance and weed interference. *Agronomy Journal*, 105(2), 539–45.

Willey, R. W. and Heath, S. B. 1969. The quantitative relationships between plant population and crop yield. In: *Advances in Agronomy* (Brady, N. C. (Ed.)), pp. 281–321. R. Halls, Cornell University, Ithaca, NY.

Zhao, D. L., Atlin, G. N., Bastiaans, L. and Spiertz, J. H. J. (2006). Cultivar weed-competitiveness in aerobic rice: Heritability, correlated traits, and the potential for indirect selection in weed-free environments. *Crop Science*, 46(1), 372–80.

Zuppinger-Dingley, D., Schmid, B., Petermann, J. S., Yadav, V., De Deyn, G. B. and Flynn, D. F. (2014). Selection for niche differentiation in plant communities increases biodiversity effects. *Nature*, 515(7525), 108–11.

Chapter 3

Invasive weed species and their effects

David R. Clements, Trinity Western University, Canada

1 Introduction

Since the late 1990s, the field of invasion biology has expanded in numerous ways. In 1999 President Bill Clinton signed an executive order calling for action against invasion of alien biological species in the United States, and this translated into action throughout the world. In 'war rooms' in the United States and elsewhere, managers strategized on how to deal with these invasive species, newly recognized for the harmful effect they may have on ecosystems (Clements and Corapi, 2005). Although Elton (1958) had first conceptualized the field much earlier, it was not until the new millennium that scientific efforts began in earnest, and many new agencies were created, often to develop cooperative efforts to manage invasive species. Weeds historically have been dealt with by agricultural agencies, but more recently the threat of invasive species was recognized by numerous other stakeholders and frequently there has been disagreement between agricultural interests and other sectors. Often it is a question of definition, and the line between agricultural weeds and invasive weeds is frequently blurred (Rejmánek, 1995; Clements et al., 2004; Smith et al., 2006; Thomas and Leeson, 2007).

http://dx.doi.org/10.19103/AS.2017.0025.03

2 What is an invasive weed?

A key element of the definition of invasive weed is whether or not the species is native to a given region. The term alien invasive weed is frequently used. This turns out to be an interesting point of correspondence between invasive and 'agricultural weeds'. Although some agricultural weeds are native to a given region, the vast majority are not. Weed scientists have been hesitant to call the plants they study invasive. They are commonly referred to as environmental weeds. However, the categories are far from mutually exclusive. Many documented cases indicate that agricultural weeds may escape into nearby natural areas and vice versa. For example, intermediate levels of disturbance compared to conventional tillage present in conservation tillage lands have fostered the invasion of cropland by biennial weeds such as horseweed (*Conyza canadensis*), bull thistle (*Cirsium vulgare*), common yarrow (*Achillea millefolium*) and others (Clements et al., 2004). Sometimes genotypic variation may facilitate invasion of cropland by roadside or waste area plants, such as creeping buttercup (*Ranunculus repens*), an ecotype better adapted to arable fields (Harris et al., 1998; Clements et al., 2004).

Thomas and Leeson (2007) examined changes in weed species composition in cropland on the Canadian prairies over the twentieth century using weed survey data (Fig. 1). They observed that the weed species composition was not constant, but that new species invaded prairie cropland, with a new species appearing every decade of the surveys. In the process the proportion of alien species continually increased, reaching 70% of the prairie species by the end of the century, and accounting for 93 to 96% of the total relative abundance index in the most recent surveys. They pointed out that the alien status of most Canadian arable weed species, by definition, makes them invasive according to some sources (e.g. Pyšek et al., 2004). Thomas and Leeson (2007) called for more research on the intersection between

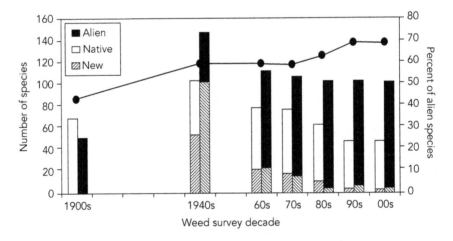

Figure 1 Number of alien and native species and percentage of aliens found in weed surveys in the Canadian prairie provinces. The sampling methodology differed among the early surveys (1900s to 1960s) and was different from the contemporary surveys (1970s to 2000s). In the contemporary survey the number of fields varied between decades. The sampling intensity differed between each survey; therefore, comparisons of total numbers of species should not be made between decades (from Thomas and Leeson, 2007).

managing agricultural and invasive species in other settings. Others agree about the benefits of sharing knowledge because the same weed species or at least closely related ones may be involved in both realms (Rejmánek, 1995; Clements et al., 2004; Smith et al., 2006).

Another key aspect of labelling a plant as invasive involves its ability to expand its local and/or geographic range. This aspect does not require the species to be alien, and thus it is quite possible for native species to be invasive, although most invasive plants are not. Rejmánek (1995) provided a conceptual model to indicate how the status of a given plant varies according to different perspectives: anthropogenic, biogeographical and ecological (Fig. 2). The anthropogenic perspective provides a challenge to defining a species as invasive, or as a weed. People have different views on what status a given species should be given, often leading to conflicts (Clements and Corapi, 2005). Even the commonly used definition of a weed as a 'plant out of place or growing where it is not wanted' (W.S. Blatchley 1912 as quoted in Zimdahl, 2007, p. 18) carries with it anthropocentric baggage. Although invasive weeds are identified by scientific disciplines (e.g. weed science, biogeography or ecology) using objective criteria, there will always be a subjective element that must be acknowledged. Figure 2 provides a helpful picture of how some of these disciplines interact.

As we move towards a definition of invasive weed, this chapter will offer and evaluate many definitions. Colautti and MacIsaac (2004) did an inventory of the literature and came up with the following list:

- a non-native species (Goodwin et al., 1999; Radford and Cousens, 2000)
- a native or non-native species that has colonized natural habitats (Burke and Grime, 1996)
- a widespread non-native species (van Clef and Stiles, 2001)
- a widespread non-native species that has a negative effect on habitat (Davis and Thompson, 2000; Mack et al., 2000)

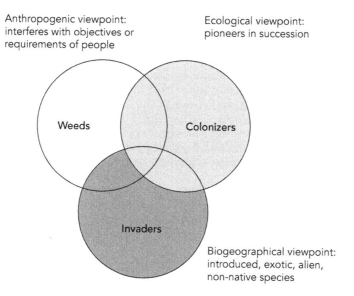

Figure 2 Weeds, colonizers and invaders from different perspectives. Redrawn from Rejmánek (1995).

The fourth definition is the most useful because it is more comprehensive, although it restricts invasive species to non-native species, and the effect on habitat is singled out, although it is widely recognized that invaders act in numerous ways.

The importance of a robust definition was emphasized when Valéry et al. (2013) argued that the discipline of invasion biology should cease to exist because it is founded on definitions that are nonsensical and counterproductive. They make their case on the fact that native species that exhibit invasive characteristics such as rapid population increase are ignored in invasion biology. In contrast, Blondel et al. (2014) argued that the discipline of invasion biology has a solid foundation and the fact that invasion biology focusses on invasive species and includes some debated concepts should not preclude it from being a *bona fide* discipline.

Blondel et al. (2014) pointed out that 'invade' is derived from the Latin term *in-vadere*, which means 'to go into', and that this must be a fundamental constituent of the science of invasion biology, making it distinct from the study of population outbreaks. They warn against definitions that are too restrictive, for example, requiring that invasions only be human-mediated. A further challenge lies in the determination of what constitutes a native species, especially considering longer timescales and the potential for long-distance dispersal by natural means (Crees and Turvey, 2015). Zimdahl (2007) held that some weeds are invasive but most are not, given the definition adopted by the International Union for Conservation of Nature and Natural Resources (IUCN):

> An alien species that becomes established in natural or seminatural ecosystems or habitats, is an agent for change, and threatens native biological diversity. (McNeely, 2001)

'Semi-natural ecosystems' are themselves difficult to define, and the relevant question here is whether or not they include agro-ecosystems. Rangelands could be described as both semi-natural and agricultural, and rangeland weeds often fit the criteria of invasive species. Although Zimdahl (2007) indicated that most weeds should not be considered invasive, he also pointed out that the 'scope of concern of weed science must expand to include invasive species.'

Legal definitions can also be helpful. For example, the legal definition in the 1999 Executive Order in the United States was: 'an alien (or non-native) species whose introduction does, or is likely to cause economic or environmental harm or harm to human health' (Executive Order 13112, 1999).

In the spirit of inclusiveness advocated by Blondel et al. (2014) and others, this chapter considers an invasive weed to be one that exhibits distribution expansion coherent with the Latin root *in-vadere*. That is, 'An invasive weed may be either native or non-native, and may have negative effects on either natural ecosystems or agroecosystems, but must clearly be invasive in that it exhibits a tendency to rapidly colonize and spread to occupy new niches.'

As mentioned earlier, the term 'weed' includes a subjective assessment, as it is seen to have negative effects from a human standpoint, but very often weeds may have innate properties that facilitate such effects, as seen in Baker's ideal weed characteristics (Baker, 1974) and other similar inventories of weed traits.

Another key question that arises from trying to define invasives is that although the effect of invasive species is a frequent subject of discussion, their effects are not always well understood or defined. Jescheke et al. (2014) developed seven questions to help define effects:

1 'Are only unidirectional changes considered or are bidirectional changes considered?'
2 'Is the definition as neutral as possible or are human values explicitly included?'
3 'Is the term *impact* only used if the change caused by a non-native species exceeds a certain threshold, or is it used for any change?'

4 'Are ecological or socio-economic changes considered, or both?'
5 'Which spatio-temporal scale is considered?'
6 'Which taxonomic or functional groups and levels of organization are considered?'
7 'Consideration of per capita change, population density, and range?'

Question 3 illustrates that the relative effect of a species may be little to none (in which case it would not be invasive) to major. The first question highlights that some invasive species may have positive effects on biodiversity, even though a negative effect of invasive species on biodiversity is held by some to be second only to habitat destruction (Wilcove et al., 1998). The following sections address some of the questions on effects of invasive species and how integrated weed management (IWM) should address them.

3 The invasion process

When an invasive species has crossed a certain abundance threshold and is clearly causing serious economic damage, standard evaluation techniques for crop pests come into play. However, there is a well-documented lag phase during most invasions (Fig. 3), which makes it challenging to assess risk of invasion and the threat posed by a given invader. The length of the lag phase can vary from a few years to centuries (Pysek and Prach, 1993; Crooks, 2005; Larkin, 2012). Various mechanisms have been postulated to explain the lag phase. Chief among these are dispersal limitations, availability of empty niches, and phenotypic or genotypic changes to the invasive species in the introduced range (Clements and DiTommaso, 2011; Perkins et al., 2013; Espeland, 2013; Murren et al., 2014). The three major phases in an invasion (Fig. 3) are introduction, expansion (after the lag phase) and saturation (Morris et al., 2013). Morris et al. found that many invaders in arid regions did not fit a logistic pattern with a lag but rather exhibited periodic spikes and crashes.

Another key consideration is that not all invasions are successful, and thus many introduced species become extinct during introduction or remain in the lag phase. Williamson's (1996) tens rule generalizes that as potential invaders enter a dispersal pathway, only 10% disperse into a new habitat, then only 10% of those establish a

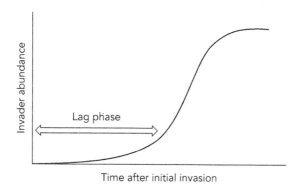

Time after initial invasion

Figure 3 Profile common to many invasions by invasive plants and other invasive species over time, illustrating the initial lag phase.

stable population in the new habitat, and finally only 10% become problematic invasive species. These problematic species are characterized by spreading widely and having serious effects, while the other 90% that form stable populations remain relatively localized.

Recent *post hoc* forensic research has attempted to provide an explanation of the invasion process with the goal of learning how to predict future invasions. Larkin (2012) utilized herbarium records to examine lag periods for invasive plant species in the U.S. Midwest, finding that 77% of species exhibited lags that ranged from 3 to 140 years. Although Larkin (2012) looked at a number of factors that could affect the length of the lag period, such as size of native range, introduction pathway, growth form and so on, no one factor provided strong predictive power. Flores-Moreno et al. (2015) also looked at herbarium specimens to trace invasion histories over 200 years, focusing on three species (*Epilobium ciliatum* Raf., *Senecio squalidus* L. and *Veronica persica* Poir.) introduced to the United Kingdom. They failed to find the expected lag in evolutionary response among the species, and concluded that evolutionary change was sufficiently fluid to occur even at the beginning of the invasion (Flores-Moreno et al., 2015).

A historical analysis of 155 tropical grasses that invaded Australia found that the long lag phase between introduction and naturalization meant that although 21 high-impact grasses found their way to Australia and escaped management, invasions of grasses are now anticipated to diminish (van Klinken et al., 2015). Morris et al. (2013) did a historic analysis of rangeland invasion in the western United States over 41–86 years to see how well the patterns of invasion corresponded to the logistic curve with three invasion phases (Fig. 3). Their results illustrated that other invasion patterns are possible, for example, 'sporadic spikes and crashes' (Morris et al., 2013). They attributed the lack of correspondence to the logistic model to the nature of the environment because the arid rangeland was subject to multi-equilibrium dynamics. Yet another forensic technique was used by Fennel et al. (2014), who analysed invasion dynamics by looking at genetic variation in seeds of *Gunnera tinctoria* buried in different soil layers. They detected a higher degree of genetic variation prior to the transition from lag to exponential phase.

An important issue generated by the lag phase is the challenge of making decisions on incipient invaders – either newly introduced or species which could potentially be introduced. Eradication is by far most feasible in early stage of invasion but it is often not apparent whether an invader is a sleeper weed that will become a major problem, or form small populations with low or no effect (Daehler, 2003; Larkin, 2012). Two key tools have been developed in response to this challenge: 1) early detection and rapid response (EDRR) and 2) weed risk assessment (WRA).

Based on the widespread occurrence of time lags during invasions, Crooks (2005) advocated taking the precautionary principle, assuming the worst of all invasive species. Regardless of initial abundance, all invaders could become problematic in the future. The precautionary approach favours EDRR in response to potential damage from waiting too long to control (Westbrooks, 2004). If invasive weeds can be contained or controlled prior to the exponential phase (see Fig. 3), the costs for management are dramatically lower. Thus, EDDR is both cost-effective and environmentally sound, because it results in relatively low levels of damage to the habitat by the invader or the control (Westbrooks, 2004). Based on an analysis of a recent deterioration in available funding for invasive species in California, Funk et al. (2014) argued that funds spent on controlling invasive weed species in the early stages of invasion result in potentially massive returns on investment. In the face of a large pool of potential invasive plants, WRA systems are a useful means to prioritize which species EDDR is to be applied to, given the assumption of limited available resources.

WRA models generally involve a questionnaire, such as the Australian WRA, often considered the standard model to use when discerning what species should be imported (McGregor et al., 2012). The Australian WRA was developed based on 49 questions, relating to the weed's performance in other areas and its potential to cause environmental or economic damage (Pheloung, 1995; Pheloung et al., 1999). These models have been used in various places for decades and have exhibited predictive accuracy (Gordon et al., 2008), but Hulme (2012) cautioned that risk assessments are not always accurate due to the difficulties inherent in measuring weed hazards objectively, predicting complex weed dynamics and uncertainty, and the subjectivity of experts. For example, when McGregor et al. (2012) tested the Australian WRA they found that although it predicted naturalization well it was not always accurate on the extent of spread. Hulme (2012) recommended augmenting the WRA approach using knowledge of experts to assess uncertainties accompanying weed population and human management dynamics (e.g. interventions to improve ecosystem resilience).

Invasive species can have tremendous economic effects, but often these occur slowly, and by the time managers begin to understand the potential it is too late (Westbrooks, 2004; Mack et al., 2000). Van Klinken et al. (2013, 2015) studied 155 tropical and subtropical grasses in Australia to determine if effects could be predicted. Major predictors were rapid rate of spread (regions invaded on a decade timescale) and semi-aquatic status (van Klinken et al., 2013). Two less robust predictors were whether or not intentional introduction occurred, such as grasses introduced for pasture, and the time elapsed since naturalization (van Klinken et al., 2013). Van Klinken et al. (2015) concluded the most important invasion pathway for Australia was intentional introduction of pasture species. This has been observed in many places for pasture grasses (Mack et al., 2000; Morris et al., 2013) and horticultural introductions (Reichard and White, 2001; Hulme, 2009; Barbier et al., 2011). These pathways are still active and globalization may increase the risk and rate of introduction.

4 Economic effects on agricultural commodities

Pimentel et al. (2000) comprehensively analysed the impact of invasive species on U.S. natural and agricultural systems, estimated to cost $138 billion per year. Weeds in crops cost $33 billion and reduced yield by 12%. Whether this cost applies to invasive weeds or simply weeds depends on one's definition. Pimentel et al. (2000) estimated that 73% of non-indigenous weeds cost $24 billion. They also noted that 45% of non-indigenous pasture weeds cost $1 billion. Furthermore, $5 billion was spent on weed management. Without a significant change in management success we must assume that costs have increased since 2000. Pimentel's cost estimates are based on numerous assumptions, which he acknowledged are difficult to make because of the lack of detailed information on invasive species and the complex relationship between the management costs and the crop-weed community. Pimentel claimed that non-indigenous weeds cause greater crop losses than natives and the estimate of $24 billion is conservative. At least 5000 non-native plants have been introduced to the United States, and have formed large populations, thus it is clear that the cost is substantial.

Duncan et al. (2004) assessed 16 invasive non-native rangeland weeds in the United States, using rate of spread, effects, area infested and other relevant data. They concluded that for many weeds there was not enough information to make accurate assessments. The total land infested by the 16 species in the United States was 51 million ha (Duncan et al., 2004).

Obviously, assessing the effects of invasive weeds on crops and rangeland in the world is challenging. Table 1 includes a few illustrative examples. Four of the species were included in Duncan et al.'s (2004) study – *Bromus tectorum*, *Carduus nutans*, *Cirsium arvense* and *Euphorbia esula*.

Table 1 Examples of effects of invasive weeds on agricultural commodities

Invasive weed species	Effect	Source
Amaranthus tuberculatus (waterhemp)	Native North American plant which evolved characteristics making it a serious agricultural pest in several states since the 1950s, reducing corn and soybean yields	Waselkov and Olsen (2014)
Bromus tectorum (cheatgrass)	Large areas of North American rangeland infested, reducing pasture quality and altering ecosystem dynamics (40 million ha in the United States)	Kelley et al. (2013)
Carduus nutans (musk thistle)	Widespread invasion of North American rangeland replacing forage species for livestock	Rauschert et al. (2015)
Centaurea solstitialis (yellow star thistle)	Covers large areas of California (6 million ha) and range extends through much of the United States and into southern Canada	DiTomaso et al. (2006)
Cirsium arvense (Canada thistle)	Serious effects on pasture systems in most temperate regions worldwide	Bourdôt et al. (2016)
Conyza canadensis (horseweed)	Establishment in agricultural fields as growers transition to minimal tillage	Clements et al. (2004)
Euphorbia esula (leafy spurge)	Established over large areas in north central North America, decreases the quality of grazing land, reducing grazing capacity	Leitch et al. (1996)
Mikania micrantha (mile-a-minute)	Invasion through the subtropics, smothering and outcompeting a variety of crops	Day et al. (2016)
Parthenium hysterophorus (parthenium weed)	Rapid invasions in >40 countries in tropics and subtropics causing yield losses in crops and pastures via competition and allelopathy	Bajwa et al. (2016)
Ranunculus repens (creeping buttercup)	Ability to invade crop environments from non-crop environments	Harris et al. (1998)

The magnitude of the effects of the ten example species (Table 1) ranges greatly. My objective is to highlight the various *types of effect* they represent. *Bromus tectorum* clearly fits the category of 'ecosystem engineer', changing the fire cycle in areas such as the Great Basin in Idaho and Utah from 60–110 years to every 3–5 years (Whisenant, 1990). The invasion of large areas of western North America by *B. tectorum* was the result of the ability of this cleistogamous winter annual to displace native perennials, native annuals and agricultural practices that favoured its spread (Mack, 1981). Although this invasion was initially facilitated by cattle ranching, it is now widespread throughout the region, regardless of disturbance level (Mack, 1981; Morrow and Stahlman, 1984; Svejcar and Tausch, 1991; Clements et al., 2007). Of the 16 species assessed by Duncan and Jachetta (2005), *B. tectorum* was the most widespread, covering 22 680 785 ha out of the total infestation of 44 197 299 ha for all species. Despite the large amount of accumulated knowledge on the effects of *B. tectorum*, Kelley et al. (2013) found that managers in Colorado and Wyoming were hesitant to devote resources to its management. Clearly there is a need for more information and better, more economically feasible methods of control (Kelley et al., 2013).

Thistle species have been particularly challenging as invaders, in pastures, crops and natural systems (Pimentel et al., 2000; Pimentel, 2009). *C. solstitalis* demonstrates how a single non-indigenous species can overrun an entire system, resulting in the loss of productive grassland in California (Pimentel et al., 2005). Young et al. (2010, 2011) conducted detailed studies on the mechanisms by which *C. solstitalis* outcompetes native plants during invasion into established communities. Communities with similar functional forms to *C. solstitialis* resisted invasion, whereas annual grass and native forb communities were ineffective in halting invasion by *C. solstitialis*, which spreads at rates from 13 to 17% per year (Duncan and Jachetta, 2005). *Carduus nutans* generally has a more localized effect and yet it is the second most frequently listed noxious weed in the United States (Skinner et al., 2000), capable of spreading at up to 22% per year (Duncan and Jachetta, 2005). Rauschert et al. (2015) sampled areas infested with *C. nutans* in Pennsylvania and found its presence greatly affected vegetation associations. Thus, as in the case of *C. solstitialis*, the nature of the invaded community is critical. As MacDougall and Turkington (2005) argued, in many cases invasive species are passengers rather than drivers of ecological change.

A thistle with effects spanning most of the temperate regions in the globe is Canada thistle, *Cirsium arvense*; in New Zealand it was estimated that the 2011–12 loss in revenue due to *C. arvense* was $685 million, affecting dairy, sheep, beef and deer production (Bourdôt et al., 2016). It is thought to have originated in the Middle East, and from there followed humanity throughout the world (Tiley, 2010). The plant tends to be a 'passenger' of highly disturbed habitats, and thrives in heavily grazed areas where competition by other species is reduced, whereas *C. arvense* itself is seldom grazed (Oswald, 1985; Casabon and Pothier, 2008; Tilley, 2010). As with *B. tectorum*, despite extensive research, there are gaps in our understanding of its effects, such as our ability to map out the seasonal growth pattern of *C. arvense*. A study by Bourdôt et al. (2016) developed a means to predict these seasonal dynamics for New Zealand which could readily be adapted to other regions affected by *C. arvense*.

Euphorbia esula (leafy spurge), though neither a thistle nor an invasive grass like *B. tectorum*, has affected large expanses of pasture in North America, at a great economic cost (Leitch et al., 1996). Duncan and Jachetta (2005) estimated that it had infested 1 487 237 acres in 17 western states and 375,154 acres in eastern states with an average annual rate of spread of 12–16% per year. A detailed report on *E. esula* (Rempel, 2010) found that there were over

1.2 million acres of leafy spurge in Manitoba, primarily affecting pastures, natural areas and waste places, with the infestation having grown 3.5 fold in 10 years. This infestation was estimated to cost the province $40.2 million per year, more than twice as much as 10 years earlier, including $10.2 million from the loss of grazing capacity, $5.8 million for chemical applications and indirect costs of $24.1 million (Rempel, 2010).

Although weeds are often segregated into two categories – 'environmental weeds' and 'agricultural weeds' – such categories may breakdown, particularly when agricultural practices change. For example, the widespread adoption of conservation tillage is promoting weeds with more perennial life histories, such as *Conyza canadensis*, horseweed (Clements et al., 2004) or *Taraxacum officinale*, dandelion (Clements et al., 1996). When soil is ploughed annually with a high disturbance implement, more annual plants with large seedbanks are promoted, whereas succession tends to progress towards biennial or perennial species in the absence of such disruptive disturbance (Swanton et al., 1993). Similarly, if there is a shift in field margins management, invasive plants that normally thrive only in natural environments may invade agricultural systems, as has been observed for *R. repens* in North America (Harris et al., 1998). Furthermore, *R. repens* appears to have gone through evolutionary changes to adapt to this new habitat (Clements et al., 2004). Similarly, Waselkov and Olsen (2014) discovered that since the 1950s *Amaranthus tuberculatus*, waterhemp, underwent evolutionary changes as this native species became an agricultural weed in North America. Even just within Illinois, 10% of weed control costs are expended on *A. tuberculatus*, amounting to $65 million per year (Waselkov and Olsen, 2014). The mechanism suggested by the study was that eastward movement of the western genetic lineage influenced by agricultural practices that provided selection pressure for the weedy form of *A. tuberculatus* to develop.

The two cosmopolitan tropical weeds (*Mikania micrantha* and *Parthenium hysterophorus*) in Table 1 are also generalists, infesting both natural and agricultural areas. *M. micrantha* is native to South America but now found throughout the Pacific Islands and southern Asia (Day et al., 2016). It infests a wide variety of crops, but is also a serious concern in native forests where it threatens rare animal species. Thus, it is a 'crossover weed' illustrating the importance of being cautious about categorizing a weed as an environmental or agricultural one. It is also capable of evolutionary adaptation (Hong et al., 2008; Li and Dong, 2009). Negative effects of *M. micrantha* are its rapid growth, high reproductive capacity and ability to re-sprout from fragments (Day et al., 2016). A bioclimatic model indicates that it is capable of invading many tropical regions where it does not yet occur, including continental Africa (Day et al., 2016).

Similarly, *P. hysterophorus* (parthenium ragweed) has many attributes that make it one of the world's worst weeds, although the details of its invasion mechanism were not well delineated until recently (Bajwa et al., 2016). Bajwa et al. (2016) explained that its success was due to morphological attributes, biological life cycle plasticity, seed biology and dispersal, competitive ability, release from natural enemies, physiological characteristics, soil relations, allelopathy, genetics and adaptability to climate change. With a considerable arsenal of attributes within each of these categories, it is little wonder that it degrades a broad array of habitats including grasslands, wastelands and cropping systems (Tamado and Milberg, 2004; Shabbir and Bajway, 2007; Dhileepan, 2009; Tadesse et al., 2010; Safdar et al., 2015). Furthermore, it is toxic to livestock and causes contact dermatitis in humans (Towers and Rao, 1992).

The species in Table 1 are examples of the kinds of effects that invasive species have on agricultural crops and rangelands, and many also degrade natural environments. The wicked problems created by invasive plants (McNeely 2013) are compounded by at least three other factors, as described in subsequent sections: indirect effects, adaptability and global forces.

5 Indirect effects

Because many of the invasive plants affecting agriculture also affect natural systems and human and/or animal health, their costs cannot be measured only in terms of crop losses (Duncan et al., 2004; Pimentel et al., 2005, Pimentel, 2009). For example, Juliá et al. (2007) found that the effect of *C. solstitialis* (yellow star thistle) on non-agricultural resources in Idaho was $4.5 million per year which was 21% of the total cost. Direct agricultural costs of $8.2 million per year included reduced grazing capacity, grower income and production outlays. However, because rangelands supply habitat for wildlife and support watershed health, *C. solstitialis* also reduced wildlife capacity, watershed quality, spending on wildlife recreation and increased water treatment costs (Juliá et al., 2007).

Some of the costs are for management of invaders: herbicides and other control methods. These costs are highly variable depending on how the invasion occurred (Fig. 3). For example, Mack et al. (2000) discussed the consequences of failing to control invasive species quickly, or failing to detect or identify future invaders. If these occur, costs will rise rapidly (Mack et al., 2000) because it is difficult 1) to determine what plants will be invasive; 2) to identify which regions are vulnerable; 3) to eradicate an established invader, and efficacy of control methods varies widely; 4) although charismatic invaders are obvious targets, ecosystem-wide management is a more strategic approach; and 5) although prevention of entry is far less costly, human nature will tend towards post-entry control.

Westbrooks (2004) has persistently argued for devoting resources to EDRR which will result in considerable savings later. To activate EDRR, early detection and reporting of plants suspected of being invasive, careful taxonomic identification and verification of suspects, detailed record keeping, assessment of new species for invasive potential and mechanisms for rapid response are mandatory (Westbrooks, 2004). A key element is WRA, which demands development of a questionnaire to assess the potential invasiveness based on known data and expert opinion (Daehler et al., 2004). Although conducting WRAs is expensive, the benefits will usually outweigh the costs (Keller et al., 2007; Hulme, 2012).

The overall effectiveness of WRA has been questioned by those who advocate a more comprehensive approach (Hulme, 2012; Leung et al., 2012). The concept of 'invasion debt' considers costs of the entire invasion cycle. The idea is that without adequate, planned protection new species inevitably invade and costs will increase. Rouget et al. (2016) tested this approach for *Acacia* species and found that although most acacias that might invade southern Africa have already invaded, there is still a potential 'establishment debt'. The addition of four new species would require a $500 million one-time management expense (clearing the *Acacia* spp.) if they were allowed to invade over the next 20 years.

Although many invasive plants are highly visible and can be controlled relatively easily, others are small and inconspicuous and produce numerous small seeds, creating challenges for EDRR approaches. Shortly after the province of British Columbia introduced EDRR, *Soliva sessilis* (carpet burweed) was used as a test case in 2005 (Polster, 2007). The first sighting in Canada was on Salt Spring Island in 1996. By 2005 the weed was found in an urban park in Victoria on Vancouver Island. Despite the expenditure of considerable time, funding and human resources, it continues to spread in British Columbia, and is still fairly abundant on Salt Spring Island (J. Brouard, pers. comm. 2015). *S. sessilis* is difficult to control because it is a very small, inconspicuous plant, which produces numerous seeds that are easily dispersed by animals, including humans (Polster, 2007).

Eriochloa villosa (woolly cupgrass) is a much more readily identified invasive plant, with stems 30–100 cm high producing conspicuous inflorescences that can be readily distinguished from other grass species, at least by experts (Darbyshire et al., 2003). It was first identified in Canada in 2000 at a site southeast of Montreal, likely introduced as a contaminant of crop seed (Darbyshire et al., 2003). *E. villosa* is a serious pest of row crops in the U.S. Corn Belt. As in the case of *S. sessilis*, an EDDR protocol was deployed, at the federal rather than provincial level; however, new populations continued to appear (Simard et al., 2015). It became a nationally regulated and quarantined weed in 2011. Seed production and dispersal from the relatively small area where it originally established is an issue, as is the development of adequate control methods. Simard et al. (2015) attempted to reduce seed production by incorporating forage legumes into crop rotation but despite the competition from legumes *E. villosa* continued to produce seeds. Since first being discovered in Canada in 2000, *E. villosa* has become well-established and it seems unlikely to be eradicated soon.

The story of woolly cupgrass in Quebec, Canada, also illustrates another type of indirect cost associated with invasive plants: the more we try to control them, the more they fight back. They evolve in response to management (Clements et al., 2004), similar to the evolution of herbicide resistance. Invasive species are capable of rapid evolution, particularly on the leading edge of an expanding range (Perkins et al., 2013). Their life history includes population viability in edge habitats and this combined with evolution of dispersal mechanisms serves to accelerate their spread (Perkins et al., 2013). Combining these selection pressures for enhanced dispersal and/or enhanced life-history attributes with management factors, climate change and globalization only serves to encourage more rapid evolutionary change in weeds (Clements and DiTomasso, 2011), unless this adaptability can be foiled by forward-thinking management strategies that anticipate evolutionary change (Clements et al., 2004).

6 Globalization and climate change effects

Although invasive plants were recognized as causing serious harm to agricultural and other commodities, in the past several decades climate change and increased global trade have served to accelerate plant invasion (Beerling et al., 1995; Bunce and Ziska, 2000; Ding et al., 2008; McDonald et al., 2009; Clements and DiTommaso, 2011; Hyvönen et al., 2012; Gallagher et al., 2013; Singh et al., 2013; Clements et al., 2014; Seebens et al., 2015). The problem can also be seen from the reverse angle, as Mack et al. (2000) referred to biotic invasions as 'agents of global change'. *All invasive species* are essentially a product of globalization, because they are introduced either intentionally or unintentionally by human agency at a global scale, just as the consensus on climate change is that it is human caused (Maibach et al., 2014). Here I highlight six species that demonstrate key responses invasive plants have to globalization and climate change, and in turn act as agents of global change themselves (Table 2).

As humanity seeks alternatives to fossil fuels, the development of biofuels has been a major strategy. Characteristics that make good biofuel crops tend to correspond very

Table 2 Examples of invasive plants associated with globalization and climate change

Invasive weed species	Effect	Source
Arundo donax (giant reed)	Cultivated as a biomass crop and although it does not produce viable seeds in cultivated areas, vegetative growth leads to risk of invasion	Barney and DiTomaso (2008)
Mikania micrantha (mile-a-minute)	Invasion through the subtropics facilitated by wind-dispersed seeds, vegetative reproduction and increased trade in South East Asia and the Pacific Islands	Day et al. (2016)
Panicum virgatum (switchgrass)	Cultivated as a biomass crop but has high seed production, rapid growth rate and broad environmental tolerance	Barney and DiTomaso (2008)
Parthenium hysterophorus (parthenium weed)	Improved water use efficiency, photosynthetic rate and higher growth rate with higher CO_2 concentrations at 25–35°C	Pandey et al. (2003)
Solidago canadensis (Canada goldenrod)	Escape from natural enemies and ability to evolve greater allelopathic and competitive ability in invaded range	Yuan et al. (2013)
Sorghum halepense (Johnson grass)	Adaptations for more northern climates for crop and non-crop environments, leading to more widespread effects	Sezen et al. (2016)

well to those that make good invasive weeds (Simberloff, 2008). Biofuels ideally can be grown with relatively few inputs, can thrive on marginal land, grow rapidly and have high reproductive capacity, mirroring some key characteristics of invasive plants (Barney and DiTomaso, 2008). Examples of global change weeds (Table 2) being cultivated as biomass crops are *Arundo donax* (giant reed) and *Panicum virgatum* (switchgrass). An obvious way of reducing the potential effects of biomass crops as potential invasive species is to curtail seed production, because *A. donax* grown in North America produces sterile seeds (Barney and DiTomaso, 2008). However, it is extremely hard to control once established and its rhizomes and fibrous roots spread and establish new plants, for example, via flooding (Barney and DiTomaso, 2008). It is being developed as a biofuel in parts of the United States, although it is legally a noxious weed in California and Texas (Barney and DiTomaso, 2008).

P. virgatum produces viable seeds in North America. Although characterized as aggressive, Barney and DiTomaso (2008) noted that there were few records of its escape from cultivation in various world regions, although this could be an artefact of relatively recent cultivation. Characteristics that would seem to predispose it to escaping cultivation include high seed production, vegetative regeneration from fragments, rapid growth rate and broad environmental tolerance (Parrish and Fike, 2005). As biofuels continue to be promoted and developed throughout the world, it is imperative that safeguards be in place to prevent any particular biofuel crop from becoming the next invasive weed problem.

The two pan-tropical invasive weeds already discussed in section 4 on agricultural effects, *M. micrantha* and *P. hysterophorus*, are also listed here because of their strong relationship to global trade and potential for range expansion under climate change (Table 2). The distributions of both species have greatly increased with expanded trade in South East Asia. Climate modelling suggests that *M. micrantha* could easily spread further (Day et al., 2016), and its spread could be exacerbated by climate change. Similarly, *P. hysterophorus* thrives in warm environments and could expand its range as tropical conditions expand poleward (Pandey et al., 2003).

Solidago canadensis (Canada goldenrod) has invaded almost every part of Asia due to increased trade, and adapted very well to its new environment, possessing enhanced allelopathy and competitive ability (Yuan et al., 2013). It is an example of a plant that in its native range (North America) is relatively innocuous and not frequently considered a weed, whereas in its invaded range it is a serious pest (Yuan et al., 2013). Two key factors are involved: 1) differences between the invaded and native environments, and 2) post-invasion evolution.

Another species exhibiting rapid evolutionary change is *Sorghum halepense* (Johnson grass) (Table 2), which shows a two-pronged invasion in North America (Sezen et al., 2016: 1)) adaptation to both non-crop and crop environments and 2) adaptation to more northern climates. *S. halepense* is a post-Columbian invader that diverged considerably in the past 200 years. Genetic analysis showed that although it initially occupied primarily agricultural habitats, *S. halepense* was able to adapt to other habitats subsequently (Sezen et al., 2016). It has also become adapted to cooler, northern climates in North America (Sezen et al., 2016), and could advance still further north (Clements and DiTommaso, 2012).

In sum, the issues surrounding three key environmental problems of our time: climate change, globalization and invasive species are clearly a wicked problem (McNeely, 2013). Wicked problems are problems wherein each component of the problem is itself highly complex and difficult to solve. Rittel and Webber (1973) listed the issues associated with wicked problems in a planning context:

1 There is no definitive formulation of a wicked problem
2 Wicked problems have no stopping rule
3 Solutions to wicked problems are not true-or-false, but good-or-bad
4 There is no immediate and no ultimate test of a solution to a wicked problem
5 Every solution to a wicked problem is a one-shot operation; because there is no opportunity to learn by trial-and-error, every attempt counts significantly
6 Wicked problems do not have an enumerable (or an exhaustively describable) set of potential solutions, nor is there a well-described set of permissible operations that may be incorporated into the plan
7 Every wicked problem is essentially unique
8 Every wicked problem can be considered to be a symptom of another problem
9 The existence of a discrepancy representing a wicked problem can be explained in numerous ways. The choice of explanation determines the nature of the problem's resolution
10 The planner has no right to be wrong

Our hope as weed scientists is to be good planners with respect to future invasions and their management, but the set of ten issues defined by Rittel and Webber (1973) apply all too well to the complicated web of questions surrounding invasive species, globalization and climate change. Will weed scientists be seen as wise planners (and managers) as the issues continue to evolve?

7 Applying IWM

Given the seriousness and complexity of invasive plant effects on agriculture and other ecosystems, the management response must be a close match in terms of seriousness and complexity if it is to succeed. IWM systems are by definition more complex than single control measures, and must consider the entire agricultural system first and foremost. Zimdahl (2007) considered the following five components essential to an IWM approach:

1 Incorporation of ecological principles
2 Use of plant interference and crop-weed competition
3 Incorporation of economic and damage thresholds
4 Integration of several weed control techniques, including selective herbicides
5 Supervised weed management frequently by a professional weed manager employed to develop a program for each crop-weed situation

Zimdahl (2007) also provided the useful metaphor for IWM as a set of tools in the carpenter's toolbox, that is, connecting with the professionalism intrinsic to component number 5. Liebman and Gallant (1997) further emphasized the value of using the tools in a truly integrated way, by avoiding too much reliance on the big hammer of chemical weed control, but instead using many little hammers.

Swanton and Weise (1991) organized IWM in crop agriculture around six major components, which can be used to illustrate applications of IWM to invasive weeds (Table 3). The five components provided by Zimdahl (2007) are intrinsic to the development of the Swanton and Weise (1991) framework.

Table 3 Applications of integrated weed management (IWM) components from Swanton and Weise (1991) to invasive species

IWM component	Application	Source
Tillage system	Zero tillage may lead to invasion of more novel invaders, particularly perennial or woody species	Ghersa et al., 2002
Critical period	Timing of grazing optimized to control *Centaurea solstitialis* (yellow star thistle) in Idaho	Wallace et al., 2008
Alternative control methods	Use of mowing to augment herbicidal control of *Peganum harmala* (African rue) in Oregon rangelands	Johnson and Davies, 2014
Crop competitiveness	Sweet potato crop utilized to compete with *Mikania micrantha* for light and nutrients in China	Shen et al., 2015
Crop-weed modelling	Use of modelling to evaluate optimal control strategies for *Andropogon gayanus* (gamba grass) in Australia	Adams and Setterfield, 2015
Crop rotation and seed bank dynamics	Invasive sunflower *Helianthus annuus* ssp. *annuus* resistant to imidazolinone herbicides requires use of crop rotation to ensure the eradication of seedbank populations	Presotto et al., 2011

In a systems approach to agriculture, one fundamental cornerstone is tillage, as producers move towards reducing the amount of soil disturbance as much as possible to provide numerous ecological benefits. In turn, ecological principles (ca. Zimdahl, 2007), such as ecological succession, come into play, whereby there is a tendency for weeds of reduced tillage systems to possess longer life cycles (e.g. biennials or perennials) as ecological succession is allowed to move further along than in high disturbance systems (Swanton et al., 1993). Thus, succession theory predicts the invasion of plants with suitable traits when zero tillage is adopted (Table 3), such as in the Rolling Pampa area of Argentina, where zero-tillage fields were invaded by seven woody species, as compared to just three woody species invading conventional tillage land (Ghersa et al., 2002). Such grassland areas requiring human intervention to ensure succession does not lead to the formation of new, undesirable communities that compromise agricultural productivity (Ghersa et al., 2002). On the other hand, reduced tillage may take advantage of the fact that later successional species tend to be less competitive, such as the shift from annuals such as *Chenopodium album* (common lambsquarters) to a greater proportion of perennial weeds such as *Taraxacum officinale* (dandelion) seen when reduced tillage is adopted in Ontario (Clements et al., 1996).

Although plant ecologists have studied plant competition extensively, there are still many aspects of plant competition that are not well understood. Knowledge of the time during the season when weeds are most competitive is valuable, so control can be focused on the critical period (Swanton and Weise, 1991). More precise knowledge of competitive interactions could similarly be used to manage invasive species. In Idaho, Wallace et al. (2008) found that if grazing by either cattle or sheep was appropriately timed, the effect on the invasive plant *Centaurea solstitialis* (yellow star thistle) was maximized (Table 3). This is just one example of numerous ways that the critical period concept may be applied to ensure that control methods target invasive plants when they are most vulnerable.

By alternative control methods, Swanton and Weise (1991) imply non-herbicidal control methods. Alternative control methods are highly favoured in many situations where invasive weeds affect natural ecosystems, because of the high likelihood of collateral damage if herbicidal control is used. Even in agricultural systems, effective alternative methods can be valuable in either replacing or simply augmenting herbicide control. Alternative methods have been particularly well-developed in management of rangeland weeds, often due to the economics of managing large land areas with minimal resources. For rangeland, Johnson and Davies (2014) found that mowing was very effective in augmenting herbicide control of *Peganum harmala* (African rue) in Oregon (Table 3).

Alternative control may not eradicate incipient invaders as effectively as herbicides, and herbicides may likewise be powerless to eradicate, and thus the question is often more on how best to manage long-term dynamics. Crop-weed modelling may provide a useful picture of these dynamics. Adams and Setterfield (2015) were able to use modelling to evaluate optimal control strategies for *Andropogon gayanus* (gamba grass) in Australia (Table 3). The exercise primarily was designed to look at different economic scenarios, but also served to examine a variety of ecological outcomes. Under a constrained budget scenario, funding control programmes annually reduced spread of *A. gayanas* by 27% annually compared to no control; investing these limited resources to fund local eradication was not as efficient (Adams and Setterfield, 2015). Given the extremely rapid spread of this species in parts of Australia, such a strategic approach could be critical in stemming the invasion.

The final components of the IWM scheme of Swanton and Weise (1991) were crop rotation and seed bank dynamics (Table 3). Presotto et al. (2011) observed the value of crop rotation and accompanying effects on seed bank dynamics for the invasive sunflower subspecies, *Helianthus annuus* ssp. *annuus* resistant to imidazolinone herbicides. This subspecies has arisen as an exoferal product of human activity (Casquero et al., 2013), with the subspecies arising as a product of evolutionary change in domestic sunflower (Ellstrand et al., 2010). The feral subspecies may interfere with its crop progenitor and be very difficult to control with herbicides. Its lower oil content leads to poorer quality harvested grains (Casquero et al., 2013). This challenging situation is further complicated by imidazolinone resistance in the wild sunflower subspecies (Presotto et al., 2011). Crop rotation serves to rotate crop-weed competition and herbicides (Swanton and Weise, 1991). Presotto et al. (2011) recommended that management practices incorporate both crop rotation and herbicide rotation. This sunflower scenario provides a lesson in the value of designing agro-ecosystems that foresee and attempt to foil the capability of invasive weeds to evolve in response to human selection pressure (Clements et al., 2004).

Frequently, one of the most important issues posed by invasive species is propagule pressure, as invasive species tend to produce more propagules in their introduced range than in their native range (Lockwood et al., 2005). Thus, managing demography must be a major focus of IWM of invasive weeds. Kerr et al. (2016) analysed control for 14 significant invasive plants throughout the globe and warned that social and economic factors must be considered.

8 Conclusion

The often explosive demography and associated rapid evolution of invasive species in response to global forces, climate change and rapid transformations in agricultural technology presents important challenges for weed scientists. In fact, this set of challenges can be defined as wicked problems that do not have easy solutions, are confounded by diverse societal values and have diverse ecological ramifications. These challenge calls for the integration of multiple approaches, well beyond any previous attempts to make IWM work.

9 Where to look for further information

The following list provides some key references on the subject of invasive weeds:

Booth, B., Murphy, S. and Swanton, C. (2010), *Invasive Plant Ecology and Natural and Agricultural Systems*. Wallingford, Oxon, UK: CABI Publishing.

Davis, M. (2009), *Invasion Biology*. New York: Oxford University Press.

Inderjit (ed.) (2009), *Management of Invasive Weeds*. Springer Science + Business Media B.V.

Perrings, C., Mooney, H., and Williamson, M. (2010), *Bioinvasions and Globalization: Ecology, Economics, Management, and Policy*. Oxford University Press.

Radosevich, S., Holt, J., and Ghersa, C. (2007), *Ecology of Weed and Invasive Plants: Relationship to Agriculture and Natural Resource Management*. John Wiley & Sons.

Shibu J., Singh, H., Batish, D. and Kohli, R. K. (2013), *Invasive Plant Ecology*. CRC Press.

10 References

Adams, V. and Setterfield, S. (2015), 'Optimal dynamic control of invasions: applying a systematic conservation approach', *Ecol. Appl.*, 25, 1131–41.

Bajwa, A., Bhagirath, S., Farooq, M., Shabbir, A. and Adkins, S. (2016), 'What do we really know about alien plant invasion? A review of the invasion mechanism of one of the world's worst weeds', *Planta*, DOI 10.1007/s00425-016-2510-x.

Baker, H. (1974), 'The evolution of weeds', *Ann. Rev. Ecol. Sys.*, 5, 1–24.

Barbier, E., Gwatipedza, J., Knowler, D. and Reichard, S. (2011), 'The North American horticultural industry and the risk of plant invasion', *Agric. Econ.*, 42(s1), 113–30.

Beerling, D., Huntley, B. and Bailey, J. (1995), 'Climate and the distribution of *Fallopia japonica*: use of an introduced species to test the predictive capacity of response surfaces', *J. Veg. Sci.*, 6, 269–82.

Blondel, B., Hoffmann, B. and Courchamp, F. (2014), 'The end of Invasion Biology: intellectual debate does not equate to nonsensical science', *Biol. Invasions*, 16, 977–9.

Bourdôt G., Hurrell G., Trolove M. and Saville, D. (2016), 'Seasonal dynamics of ground cover in *Cirsium arvense* – a basis for estimating grazing losses and economic impacts', *Weed Res.*, 56, 179–91.

Bunce, J. and Ziska, L. (2000), 'Crop ecosystem responses to climatic change: crop/weed interactions'. In: Reddy, K. and Hodges, H. (eds), *Climate Change and Global Crop Productivity*. New York, NY: CABI, pp. 333–48.

Burke, M. and Grime, J. (1996), 'An experimental study of plant community invasibility', *Ecology*, 77, 776–90.

Casabon, C. and Pothier, D. (2008), 'Impact of deer browsing on plant communities in cutover sites on Anticosti Island', *Ecoscience*, 15, 389–97.

Casquero, M., Presotto, A. and Cantamutto, M. (2013), 'Exoferaliy in sunflower (*Helianthus annuus* L.): A case study of intraspecific/interbiotype interference promoted by human activity.' *Field Crops Res.*, 142, 91–101.

Clements, D. and Corapi, W. (2005), 'Paradise lost? Setting the boundaries around invasive species', *Pers. Sci. Christian Faith*, 57, 44–54.

Clements, D. and DiTommaso, A. (2011), 'Climate change and weed adaptation: can evolution of invasive plants lead to greater range expansion than forecasted?' *Weed Res.*, 51(3), 227–40.

Clements, D. and DiTommaso, A. (2012), 'Predicting weed invasion in Canada under climate change: evaluating evolutionary potential', *Can. J. Plant Sci.*, 92, 1013–20.

Clements, D., Benoit, D., Murphy, S. and Swanton, C. (1996), 'Tillage effects on weed seed return and seedbank composition', *Weed Sci.*, 44, 314–22.

Clements, D., DiTommaso, A., Jordan, N., Booth, B., Murphy, S., Cardina, J., Doohan, D., Mohler, C. and Swanton, C. (2004), 'Adaptability of plants invading North American Cropland', *Agric. Ecosys. Environ.*, 104, 379–98.

Clements, D., Krannitz, P. and Gillespie, S. (2007), 'Seed banks of invasive and native plant species in a semi-desert shrub-steppe with varying grazing histories', *Northwest Sci.*, 81, 37–49.

Clements, D., DiTommaso, A. and Hyvönen, T. (2014), 'Ecology and management of weeds in a changing climate', In: Chauhan, B. and Mahajan G. (eds), *Recent Advances in Weed Management*. New York, NY: Springer, pp. 13–37.

Crees, J. and Turvey S. (2015), 'What constitutes a "native" species? Insights from the Quaternary faunal record', *Biol. Cons.*, 186, 143–8.

Crooks, J. (2005), 'Lag times and exotic species: the ecology and management of biological invasions in slow-motion', *Ecoscience*, 12, 316–29.

Daehler, C. (2003), 'Performance comparisons of co-occurring native and alien invasive plants: implications for conservation and restoration', *Annu. Rev. Ecol. Evol. Syst.*, 34, 183–211.

Daehler, C., Denslow, J., Ansari, S. and Kuo, H. (2004), 'A risk-assessment system for screening out invasive pest plants from Hawaii and other Pacific Islands', *Cons. Biol.*, 18(2), 360–8.

Darbyshire, S., Wilson, C. and Allison, K. (2003), 'The biology of invasive alien plants in Canada. 1. *Eriochloa villosa* (Thunb.) Kunth', *Can. J. Plant Sci.*, 83(4), 987–99.

Davis, M. and Thompson, K. (2000), 'Eight ways to be a colonizer; two ways to be an invader: a proposed nomenclature scheme for invasion ecology', *Bull. Ecol. Soc. Am.*, 81, 226–30.

Day, M., Clements, D., Gile, C., Shen, S., Weston, L. and Zhang, F. (2016), 'Biology and impacts of Pacific Island invasive species. 13. *Mikania micrantha*, mile-a-minute (Magnoliopsida: Asteraceae)', *Pacific Sci.*, 70(3), 257–85.

Dhileepan, K. (2009), 'Managing parthenium weed across diverse landscapes: prospects and limitations', In: Inderjit, (ed.), *Management of invasive weeds*. Netherlands: Springer, pp. 227–59.

Ding, J., Mack, R., Lu, P., Ren, M. and Huang, H. (2008), 'China's booming economy is sparking and accelerating biological invasions', *BioScience*, 58, 317–24.

DiTomaso, J., Kyser, G. and Pitcairn, M. (2006), *Yellow Starthistle Management Guide*. Berkeley, CA: California Invasive Plant Council Publication 2006-03. 78 pp.

Duncan, C., Jachetta, J., Brown, M., Carrithers, V., Clark, J., DiTomaso, J., Lym, R., McDaniel, K., Renz, M. and Rice, P. (2004), 'Assessing the economic, environmental, and societal losses from invasive plants on rangeland and wildlands', *Weed Technol.*, 18(sp1), 1411–16.

Duncan, C. and Jachetta, J. (2005), 'Introduction', In: Duncan, C. and Clark, J. (eds), *Invasive Plants of Range and Wildlands and Their Environmental, Economic, and Societal Impacts*. Weed Science Society of America, Lawrence, KS, pp. 1–7.

Ellstrand, N., Heredia, S., Leak-Garcia, J., Heraty, J., Burger, J., Yao, L., Nohzadeh-Malakshah, S. and Ridley, C. (2010), 'Crops gone wild: evolution of weeds and invasives from domesticated ancestors', *Evol. Appl.*, 3(5–6), 494–504.

Elton, C. (1958), *The Ecology of Invasions by Plants and Animals*. London: Methuen.

Espeland, E. (2013), 'Predicting the dynamics of local adaptation in invasive species', *J. Arid Land*, 5(3), 268–74.

Executive order 13112. 1999. Presidential documents - invasive species, by President Clinton, W. J. on 3 Feb. *Federal Register*, 64(25), 6183–6.

Fennell, M., Gallagher, T., Vintro, L. and Osborne, B. (2014), 'Using soil seed banks to assess temporal patterns of genetic variation in invasive plant populations', *Ecol. Evol.*, 4(9), 1648–58.

Flores-Moreno, H., García-Treviño, E., Letten, A. and Moles, A. (2015), 'In the beginning: phenotypic change in three invasive species through their first two centuries since introduction', *Biol. Invasions*, 17(4), 1215–25.

Funk, J., Matzek, V., Bernhardt, M. and Johnson, D. (2014), 'Broadening the case for invasive species management to include impacts on ecosystem services', *BioScience*, 64, 58–63.

Gallagher, R., Duursma, D., O'Donnell, J., Wilson, P., Downey, P., Hughes, L. and Leishman, M. (2013), 'The grass may not always be greener: projected reductions in climatic suitability for exotic grasses under future climates in Australia', *Biol. Invasions*, 15, 961–75.

Ghersa, C., de la Fuente, E., Suarez, S. and Leon, R. (2002), 'Woody species invasion in the Rolling Pampa grasslands, Argentina', *Agric. Ecosys. Environ.*, 88, 271–8.

Goodwin, B., McAllister, A. and Fahrig, L. (1999), 'Predicting invasiveness of plant species based on biological information', *Cons. Biol.*, 13, 422–6.

Gordon, D., Onderdonk, D., Fox, A. and Stocker, R. (2008). 'Consistent accuracy of the Australian weed risk assessment system across varied geographies', *Diversity Distrib.*, 14, 234–42.

Harris, S., Doohan, D., Gordon, R. and Jensen, K. (1998), 'The effect of thermal time and soil water on emergence of *Ranunculus repens*', *Weed Res.*, 38, 405–12.

Hong, L., Niu, H., Shen, H., Ye, W. and Cao, H. (2008), 'Development and characterization of microsatellite markers for the invasive weed *Mikania micrantha* (Asteraceae)', *Molecular Ecology Resources*, 8(1), 193–5.

Hulme, P. (2009). 'Trade, transport and trouble: managing invasive species pathways in an era of globalization', *J. Appl. Ecol.*, 46(1), 10–18.

Hulme, P. (2012). 'Weed risk assessment: a way forward or a waste of time?', *J. Appl. Ecol.*, 49(1), 10–19.

Hyvönen, T., Luoto, M. and Uotila, P. (2012), 'Assessment of weed establishment risk in a changing European climate', *Agric. Food Sci.*, 21, 348–60.

Jeschke, J., Bacher, S., Blackburn, T., Dick, J., Essl, F., Evans, T., Gaertner, M., Hulme, P., Kühn, I., Mrugala, A. and Pergl, J. (2014) 'Defining the impact of non-native species', *Cons. Biol.*, 28(5), 1188–94.

Johnson, D., and Davies, K. (2014), 'Effects of integrating mowing and Imazapyr application on African rue (*Peganum harmala*) and native perennial grasses', *Invasive Plant Sci. Manage.*, 7(4), 617–23.

Juliá, R., Holland, D. and Guenthner, J. (2007), 'Assessing the economic impact of invasive species: The case of yellow starthistle (*Centaurea solsitialis* L.) in the rangelands of Idaho, USA', *J. Environ. Manage.*, 85(4), 876–82.

Keller, R., Lodge, D. and Finnoff, D. (2007), 'Risk assessment for invasive species produces net bioeconomic benefits', *Proc. Nat. Acad. Sci.*, 104(1), 203–7.

Kelley, W., Fernandez-Gimenez, M. and Brown, C. (2013), 'Managing downy brome (*Bromus tectorum*) in the central rockies: land manager perspectives', *Invasive Plant Sci. Manage.*, 6, 521–35.

Kerr, N., Baxter, P., Salguero-Gómez, R., Wardle, G. and Buckley, Y. (2016), 'Prioritizing management actions for invasive populations using cost, efficacy, demography and expert opinion for 14 plant species world-wide', *J. Appl. Ecol.*, DOI 10.1111/1365-2664.12592.

Larkin, D. (2012). 'Lengths and correlates of lag phases in upper-Midwest plant invasions', *Biol. Invasions*, 14(4), 827–38.

Leung, B., Roura-Pascual, N., Bacher, S., Heikkilä, J., Brotons, L., Burgman, M., Dehnen-Schmutz, K., Essl, F., Hulme, P., Richardson, D., Sol, D., Vilà, M. and Rejmánek, M. (2012), 'TEASIng apart alien species risk assessments: a framework for best practices', *Ecol. Lett.*, 15, 1475–93.

Li, J. and Dong, H. (2009), 'Fine scale clonal structure and diversity of invasive plant *Mikania micrantha* H. B. K. and its plant parasite *Cuscata campestris* Yunker', *Biol. Invest.*, 11, 687–95.

Liebman, M., and Gallandt, E. (1997), 'Many little hammers: ecological management of crop-weed interactions', *Ecology in Agric.*, 291–343.

Lockwood, J., Cassey, P. and Blackburn, T. (2005), 'The role of propagule pressure in explaining species invasions', *Trends Ecol. Evol.*, 20(5), 223–8.

Mack, R. (1981), 'Invasion of *Bromus tectorum* L. into Western North America: an ecological chronicle', *Agro-Ecosystems*, 7, 145–65.

Mack, R., Simberloff, D., Lonsdale, W., Evans, H., Clout, M. and Bazzaz, F. (2000), 'Biotic invasions: causes, epidemiology, global consequences and control,' *Ecol. Appl.*, 10, 689–710.

MacDougall, A. and Turkington, R. (2005), 'Are invasive species the drivers or passengers of change in degraded ecosystems?' *Ecology*, 86, 42–55.

Maibach, E., Myers, T. and Leiserowitz, A. (2014). 'Climate scientists need to set the record straight: There is a scientific consensus that human-caused climate change is happening', *Earth's Future*, 2(5), 295–8.

McDonald, A., Riha, S., DiTommaso, A. and DeGaetano, A. (2009), 'Climate change and the geography of weed damage: analysis of US maize systems suggests the potential for significant range transformations', *Agric Ecosys. Environ.*, 130, 131–40.

McGregor, K., Watt, M., Hulme, P. and Duncan, R. (2012), 'How robust is the Australian Weed Risk Assessment protocol? A test using pine invasions in the Northern and Southern hemispheres', *Biol. Invasions*, 14(5), 987–98.

McNeely, J., (ed.) (2001), *The Great Reshuffling: Human Dimensions of Invasive Alien Species.* IUCN, Gland, Switzerland and Cambridge, UK., 242pp.

McNeely, J. (2013), 'Global efforts to address the wicked problem of invasive alien species', In: *Plant Invasions in Protected Areas*, Netherlands: Springer, pp. 61–71.

Morris, C., Morris, L., Leffler, A., Collins, C., Forman, A., Weltz, M. and Kitchen, S. (2013), 'Using long-term datasets to study exotic plant invasions on rangelands in the western United States', *J. Arid Environ.*, 95, 65–74.

Morrow, L. and Stahlman, P. (1984). 'The history and distribution of downy brome (*Bromus tectorum*) in North America', *Weed Sci.*, 32(Suppl. 1), 2–6.

Murren, C.; Purvis, K., Glasgow, D., Messervy, J., Penrod, M. and Strand, A. (2014), 'Investigating lag phase and invasion potential of *Vitex rotundifolia*: a coastal dune exotic', *J. Coastal Res.*, 30(4), 815–24.

Oswald, A. (1985), 'Impact and control of thistles in grassland', In: Brockman, J. (ed.), *Weeds, Pest and Diseases of Grassland and Herbage Legumes*, British Grassland Society, Croydon, UK., pp. 128–36.

Pandey, D., Palni, L. and Joshi, S. (2003), 'Growth, reproduction, and photosynthesis of ragweed parthenium (*Parthenium hysterophorus*)', *Weed Sci.*, 51, 191–201.

Parrish, D. and Fike, J. (2005), 'The biology and agronomy of switchgrass for biofuels', *Crit. Rev. Plant Sci.*, 24, 423–59.

Perkins, T., Phillips, B., Baskett, M. and Hastings, A. (2013), 'Evolution of dispersal and life history interact to drive accelerating spread of an invasive species', *Ecol. Lett.*, 16(8), 1079–87.

Pheloung, P. (1995), *Determining the weed potential of new plant introductions to Australia: a report on the development of a weed risk assessment system commissioned by the Australian Weeds Commission and the Plant Industries Committee*, Agriculture Protection Board Western Australia.

Pheloung, P., Williams, P. and Halloy, S. (1999), 'A weed risk assessment model for use as a biosecurity tool evaluating plant introductions', *J. Environ. Manage.*, 57(4), 239–51.

Pimentel, D., (2009), 'Invasive plants: their role in species extinctions and economic losses to agriculture in the USA', In: Interjit, (ed.), *Management of invasive weeds*, Netherlands: Springer, pp. 1–7.

Pimentel, D., Lach, L., Zuniga, R. and Morrison, D. (2000), 'Environmental and economic costs of nonindigenous species in the United States', *BioScience*, 50(1), 53–65.

Pimentel, D., Zuniga, R. and Morrison, D. (2005), 'Update on the environmental and economic costs associated with alien-invasive species in the United States', *Ecol. Econ.*, 52(3), 273–88.

Polster, D. (2007), 'Eradicating carpet burweed (*Soliva sessilis* Ruiz & Pavón) in Canada', In: Clements, D. and Darbyshire, S. (eds), *Invasive plants: Inventories, strategies and action*. Topics in Canadian Weed Science, Volume 5. Sainte Anne de Bellevue, Québec: Canadian Weed Science Society – Société canadienne de malherbologie, pp. 71–82.

Presotto, A., Ureta, M., Cantamutto, M. and Peverence, M. (2011), 'Effects of gene flow from IMI resistant sunflower crop to wild *Helianthus annuus* populations', *Agric. Ecosys. Environ.*, 146, 153–61.

Pysek, P. and Prach, K. (1993), 'Plant invasions and the role of riparian habitats: a comparison of four species alien to central Europe', *J. Biogeog.*, 20, 413–20.

Pyšek, P., Richardson, D., Rejmánek, M., Webster, G., Williamson, M. and Kirschner, J. (2004), 'Alien plants in checklists and floras: towards better communication between taxonomists and ecologists', *Taxon*, 53, 131–43.

Radford, I. and Cousens, R. (2000), 'Invasiveness and comparative life-history traits of exotic and indigenous *Senecio* species in Australia', *Oecologia*, 125, 531–42.

Rauschert, E., Shea, K. and Goslee, S. (2015), 'Plant community associations of two invasive Thistles', *Ann Bot-London Plants*, 7, plv065, DOI 10.1093/aobpla/plv065.

Reichard, S. and White, P. (2001), 'Horticulture as a pathway of invasive plant introductions in the United States: Most invasive plants have been introduced for horticultural use by nurseries, botanical gardens, and individuals', *BioScience*, 51(2), 103–13.

Rejmánek, M. (1995), 'What makes a species invasive?' In: Pyšek, P., Prach, K., Rejmánek, M. and Wade, M. (eds), *Plant Invasions: General Aspects and Special Problems*. Amsterdam: SPB Academic Publishing, pp. 3–13.

Rempel, K. (2010), *Economic Impact Assessment of Leafy Spurge in Southern Manitoba: Final Report.* Rural Development Institute, Brandon University.

Rittel, H. and Webber, M. (1973), 'Dilemmas in a general theory of planning', *Policy sciences*, 4(2), 155–69.

Safdar, M., Tanveer, A., Khaliq, A. and Riaz, M. (2015), 'Yield losses in maize (*Zea mays*) infested with parthenium weed (*Parthenium hysterophorus* L.)', *Crop Prot.*, 70, 77–82.

Seebens, H, Essl, F, Dawson, W., Fuentes, N., Moser, D., Perg, J., Pyšek, P., van Kleunen, M., Weber, E., Winter, M. et al. (2015), 'Global trade will accelerate plant invasions in emerging economies under climate change', *Glob. Change Biol.*, 21, 4128–40.

Sezen, U., Barney, J., Atwater, D., Pederson, G., Pederson, J., Chandler, J. M. et al. (2016), Multi-Phase US Spread and Habitat Switching of a Post-Columbian Invasive, *Sorghum halepense*. *PLoS ONE*, 11(10), e0164584, DOI 10.1371/journal.pone.0164584.

Shabbir, A. and Bajwa, R. (2007), 'Parthenium invasion in Pakistan – a threat still unrecognized', *Pak. J. Bot.*, 39, 2519–26.

Shen, S., Xu, G., Clements, D., Chen, A., Zhang, F., Jin, G. and Kato-Noguchi, H. (2015), 'Suppression of the invasive plant mile-a-minute (*Mikania micrantha*) by local crop sweet potato (*Ipomoea batatas*) by means of higher growth rate and competition for soil nutrients', *BMC Ecology*, 15, 1–10. DOI 10.1186/s12898-014-0033-5.

Simard, M., Nurse, R. and Darbyshire, S. (2015), 'Emergence and seed production of woolly cupgrass (*Eriochloa villosa*) in legume forage crops', *Can. J. Plant Sci.*, 95(3), 539–48.

Simberloff, D. (2008), 'Invasion biologists and the biofuels boom: Cassandras or colleagues', *Weed Sci.*, 56(6), 867–72.

Singh, R., Prasad, P. and Reddy K. (2013), 'Impacts of changing climate and climate variability on seed production and seed industry', *Adv. Agron.*, 118, 49–110.

Skinner, K., Smith, L. and Rice, P. (2000), 'Using noxious weed lists to prioritize targets for developing weed management strategies', *Weed Sci.*, 48, 640–4.

Smith, R., Maxwell, B., Menalled, F. and Rew, L. (2006). 'Lessons from agriculture may improve the management of invasive plants in wildland systems', *Front. Ecol. Environ.*, 4, 428–34.

Svejcar, T. and Tausch, R. (1991), 'Anaho Island, Nevada: a relict area dominated by annual invader species', *Rangelands*, 13, 233–6.

Swanton, C. and Weise S. (1991), 'Integrated weed management: the rationale and approach', *Weed Technol.*, 5, 657–63.

Swanton, C., Clements, D. and Derksen, D. (1993), 'Weed succession under conservation tillage: A hierarchical framework for research and management', *Weed Technol.*, 7, 286–97.

Tadesse, B., Das, T. and Yaduraju, N. (2010), 'Effects of some integrated management options on parthenium interference in sorghum', *Weed Biol. Manage.*, 10, 160–9.

Tamado, T. and Milberg, P. (2004), 'Control of Parthenium (*Parthenium hysterophorus*) in grain sorghum (*Sorghum bicolor*) in the smallholder farming system in eastern Ethiopia', *Weed Technol.*, 18, 100–5.

Thomas, G. and Leeson, J. (2007), 'Tracking long-term changes in the arable weed flora of Canada', In: Clements, D. and Darbyshire, S. (eds), *Invasive plants: Inventories, strategies and action. Topics in Canadian Weed Science, Volume 5*. Sainte Anne de Bellevue, Québec: Canadian Weed Science Society – Société canadienne de malherbologie, pp. 43–69.

Tiley, G. (2010). Biological Flora of the British Isles: *Cirsium arvense* (L.) Scop. *J. Ecol.*, 98(4), 938–83.

Towers, G. and Subba Rao, N. (1992), 'Impact of the pan-tropical weed, *Parthenium hysterophorus* L. on human affairs', In: *Proceedings of the first international weed control congress* 2, 134–8.

Valéry, L., Fritz, H. and Lefeuvre, J. C. (2013), 'Another call for the end of invasion biology', *Oikos*, 122(8), 1143–6.

van Clef, M. and Stiles, E. (2001), 'Seed longevity in three pairs of native and non-native congeners: assessing invasive potential', *Northeastern Naturalist*, 8, 301–10.

van Klinken, R., Panetta, F., Coutts, S. (2013), 'Are high-impact species predictable? An analysis of naturalised grasses in Northern Australia', *PLoS ONE* 8(7), e68678, DOI 10.1371/journal.pone.0068678.

van Klinken, R., Penetta, F., Coutts, S. and Simon, B. (2015), 'Learning from the past to predict the future: an historical analysis of grass invasions in northern Australia', *Biol. Invasions*, 17, 565–79.

Wallace, J., Wilson, L. and Launchbaugh, K. (2008), 'The effect of targeted grazing and biological control on yellow starthistle (*Centaurea solstitialis*) in canyon grasslands of Idaho', *Rangeland Ecol. Manage.*, 61, 314–20.

Waselkov, K. and Olsen, K. (2014), 'Population genetics and origin of the native North American agricultural weed waterhemp (*Amaranthus tuberculatus*; Amaranthaceae)', *Am. J. Bot.*, 101(10), 1726–36.

Westbrooks, R. (2004), 'New approaches for early detection and rapid response to invasive plants in the United States', *Weed Technol.*, 18, 1468–71.

Whisenant, S. (1990), Changing fire frequencies on Idaho's Snake River Plain: Ecological and Management Implications. Ogden, Utah: General Technical Report INT—U.S. Department of Agriculture, Forest Service, Intermountain Research Station.

Wilcove, D., Rothstein, D., Dubow, J., Phillips, A. and Losos, E. (1998), 'Quantifying threats to imperiled species in the United States', *BioScience*, 48, 607–15.

Williamson, M. (1996), *Biological Invasions (Population and Community Biology Series)*. New York: Chapman and Hall.

Young, S., Kyser, G., Barney, J., Claassen, V. and DiTomaso, J. (2010), 'Spatio-temporal relationship between water depletion and root distribution patterns of *Centaurea solstitialis* and two native perennials', *Restor. Ecol.*, 18, 323–33.

Young S., Kyser, G., Barney J., Claassen, V. and DiTomaso, J. (2011), 'The role of light and soil moisture in plant community resistance to invasion by yellow starthistle (*Centaurea solstitialis*)', *Restor. Ecol.*, 19, 599–606.

Yuan, Y., Wang, B., Zhang, S., Tang, J., Tu, C., Hu, S., Yong, J. W. and Chen, X. (2012), Enhanced allelopathy and competitive ability of invasive plant *Solidago canadensis* in its introduced range. *J. Plant Ecol.*, 6 (3), 253–263.

Zimdahl, R. (2007), *Fundamentals of Weed Science*. Elsevier, Amsterdam. 666 pp.

Part 2

IWM principles

Key issues and challenges of integrated weed management

C. J. Swanton and T. Valente, University of Guelph, Canada

1 Introduction

One of the greatest challenges that we face in agriculture is how to intensify agricultural productivity while at the same time enhance ecosystem services. Intensifying production while enhancing ecosystem services such as water quality, plant and animal diversity, pollinator habitat, etc., may at first glance appear to be two diametrically opposed concepts. Is it possible, through innovative research, to consider this goal to be achievable?

To address this goal, we will need to evaluate carefully how we use our current weed-management technology such as herbicides and herbicide-resistant crops. Herbicide chemistries and herbicide-resistant crops have provided excellent technologies that have resulted in significant changes to the way that weeds can be controlled. One could argue that the success of this technology has led to the development of cropping systems that rely primarily on these technologies rather than using them as a component of an integrated weed-management (IWM) programme.

IWM is based on sound agronomic principles. It involves the use of many little hammers, that is, selection pressures (Liebman and Gallandt, 1997) that on their own are not sufficient control measures, but if taken together into a system approach can represent a

http://dx.doi.org/10.19103/AS.2017.0025.04

powerful selection pressure for the control of weeds. Knowledge of these varied selection pressures is essential for growers to make informed decisions regarding their weed-management programme. An effective IWM strategy involves the systematic integration of selection pressures created by choice of tillage, crop density and row width, nutrient management, an understanding of the applicability of the critical period for weed control and the importance of a diverse crop rotation (Swanton and Murphy1996; Swanton and Weise1991). These agronomic practices need to be developed into a logical sequence with a focus on the enhancement of crop health. IWM must not be seen as increasing risk to management, but rather as useful knowledge for the development of an effective weed-management strategy. In this chapter, we highlight several key components that must form the bases for an effective IWM strategy.

2 Tillage

Disturbance caused by tillage is a powerful selection pressure imposed upon weed communities. Disturbance via tillage can be characterized by magnitude, area affected, frequency (Blackshaw et al. 1994, 2001; Booth, Murphy and Swanton 2010; Buhler and Mester 1991; Walker and Willig 1999) and timing (Smith 2006). Magnitude is a combination of physical force and severity, for example, deep ploughing. The area affected will determine the heterogeneity of the environment which will influence the availability of safe sites for seedling germination and establishment. Frequency and timing of disturbance will play a key role in determining the life cycle characteristics of species that are able to survive. For example, Derksen et al. (1993) found wind-dispersed species and volunteer crops were generally associated with reduced tillage and summer annual dicots with conventional tillage. In a long-term tillage and cropping system experiment, Buhler et al. (1994) found greater and more diverse populations of perennial weeds in reduced tillage systems compared with the mouldboard plough system. Research by Smith (2006) reported that spring tillage created weed communities dominated by annual broadleaf weeds and C_4 grasses, while fall tillage led to communities dominated by late-emerging broadleaf weeds and C_3 grasses. It should be noted that the impact of tillage on the extant weed population will also be an interaction with location, environment and the distribution of seeds within the soil profile.

   Distribution of seeds within the soil profile is a function of tillage depth and soil type. For example, Clements et al. (1996) reported that on a silt loam soil, the top 5 cm of soil contained 37 and 33% of the total seedbank in mouldboard and ridge-till systems, respectively, compared with 74 and 61% in no-till and chisel ploughed systems. Seed distribution within the mouldboard ploughed treatment was more uniformly distributed over the soil-sampling depth of 15 cm. In tillage studies initiated in 1962, in Ohio, Cardina et al. (1991) reported that 72–47% of seeds were found in the top 5 cm in a silt loam and silty clay loam soils, respectively. This study suggested that seed distribution within the seedbank differed among soil types receiving the same tillage treatment. They found a more uniform distribution of seeds for each tillage type within the soil profile of a silty clay loam soil. They suggested that seeds shedding from the parent plant may be deposited deeper within the soil profile as a result of large soil cracks that form when the soil dries. Studies conducted on a sandy soil (Barberi and Lo Cascio 2001; Swanton et al. 2000) found differences in the vertical seed distribution notably with chisel and mouldboard ploughed treatments compared with studies conducted on a finer texture soil (see also Ball 1992; Clement et al. 1996; Yenish, Doll and Buhler, 1992). This research suggested that soil texture, particularly, infiltration characteristics, soil strength and larger pore size distribution may allow seeds to penetrate deeper into the soil profile.

Keeping the vast majority of the weed seedbank close to the surface of the soil is an important IWM strategy. This is the harshest environment for the seeds in terms of environmental conditions of moisture fluctuations, temperature extremes, seed pathogens and predation (Davis and Raghu 2010; Murphy et al. 2006; Yenish, Doll and Buhler 1992; Clements et al. 1996). Keeping seeds closer to the soil surface facilitates the ability of these natural processes to contribute to the overall reduction in seed number within the seedbank. For example, daily seed predation rates for seeds on the soil surface range from 2.9% (Cromar, Murphy and Swanton 1999), 4.2 to 4.8% (Brust and House 1988) and 11% per day (Cardina and Sparrow 1996). In addition, soilborne microorganisms such as bacteria, fungi and viruses can cause seed decay, reduce seed germination and enhance seedling death (Kremer 1993; Swanton and Booth 2009). Soilborne microorganisms are generally more abundant and active near the soil surface and in tillage systems where minimum soil disturbance occurs (Swanton and Booth 2009). Over a 6-year period, Murphy et al. (2006) found that total seedbank losses were highest in no-till compared with chisel and mouldboard plough systems declining from an initial count of 41 000 seeds to 8 000 seeds per m^3 in year six of the study. This reduction in seed number is an important strategy for long-term weed management; however, in any given year, the timing of weed-seedling emergence from varying depths within the soil seedbank will influence the outcome of weed and crop competition.

3 Time of weed emergence relative to the crop

Knowledge of the time of weed emergence relative to the crop is central to our understanding of the impact of weeds on crop yield. The early-emerging weed seedlings are by far the most competitive; as the crop develops later, emerging weeds are much less competitive (Dew 1972; Hagood et al. 1981, O'Donovan et al. 1985; Kropff 1988). For example, data collected from threshold studies on redroot pigweed (*Amaranthus retroflexus* L.) in corn reported a yield loss of 5–34% caused by pigweed densities of 0.5–8 plants per metre of row, if emerging up to the fourth leaf stage of corn (Knezevic et al. 1994). Pigweed emerging after the seventh leaf stage of corn, however, did not cause any yield loss. Maximum yield loss caused by early-emerging barnyard grass seedlings (*Echinochloa crus-galli* (L.) Beauv.) ranged from 26 to 35% compared with less than 6% after the fourth leaf stage of corn growth (Bosnic and Swanton 1997). A similar response to time of weed emergence relative to the crop was reported by Cardina, Sparrow and McCoy (1995), Chikoye, Weise and Swanton (1995) and Dieleman et al. (1995, 1996). In all of these studies, there was a clear response to duration of weed interference and weed-seedling density. Knowledge of the time of weed-seedling emergence relative to the crop is an underlying principle of IWM. This fundamental principle would suggest that good weed management should be judged on decisions made in the early stage of crop development rather than solely on appearance at harvest.

4 Critical periods for weed control

The critical period for weed control is directly influenced by the time of weed emergence relative to the crop. It defines the crop growth stages during which weeds must be controlled in order to prevent unacceptable yield loss. It provides growers with knowledge of the importance of early weed control and is a guide from which the cost of weed control can be measured against yield (Norsworthy et al. 2004; Swanton and Weise 1991; Knezevic et al. 2002). The critical period for weed control describes a yield loss window and a harvest window. For example, Hall et al. (1992) defined a yield loss window for weed control to occur between the 3rd and 14th leaf tip of corn in order to maintain a yield loss of less than 2.5%. This critical period was further refined to occur between the 3rd and 10th leaf tip of corn growth for a 5% yield loss level (Swanton, unpublished data). In addition to defining the importance of early season weed control, the critical weed-free period provides a window for mechanical weed control, crop scouting and guidance for the introduction of a cover crop.

The harvest window occurs from the end of the critical period until the crop is harvested. Weeds that emerge within this window are not likely to affect crop yield, unless soil nutrient status is low, but may influence crop quality or interfere with harvest ability. It is important to understand that the cost of weed control between the yield window and the harvest window is very different. If weeds are to be controlled in the harvest window, then growers must be confident that money will be lost because of a down grading in crop quality or that the weeds will interfere with harvest operations or there will be a delay in harvest because of an increase in crop seed moisture content caused by the presence of weeds.

The duration of the critical period for weed control, however, will vary with crop (Martin, Van Acker and Friesen, 2001; Mohammadi et al., 2005; Weaver and Tan 1983) and the environment and cultural practices (Knezevic et al. 2002). For example, the critical period for soybean was estimated to occur from soybean emergence to the V4 stage of growth in order to prevent a yield loss of more than 2.5% (Van Acker et al. 1993), whereas in white bean (*Phaseolus vulgaris* L.), weed control was critical between the 2nd trifoliate and the first flower stage of growth in order to prevent a yield loss of 3% or more (Woolley et al. 1993). The environmental conditions of soil moisture, temperature and rainfall will invariably influence weed-seedling emergence patterns and thus the onset of the critical period. Cultural practices such as reducing row widths and nutrient management may enhance crop competitiveness through early canopy closure. Studies conducted in corn, however, concluded that the critical period for weed control did not differ between corn rows of 97 vs 48 cm as light interception between the rows did not vary throughout the growing season (Norsworthy and Oliviera 2004). Row spacing, however, was found to influence the critical time for weed removal in soybeans (Knezevic, Evans and Mainz 2003). The results from this study suggested that increasing row width from 19 to 76 cm reduced the tolerance of soybean to the presence of early-emerging weeds. Initiation of yield loss was detected at the first trifoliate in the 76 cm row width compared with the third trifoliate for soybeans planted in 19 cm row spacing. In addition, the amount and timing of nitrogen applied to the crop can have a significant bearing on the timing of the critical period. In the study by Evans et al. (2003), the onset of the critical period occurred earlier in corn that received 0 and 60 kg N ha^{-1} nitrogen rates compared with corn that received 120 kg N ha^{-1}. An increase in nitrogen generally decreased the length of the critical period for weed control possibly as a result of a more rapid development of the crop canopy and leaf area index.

5 Crop morphology

The concept of utilizing crop competitive morphology to suppress weeds has been a research goal for many years in weed science. One would anticipate that the morphological features required to suppress weeds would be intuitively obvious. Morphological crop features such as rapid early-seedling growth, fast rate of leaf appearance and early canopy fill are examples of plant characteristics that should provide a consistent competitive advantage (Begna et al. 2001; Buhler 2002). So, the fundamental question then becomes, why have we not made progress in breeding of weed-competitive crop cultivars and hybrids?

This lack of progress in breeding a weed-competitive crop is a result of the lack of meaningful correlations with the selected morphological traits and weed suppression. Staniforth (1962) found no differences in competitive ability between four soybean varieties under varying levels of soil fertility, planting density and weed infestation. A study conducted by Burnside (1972) compared ten soybean varieties and found three weed-competitive cultivars. Harosoy 63', Amsoy and Corsoy did not identify any unique morphological feature associated with this ability. A study conducted by Garrity et al. (1992) compared the weed suppressive and competitive abilities of twenty-five upland rice cultivars and found the characteristic that most strongly correlated with reduced weed biomass to be crop plant height. Malik et al. (1993) found that white bean varieties differed in their competitiveness with weeds with some white bean varieties significantly reducing weed biomass; late maturing white bean cultivars were found to be more competitive against weeds. Dingkuhn et al. (1999) compared sixteen lines of rice cultivars to determine their competitiveness both intra-and inter-specifically. They found a strong correlation between leaf area index, specific leaf area and tillering and competitiveness, regardless of the competitor. Watson et al. (2006) compared the ability of twenty-nine barley cultivars varying in height, seed morphology and class in their ability to compete with and withstand weed competition. Generally tall, hulled cultivars were more competitive with weeds than semi-dwarf and hull-less cultivars. This study, however, concluded that the correlation coefficients (for specific morphological traits) were not strong enough to be useful within a breeding programme. Considering this lack of consistency, the best recommendation may be to follow the simple rule that the best-yielding crop cultivars and hybrids under well-managed conditions should also be the best yielding under weedy conditions.

6 Row width and seeding density to reduce weed competitiveness

An alternative approach to enhance crop-competitive ability is to adjust row width and seeding density. The goal of adjusting row widths and seeding densities is to increase the rate and timing of crop canopy closure, thereby providing a biological control mechanism that suppresses above-ground weed biomass and growth by changing the light quantity and quality (Arce et al. 2009; Harder et al. 2007; Harker et al. 2003; Johnson et al. 1998; McLachlan et al. 1993; Tollenaar et al. 1994). For example, Wax and Pendleton (1968) attributed the observed decrease in above-ground weed biomass in narrow soybean rows (25 cm) compared with wide rows planted at 75 cm spacing to increased shading from crop plants in narrow rows. Similar results were found by Légère and Schreiber (1989) with narrow rows (25 cm) resulting in reduced redroot pigweed biomass and improved soybean leaf area production. Malik et al. (1993) evaluated the ability of row spacing and seeding rate of white beans to suppress the competitive ability of weeds; in this study, weed-competitive ability was not reduced. Similarly, Valente (2017) investigated the impact of differing soybean cultivar morphology and seeding rate as a tool to suppress weed growth. The results of this study found that increasing the number of plants per hectare did not increase yield nor improve weed control. Adjusting row width and seeding densities is definitely a valuable strategy to assess for weed suppression; however, the results may vary with crop type and growing conditions.

7 Nutrient management

The principle behind nutrient management studies has been to enhance crop competitiveness with weeds by selective placement and timing of nutrients, primarily nitrogen. The ability to utilize nitrogen placement and timing as an effective IWM strategy, however, has not resulted in consistent results (see Table 1). Of the studies cited in Table 1, 40% reported a decrease in weeds, 40% reported an increase in weeds and 20% reported variable results as a result of selective placement and timing of nitrogen. For example, Carlson and Hill (1986) attributed an increase in wild oat density and subsequent competitive ability to the addition of varying nitrogen rates and timing in wheat. Kirkland and Beckie (1998) reported a reduction in green foxtail (*Setaria viridis* L. (Beauv.)) density, regardless of nitrogen placement, in spring wheat. Nitrogen applied as a side-banded or broadcast treatment increased crop competitiveness with weeds, resulting in a reduction in weed biomass and density and increased grain yield. In a study utilizing both cultivated and weedy rice (or red rice) (*Oryza sativa*) Burgos et al. (2006), concluded that the competitive ability of weedy rice was enhanced by greater nitrogen use efficiency, resulting in a more rapid biomass accumulation compared with cultivated rice.

Research on nitrogen management as it applies to weed management begs the question whether higher rates of nitrogen are required to enhance the competitive ability of high-yielding hybrids and cultivars? The alternative view would suggest that modern stress-tolerant hybrids and cultivars are higher yielding under both low and high nitrogen conditions when compared with older hybrids and cultivars. In a study conducted by Tollenaar et al. (1994), old corn hybrids (Pride 5, released in 1959) and more recently released hybrids were compared under conditions of high and low nitrogen and high and low weed pressure. The effect of weed competition was greatest under conditions of low nitrogen and high weed pressure. Importantly, the old hybrid suffered the greatest yield loss under these conditions compared with the newer hybrids. These results would suggest that new high-yielding, stress-tolerant hybrids and cultivars do not require higher levels of nitrogen to reduce the competitive effects of weeds.

Table 1 Impact of nitrogen application on weed population and biomass in various cropping systems (1962–2006)

Author/year	Crop	Nitrogen application method	Timing	Impact on weed population/biomass (+/−)
Sexsmith and Pittman, 1963	Wild oat (*Avena sterilis* (L.))	Broadcast	Pre-plant	−
Reinertsen Cochran and Morrow, 1984	Wheat (*Triticum aestivum* (L.))	Surface and banded	At planting	Surface (+) Banded (−)
Carlson and Hill, 1986	Spring wheat (*Triticum aestivum* (L.))	Broadcast	Pre-plant	+
Cochran, Morrow and Schirman, 1990	Winter wheat (*Triticum aestivum* (L.))	Broadcast and deep-banded	Fall application	−
Tollenaar et al., 1994	Corn (*Zea mays* (L.))	Broadcast	Pre-plant	+
Jornsgard et al., 1996	Spring and winter wheat (*Triticum aestivum* (L.))	N/A	At planting	−
Kirkland and Beckie, 1998	Spring wheat (*Triticum aestivum*)	Side banded and broadcast	Pre-plant	−
Blackshaw, Semach and Janzen, 2002	Spring wheat (*Triticum aestivum* (L.))	Broadcast, surface pools, point injected	Pre-plant	Broadcast (+) Surface pools (+) Point injected (−)
Blackshaw et al., 2003	Various weed species	Surface applied	After planting	+
Evans et al., 2003	Corn (*Zea mays* (L.))	Broadcast	Pre-plant	−
Blackshaw, Molnar and Janzen, 2004	Spring wheat (*Triticum aestivum* (L.))	Broadcast, banded, point injected	Fall or spring application	Broadcast (+) Banded (+) Point injected (−)
Brainard, DiTommaso and Mohler, 2006	Vegetable farms (Powell amaranth (*Amaranthus powellii*))	N/A	Early season or split	+
Burgos et al., 2006	Rice (*Oryza sativa* (L.))	Hand applied	Post emergence	+
Hungria et al., 2006	Soybean (*Glycine max* (L.))	Broadcast	At planting and/or R2 or R4	+
Osborne and Riedell, 2006	Soybean (*Glycine max* (L.))	Banded	At planting	−

+ increased weed populations and biomass; − decreased weed populations and biomass.

8 Crop rotation

Intuitively, the more diverse a crop rotation is, the greater is the opportunity for variable selection pressures such as planting date, crop ground cover and canopy closure to influence weed population dynamics. Liebman and Dyck (1993) suggested that the use of varying crop sequences, within a rotation, may alter periods of direct resource competition, thereby reducing the competitive effects of weeds on crops. A diverse crop rotation reduces the ability of the weed population to adapt to similar growth habits and nutrient requirements of the crop (Liebman and Dyck 1993). It is, however, difficult to isolate only the effect of crop rotation on weed population dynamics. Most studies involving crop rotation are usually conducted under a cropping system perspective, utilizing additional variables of tillage and herbicides. For example, Schreiber (1992) reported that crop rotation compared with continuous corn reduced giant foxtail (*Setaria faberii* (Herrm.)) populations with the greatest reduction occurring as a result of the most diverse rotation (i.e. soybean/wheat/corn rotation) in mouldboard, chisel and no-till systems. In western Canada, weed communities in continuous cropping systems tended to have greater total weed densities and were more similar in composition than cropping systems that incorporated a fallow treatment (Derksen et al. 1993). Doucet et al. (1999), however, concluded that weed management was more effective than crop rotation in determining weed density and weed species diversity, whereas Cardina, Herms and Doohan (2002) concluded that the interaction of both crop rotation and tillage was critical in influencing weed seed abundance and composition in the soil seedbank. Multiple studies have reported similar results, supporting the general principle that in all tillage systems, diverse cropping systems provide enhanced opportunities for an effective weed-management strategy which then resulted in a reduction in both weed abundance and impact on crop yield (Ball 1992; Booth, Murphy and Swanton 2003; Cardina, Herms and Doohan 2002; Derksen et al. 2002; Liebman and Davis 2000; Schreiber 1992; Stevenson et al. 1997; Westerman et al. 2005).

9 Future trends and conclusion

The components of IWM have received considerable attention in terms of research (see Buhler, 2002; Mortensen, Bastiaans and Sattin 2000). Although growers may not identify with IWM, we would suggest that many growers already apply several of the key components as part of their agronomic programme, that is, reduced tillage, seeding density and critical periods for weed control. These current practices have been field tested and proven under variable environmental conditions. The next advancement for IWM may well be dependent upon the environmental and quality demands placed on the commodities that we produce and trade on a global basis. A common demand within the global market may be 'how green is your commodity?' 'What is the carbon footprint of your production system?' 'Have you completed a life cycle analysis?' (See also Buhler 2002; Swanton et al. 2008; Wall et al. 2001).

Deytieux et al. (2012) asked the fundamental question 'Is IWM efficient for reducing environmental impacts of cropping systems?' In this study, a life cycle assessment was conducted on five different cropping systems involving differences in crop rotation sequencing, fertilization, pesticide use and soil tillage. Their analysis suggested that a cropping system implementing key components of IWM improved 'energy input, greenhouse gas emission, eutrophication and acidification and eco-toxicity and human toxicity'. Indicators of biodiversity and soil quality, however, were not consistently different among systems.

Global concerns regarding the environmental impact of intensive agricultural production, loss of biodiversity, pesticide residues and climate change will invariably influence our future research direction. The challenge for weed-science research will be to anticipate international demands regarding environmental impact and food quality that will influence our ability to trade commodities at a premium price. In order to do this, IWM research must change from being a descriptive science to a predictive science. We continue to need improved decision support systems, risk-management analyses and the knowledge of the interactions of cropping systems and weed management. Precision agriculture tools may provide us with the ability to manage spatial and temporal variabilities of weed populations. Advancements in precision sprayer technology, the detection and mapping of weed populations and the development of self-propelled robotic cultivators will continue. The application of these new technologies within our cropping system will need to be thoroughly tested in terms of profitability and environmental stewardship. Can precision agriculture reduce the amount of herbicide applied into the environment and, as a result, enhance ecosystem services such as water quality? The enhancement of ecosystem services is clearly an expectation for the future of IWM.

10 Where to look for further information

In addition to the references listed in this chapter, further readings may include:

- The United Nations Food and Agriculture Organization publication entitled, "Conservation of natural resources for sustainable agriculture. What you should know about Integrated Weed Management. http://www.fao.org/ag/ca/training_materials/cd27-english/wm/weeds.pdf
- The United Nations Food and Agriculture Organization publication entitled, "What is Integrated Weed Management", http://www.fao.org/agriculture/crops/thematic-sitemap/theme/spi/scpi-home/managing-ecosystems/integrated-weed-management/iwm-what/en/.

11 References

Arce, G. D., Pedersen, P. and Hartzler, R. G., 2009. Soybean seeding rate effects on weed management. *Weed Technology*, 23(1), pp. 17–22.

Ball, D. A., 1992. Weed seedbank response to tillage, herbicides, and crop rotation sequence. *Weed Science*, 40, pp. 654–9.

Barberi, P. and Lo Cascio, B., 2001. Long-term tillage and crop rotation effects on weed seedbank size and composition. *Weed Research*, 41(4), pp. 325–40.

Begna, S. H., Hamilton, R. I., Dwyer, L. M., Stewart, D. W., Cloutier, D., Assemat, L., Foroutan-Pour, K. and Smith, D. L., 2001. Morphology and yield response to weed pressure by corn hybrids differing in canopy architecture. *European Journal of Agronomy*, 14(4), pp. 293–302.

Blackshaw, R. E., Brandt, R. N., Janzen, H. H., Entz, T., Grant, C. A. and Derksen, D. A., 2003. Differential response of weed species to added nitrogen. *Weed Science*, 51(4), pp. 532–9.

Blackshaw, R. F., Larney, F. O., Lindwall, C. W. and Kozub, G. C., 1994. Crop rotation and tillage effects on weed populations on the semi-arid Canadian prairies. *Weed Technology*, 8, pp. 231–7.

Blackshaw, R. E., Larney, F. J., Lindwall, C. W., Watson, P. R. and Derksen, D. A., 2001. Tillage intensity and crop rotation affect weed community dynamics in a winter wheat cropping system. *Canadian Journal of Plant Science*, 81(4), pp. 805–13.

Blackshaw, R. E., Semach, G. and Janzen, H. H., 2002. Fertilizer application method affects nitrogen uptake in weeds and wheat. *Weed Science*, 50(5), pp. 634–41.

Blackshaw, R. E., Molnar, L. J. and Janzen, H. H., 2004. Nitrogen fertilizer timing and application method affect weed growth and competition with spring wheat. *Weed Science*, 52(4), pp. 614–22.

Booth, B. D., Murphy, S. D. and Swanton, C. J., 2003. *Weed Ecology in Natural and Agricultural Systems*. CABI Publishing, UK, ISBN 0 85199 528 4, 303pp.

Booth, B. D., Murphy, S. D. and Swanton, C. J., 2010. *Invasive Plant Ecology in Natural and Agricultural Systems*. CABI Publishing, UK, ISBN 13: 978 1 84593 605 1, 214pp.

Bosnic, A. C. and Swanton, C. J., 1997. Influence of barnyard grass (*Echinochloa crus-galli*) time of emergence and density on corn (*Zea mays*). *Weed Science*, 45, pp. 276–282.

Brainard, D. C., DiTommaso, A. and Mohler, C. L., 2006. Intraspecific variation in germination response to ammonium nitrate of Powell amaranth (*Amaranthus powellii*) seeds originating from organic vs. conventional vegetable farms. *Weed Science*, 54(3), pp. 435–42.

Brust, G. E. and House, G. J., 1988. Weed seed destruction by arthropods and rodents in low-input soybean agroecosystems. *American Journal of Alternative Agriculture*, 3(1), pp. 19–25.

Buhler, D. D. and Mester, T. C., 1991. Effect of tillage systems on the emergence depth of giant (*Setaria faberii*) and green foxtail (*Setaria viridis*). *Weed Science*, 39, pp. 200–203.

Buhler, D. D., 1992. Population dynamics and control of annual weeds in corn (*Zea mays*) as influenced by tillage systems. *Weed Science*, 40, pp. 241–48.

Buhler, D. D., Gunsolus, J. L. and Ralston, D. F., 1992. Integrated weed management techniques to reduce herbicide inputs in soybean. *Agronomy Journal*, 84(6), pp. 973–8.

Buhler, D. D., Stoltenberg, D. E., Becker, R. L. and Gunsolus, J. L., 1994. Perennial weed populations after 14 years of variable tillage and cropping practices. *Weed Science*, 42, pp. 205–9.

Buhler, D. D., 2002. 50th Anniversary – Invited Article: Challenges and opportunities for integrated weed management. *Weed Science*, 50(3), pp. 273–80.

Burgos, N. R., Norman, R. J., Gealy, D. R. and Black, H., 2006. Competitive N uptake between rice and weedy rice. *Field Crops Research*, 99(2), pp. 96–105.

Burnside, O. C., 1972. Tolerance of soybean cultivars to weed competition and herbicides. *Weed Science*, 20, pp. 294–7.

Cardina, J., Regnier, E. and Harrison, K., 1991. Long-term tillage effects on seed banks in three Ohio soils. *Weed Science*, 39, pp. 186–94.

Cardina, J., Sparrow, D. H. and McCoy, E. L., 1995. Analysis of spatial distribution of common lambsquarters (*Chenopodium album*) in no-till soybean (*Glycine max*). *Weed Science*, 43, pp. 258–268.

Cardina, J. and Sparrow, D. H., 1996. A comparison of methods to predict weed seedling populations from the soil seedbank. *Weed Science*, 44, pp. 46–51.

Cardina, J., Herms, C. P. and Doohan, D. J., 2002. Crop rotation and tillage system effects on weed seedbanks. *Weed Science*, 50(4), pp. 448–60.

Carlson, H. L. and Hill, J. E., 1986. Wild oat (*Avena fatua*) competition with spring wheat: Effects of nitrogen fertilization. *Weed Science*, 34, pp. 29–33.

Chikoye, D., Weise, S. F. and Swanton, C. J., 1995. Influence of common ragweed (*Ambrosia artemisiifolia*) time of emergence and density on white bean (*Phaseolus vulgaris*). *Weed Science*, 43, pp. 375–80.

Clements, D. R., Benott, D. L., Murphy, S. D. and Swanton, C. J., 1996. Tillage effects on weed seed return and seedbank composition. *Weed Science*, 44, pp. 314–22.

Cochran, V. L., Morrow, L. A. and Schirman, R. D., 1990. The effect of N placement on grass weeds and winter wheat responses in three tillage systems. *Soil and Tillage Research*, 18(4), pp. 347–55.

Cromar, H. E., Murphy, S. D. and Swanton, C. J., 1999. Influence of tillage and crop residue on postdispersal predation of weed seeds. *Weed Science*, 47, pp. 184–94.

Davis, A. S. and Raghu, S., 2010. Weighing abiotic and biotic influences on weed seed predation. *Weed Research*, 50(5), pp. 402–12.

Derksen, D. A., Lafond, G. P., Thomas, A. G., Loeppky, H. A. and Swanton, C. J., 1993. Impact of agronomic practices on weed communities: Tillage systems. *Weed Science*, 41, pp. 409–17.

Derksen, D. A., Anderson, R. L., Blackshaw, R. E. and Maxwell, B., 2002. Weed dynamics and management strategies for cropping systems in the northern Great Plains. *Agronomy Journal*, 94(2), pp. 174–85.

Dew, D. A., 1972. An index of competition for estimating crop loss due to weeds. *Canadian Journal of Plant Science*, 52(6), pp. 921–7.

Deytieux, V., Nemecek, T., Knuchel, R. F., Gaillard, G. and Munier-Jolain, N. M., 2012. Is integrated weed management efficient for reducing environmental impacts of cropping systems? A case study based on life cycle assessment. *European Journal of Agronomy*, 36(1), pp. 55–65.

Dieleman, A., Hamill, A. S., Weise, S. F. and Swanton, C. J., 1995. Empirical models of pigweed (*Amaranthus* spp.) interference in soybean (*Glycine max*). *Weed Science*, 43, pp. 612–18.

Dieleman, A., Hamill, A. S., Fox, G. C. and Swanton, C. J., 1996. Decision rules for postemergence control of pigweed (*Amaranthus* spp.) in soybean (*Glycine max*). *Weed Science*, 44, pp. 126–32.

Dingkuhn, M., Johnson, D. E., Sow, A. and Audebert, A. Y., 1999. Relationships between upland rice canopy characteristics and weed competitiveness. *Field Crops Research*, 61(1), pp. 79–95.

Doucet, C., Weaver, S. E., Hamill, A. S. and Zhang, J., 1999. Separating the effects of crop rotation from weed management on weed density and diversity. *Weed Science*, 47, pp. 729–35.

Evans, S. P., Knezevic, S. Z., Lindquist, J. L., Shapiro, C. A. and Blankenship, E. E., 2003. Nitrogen application influences the critical period for weed control in corn. *Weed Science*, 51(3), pp. 408–17.

Garrity, D. P., Movillon, M. and Moody, K., 1992. Differential weed suppression ability in upland rice cultivars. *Agronomy Journal*, 84(4), pp. 586–91.

Hagood Jr, E. S., Bauman, T. T., Williams Jr, J. L. and Schreiber, M. M., 1981. Growth analysis of soybeans (*Glycine max*) in competition with jimsonweed (*Datura stramonium*). *Weed Science*, 29, pp. 500–4.

Hall, M. R., Swanton, C. J. and Anderson, G. W., 1992. The critical period of weed control in grain corn (*Zea mays*). *Weed Science*, 40, pp. 441–7.

Harder, D. B., Sprague, C. L. and Renner, K. A., 2007. Effect of soybean row width and population on weeds, crop yield, and economic return. *Weed Technology*, 21(3), pp. 744–52.

Harker, K. N., Clayton, G. W., Blackshaw, R. E., O'Donovan, J. T. and Stevenson, F. C., 2003. Seeding rate, herbicide timing and competitive hybrids contribute to integrated weed management in canola (*Brassica napus*). *Canadian Journal of Plant Science*, 83(2), pp. 433–40.

Hungria, M., Franchini, J. C., Campo, R. J. and Graham, P. H., 2005. The importance of nitrogen fixation to soybean cropping in South America. In: *Nitrogen Fixation in Agriculture, Forestry, Ecology, and the Environment* (pp. 25–42). Springer Netherlands.

Johnson, G. A., Hoverstad, T. R. and Greenwald, R. E., 1998. Integrated weed management using narrow corn row spacing, herbicides, and cultivation. *Agronomy Journal*, 90(1), pp. 40–6.

Jørnsgård, B., Rasmussen, K., Hill, J. and Christiansen, J. L., 1996. Influence of nitrogen on competition between cereals and their natural weed populations. *Weed Research*, 36(6), pp. 461–70.

Kirkland, K. J. and Beckie, H. J., 1998. Contribution of nitrogen fertilizer placement to weed management in spring wheat (*Triticum aestivum*). *Weed Technology*, 12, pp. 507–14.

Knezevic, S. Z., Weise, S. F. and Swanton, C. J., 1994. Interference of redroot pigweed (*Amaranthus retroflexus*) in corn (*Zea mays*). *Weed Science*, 42, pp. 568–73.

Knezevic, S. Z., Evans, S. P., Blankenship, E. E., Van Acker, R. C. and Lindquist, J. L., 2002. Critical period for weed control: The concept and data analysis. *Weed Science*, 50(6), pp. 773–86.

Knezevic, S. Z., Evans, S. P. and Mainz, M., 2003. Row spacing influences the critical timing for weed removal in soybean (*Glycine max*) 1. *Weed Technology*, 17(4), pp. 666–73.

Kremer, R. J., 1993. Management of weed seed banks with microorganisms. *Ecological Applications*, 3(1), pp. 42–52.

Kropff, M. J., 1988. Modelling the effects of weeds on crop production. *Weed Research*, 28(6), pp. 465–71.

Légère, A. and Schreiber, M. M., 1989. Competition and canopy architecture as affected by soybean (*Glycine max*) row width and density of redroot pigweed (*Amaranthus retroflexus*). *Weed Science*, 37, pp. 84–92.

Liebman, M. and Dyck, E., 1993. Crop rotation and intercropping strategies for weed management. *Ecological Applications*, 3(1), pp. 92–122.

Liebman, M. and Gallandt, E. R., 1997. Many little hammers: Ecological management of crop-weed interactions. In: *Ecology in Agriculture*, pp. 291–343.

Liebman, M. and Davis, A. S., 2000. Integration of soil, crop and weed management in low-external-input farming systems. *Weed Research (Oxford)*, 40(1), pp. 27–47.

Malik, V. S., Swanton, C. J. and Michaels, T. E., 1993. Interaction of white bean (*Phaseolus vulgaris* L.) cultivars, row spacing, and seeding density with annual weeds. *Weed Science*, 41, pp. 62–8.

Martin, S. G., Van Acker, R. C. and Friesen, L. F., 2001. Critical period of weed control in spring canola. *Weed Science*, 49(3), pp. 326–33.

McLachlan, S. M., Tollenaar, M., Swanton, C. J. and Weise, S. F., 1993. Effect of corn-induced shading on dry matter accumulation, distribution, and architecture of redroot pigweed (*Amaranthus retroflexus*). *Weed Science*, 41, pp. 568–73.

Mohammadi, G., Javanshir, A., Khooie, F. R., Mohammadi, S. A. and Zehtab Salmasi, S., 2005. Critical period of weed interference in chickpea. *Weed Research*, 45(1), pp. 57–63.

Mortensen, D. A., Bastiaans, L. and Sattin, M., 2000. The role of ecology in the development of weed management systems: An outlook. *Weed Research (Oxford)*, 40(1), pp. 49–62.

Murphy, S. D., Clements, D. R., Belaoussoff, S., Kevan, P. G. and Swanton, C. J., 2006. Promotion of weed species diversity and reduction of weed seedbanks with conservation tillage and crop rotation. *Weed Science*, 54(1), pp. 69–77.

Norsworthy, J. K. and Oliveira, M. J., 2004. Comparison of the critical period for weed control in wide- and narrow-row corn. *Weed Science*, 52(5), pp. 802–7.

O'Donovan, J. T., Remy, E. A. D. S., O'Sullivan, P. A., Dew, D. A. and Sharma, A. K., 1985. Influence of the relative time of emergence of wild oat (*Avena fatua*) on yield loss of barley (*Hordeum vulgare*) and wheat (*Triticum aestivum*). *Weed Science*, pp. 498–503.

Osborne, S. L. and Riedell, W. E., 2006. Starter nitrogen fertilizer impact on soybean yield and quality in the Northern Great Plains. *Agronomy Journal*, 98(6), pp. 1569–74.

Reinertsen, M. R., Cochran, V. L. and Morrow, L. A., 1984. Response of spring wheat to N fertilizer placement, row spacing, and wild oat herbicides in a no-till system. *Agronomy Journal*, 76(5), pp. 753–6.

Schreiber, M. M., 1992. Influence of tillage, crop rotation, and weed management on giant foxtail (*Setaria faberi*) population dynamics and corn yield. *Weed Science*, 40, pp. 645–53.

Sexsmith, J. J. and Pittman, U. J., 1963. Effect of nitrogen fertilizers on germination and stand of wild oats. *Weeds*, 11, pp. 99–101.

Smith, R. G., 2006. Timing of tillage is an important filter on the assembly of weed communities. *Weed Science*, 54(4), pp. 705–12.

Staniforth, D. W., 1962. Responses of soybean varieties to weed competition. *Agronomy Journal*, 54(1), pp. 11–13.

Stevenson, F. C., Legere, A., Simard, R. R., Angers, D. A., Pageau, D. and Lafond, J., 1997. Weed species diversity in spring barley varies with crop rotation and tillage, but not with nutrient source. *Weed Science*, pp. 798–806.

Swanton, C. J. and Weise, S. F., 1991. Integrated weed management: The rationale and approach. *Weed Technology*, 5, pp. 657–63.

Swanton, C. J. and Murphy, S. D., 1996. Weed science beyond the weeds: The role of integrated weed management (IWM) in agroecosystem health. *Weed Science*, 44, pp. 437–45.

Swanton, C. J., Shrestha, A., Knezevic, S. Z., Roy, R. C. and Ball-Coelho, B. R., 2000. Influence of tillage type on vertical weed seedbank distribution in a sandy soil. *Canadian Journal of Plant Science*, 80(2), pp. 455–7.

Swanton, C. J., Mahoney, K. J., Chandler, K. and Gulden, R. H., 2008. Integrated weed management: Knowledge-based weed management systems. *Weed Science*, 56(1), pp. 168–72.

Swanton, C. J. and Booth, B. D., 2009. Management of weed seedbanks in the context of populations and communities. *Weed Technology*, 18, pp. 1496–502.

Tollenaar, M., Nissanka, S. P., Aguilera, A., Weise, S. F. and Swanton, C. J., 1994. Effect of weed interference and soil nitrogen on four maize hybrids. *Agronomy Journal*, 86(4), pp. 596–601.

Valente, T., 2017. Evaluation of Cultural Weed Control Methods in Soybean *(Glycine max)* to Improve Crop Competitiveness, University of Guelph, pp. 1–62.

Van Acker, R. C., Swanton, C. J. and Weise, S. F., 1993. The critical period of weed control in soybean [*Glycine max* (L.) Merr.]. *Weed Science*, 41, pp. 194–200.

Wall, E., Weersink, A. and Swanton, C., 2001. Agriculture and ISO 14000. *Food Policy*, 26(1), pp. 35–48.

Watson, P. R., Derksen, D. A. and Van Acker, R. C., 2006. The ability of 29 barley cultivars to compete and withstand competition. *Weed Science*, 54(4), pp. 783–92.

Wax, L. M. and Pendleton, J. W., 1968. Effect of row spacing on weed control in soybeans. *Weed Science*, 16, pp. 462–5.

Weaver, S. E. and Tan, C. S., 1983. Critical period of weed interference in transplanted tomatoes (*Lycopersicon esculentum*): Growth analysis. *Weed Science*, 31, pp. 476–81.

Westerman, P. R., Liebman, M., Menalled, F. D., Heggenstaller, A. H., Hartzler, R. G. and Dixon, P. M., 2005. Are many little hammers effective? Velvetleaf (*Abutilon theophrasti*) population dynamics in two-and four-year crop rotation systems. *Weed Science*, 53(3), pp. 382–92.

Willig, M. R. and Walker, L. R., 1999. Disturbance in terrestrial ecosystems: Salient themes, synthesis, and future directions. In: *Ecosystems of the World*, pp. 747–68.

Woolley, B. L., Michaels, T. E., Hall, M. R. and Swanton, C. J., 1993. The critical period of weed control in white bean (*Phaseolus vulgaris*). *Weed Science*, pp. 180–84.

Yenish, J. P., Doll, J. D. and Buhler, D. D., 1992. Effects of tillage on vertical distribution and viability of weed seed in soil. *Weed Science*, 40, pp. 429–33.

Chapter 5

Ethical issues in integrated weed management[1]

Robert L. Zimdahl, Colorado State University, USA

1 Introduction

Regardless of whether one lives in a developed or developing country and whether one is rich or poor, male or female, educated or not, we all live in a post-industrial, information-age society. We are also fortunate to live in an era of scientific achievement and technological progress, perhaps unequalled in human history, which has not only created the good standard of life many of us enjoy but also some of the problems we face. The achievements include:

- Waking up this morning to music from your cell phone.
- Preparing breakfast in your microwave as you review the news on your computer, which gives you nearly instant access to information that is orders of magnitude greater than the resources of most of the world's libraries.
- Medical advances that cure what used to kill or cripple.
- Immunization to prevent childhood diseases.
- Elimination of smallpox and possibly polio in the near future.
- Vastly improved detection and control of some diseases.
- Travel at speeds and convenience unknown to our grandparents, across oceans and mountains that were once formidable barriers.
- Finally, because of agricultural research and technology abundance of food for many, though, sadly not for all.

[1] This chapter was first published in the 50th Anniversary Celebratory Volume of the Asia-Pacific Weed Science Society, 2017. Nimal Chandrasena and Adusumilli Naraayana Rao (Eds.), *26th Asian-Pacific Weed Science Society Conference: Golden Jubilee Conference, Kyoto, Japan.* 19–22 September 2017, Balaji Scan Ltd, Lakadikapul, India (ISBN: 978-81-931978-5-1).

http://dx.doi.org/10.19103/AS.2017.0025.24

Particular problems facing agriculture include climate change, global warming, pollution of all forms, social inequality, environmental degradation and soil erosion. Weed science, a subdivision of agriculture, poses additional problems: herbicide resistance, invasive species, biotech/genetically modified organisms (GMOs) and concerns about sustainability. Many citizens of developed countries know and benefit from the achievements of agricultural science and technology and are concerned about the problems science and technology have wrought.

We live in a world where progress, which is frequently equated with growth, is expected and generally regarded as good. Many want more of the good things of life and expect the future to be bigger, better, easier and arrive faster. We exult in the good and lament the bad. So many aspects of our life change faster than we are able to keep up. We may not always know our destination, but we are going there in a hurry. We are beneficiaries and believers in the efficacy of technology, which promises to solve the problems of society, agriculture and our extractive, industrial economy. Those involved in agriculture believe that development and use of more and more sophisticated, high-energy, advanced technology is always good and more technology will be better. The agricultural problems caused by the unintended consequences of technological solutions will, many are certain, be solved by more high-tech solutions – by improved technology. Absolutism, and its intellectual cousin certitude, are the great diseases of philosophical thought about ideas and ideals.

I do not mean or intend to imply that we should abandon science and its resultant technology. I do assert that we need 'to abandon the narcissistic illusion that we can control our interventions in an infinitely complex world' (Jensen 2016). We humans, earth's dominant species, are not just figures in the landscape – we are shapers of the landscape (Bronowski 1973, p.1). It is my view that we should carefully think about whether our shaping of the landscape is desirable and sustainable. Although we may always know what we are doing, we should and are obligated to consider what we may be undoing. A degree of intellectual humility might compel us to be more careful with our tinkering (Jensen). We need to cultivate in ourselves and our students the intellectual humility that helps us be more careful with our science and our technology and leads to thought about the moral dimension of what we do and undo.

2 Ethical principles

With that brief introduction, it is my intent to ask two questions (Zimdahl 2012):

- How do you *decide* what you choose to do is the right thing to do?
- How do you *know* what is the right thing to do?

We all have a sense of what is right and wrong, which is often unexamined and not supported by careful reasoning. A guide towards helping decide what one ought to do is found in our ethical principles, which are guides not answers that help us decide and may govern what is right and wrong.

Understanding and using ethical principles, our invisible guides, is often complicated by confusion over what ethics is and is not. Ethics is not four things:

- It is not a set of prohibitions – ethics is not a set of do not rules concerned with our behaviour or religion.
- Ethics is not an ideal system that is noble in theory but useless in practice. The reverse is true. An ethical judgement that is no good in practice has serious theoretical faults.
- Ethical principles are not relative to time, culture, society or parents. They are influenced by, but are not necessarily determined by, them.
- Ethics is not just subjective. Many people think that an ethical act is always deemed to be right or wrong based on one's feelings and nothing more. Such an attitude means there is no such thing as an objective right or wrong.

Critical thinking, an intellectually disciplined process of conceptualizing, analysing and evaluating information gathered from observation and experience and using it as a guide to belief and action, is required. It employs universal intellectual values: clarity, accuracy, relevance, sound evidence, good reasons and fairness. It is difficult.

We all have personal ethics which guide our daily behaviour. We are subject to social ethical expectations about torture, pornography, civil rights and treatment of children. Scientists also have professional ethics: do not fabricate data, give proper credit, be honest, include opposing results and disclose conflicts of interest.

When ethical standards go wrong, as they do, they are a prelude to sweatshops, mistreatment of women, concentration camps, child labour and torture. We also know that ethical standards change. Examples include business hours, smoking, women's rights and treatment of animals and the environment.

3 Ethics in agriculture

All academic disciplines have an ethical component. I suggest that the truest test of the moral condition of any scientific or other discipline, indeed of one's life, is a willingness to examine its moral condition. In agriculture, we have not examined our ethical base or the reasons for it. We have assumed that agriculture had an adequate ethical foundation. That assumption was not questioned. We all make assumptions and we do not want them questioned, we want to use them. In natural resources and environmental study, examination of the ethical base has occurred because public pressure demanded.

Philosophers study ethics. They do not tell us what is right and wrong, they show us how to think about what is right and wrong. Change occurs because an unexamined principle, once articulated and brought into the light for examination, is found lacking and abandoned in favour of other more carefully constructed thinking. The US culture provides several examples of change: civil rights, women's rights, environmental rights and animal rights. But there has been no comparable change and little critical thinking in agriculture about our ethical foundation.

Scientists claim that in the scientific realm there are answers which can be defined mathematically and are publicly verifiable, literal, definitive, precise and falsifiable. In contrast, as stated above, it is common to believe that ethical positions are purely subjective – they are only opinions and lack a rational justification. That is false. When I say slavery, Nazi Germany or female genital mutilation are wrong, is this merely my view, that I, at this time and place in my society, say they are wrong? No! There is widespread, perhaps near universal agreement that these things are wrong and the reasons provided across cultures will be similar. Ethical claims are supported by careful, logical reasoning. The normative, descriptive language speaks of what is most important and why it is or ought to be valued. Moral/ethical reasoning reflects a long, distinguished history of rational public discourse.

The ethical position that characterizes agriculture is productionism. It is the central, indeed often the only norm, of agriculture. The moral imperative is to produce food and fibre to benefit all humanity. It is what must be sustained. Those involved in agriculture, whether they are producers, suppliers or researchers, regardless of their employer should ask and debate if production is a sufficient criterion for judging all agricultural activities. Does it justify everything? What about other specific responsibilities: achieving sustainable production practices, decreasing pollution, eliminating soil erosion, eliminating harm to other plant and animal species, ending habitat destruction and ending water pollution and mining of water for irrigating agricultural crops. All segments of the agricultural enterprise ought to work towards accomplishing these equally worthy, morally good goals. Developing integrated weed management systems should, at a minimum, consider these goals.

Agricultural scientists have assumed that as long as their research and the resultant technology increase food production and availability, they and the end users are somehow exempt from negotiating and re-negotiating the moral bargain that is the foundation of the modern democratic state (Thompson 1989). It is unquestionably a moral good to feed people. Therefore it is assumed that anyone who questions agriculture's morality or the results of its technology simply does not understand the importance of what is being done. It is assumed that agricultural practitioners are technically capable and that the good results of their technology make them morally correct. We are obliged to question that assumption:

> We have lived by the assumption that what was good for us would be good for the world. We have been wrong. For I do not doubt that it is only on the condition of humility and reverence before the world that our species will be able to remain in it.
>
> (Berry 1999)

The public is concerned about pesticides in soil, water and food; cruelty to animals; biotech/GMOs; corporate agriculture; mining of water; loss of small farms and rural communities; loss of genetic diversity; pollution by animal factory wastes; exploitation of and cruelty to agricultural labour; and soil erosion. These are not just concerns of a radical fringe of society. They are general societal concerns and an agricultural system that justifies everything because it increases production faces ethical challenges, which should not be ignored.

Agriculture is the essential human activity. It is the largest, most widespread and most important human interaction with the environment. Because agriculture has well-defined, unavoidable, negative environmental consequences, it is my view that agriculture must develop and have a firm ethical foundation. It is not just about results. We should not assume that because those in agriculture believe in what they do, and the results have been mostly good – more people are fed than ever before – that those who practise and support agriculture automatically have societal acceptability.

Those in agriculture and in weed science are certain about the moral correctness, the goodness, of their activity. The basis of that moral certainty is not clear to those who have it. Therefore, agriculture's moral certainty is potentially harmful because it is unexamined by most of its practitioners. Moral certainty and the absence of reasoned discourse and debate inhibit discussion about what sustainable integrated weed management systems ought to do or be. Debate will uncover the foundational moral theories, the often invisible foundation on which actions rest. Debate will reveal the reasons, the justification for deciding that what one does is what one ought to do. Exploration of agriculture's moral foundation will not reveal a single guiding principle that will solve all agricultural dilemmas. It will reveal several principles that will be useful as alternative production technologies are explored.

Western agriculture is a productive marvel which is envied by many societies where hunger rather than abundance dominates. Science and technology have created steady yield increases by development of higher yielding cultivars, synthetic fertilizers, improved soil management and mechanization, and improved weed control. Without the yield increases that have occurred since 1960, the world would now require an additional 10–12 million square miles [roughly equal to the land area of the United States, the European Union and Brazil combined (Avery 1997)] for producers to achieve present levels of food production. Modern high-yield agriculture may not be one of the world's problems, but rather the solution to providing sufficient food for all, sufficient land for wildlife and protecting the environment. But there are risks.

The technology required to feed the world has always exposed people to risk. In the past most of the risk was borne by users of the technology. Now many risks of agricultural technology are borne by others. Technology developers, sellers, regulators and users, in their moral certainty (Zimdahl 2002) have not secured or even considered how to secure the public's consent to use technology that exposes people to involuntary risk.

Agricultural producers and those who support them with technology have been seduced into thinking that, as long as they increase food availability, they are exempt from seeking societal approval for employing the technology that modern agriculture requires, which exposes people to involuntary risks. That is not how modern democracies are supposed to work. A result is that citizens of democratic societies have become reluctant to entrust their water, their diets or their natural resources blindly into the hands of farmers, agribusiness firms and agricultural scientists. Another result is development of small-scale farmers' markets where consumers trust the food and those who produce it and demand more governmental regulation of agricultural and weed management practices.

Agricultural people must participate in the dialogue that leads to social consensus about risks, and they must be willing to contribute the time and resources required to understand the positions of their fellow citizens. For most non-agricultural segments of society, these are not new demands. For agriculture they are. Agriculturalists have been so certain of the moral correctness of their pursuit of increased production that they have failed to listen to and understand the positions of other interest groups (e.g. environmental, organic). Agriculturalists have not articulated any primary value position other than the value of production and have not offered reasons why production ought to retain its primacy.

4 Sustainability as an ethical goal

What is the primary agricultural problem? Is it production? Of course it is. However, problems related to distribution, waste (Institution of Mechanical Engineers 2013) and poverty must also be considered. Production of abundant food and fibre must remain a goal of agriculture. A morally pluralistic world compels us to ask if the endless pursuit of more production is the right answer to the many ethical dilemmas agriculture faces. We are encouraged to explore other goals that ought to be considered by agriculture and ask when and why one or more of them should take precedence over production. For example: sustainable, environmentally safe production that meets human needs and contributes to a just social order may be of greater moral importance than profitable production. That is not the dominant agricultural view. Sustainability is regarded by those in agriculture as primarily a production and secondarily an environmental goal. In weed management, to sustain usually means protecting the productive resource (soil, water, gene pools) to maintain production. Others argue the productive resource is important, but ranks below sustaining environmental quality. This debate goes to the heart of what agriculture ought to be. Agriculture has a major responsibility because it is so widespread and has the potential to care for or harm so much land. This is a different view from protecting only the productive ability of land. Land is not simply a productive resource. It is the basis of life. Without the land there will be no agriculture, so land must be regarded as something more than one of a number of other productive resources (e.g. fertilizer, machines, irrigation water, herbicides, or seed). To harm or destroy the land is to destroy something essential to life, and that certainly raises a moral question.

The challenge of achieving agricultural sustainability is that it involves values. It is generally not acknowledged in agricultural science that values are not external to the science and technology but its basis. Scientists know they are responsible for the scientific integrity of their work and for its intellectual contribution. They do not as readily assume responsibility for the moral aspects of their work. All of science and all of agricultural science is involved in moral/value questions. Science is not value-free, it is value-laden.

The research and teaching we do now involve assumptions and a view of a future we expect, desire or fear. As weed scientists proceed towards truly integrated weed management systems there will be conflicting interests, incompatible analyses based on different views of the nature of the problem, rising material expectations and different views of sustainability. It is unusual to find anyone opposed to sustainability. It is equally clear that there are many views of what ought to be sustained and how to achieve sustainability.

I know that weed and other agricultural scientists are ethical in the conduct of their science and in their personal lives, but they do not extend ethics into their work. They are realists, not idealists. Realists run agricultural research and the world; idealists do not. Idealists attend academic conferences and may write thoughtful articles. But the action is elsewhere. The reality is produce profitably or perish in the real agricultural world. Realism rules, and philosophical and ethical correctness may be interesting but they are not necessary for useful work in agriculture or other scientific disciplines.

Such a position needs to be called into question. We need to accept the difficult task of conducting an ethical analysis of weed science and its results. We must strive for an analysis of what it is about agricultural practices that limits our aspirations and needs modification. The analysis must include departments of agriculture, university departments, scientific societies, research institutions and commercial organizations that serve and profit from agriculture. We must strive to strengthen those features that are beneficial to society and change those that are not. We must be sufficiently confident to study ourselves and our institutions and dedicated to the task of modifying both.

To preserve what is best about modern agriculture and to identify the abuses modern technology has wrought on our land, our people and other creatures, and begin to correct them will require many lifetimes of work. Agriculturalists must see agriculture in its many forms – productive, scientific, environmental, economic, social, political and moral. It is not sufficient to justify all management activities on the basis of increased production. Other criteria, many with a clear moral foundation, should be included. We do not and no one ever will live in a post-agricultural society. All societies have an agricultural foundation within their borders or elsewhere. Those in agriculture must strive to assure all that the ethical foundation of the largest and most important human interaction with the environment is secure.

5 Conclusion

I encourage my colleagues and friends in all phases of agriculture to think about the ethical aspects of agriculture - the essential human endeavour. Increasing production should not be regarded as justification for all agricultural technology which imposes risks on the environment and on all who must eat. Failure to consider the ethical dimension of the agricultural enterprise and engaging in realistic discussions about its successes and failures leads to the risk of losing public support and understanding. All of those engaged in agriculture have personal and professional ethics which guide their behaviour. Extending their ethical thought to the larger agricultural realm will lead to and ensure a strong ethical foundation for agriculture.

6 Where to look for further information

Interested readers are directed to the *Journal of Agricultural and Environmental Ethics*, as well to following:

Wendell Berry and Wes Jackson, and Kirschenmann, F. 2010. *Creating an ecological conscience.* Counterpoint Press. Berkeley, CA. 403 pp.
Thompson, P. B. 1995. *The spirit of the soil.* Routledge. 196 pp.
Thompson, P. B. 1998. *Agricultural ethics.* Iowa State University Press, Ames, IA.
Rachels, J. and Rachels, S. 2010. *Elements of moral philosophy*, 6th. Ed., McGraw Hill. 203 pp.
Zimdahl, R. L. 2012. *Agriculture's ethical horizon*, 2nd Ed. Elsevier. 274 pp.
Zimdahl, R. L. 2012. *Weed science: a plea for thought-revisited.* Springer. London. 73 pp.

7 References

Avery, D. 1997. Saving the planet with pesticides and biotechnology and European farm reform, pp. 3–18. In: *Proceedings of the British Crop Protection Conference – Weeds.* British Crop Protection Council, Croydon.
Berry, W. 1999. In distrust of movements. The land report 65 (Fall), pp. 3–7. The Land Institute, Salina, KS.
Bronowski, J. 1973. *The Ascent of Man.* Little, Brown and Company, Boston, MA, 448p.
Institution of Mechanical Engineers. 2013. www.imeche.org/archives/13-01-10/New_report... Accessed May 2013.

Jensen, R. 2016. What is the world? Who are we? What are we going to do about it? The land report 16 (Fall), pp. 22–26. The Land Institute, Salina, KS.

Thompson, P. B. 1989. Values and food production. *Journal of Agricultural Ethics* 2:209–23.

Zimdahl, R. L. 2002. Moral confidence in agriculture. *American Journal of Alternative Agriculture*. 17(1):44–53.

Zimdahl, R. L. 2012. *Agriculture's Ethical Horizon*. Elsevier, Inc., London, UK, 274p.

Chapter 6

Surveillance and monitoring of weed populations

Anita Dille, Kansas State University, USA

1 Introduction

It is important to survey and monitor weed populations in agricultural fields because a farmer needs to know what weed species and communities occur and where in their field to make appropriate and effective weed management decisions for the current growing season and for future crop production. A weed map is most useful when it is a complete, accurate and unbiased assessment of the weed populations for a given field, farm or region. It is not easy to achieve because of the time and labour required, the complexity of the scouting information, the ability to identify each species, the spatial and temporal variation in weed density and distribution and the scale at which the information is collected.

Weed management is a key input for successful crop production. There is no question that weeds occur in crop fields, so why would there be an emphasis on surveying and monitoring weed populations? An integrated weed management (IWM) programme requires understanding the cropping (crop choice and rotational sequence) and tillage system for a farm. Regular scouting to determine if weed control is economically justified is a fundamental principle of IWM (Swanton and Weise 1991; Wiles 2005).

Field scouting (surveying and monitoring) weed populations requires a pre-planned strategy for travelling across a field, the size and shape of sample units, how many samples to observe, types of information to gather about the weed population in each sample unit and a method for recording that information (Doll et al. 1998; Fishel et al. 2009; Wiles 2005). If one was not concerned over the spatial distribution of weed populations in a given field, management decisions could be based on a general assessment of last year's weed problems, information gleaned by walking along the field edges or driving a zigzag pattern across the field in the spring to document the weed species that occur.

http://dx.doi.org/10.19103/AS.2017.0025.05

Depending on the purpose of a given weed map, the amount of time and labour able to be dedicated to generating the map will affect the desired level of detail that can be obtained. The approach to sampling is different if the purpose of scouting a field is to determine the average weed population, and to select the best uniformly applied weed control recommendation. However, if the goal is site-specific weed management, which could be viewed as general weed management, optimizing management for a particular weed species or finding small patches of a weed, then the sampling plan must be different (Wiles 2005). As more sophisticated technologies are becoming available, remote sensing tools can be used to 'cover' the field in less time and reduce the labour and time required, and shifting from paper and pencil notes to computer-based scouting tools. A Global Navigation Satellite System (GNSS) is used to record observations, as new approaches to field scouting will produce better, more accurate information about the weed populations in a field throughout the year.

Challenges to regular weed scouting are a) when and how often should the field be scouted (temporal variation in weed populations), b) what level of detail and how many samples to take (spatial variation in weed populations), c) what scale of information is needed for a farm or for a broader perspective, d) what tools to use when scouting and e) what is the right equipment that can manage the variation in weed population occurrence.

2 Temporal and spatial variation

2.1 Temporal variation: when and how often to scout a field

Successful IWM strategies include varying the choice of crops, crop production practices, weed control tactics and tillage systems from year to year to keep weed populations manageable (Norsworthy et al. 2012; Swanton and Weise 1991; Swanton et al. 2008). Utilizing diverse practices introduces temporal variation into the cropping system as well as suggesting optimal times to survey and monitor the weed populations, and thus needing to recommend different weed control tools that are effective on different weed species and on plants at different growth stages. Temporal variation can be described as year-to-year changes that are expected to occur in weed populations, but also changes within and through a given growing season, depending on the life cycle of the weed and the duration of the cropping system.

If weed populations were consistent from year-to-year in their location and density, they would be much easier to scout for and to manage. Unfortunately, they are not. Colbach et al. (2000) sampled annual and perennial weeds in a small no-till field that was planted to row crops over five years. Patchiness of weed seedlings was strongest for annuals and, more generally, for species that had seed dispersed by the combine harvester, such as wild oats (*Avena fatua, Avena sterilis*) (Barroso et al. 2006) and Johnsongrass (*Sorghum halepense*) (Barroso et al. 2012). Patches of perennials were persistent across years as were species that dispersed seed prior to combine harvest (Colbach et al. 2000). Gerhards et al. (1997) also reported higher patch persistence for perennials. The most abundant weed species in the no-till field situation was green foxtail (*Setaria viridis* L.), but predictions of its occurrence over time were only valid for the following year (Colbach et al. 2000). Wilson and Brain (1991) reported that blackgrass (*Alopecurus myosuroides* Huds.), an annual grass with high seed production, was spatially stable which was attributed to effective herbicides that limited spread across the field.

Understanding the temporal and spatial dynamics of perennial weed species that occur in reduced- or no-tillage systems, which are no longer disturbed by ploughing and cultivation, is important for better weed control decisions (Webster et al. 2000). Patches of hemp dogbane (*Apocynum cannabinum* L.) were at 50% of final occupied area early in the growing season, when only 22% of the shoots had emerged, highlighting that early weed scouting would be useful, and understanding patch dynamics would lead to more effective weed management (Webster et al. 2000). An annual weed species with a persistent seedbank such as common sunflower (*Helianthus annuus* L.) or velvetleaf (*Abutilon theophrasti* Medik) would be present in the same field annually (Burton et al. 2004; Dieleman and Mortensen 1999; Gerhards et al. 1997). Imposed weed management also influences the level of temporal stability observed for different weed species. Annual grass weeds and common ragweed (*Ambrosia artemisiifolia* L.) infestations were mapped for nine years in a corn/soybean production field (Clay et al. 2006). A higher mean density of annual grass weeds (12–133 plants m^{-2}) resulted in a more stable distribution in time, space and density than the overall lower density of common ragweed (<1–37 plants m^{-2}). With stable populations a scout could return to the same location and find only moderate changes in density and patch margins. However, when effective weed control practices are used, weed populations are expected to change, highlighting the need for continuous scouting.

It is known that weeds present at the time of crop establishment and those that emerge with the crop cause the most yield loss if not controlled (Zimdahl 1980, 2004). Effective techniques need to be developed to scout for and detect these weeds early in the growing season, so proper control can be initiated. The right time to control weeds is early in the crop life cycle, during the critical period of weed control, so that the crop can establish and outcompete any low-density or late-emerging weeds (Dille 2014). Based on previous year's maps of weed problems, one would know what weed species to expect, know the potential emergence timing of the key weeds (Forcella 2000) and be able to scout for the expected weeds that will guide the selection of the best control tactic. For example, Lundy et al. (2014) used site-specific real-time temperatures to improve the accuracy of weed emergence predictions in direct-seeded rice systems. This would guide the farmer to make real-time weed management decisions.

If the IWM programme requires a decision about a post-emergence weed control application (chemical or mechanical), scouting should occur again soon after crop emergence to determine if there were any escapes from initial weed control practices, what new emerging weed populations are present and where they are located. Opportunities for more scouting would occur after the application of post-emergence weed control to evaluate its success, until no more in-crop weed control practices can be implemented. For planning for future years, surveying weed populations during the harvesting operation is also possible as well as after harvest and during fallow periods.

The focus here has been on how to detect the weed species and communities growing in the crop. Much recent detection research has been using some type of remote sensing technology, initially with collecting and processing spatial data from sensors mounted on satellite or airplane platforms. These have successfully been used to map a variety of factors in crop fields, including weed distributions. However, the spatial and temporal resolution of the images from such platforms has been limited for practical weed management decision-making (see Case Study 2).

2.2 Spatial variation: where to scout and how many samples to take

The ultimate weed map would show where every individual weed by species was located, including its size. To achieve this will be expensive, and this level of detail is more than needed because herbicide application is done primarily with wide spray booms and long field passes. When farmers have access to precision agriculture tools to manage the weed populations based on a weed map, the detail needed in such a map will increase. For example, where smaller areas of a field can be treated within the width of a spray boom (e.g. boom section or individual nozzle control).

A quick visual assessment of weed populations within a field shows that weeds are not distributed regularly (unless seeded with crop or extremely weedy) nor randomly (unless newly infested by windblown seed) (Firbank 1993; Clay and Johnson 1999) (Fig. 1). Weeds are commonly aggregated in patches, circular clumps or strips (Cardina and Doohan 2000; Cardina et al. 1997; Firbank 1993; Johnson et al. 1996a; Thornton et al. 1990). Sometimes the cause of the spatial pattern is obvious, for example, invasion from the field edge (Marshall 1989; Marshall and Arnold 1995; Wilson and Aebischer 1995), areas where no crop was able to establish due to drought or flooding, strips that were missed by weed control application (non-overlapping sprayer passes), inter-row cultivator injury to crop, or seed dispersal by tillage equipment (Rew and Cussans 1997) or a combine harvester

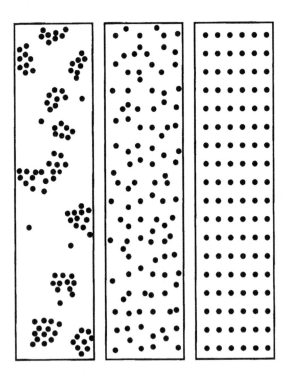

Figure 1 Three types of weed spatial distributions from left to right: patchy or aggregated, random and regular or uniform.

(Barroso et al. 2006; Colbach et al. 2000). Sometimes the reason for the pattern is not obvious and may be the result of variation in soil properties (Andreasen et al. 1991; Burton et al. 2004; Dieleman et al. 2000a,b) or dispersal from an isolated colonist in the past (Colbach et al. 2000; Dieleman and Mortensen 1999).

An annual weed species occurs because at least one seed was present and, if left uncontrolled, will grow and disperse its seed locally or to some distance from the parent plant. Through time, this could develop into a patch of weeds. Detailed spatial observations of individual species have shown that they can form high-density patch centres that decrease in density as one moves to the patch edge (Dieleman and Mortensen 1999; Webster et al. 2000) or that they follow a spatial characteristic at the field level such as depressions (Burton et al. 2004; Humston et al. 2005).

Based on detailed surveys of weed populations using grid-based sampling quadrats, with no regard to spatial distribution, the frequency or numerical distribution of species tends to follow a negative binomial distribution rather than a normal distribution (Berti et al. 1992; Johnson et al. 1996a,b; Wiles 2005; Wiles et al. 1992). This distribution is usually characterized by a large number of weed-free (zero) and/or low-density counts (between patches), and a few very high-density counts (inside patches) (Wiles et al. 1992). To utilize a statistical model to sample a population, with a goal of determining the 'average' weed density across a field when one was not going to manage spatially, the number of samples needed needs to be very high, that is >300 samples/field, to obtain a reasonable average of weed density (Berti et al. 1992; Gold et al. 1996; Johnson et al. 1996b). Alexander et al. (2005) evaluated the performance of different sampling strategies when scouting for aphids in a winter wheat field. Theoretically, taking random samples would be most appropriate but, to address cost and efficiency, one most follow a path through the field and sample along the path. In crop protection, field monitoring or 'scouting' schemes generally avoid random or grid samples and use contiguous sample units. The position of the sample unit on the path is expected to represent the area surveyed, usually at the centre of the area, and its extent covers the area represented (Alexander et al. 2005). If one knew the spatial pattern of the pest, this would determine the choice of a sampling scheme and improve accuracy of the sample. Line transects longer than the grain (distance among patches) of the spatial pattern, up to twice the size of the patch, are recommended when weed scouting (Wiles 2005).

Another question is the appropriate quadrat size to sample weeds and how far apart each quadrat should be (Dille et al. 2002). Different quadrat sizes detect different scales of spatial pattern (Firbank 1993; Kershaw and Looney 1985). Quadrat size should be influenced by the purpose of the data collected. However, in practice this is rarely the case; the choice of quadrat size is often dependent on the practical constraints of time and labour. Separation distance needs to capture the spatial variation of each species. To create a complete spatial map of weed species, there is an interpolation step needed to describe the weed distribution between sampled points (Dille et al. 2002).

There is a great opportunity to generate spatial distribution maps of weed communities by combining the fact that weed counts tend to follow this negative binomial distribution if patchy or aggregated in a field, and by adding the spatial component through documenting the specific GNSS location information together with the weed count. If the occurrence of weeds is generally low, they tend to occur in patches in the field, with large portions of the field having no weeds, and others with high-density patch centres decreasing to lower density margins (Dieleman and Mortensen 1999).

3 Monitoring weed populations

3.1 Types of weed information

Monitoring involves the assessment of the change in weed species occurrence, density (#/m²), patch margins (m² occupied) and locations of weeds over time in a field. New weed species could be identified soon after the arrival in order to control them effectively before they spread. If a current weed population was increasing in density and patch area from one year to the next, it could be because:

A the population developed herbicide resistance to a previously used herbicide, or
B an ineffective control practice was used. For example, excellent grass control was obtained, but broadleaf weed species are released from competition, or
C poor application or environmental conditions prevented effective control.

Through monitoring, early detection of weed problems can be identified before they become unmanageable.

If a given weed species consistently changes locations across a given field, scouting for the population should be done each year, along with determining why it is moving. Some weed species have short-lived seed in the soil and, with changing crop production practices, could find optimal growing environments in different parts of the field. If movement of the weed is due to a crop production practice, such as tillage that moves seed or vegetative propagules (Rew and Cussans 1997), or a combine-harvester that distributes seed in the chaff during harvest (Barroso et al. 2006, 2012; Colbach et al. 2000), these practices could be changed.

One can do a better job of long-term IWM planning once the distribution of weed populations is known, such as designing the best sequence of crops, adjusting time of planting or modifying methods of fertilizing. The science and technological sophistication of sampling for weed populations in farmers' fields has advanced significantly over the past 30 years, with advancements in computer-based tools, such as more rapid processing of data with increased computing power, and remote sensing with UAV platforms versus satellites or airplanes, that have increasing sophistication and resolution in cameras from RGB to multi-spectral to hyperspectral images. Approaches to combine all these different types of data were proposed by Bourgeois et al. (2012) by using a hierarchical Bayesian model to more accurately describe the spatial weed distributions.

3.2 Regional- and global-level surveillance and monitoring of weed populations

So far this chapter has focused on generating maps for in-field management of weed populations in agronomic crops. However, one might be interested in knowing the overall regional, national or global distribution of a new weed species, documenting shifts in weed species distribution through time across a wide geographic area because of changing crop production practices, or tracking the spread of invasive weed species. As the timescale increases, effects of larger-scale factors become more apparent. These may involve changing patterns of farming (e.g. more diverse crop rotations, incorporating cover crops, changing tillage systems from mouldboard ploughing to no-tillage, diversifying chemical control options, etc.), and an ever-changing climate. These processes may seem to be too slow to be readily observed, but there are long-term data from other sources – historical surveys, herbarium records, published accounts – which can be used to infer long-term changes in at least some weed species (Firbank 1993).

At the level of individual farmers and their fields, one can see very different weed populations on adjacent farms due to the management choices made by farmers: crop rotation, timing of weed control and tillage systems. There is some interaction between farms with weeds that do not 'respect' fence lines, landownership or property boundaries. For example, species that can be distributed by tumbling such as kochia or Russian thistle can move several miles (Stallings et al. 1995). Wind dispersal of pollen and seed also moves species across large distances, for example, horseweed (*Conyza canadensis* L.) (Dauer et al. 2006).

Across a region, surveys on changing weed abundance and distribution in response to a variety of tillage systems were useful in identifying future weed issues in southwestern Ontario, Canada (Frick and Thomas 1992). The occurrence and distribution of herbicide-resistant wild oat (*A. fatua*) (Beckie et al. 1999b) and green foxtail (*S. viridis*) (Beckie et al. 1999a) in small-grain production areas of the Canadian prairies guided weed control efforts and educational programmes for farmers. Province-wide surveys on a decadal scale in western Canada have provided insight into the shift of key weed species (Thomas 1985). Firbank et al. (1998) mapped weeds across the United Kingdom. Dauer et al. (2009) developed a simulation model to evaluate how heterogeneity of the landscape and choices of crops could influence the establishment and spread of a weedy species resistant to glyphosate, specifically horseweed. As the landscape became more diverse, there were fewer opportunities for horseweed to increase. Based on these results, more effective options for weed control can be proposed (Dauer et al. 2009).

The goals and objectives for surveying and monitoring non-indigenous or invasive plant species for management in a rangeland or on public lands will be different than the information needed about weed populations found in a corn or soybean field. Rew and Pokorny (2006) described procedures for mapping non-indigenous plant species.

Finally, at the national- or global–spatial scale, we encounter limits to the occurrence of weed species, likely due to the availability of suitable habitat provided by climate conditions (Firbank 1993). However, with anticipated changes in the climate, range expansion of weed species that are adaptable can occur (Bradley et al. 2010; Clements and DiTommaso 2011; Hellmann et al. 2008; McDonald et al. 2009). Ziska et al. (2011) proposed potential effects of changes in climate on the vulnerability of current agricultural systems to invasion by pests. With changes in temperature, rainfall and carbon dioxide levels, there is the potential for weed species to move and spread into new areas or be more competitive.

4 Case studies: how research has been used to improve practice

4.1 Case study 1

Based on USDA-NASS summaries of pest management activities that farmers employ, monitoring that includes weed scouting (deliberately or part of general observations while performing other tasks) typically ranges from 88 to 99% of planted acres for wheat, soybean, cotton and corn (USDA-NASS 2017). Scouting and monitoring of weed species and communities is one of the top recommendations for herbicide resistance management

(Norsworthy et al. 2012). Depending on the weed species, occurrence and patches of herbicide-resistant weeds could be detected before the problem becomes widespread. There are many crop consulting and scouting firms located in Kansas, other central Great Plains states and throughout the United States, that provide regular surveying and monitoring of whole fields to provide recommendations for in-season weed control to their customers.

Several US university extension publications describe the scouting patterns recommended for entering a field and documenting different pest problems, for example, the University of Wisconsin-Extension for corn (Doll et al. 1998) and the University of Missouri-Extension for field crops (Fishel et al. 2009). The most common and efficient scouting pattern for a square or rectangular field is to walk along a predetermined zigzag or M-shaped route through a field (Doll et al. 1998; Fischel et al. 2009). If the field is irregular in shape, the scout needs to keep in mind that they must cover a representative area of the field and in some type of zigzag pattern. It is important to recognize that one cannot scout the field edge only and expect pest populations to be the same in other areas (Doll et al. 1998). However, focusing on edges might be appropriate for certain weed problems such as perennial species that invade from wooded or fenced edges. For large fields (>50 acres), the scout's accuracy diminishes and it is suggested that the area be broken into separate field units with their own records. Fields could be separated into units based on geography, previous cropping history or changes in soil types (Doll et al. 1998; Fischel et al. 2009).

The methods of recording observations range from use of pencil and paper maps through to handheld, mobile mapping tools with GNSS capabilities and field scouting software programmes. Numerous scouting and consulting firms have generated their own computer-based 'apps' for their customers to access. For example, Hopkins (2017) summarized at least 17 different field scouting 'apps' that are currently available, often as a free download but tied to a specific company and needing a username and password.

For weed scouting, and for providing valuable information to the farmer, it is recommended that the first survey occurs shortly after crop emergence and continues at weekly intervals until no more control options are available (Doll et al. 1998). When returning to visit a field, change the walking path to ensure that all areas of the field are covered. Return to the same areas of the field to ensure adequate weed control was obtained. The number of stops should be a minimum of ten per 50-acre field, occurring along the arms of the M- or W-path, as well as identifying pockets of weed infestation problems as they move between stops. At each stop in the field, the scout should record the weed species, its relative abundance and growth stages, along with other crop and pest information that occurs in whatever format they are available. There may always be an opportunity to do site-specific weed management. Therefore, weed abundance might be classified as scattered, slight, moderate and severe. Relative to corn, 'scattered' would mean that weeds were present but with few individuals, probably will produce weed seed, but will not cause economic yield losses in the current season. 'Slight' would mean weeds scattered throughout the field, but on average <1 per m-row, and likely to cause economic yield losses. A 'moderate' infestation would be a fairly uniform number across the field with an average of five plants per m-row or scattered dense weed patches, while a 'severe' infestation could be >3 broadleaf weeds per m-row and >10 grass weed plants per m-row, or large areas of dense infestations (Doll et al. 1998).

4.2 Case study 2

Another approach to surveying and visualizing weed species and communities across whole fields or larger land areas is remotely sensed images. The technology has changed and improved over the past 30 years. The detection and identification of weeds growing in a field (with or without a crop) is still a challenge but new technologies are improving to allow for better spatial data at a finer resolution based on improved sensors and rapid data processing.

Aerial and satellite imageries were used to identify patches of weeds, often at a later growth stage such as flowering, when the weed was at a different growth stage than the surrounding vegetation and could be detected more easily (Clay et al. 2004; Everitt et al. 1993). This was effective for detecting invasive plant species in more remote geographical areas that were difficult to physically scout (Everitt et al. 1993; Lass and Callihan 1993; Prather and Callihan 1993). Another limiting factor was the resolution of the images obtained from these satellite or airplane views such as 30 m by 30 m pixels on the ground, where the occurrence and thus the reflectance within this image needed to be dominated by the weeds. Often this meant that the weeds had to be large, flowering and after the optimal time for weed control.

If the goal of such imagery is to make early season weed management decisions, imagery needs to identify smaller weeds soon after crop emergence (Dille 2014; Torres-Sánchez et al. 2013). Mapping weeds earlier in the growing season presents some difficulties for three main biological reasons:

1 Weeds are small, and generally distributed in small patches, which makes it necessary to work with remote images of very small pixel sizes, often on the order of centimetres (Dille 2014).
2 Grass weeds and monocotyledonous crops (e.g. wild oat (*Avena* sp.) in wheat) or broadleaved weeds and many dicotyledonous crops (e.g. common lambsquarters (*Chenopodium album*) in sunflower), generally have similar reflectance properties early in the season, which decreases the possibility of discriminating between vegetation classes using only spectral information.
3 Soil background reflectance may interfere with detection (Thorp and Tian 2004; Torres-Sánchez et al. 2013).

Imagery from UAVs has the potential:

1 to be used to determine if and where weeds are present earlier in the growing season and eventually could be processed to separate monocot versus dicot plants, and even to identify individual species.
2 to be used to evaluate efficacy of imposed weed control practices (relative to an untreated area, or scouting after weed control applied), and determine the need for subsequent control (Dille et al. 2017; Gundy et al. 2017).
3 to be used for surveying of new, resistant weeds (after herbicide application, those that are surviving the treatment could be controlled with a different practice quickly).
4 to be used for surveying for new invading weeds.

Researchers continue to evaluate what type of camera and what wavelengths are best to capture the most useful data to create weed maps (Chang et al. 2004; Keller et al.

2014; Peña et al. 2015). More and more examples of using UAVs in agricultural fields are being reported. Zhang and Kovacs (2012) reviewed the advances in UAV platforms for precision agriculture applications, and highlighted steps in producing useful data from these images. Torres-Sánchez et al. (2013) outlined the series of steps to obtain images using UAVs and to discriminate weed infestations in a sunflower crop in the early growing season for post-emergence treatments. Peña et al. (2013) evaluated the same series of steps with UAV imagery to generate an early season weed map in a corn crop with 86% overall accuracy when categorizing the image into three categories of weed-free, low-weed coverage (<5% weeds) and weedy.

5 Summary and future trends

As technological advancements in weed control techniques continue to be developed, opportunities arise to manage individual weed species and communities on a finer scale than a whole field. Understanding the biological, ecological and human causes for different numerical and spatial distributions of individual weeds in fields could assist in the development of more appropriate technology. If individual weeds could be detected and identified (by colour or shape features) and separated from the crops, more targeted and specific weed control could be possible. The next step could be to link the identification of individual weed species to other field-level characteristics, such as soil physical and chemical properties, elevation, relative position to field boundaries and position in the landscape of other crops and fields, to direct future scouting to where these weeds will likely be found. Through surveillance and monitoring of weed species and communities, their effect on crop growth and development could be reduced and production could be maximized.

Conducting surveys to identify new herbicide-resistant populations and invading weed species are key to understanding the regional and global distributions of these weeds. Once known, continuous monitoring and appropriate weed control actions could minimize future problems.

6 Where to look for further information

Additional texts provide more background on surveillance and monitoring of weeds. For non-indigenous plant species (invasive plans), Rew and Pokorny (2006) provided a detailed description of inventory and survey methods. With rapid technological advances in weed control, Young and Pierce (2014) edited a book that describes many facets and considerations for weed control automation. Perspectives on implementing and adopting integrated pest management are reported by Norris et al. (2003). A national survey of common and troublesome weed species is conducted and reported online by Lee Van Wychen, Director of Science Policy with the Weed Science Society of America (www.wssa.net).

Key professional societies and regular conferences:

- Weed Science Society of America annual meeting – February of each year (www.wssa.net)

- European Conference on Precision Agriculture (ECPA) – meets in alternate years; published proceedings
- International Society of Precision Agriculture (ISPA) (www.ispag.org)
- International Conference on Precision Agriculture (ICPA) – meets in alternate years; publishes proceedings

7 References

Alexander, C. J., Holland, J. M., Winder, L., Woolley, C. and Perry, J. N. (2005) Performance of sampling strategies in the presence of known spatial patterns. *Ann. Appl. Biol.* 146:361–70.

Andreasen, C., Streibig, J. C. and Haas, H. (1991) Soil properties affecting the distribution of 37 weed species in Danish fields. *Weed Res.* 31:181–7.

Barroso, J., Andújar, D., San Martín, C., Fernández-Quintanilla, C. and Dorado, J. (2012) Johnsongrass (*Sorghum halepense*) seed dispersal in corn crops under Mediterranean conditions. *Weed Sci.* 60:34–41.

Barroso, J., Navarrete, L., Sánchez del Arco, M. J., Fernández-Quintanilla, C., Lutman, P. J. W., Perry, N. H. and Hull, R. I. (2006) Dispersal of *Avena fatua* and *Avena sterilis* patches by natural dissemination, soil tillage and combine harvesters. *Weed Res.* 46:118–28.

Beckie, H. J., Thomas, A. G. and Légère, A. (1999a) Nature, occurrence, and cost of herbicide-resistant green foxtail (*Setaria viridis*) across Saskatchewan ecoregions. *Weed Technol.* 13:626–31.

Beckie, H. J., Thomas, A. G., Légère, A., Kelner, D. J., Van Acker, R. C. and Meers, S. (1999b) Nature, occurrence and cost of herbicide-resistant wild oat (*Avena fatua*) in small-grain production areas. *Weed Technol.* 13:612–25.

Berti, A., Zanin, G., Baldoni, G., Grignani, C., Mazzondini, M., Montenmurro, P., Tei, F., Vazzanan, C. and Biggianti, P. (1992) Frequency distribution of weed counts and applicability of a sequential sampling method to integrated weed management. *Weed Res.* 32:39–44.

Bourgeois, A., Gaba, S., Munier-Jolain, N., Borgy, B., Monestiez, P. and Soubeyrand, S. (2012) Inferring weed spatial distribution from multi-type data. *Ecol. Model.* 226:92–8.

Bradley, B. A., Wilcove, D. S. and Oppenheimer, M. (2010) Climate change increases risk of plant invasion in the Eastern United States. *Biol. Invasions* 12:1855–72.

Burton, M. G., Mortensen, D. A., Marx, D. B. and Lindquist, J. L. (2004) Factors affecting the realized niche of common sunflower (*Helianthus annuus*) in ridge-tillage corn. *Weed Sci.* 52:779–87.

Cardina, J., Johnson, G. A. and Sparrow, D. H. (1997) The nature and consequence of weed spatial distribution. *Weed Sci.* 45:364–73.

Cardina, J. and Doohan, D. J. (2000) Weed biology and precision farming. SSMG-25. International Plant Nutrition Institute. Site Specific Management Guidelines. www.ipni.net/ssmg.

Chang, J., Clay, S. A., Clay, D. E. and Dalsted, K. (2004) Detecting weed-free and weed-infested areas of a soybean field using near-infrared spectral data. *Weed Sci.* 52:642–8.

Clay, S. A., Chang, J., Clay, D. E., Reese, C. L. and Dalsted, K. (2004) Using remote sensing to develop weed management zones in soybeans. SSMG-42. International Plant Nutrition Institute. Site Specific Management Guidelines. www.ipni.net/ssmg.

Clay, S. and Johnson, G. (1999) Scouting for weeds. SSMG-15. International Plant Nutrition Institute. Site Specific Management Guidelines. www.ipni.net/ssmg.

Clay, S. A., Kreutner, B., Clay, D. E., Reese, C., Kleinjan, J. and Forcella, F. (2006) Spatial distribution, temporal stability, and yield loss estimates for annual grasses and common ragweed (*Ambrosia artemisiifolia*) in a corn/soybean production field over nine years. *Weed Sci.* 54:380–90.

Clements, D. R. and DiTommaso, A. (2011) Climate change and weed adaptation: can evolution of invasive plants lead to greater range expansion than forecasted? *Weed Res.* 51:227–40.

Colbach, N., Forcella, F. and Johnson, G. A. (2000) Spatial and temporal stability of weed populations over five years. *Weed Sci.* 48:366–77.

Dauer, J. T., Mortensen, D. A. and Humston, R. (2006) Controlled experiments to predict horseweed (*Conyza canadensis*) dispersal distances. *Weed Sci.* 54:484–9.

Dauer, J. T., Luschei, E. C. and Mortensen, D. A. (2009) Effects of landscape composition on spread of an herbicide-resistant weed. *Landsc. Ecol.* 24:735–47.

Dieleman, J. A. and Mortensen, D. A. (1999) Characterizing the spatial pattern of *Abutilon theophrasti* seedling patches. *Weed Res.* 39:455–67.

Dieleman, J. A., Mortensen, D. A., Buhler, D. D., Cambardella, C. A. and Moorman, T. B. (2000a) Identifying associations among site properties and weed species abundance. I. Multivariate analysis. *Weed Sci.* 48:567–75.

Dieleman, J. A., Mortensen, D. A., Buhler, D. D. and Ferguson, R. B. (2000b) Identifying associations among site properties and weed species abundance. II. Hypothesis generation. *Weed Sci.* 48:576–87.

Dille, J. A. (2014) Chapter 4: Plant morphology and critical period of weed control. In: *Automation: The Future of Weed Control in Cropping Systems* (S. L. Young and F. J. Pierce (Eds)). Springer Science+Business Media, Dordrecht, pp. 51–69.

Dille, J. A., Asebedo, A. R. and Gundy, G. J. (2017) Detecting emerging weeds in the field with UAV-based imagery. In: *Proceedings of ECPA*, Edinburgh, Scotland, 16–20 July 2017.

Dille, J. A., Milner, M., Groeteke, J. J., Mortensen, D. A. and Williams II, M. M. (2002) How good is your weed map? A comparison of spatial interpolators. *Weed Sci.* 51:44–55.

Doll, J., Grau, C., Jensen, B., Wedberg, J. and Meyer, J. (1998) Scouting corn: A guide for Wisconsin corn production. Publication A3547. University of Wisconsin-Extension, Madison, WI, 15p.

Everitt, J. H., Escobar, D. E., Billarreal, R., Alaniz, M. A. and Davis, M. R. (1993) Integration of airborne video, global positioning system and geographic information system technologies for detecting and mapping two woody legumes on rangelands. *Weed Technol.* 7:981–7.

Firbank, L. G. (1993) Chapter 9: The implications of scale on the ecology and management of weeds, pp. 91–104. In: *Landscape Ecology and Agroecosystems* (R. G. H. Bunce, L. Ryszkowski and M. G. Paoletti (Eds)). Lewis Publishers, CRC Press, Inc., Boca Raton, FL.

Firbank, L. G., Ellis, N. E., Hill, M. O., Lockwood, A. K. and Swetnam, R. D. (1998) Mapping the distribution of weeds in Great Britain in relation to national survey data and to soil type. *Weed Res.* 38:1–10.

Fishel, F., Bailey, W., Boyd, M., Johnson, B., O'Day, M., Sweets, L. and Wiebold, B. (2009) Integrated pest management: Introduction to Crop Scouting. Publication IPM1006. University of Missouri-Extension, Columbia, MO, 24p.

Forcella, F. (2000) Estimating the timing of weed emergence. SSMG-20. International Plant Nutrition Institute. Site Specific Management Guidelines. www.ipni.net/ssmg.

Frick, B. and Thomas, A. G. (1992) Weed surveys in different tillage systems in southwestern Ontario field crops. *Can J. Plant Sci.* 72:1337–47.

Gerhards, R., Wyse-Pester, D. Y., Mortensen D. A. and Johnson G. A. (1997) Characterizing spatial stability of weed populations using interpolated maps. *Weed Sci.* 45:108–19.

Gold, H. J., Bay, J. and Wilkerson, G. G. (1996) Scouting for weeds, based on the negative binomial distribution. *Weed Sci.* 44:504–10.

Gundy, G. J., Dille J. A. and Asebedo, A. R. (2017) Efficacy of variable rate soil-applied herbicides based on soil electrical conductivity and organic matter differences. In: *Proceedings of ECPA*, Edinburgh Scotland, 16–20 July 2017.

Hellmann, J. J., Byers, J. E., Bierwagen, B. G. and Dukes, J. S. (2008) Five potential consequences of climate change for invasive species. *Conserv. Biol.* 22:534–43.

Hopkins, M. (2017) 17 Field Scouting Apps for Precision Agriculture. http://www.precisionag.com/professionals/tools-smart-equipment/17-field-scouting-apps-for-precision-agriculture/. Accessed 29 April 2017.

Humston, R., Mortensen, D. A. and Bjørnstad, O. N. (2005) Anthropogenic forcing on the spatial dynamics of an agricultural weed: the case of the common sunflower. *J. Appl. Ecol.* 42:863–72.

Johnson, G. A., Mortensen, D. A. and Gotway, C. A. (1996a) Spatial and temporal analysis of weed seedling populations using geostatistics. *Weed Sci.* 44:704–10.

Johnson, G. A., Mortensen, D. A., Young L. J. and Martin, A. R. (1996b) Parametric sequential sampling based on multistage estimation of the negative binomial parameter k. *Weed Sci.* 44:555–9.

Keller, M., Gutjahr, C., Möhring, J., Weis, M., Sökefeld, M. and Gerhards, R. (2014) Estimating economic thresholds for site-specific weed control using manual weed counts and sensor technology: An example based on three winter wheat trials. *Pest Manag. Sci.* 70:200–11. Doi:10.1002/ps.3545.

Kershaw, K. A. and Looney J. H. H. (1985) Chapter 7: The Poisson series and the detection of non-randomness. In: *Quantitative and Dynamic Plant Ecology.* 3rd Ed., Edward Arnold, Baltimore, MD, pp. 121–37.

Lass, L. W. and Callihan, R. H. (1993) GPS and GIS for weed surveys and management. *Weed Technol.* 7:249–54.

Lundy, M. E., Hill, J. E., van Kessel, C., Owen, D. A., Pedroso, R. M., Boddy, L. G., Fischer, A. J. and Linquist, B. A. (2014) Site-specific, real-time temperatures improve the accuracy of weed emergence predictions in direct-seeded rice systems. *Agr. Syst.* 123:12–21.

Marshall, E. J. P. (1989) Distribution patterns of plants associated with arable field edges. *J. Appl. Ecol.* 26:247–57.

Marshall, E. J. P. and Arnold, G. M. (1995) Factors affecting field weed and field margin flora on a farm in Essex, UK. *Landsc. Urban Plan.* 31:205–16.

McDonald, A., Riha, S., DiTommaso, A. and DeGaetano, A. (2009) Climate change and the geography of weed damage: Analysis of U.S. maize systems suggests the potential for significant range transformations. *Agr. Ecosys. Environ.* 130:131–40.

National Research Council (1997) *Precision Agriculture in the 21st Century.* National Academy Press, Washington, DC, 149p.

Norris, R. F., Caswell-Chen, E. P. and Kogan, M. (2003) *Concepts in Integrated Pest Management.* Pearson Education, Inc., Upper Saddle River, NJ, 586p.

Norsworthy, J. K., Ward, S. M., Shaw, D. R., Llewellyn, R. S., Nichols, R. L., Webster, T. M., Bradley, K. W., Frisvold, G., Powles, S. B., Burgos, N. R., Witt, W. W. and Barrett, M. (2012) Reducing the risks of herbicide resistance: Best management practices and recommendations. *Weed Sci.* 60(special issue):31–62.

Peña, J. M., Torres-Sánchez, J., de Castro, A. I., Kelly, M. and López-Granados, F. (2013) Weed mapping in early-season maize fields using object-based analysis of unmanned aerial vehicle (UAV) images. *PLoS ONE* 8(10): e77151. doi:10.1371/journal.pone.0077151.

Peña, J. M., Torres-Sánchez, J., Serrano-Pérez, A., de Castro, A. I. and López-Granados, F. (2015) Quantifying efficacy and limits of unmanned aerial vehicle (UAV) technology for weed seedling detection as affected by sensor resolution. *Sensors* 15:5609–26.

Prather, T. S. and Callihan, R. H. (1993) Weed eradication using geographic information systems. *Weed Technol.* 7:265–9.

Rew, L. J. and Cussans, G. W. (1997) Horizontal movement of seeds following tine and plough cultivation: implications for spatial dynamics of weed infestations. *Weed Res.* 37:247–56.

Rew, L. J. and Pokorny, M. L. (Eds) (2006) Inventory and Survey Methods for Nonindigenous Plant Species. Montana State University Extension, Bozeman, MT, 75p.

Stallings, G. P., Thill, D. C., Mallory-Smith, C. A. and Lass, L. W. (1995) Plant movement and seed dispersal of Russian thistle (*Salsola iberica*). *Weed Sci.* 43:63–9.

Swanton, C. J., Mahoney, K. J., Chandler, K. and Gulden, R. H. (2008) Integrated weed management: Knowledge-based weed management systems. *Weed Sci.* 56:168–72.

Swanton, C. J. and Weise, S. F. (1991) Integrated weed management: The rationale and approach. *Weed Technol.* 5:657–63.

Thomas, A. G. (1985) Weed survey system used in Saskatchewan for cereal and oilseed crops. *Weed Sci.* 33:34–43.

Thornton, P. K., Fawcett, R. H., Dent, J. B. and Perkins, T. J. (1990) Spatial weed distribution and economic thresholds for weed control. *Crop Protect.* 9:337–42.

Thorp, K. R. and Tian, L. F. (2004) A review on remote sensing of weeds in agriculture. *Prec. Agr.* 5:477–508.

Torres-Sánchez, J., López-Granados, F., de Castro, A. I. and Peña-Barragán, J. M. (2013) Configuration and specifications of an unmanned aerial vehicle (UAV) for early site specific weed management. *PLoS ONE* 8(3): e58210. doi:10.1371/journal.pone.0058210.

[USDA-NASS] US Department of Agriculture – National Agricultural Statistics Service (2017) Agricultural Chemical Use Program. https://www.nass.usda.gov/Surveys/Guide_to_NASS_Surveys/Chemical_Use/. Accessed 30 May 2017.

Webster, T. M., Cardina, J. and Woods, S. J. (2000) Spatial and temporal expansion patterns of *Apocynum cannabinum* patches. *Weed Sci.* 48:728–33.

Wiles, L. J. (2005) Sampling to make maps for site-specific weed management. *Weed Sci.* 53:228–35.

Wiles, L. J., Wilkerson, G. G., Gold, H. J. and Coble, H. D. (1992) Modeling weed distribution for improved postemergence control decisions. *Weed Sci.* 40:546–53.

Wilson, P. J. and Aebischer, N. J. (1995) The distribution of dicotyledonous arable weeds in relation to distance from the field edge. *J. Appl. Ecol.* 32:295–310.

Wilson, B. J. and Brain, P. (1991) Long-term stability of distribution of *Alopecurus myosuroides* Huds. within cereal fields. *Weed Res.* 31:367–73.

Young, S. L. and Pierce, F. J. (Eds) (2014) *Automation: The Future of Weed Control in Cropping Systems.* Springer Science+Business Media, Dordrecht, 265p.

Zhang, C. and Kovacs, J. (2012) The application of small unmanned aerial systems for precision agriculture: A review. *Prec. Agric.* 13:693–712.

Zimdahl, R. L. (1980) *Weed-Crop Competition: A review.* International Plant Protection Center, Oregon State University, 196p.

Zimdahl, R. L. (2004) *Weed-Crop Competition: A Review.* 2nd Ed., Blackwell Publishing Professional, Ames, IA, 219p.

Ziska, L. H., Blumenthal, D. M., Runion, G. B., Hunt Jr., E. R. and Diaz-Soltero, H. (2011) Invasive species and climate change: An agronomic perspective. *Clim. Change* 105:13–42.

Part 3

Using herbicides in IWM

Chapter 7

Site-specific weed management

S.A. Clay and S.A. Bruggeman, South Dakota State University, USA

1 Introduction

Since the advent of agriculture, weed management has been a critical component of the production system. However, weed management strategies are culture specific. For example, in the Central United States, Native Americans used squash and beans planted between corn plants to suppress weeds, whereas the settlers used tillage (Clay et al., 2017). Anyone who has tended a garden knows about weeds. When scaling up from garden plots to agronomic fields, a single season of mismanagement can turn a relatively weed-free field into a patch of weeds. This is because the soil contains millions of viable weed seeds of multiple species per acre. If there is emergence and survival of only 1% of 1 000 000 seeds, the result is 10 000 plants per acre. Depending on the species, plant competition and time of emergence, a single new plant can produce 500 to over 1 000 000 seeds during a growing season (Stevens, 1932; Clay et al., 2005; Uscanga-Mortera et al., 2007) that replenish the soil seed bank, which perpetuates the weed problem and increases the need for management.

'One year's seeding is seven years' weeding' is a phrase often used to describe the long-term consequences of poor weed management (Stephens, 1982). Unlike crop seed that is selected to have uniform germination, seeds of most weed species have dormancy, either innate or acquired, by exposure to and endurance of environmental conditions. There are numerous causes of seed dormancy. These include chemical suppression, mechanical suppression due to hard seed coats, and temperature requirements, among

http://dx.doi.org/10.19103/AS.2017.0025.09

others. No matter the cause, the result is that not all weed seeds germinate at the same time within a season or in a year, but in weed flushes throughout a season, giving temporal in-season variability. Seeds produced may remain viable for 20 or more years (Toole and Brown, 1946; Darlington, 1951; Egley and Williams, 1990; Lewis, 1973), leading to among-season variability. Current research also is revealing that some weeds begin to change the soil's microbial populations to ultimately enhance or impair their own growth (i.e. plant–soil microbial feedback loops) (Lou et al., 2014). While the magnitude of microbial change may not be of concern in the short term, the legacy effects may hamper future crop production.

Weeds cause plants to alter gene expression very early in the crop's life cycle, especially down-regulation of photosynthetic genes (Clay et al., 2009; Moriles et al., 2012; Rajcan et al., 2004). Later weeds may compete for resources (light, nutrients and water), which ultimately results in crop yield reduction. Clearly, undesirable species need to be controlled in the near term and managed over the long term. But what control methods should be used? Prior to the introduction of selective herbicides (starting with 2,4-D in the 1950s), producers may have used hand labour (hoeing, pulling), cultivation or applying inorganic chemicals (such as table salt) as spot treatments to weedy areas. As farms grew larger by field consolidation (Christensen et al., 1998), less hand labour was available or became too expensive, and technology advanced (e.g. larger tractors, spray equipment, selective herbicides). These changes led to modifications in control methods, from spot treatments to broadcast, uniform applications of selective herbicides or cultivation. However, site-specific weed management can have positive effects on economics and environmental quality (Gerhards and Christensen, 2006) if variability in weed species and/or densities is great enough.

2 Site-specific weed management

2.1 Definition of site-specific management

Starting in the late 1980s, Precision Agriculture, or site-specific farming, first coined as 'farming by soil type' (Robert et al., 1996), has been an ongoing topic for researchers, industry and producers. Site-specific farming implies that management is tailored to a specific site for specific problems. The impetus for site-specific management has been economical (expense of inputs vs. commodity return) and environmental (e.g. unused chemical affecting water quality). Site-specific management is often thought of as nutrient management, with large investments in fertilizer management and application. However, research and development for site-specific weed, and other pest management strategies, were initiated about the same time (Clay, 2011; Christensen et al., 2014).

Prior to all the 'gadgets' and 'gizmos' associated with today's thinking about site-specific management (e.g. GIS, drones, robotics and global positioning system (also known as GNSS, Global Network Navigator Service)), producers whose main source of income was their farm knew that there were high- and low-producing areas. By observing cropping patterns over several seasons, farmers realized that some field areas needed extra attention and gained knowledge about best management of these areas by trial and error. Today, the farming situation has changed. There are more absentee landlords, who often want maximum yields to maximize profit (Clay et al., 2017; Reitsma et al., 2016). Farmers may not spend the cropping season in their fields, but have income derived from off-the-farm jobs, with many operations handed off to custom crews. Consultants, who may have responsibilities for 50 000 acres or more, can be hired to provide the best agronomic decisions, but may be employed for only a few cropping seasons, so that long-term continuity about problems and solutions is lacking.

Precision agriculture and site-specific management can use information technologies, such as GPS, remote sensing, tractor guidance systems and other equipment (Gerhards, 2013; Humburg, 1999). Ground-based field scouting (Wiles et al., 2007; Mandal et al., 2009) provides only a few points of data when compared with a remote sensed image. However, if points are carefully chosen, the collected data can be useful to detect and manage spatial and temporal variability of agronomic problems within and across fields (Clay et al., 1999). Adjusting inputs based on field variability facilitates management to optimize (rather than maximize) growth, which should result in greater economic return. 'Exploiting the potential of precision agriculture technologies in sustainable ways depends on whether we first ask, "Are we doing the right thing?" (strategic approach) as opposed to, "Are we doing it right?" (tactical approach)' (Melakeberhan, 2002).

A guiding principle of sustainable agricultural production is to produce a profit while minimizing long-term environmental risks (Clay and Kitchen, forthcoming). Several contrasting risks need to be addressed. Farmers must make a profit. In addition, if the land is not maintained sustainably, it will not be productive. When a field area fails to provide yields that bring in revenue that is in excess of labour, land, equipment and crop input costs, this area may be called 'marginal' land, especially if yields are low year-after-year (Massey et al., 2008). These areas should be considered for land use change, usually from row crop to pasture-type production.

Risks about agricultural production cited by the general public typically include concerns about food security and environmental quality. Food security includes concern about chemical residues in food, and having enough safe, nutritious food available at a reasonable cost. Environmental risks include preserving water, soil and air quality, and off-site movement of inputs that may harm animal habitat or non-target vegetation. Precision agriculture aims to refine larger-scale regional predictions of both abiotic and biotic stress variability into field-scale and within-field information to better target management decisions. Precision agriculture will reduce production agriculture's risks and the concerns of the public about agriculture.

2.2 Uniform management vs. site-specific management

When exploring the feasibility of site-specific weed management, it is important to consider the variability in a field. The crux of site-specific management is 1) the presence of variability (the quality of being uneven or lacking uniformity) that influences weed management and 2) variability great enough to influence the choice of inputs or the success of management practices (Adamchuk et al., 2010; Clay and Kitchen, forthcoming).

Planning and execution of site-specific management includes: 1) determining site variability, 2) deciding if variability is great enough to establish multiple management approaches (zones) and 3) assessing if available equipment has the capacity to achieve the goal. Considerations for variable site-specific weed management include: 1) identifying weed species and densities, 2) locating where they are, 3) determining the availability of techniques for acceptable control, 4) determining the optimal times for control and 5) determining the economic consequences of using a single action versus integrated approaches. Additional considerations may be appropriate. Identifying the weed species and density may not be sufficient, as resistance to herbicides may be an unforeseen problem in the population (Heap, 2016). For example, if a weed is resistant to one or more modes of action, this would change herbicide control options and would affect treatment cost. The best answer(s) may conflict with practical realities of farming. For example, inclement weather may not allow control to be conducted when the weeds are most susceptible, and by the time control can be performed, the weeds would have grown (and are no longer susceptible, or as well controlled), so the planned control method is not possible and irreversible yield loss may occur.

In precision agriculture, information regarding spatial and temporal variability needs to be converted into knowledge identifying problems and possible solutions. This implies that the manager understands the limitations of a field (e.g. areas of droughty soils, salinity problems, flood-prone areas) and that he or she applies agronomic knowledge (e.g. will the area respond to additional fertilizer or will water be the limiting factor?) to optimize yield potential for each site. Applying the knowledge may mean choosing more than one hybrids for a field, modifying plant density by field area, applying different fertilizers or pesticides by area, or changing rates or methods for control on the go. This process of thinking through the data is the single most important step in site-specific management, but often is overlooked, because available technology is the centre of attention.

3 Weed variability and its influence on weed management

Site-specific weed management implies that an area will be divided into two or more areas based on the variability of characteristics that influence weed management. Site-specific assessment may conclude that the best approach is to apply a single herbicide at a single rate across a field. As mentioned above, the variability of the characteristic, that is, the quality of being uneven or lacking in uniformity, has to be great enough in degree to influence the input(s), outcome(s) or effects(s) of the management practices (Clay and Kitchen, forthcoming). The desired outcomes may be based on agronomic considerations (e.g. greater yield, better weed control) and economic returns (e.g. greater return on investment), or may be chosen to lessen environmental effect in sensitive areas (e.g. changing herbicide rate due to soil characteristics, managing areas near homesteads, water ways, or in areas close to endangered species in a different manner than other areas). Both spatial and temporal variabilities are considered most often when exploring if site-specific management is appropriate.

3.1 Spatial variability

Spatial variability is when a measured quantity or quality differs from one physical location to another. Spatial variability across a site may be due to differences in the five soil-forming factors: climate, living organisms, relief, parent material and time. These factors influence soil chemical (pH, EC, CEC), physical (bulk density, water holding capacity and movement) and biological properties (e.g. organic matter and microbial activity). Monitoring and modifying some of the characteristics were noted and acted upon (specifically soil pH and liming) to improve farming early in the twentieth century (Linsley and Bauer, 1929). Soil characteristics (pH, texture, organic matter) can influence the application rate and efficacy of soil-active herbicides.

Weeds often occur in patches across agronomic fields (Marshall, 1988; Brain and Cousens, 1990; Cardina et al., 1997; Chang et al., 2004; Clay et al., 1999, 2006 a and b;

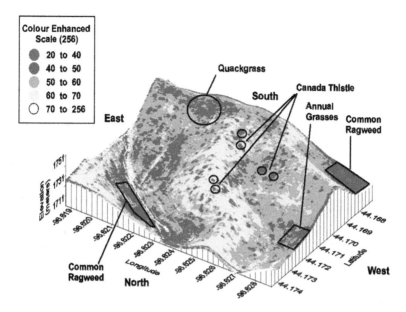

Figure 1 Georeferenced colour-enhanced satellite image of a 160 acre (63 ha) South Dakota field. Note that all the small red dots in the middle of the field, although not circled, are Canada thistle patches (Dalsted and Queen, 1999).

Gehards and Christensen, 2006) (Fig. 1). The presence, abundance or absence of some weed species can also be influenced by the spatial soil characteristics (Dieleman et al., 2000; Dille et al., 2002; Reisinger et al., 2005; Walter et al., 2002). As examples, common cocklebur (*Xanthium strumarium*) and field horsetail (*Equisetum arvense*) are typically found in wetter areas, whereas field sandbur (*Cenchrus pauciflorus*) and puncturevine (*Tribulus terrestris*) are often found in sandy, drier sites. Weed species and density, the presence of herbicide-resistant species, or crop growth stage may also vary spatially across a field and may modify the choice of treatment. In addition, attributes that could influence the environmental effect of management decisions would be distance to streams or sensitive areas or depth to groundwater.

3.2 Temporal variability

Temporal variability examines the changes to parameters of interest over time. Changes may occur over the short term (e.g. weeds growing from 2 to 8 cm in a few days) or over the long term (e.g. different weed species becoming dominant over several growing seasons). Microclimate conditions within a year can affect the appearance of pests across a field (Waggoner and Aylor, 2000). Temporal variability across years has been shown to influence the expansion and contraction of weed patches in specific fields (Johnson et al., 1995; Johnson et al., 1996; Clay et al., 2006b) (Fig. 2). Therefore, a single year of observation is not sufficient for multiyear recommendations.

Characteristics that may be of interest for site-specific weed management include water availability, soil temperature, growing degree days (GDD), crop type and crop and weed growth stages. For exterminating a weed at a specific time during a season,

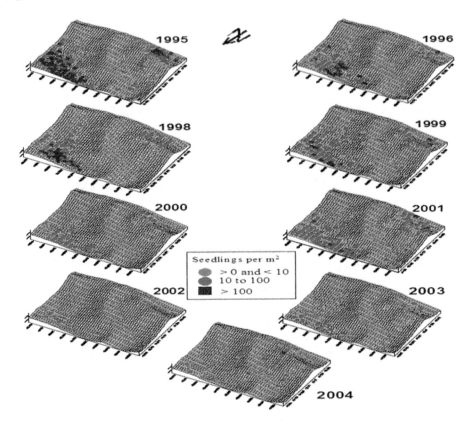

Figure 2 Example of historic record of common ragweed seedlings scouted in May of each year by grid sampling (modified from Clay et al., 2006b).

temporal variability is often examined within a single growing season. Models that use abiotic weather parameters to predict the likely occurrence of crop stress and growth of biotic vectors have been developed. For example, temperature (based on GDD) alone, or combined with soil water conditions (based on hydrothermal time, HTT), can be used to predict weed emergence (Vleeshouwers and Kropff, 2000; Masin et al., 2010; Gonzalez-Andujar et al., 2016; Davis et al., 2013) and growth rates (Forcella, 1998; Forcella et al., 2000). The GDD and HTT information can help scouts anticipate when and which problems will be observed and when to schedule weed management. For example, cool season weeds emerge in early spring and cause problems during emergence of spring crops, whereas warm season weeds emerge later, sometimes after all control has been done. Winter annual weeds emerge in late fall and often cause problems with harvest or with fall planted crops, such as winter wheat.

It should be noted that multiple in-season observations may be warranted due to differences in emergence patterns of weeds (Clay et al., 2006a) (Fig. 3 and 4). A single in-season observation may not provide all the information needed for post-emergent weed control decisions. In addition, even though GDD are the same within an area, spatial microsite characteristics may also influence weed emergence patterns and differ

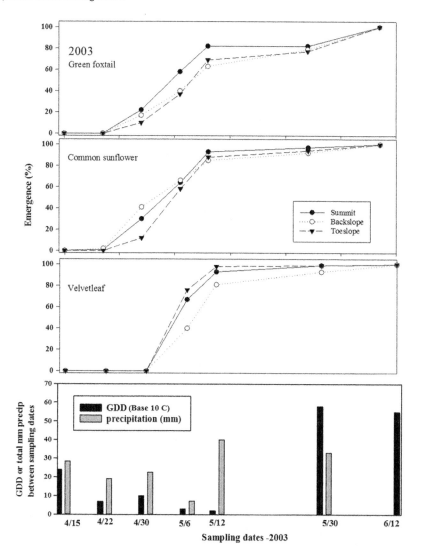

Figure 3 Weed species cumulative emergence by date and landscape position in 2003 near Brookings, South Dakota. Weeds in replicated m² areas were enumerated by species every 7–10 d until no additional plants emerged, and the per cent emergence by sampling date was back calculated from the total number observed. GDD (base 10 C) and precipitation (mm) are provided for each sampling period. The GDD and precipitation for April 15 are cumulative from 1 January 2003 (Author's unpublished data).

by species. The 2003 data (Fig. 3) imply that a post-emergence herbicide application may be unneeded for velvetleaf at any field location on April 30, but 40% of common sunflower had emerged and was growing. Waiting until nearly all weeds have emerged creates a weed population of varying growth stages, with the larger weeds becoming extremely difficult to control, and may also overlap into the crop's critical weed-free

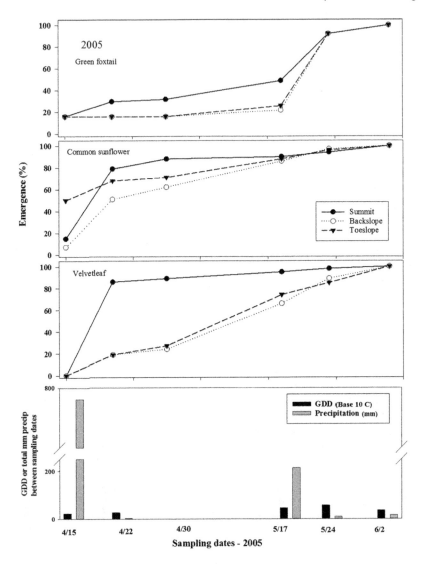

Figure 4 Weed species cumulative emergence by date and landscape position in 2005 near Brookings, South Dakota. Weeds in replicated m² areas were enumerated by species every 7–10 d until no additional plants emerged, and the per cent emergence by sampling date was back calculated from the total number observed. GDD (base 10 C) and precipitation (mm) are provided for each sampling period. The GDD and precipitation for April 15 are cumulative values from 1 January 2005 (Author's unpublished data).

period, generating unrecoverable yield loss. The best management will consider varying the crop over multiple seasons. For example, the same species were tracked in 2003 and 2005 with different emergent patterns (Fig. 4). GDD (base 10 C) leading up to May crop planting were similar in both years. However, spring 2003 was very dry, whereas spring 2005 was very wet. This difference contributed to earlier emergence in 2005.

4 Field scouting: measuring spatial and temporal variabilities of weeds

To create an effective application map, spatial and temporal variabilities of all weeds and species distribution must be measured (Gerhards, 2013). In the past, this information has been collected by field scouting, which created hand-drawn weed maps (Wiles et al., 2007). Other more formal scouting operations may be carried out with coordinates saved for crucial information logged using GPS or differential GPS (DGPS) (which is a corrected satellite signal to give positional sub-meter accuracy) (Clay and Kitchen, forthcoming). When crop, pest or soil parameters are measured and recorded in association with GPS coordinates, it is referred to as *georeferenced* data because the data have a spatial context. These data can be evaluated by software that analyses spatial relationships (*geospatial analysis* or *spatial analysis*), from which maps can be prepared. Software packages used to analyse, manage and map spatial data for a wide range of applications include Geographic Information Systems (GIS) and Farm Management Information Systems.

Clay and Johnson (1999) described several sampling schemes for a field to guide scouting activities. The on-the-ground methods included random sampling, alternative assessment such as a W pattern (Fig. 5) or dividing a field into areas of similar topography, grid sampling with geostatistical interpretation (Donald, 1994; Clay et al., 1999) or observation at harvest (Wiles et al., 2007). When examining the population distribution of weeds, the density data are typically bimodal, with a lot of areas having very low densities and some having very high densities, even in restricted field sizes (Marshall, 1998; Colbach et al., 2000 and 2011). Ground-based sampling techniques provide good information on the weed species present and growth stages. However, ground-based techniques have limited value when developing site-specific recommendations. Due to patchiness of weed infestations, developing spatially accurate weed maps through kriging or nearest neighbour estimates is difficult, even if grid sampling is used, due to the limited number of ground-based observations obtained during field scouting (Clay et al., 1999). This is exacerbated because methods are labour intensive [e.g. 72 ¼-m^2 quadrat samples were needed per ha to provide good approximations of grass weed populations (Marshall, 1988)], and with limited time, large areas of fields are not inspected, and minimal information for decisions is available.

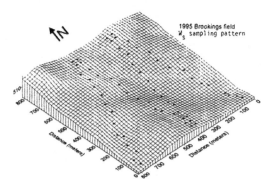

Figure 5 Example of a 'W' pattern for weed scouting (modified from Clay et al., 1999).

In today's communication and technology age, the information that needs to be collected to know field variability and to prescribe the best solutions is expected to be obtained routinely in a near instantaneous, inexpensive manner. The information has to be available in real time (not days or weeks later), and the resulting plan of action must be accomplished with minimal changes or added costs to the norm of a one-management-type fits the whole field. The challenges with information collection, economic uncertainty and implementation complexities have been pointed to as slowing the adoption of site-specific management (Zhang and Kovacs, 2012).

Adaptive sampling could use aerial imagery from satellite or airplane platforms, or from unmanned aerial vehicles (UAVs) (Fig. 6 and 7) equipped with cameras and sensors that have multispectral capabilities to examine spectral properties of an area (Fig. 1, Fig. 8–10) (Scotford and Miller, 2005; Zwiggelaar, 1998; Clay, 2006b). The imagery is based on reflected radiance from objects, which can help identify problem areas. The spectral data may be detected as broadband wavelength intervals (usually 4–8 bands with intervals ranging from 0.4 to 10 um) or, in some cases, hyperspectral wavelengths with 200 or more intervals with bands ranging from 0.4 to 2.5 um (O'Neill and Dalsted, 2011). While the aerial images can provide field-wide information, the resolution is not fine enough to distinguish specific problems, and sorting multiple plant species in a mixed population using hyperspectral signatures has not been resolved at this time, but is a popular research topic. Typically, what is detected in the image are anomalous areas (e.g. irregularities in the image due to a wide array of causes) (Fig. 1, 6 and 7) that must be field scouted to understand the specific reason for the anomaly (weeds, insects, diseases, poor soil conditions, etc.). When examining an image, this information may be paired with prior knowledge to aid in weed detection. For example, weed presence may be implied if row crop imagery has vegetation between rows, or imagery may be combined with environmental gradients to develop predictive models and set boundaries for different species based on specific field types or species resource needs (McGowen et al., 2001; Lass et al., 2011). Ultimately, these images can be used to help with decisions for spatial extent of treatments (Clay et al., 2004).

There are other concerns with remote sensed images (O'Neill and Dalsted, 2011). The geometric accuracy of a remote sensed image is based on the relation of the image to its true shape which can be distorted, especially in areas not directly below the sensor.

Figure 6 Image of a rotary-winged drone with a camera as a payload (Image courtesy of J.C. Streibig, University of Copenhagen).

Figure 7 Image of a fixed-wing unmanned aerial vehicle (Image courtesy of AgEagle, Neodesha, KS, and Raven Industries, Sioux Falls, South Dakota).

Figure 8 False colour image of a South Dakota soybean field. Although the irregular areas on the left and right sides of the image look nearly identical, after ground scouting, the area on the left was determined to be an infestation of waterhemp (*Amaranthus rudis*). The area on the right was a combination of bean leaf beetle and grasshopper damage (Image courtesy of the author).

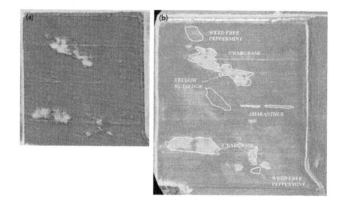

Figure 9 (a) and (b). Aerial field image of a 14 ha field obtained with aircraft flying 1800 m above the target and using a multispectral band camera (a) and identification of field problems (b) using ground-truth information (modified from Gumz and Weller, 2011).

Images are best when the sensor is positioned directly above the target (nadir position), with increasingly less accuracy as the angle increases from the centre point. The field of view of a sensor is the solid angle through which a detector element (pixel sensor) is sensitive to electromagnetic radiation at any one time, which is dependent on the camera's field of view (in degrees) and is the dimension of visible ground for a known

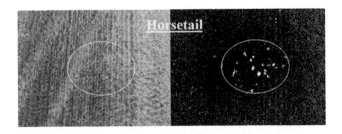

Figure 10 Drone imagery of wheat stubble field (left) and enhanced black and white imagery of the same area (right) with weed infestation circled on each. The weed of interest was horsetail (*Equisetum* sp.) (Images courtesy of J. Streibig, University of Copenhagen).

altitude. The instantaneous field of view is determined by the size of the detector element and the camera's focal length. The accuracy of the images becomes extremely important if overlaying two different types of images (e.g. a satellite image of a field with a yield map generated during harvest). The spatial resolution of the image needs to be known. A single pixel may represent small (<1 m² areas) to very large (30 m²) areas. With finer resolution, the image will have more information and better accuracy. The larger the pixel size, the more dissimilar things are represented in the pixel, which can lead to faulty interpretation. For example in an image with a resolution of 1 m² pixels, a 40 ha field is represented by 400 000 pixels. This fine resolution shows individual crop rows, as well as vegetation between rows. However, a 40 ha field represented by an image with a resolution of 30 m² pixels will contain only 440 pixels, with each pixel blending the signals from an area 900 times larger. Using this coarse resolution image, the interpretation may be that the field does not have a weed problem.

Radiometric resolution defines the level of brightness of the image and, depending on the sensor, can be 8 bit (recording 256 levels of brightness), 10 bit (1024 levels of brightness) or 11 bit (2048 levels of brightness). For satellite images, the temporal resolution should be considered. The return time of satellites to cover a given area may be days or weeks, although with the multitude of satellites taking images in 2016, the return intervals have been shortened. The cost of satellite imagery varies from free (example USGS EarthExplorer http://lpdaac.usgs.gov/data_access/usgs_earthexplorer) to expensive (hundreds of dollars per image with Quickbird technology, Harris MapMart, http://cms.mapmart.com/Products/DigitalElevationModel/Intermap.aspx), depending on resolution, provider and other factors.

Image quality from satellites (160–35 000 km above the earth) and airplane (1–10 km above the earth) platforms is subject to atmospheric conditions (e.g. cloudiness, haze due to smoky conditions or water vapour) (Chang and Coakley, 1993; Chang et al., 2005) and the time of day. Correction of images may be needed as radiance from an object is scattered by gases and aerosols in the atmosphere, and surface glint may also interfere with the detected radiance (Gao et al., 2009). The best images are taken between 10 am and 2 pm, on bright, sunny, cloud-free days. The height of an aerial platform and the sensor type will determine the resolution of the image and the area of interest will be recorded as a single image or as multiple images. If several flight paths (guided by GPS) need to be flown over the area, the individual images need to be 'stitched together' to provide mosaic coverage (Zhang and Kovacs, 2012). The flight paths should overlap (by 30%) and have the sensor as parallel to the ground as possible so that there is no yawing (tilting) of the aircraft. There are computer programmes that will process this type of information within an hour. Using this information for application control maps requires that the data output format be compatible with the controller input format.

Field imagery can be obtained in real time or near real time from unmanned aircraft systems (UAVs) such as rotary wing (Fig. 6) or fixed-wing aircraft (Fig. 7), or from drones (Hugenholtz et al., 2013; Pena et al., 2015; Rasmussen et al., 2013; Zhang and Kovacs, 2012). UAVs equipped with sensors provide colour or false-colour images (Fig. 9) or other information of interest, such as multispectral images, gas sensing or thermal information (Roldan et al., 2015). Because of the lower flight altitudes (125 m or less), atmospheric correction is not needed. GPS receivers are part of the components used on the UAV to programme the flight path so data obtained can be georeferenced (Watts et al., 2010). There are numerous advantages to using UAVs for field scouting. The opportunities to obtain images are not limited by satellite overflight times or dependence on an aircraft. The cost of purchasing a UAV is low compared to that of other remote sensing instruments. The flight time is relatively fast (cruising speeds of 90 km/hr for some models) and the take-off and landing distances are minimal (Watts et al., 2010). The course can be preprogrammed, so that there is little operator error. The altitude of the flight path can be lower than a manned aircraft. This can result in greater pixel resolution, but must be balanced against the number of flight paths and time needed for data collection and processing. Disadvantages of using UAVs include limited weight payload. Sensors and batteries , which determine flying time, must be light weight. Unfortunately, small light weight UAVs have limited flight time and high winds can challenge the stability of the flight path and yawing of the machine. Poor stability and yawing may lead to distorted images and make fusion into a single image difficult. Flying and operating UAVs in the United States is regulated by the Federal Aviation Administration to prevent interference with commercial aircraft and to maintain safety and security. Rules for UAV operation other than for hobbyist activities were revised in August 2016, and regulations should be consulted prior to use (https://www.faa.gov/uas/).

If remote sensing is used, the scale for the image (resolution) and the time when the image was obtained will define how sensitive the image is and its capability to detect different problems. For example, a very dense, large (20 m by 20 m) weed patch present before crop emergence should be easily identified, even if the image resolution is fairly coarse (e.g. 10 m pixels). However, if weeds are sparse, the area of infestation is small, and the crop is close to canopy closure, the weed patch may not be detected. If the image is entered into a controller as an after-mapping application, the area may be missed. A species that does not influence yield at low densities is not important. However, missing certain species of weeds at low densities can result in high yield loss [e.g. giant ragweed (*Ambrosia trifida*) or Palmer amaranth (*Amaranthus palmeri*)] and is unacceptable. Treating small, sparsely populated patches of these species may be appropriate.

5 Other sensing methods and controlling weeds based on spatial variability

Agricultural engineers have been working, independently and with weed scientists, to improve sensing of weeds for site-specific management and application control technologies. To determine the presence of weeds, machine vision and machine learning

Figure 11 Dual nozzles of the Hawkeye™ sprayer system (Image courtesy of Raven Industries, Sioux Falls, South Dakota).

Figure 12 Robovator™, a mechanical tillage system with static knives close to the row and mobile knives for between-row cultivation. The mobile knives are controlled by a system that has cameras and interpreting software to move the knives to weedy areas. More information is available at http://www.visionweeding.com/robovator-mechanical/ (Image courtesy F. Poulsen Engineering, Denmark).

techniques can be used. This technology is based on understanding spectral signatures and plant shapes (Li, 2014; Robbins, 1998; Astrand and Baerveldt, 2005; Du et al., 2007; Slaughter et al., 2008a; Downey et al., 2004).

Some sensing devices have been mounted in front of sprayer nozzles (Fig. 11) or tillage implements (Fig. 12), with computers having specific software capable of distinguishing crops from weeds to determine where control is needed. The information may trigger action: application of an herbicide by turning on individual nozzles or sections of booms on a sprayer, or movement of mobile knives to disrupt the area of concern (Fig. 12). Differential herbicide application may also be attained by using a double boom system (Fig. 11) that is plumbed to have different herbicides, and depending on the weed problem, the booms may be on or off in tandem or separately. Alternatively, direct injection systems may change the herbicide on the go to apply different herbicides where needed.

Sensors have also been mounted on autonomous robots (Slaughter et al., 2008b; Concurrent Solutions, 2004). There are numerous designs on internet sites, some theoretical, and others as prototype designs or working models (Fig. 13 and 14). All of the robots have some kind of device, either a sensor or a camera, to detect weed presence or to differentiate an unwanted plant from the crop. Some machines have artificial lighting, so that natural sunlight (if limited due to clouds or sun angle) is not limiting (Hemming and Rath, 2001). Depending on the intended use, the robots are equipped with mechanical methods of control (e.g. BoniRob, which identifies non-crop plants using machine vision and learning and rams the unwanted plant into the ground or the Tertill that, when it detects a weed, uses a weed-whipper action to destroy the plant) (Fig. 13 and 14), whereas other types of robots have been developed with herbicide application capabilities (Sogaard and Lund, 2007; Concurrent Solutions, 2004).

For larger areas, direct injection pumps may inject different herbicides from different reservoirs or may apply a single herbicide at varying rates when weeds are present

Figure 13 BoniRob™, an autonomous robot that detects weed presence and pushes the unwanted plant into the ground. Intelligence for the fields: Bosch robot gets rid of weeds automatically and without herbicides: New applications for sensor technology and algorithms. Bosch press release, 8 October 2015. https://www.deepfield-robotics.com/en/News-Detail_151008.html (Image courtesy of Robert Bosch Start-up GmbH, Germany, http://www.bosch-startup.com/).

Figure 14 The Tertill™ robot, a prototype of Franklin Robotics (Image courtesy of Rory MacKean rory@franklinrobotics.com).

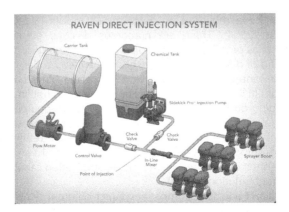

Figure 15 Sidekick Pro™ illustration showing the direct injection method of supplying different herbicides into a boom based on prescription maps (Image courtesy of Raven Industries, Sioux Falls, South Dakota).

(Humburg, 1999; Sokefeld et al., 2005) (Fig. 15). The computer-controlled injection system may have the capacity to adjust the application map using sensors with look-ahead capability. The response time of changing a herbicide application based on direct injection depends on how close the herbicide reservoir is to the nozzle (Sokefeld et al., 2005), with lags as short as 2.5 s for a 10-nozzle boom (El Aissaoui, 2016). Tractor speed and boom width determine the size of an area that can be treated at one time. A tractor travelling at 15 km/h will cover 10 m in 2.5 s. If the swath is 30 m, then the smallest area for treatment will be 300 m². However, changing to a different herbicide will require a lead time to switch and recharge the nozzles, so either speed must be decreased or boom length must be shortened to obtain the desired application precision.

6 Results, interpretation and management decisions

6.1 Results and interpretation of field scouting data to inform site-specific management decisions

The parameters investigated by scouting before site-specific management decisions are implemented should include 1) determining what pest is present, 2) suitable timing to observe the pest (temporal timing), 3) pest biology and sampling method (Can the pest be identified by the chosen method?), 4) pest location (spatial position) and 5) the density of the infestation (Is the pest at a high enough level at the location to cause yield loss or to injure the crop?). After understanding these parameters, the decision is made about where and when to treat with the desired method.

The level of weed infestation can be described in terms of economics, where control is not implemented until yield loss is greater than the economics of the control method. There are numerous reports on expected yield loss due to weed presence, based on crop, weed species, weed density and timing of infestation (Zimdahl, 2004). The expected yield loss can be monetized and compared with the treatment cost. If treatment costs are not recovered, then a no-treatment decision is logical. The type of scouting method and interpretation are important determinants of treatments (Clay, 2006a). For example, Fig. 2 represents the results of grid sampling data for common ragweed densities at 1300 points in a 63 ha field over a series of years. Based on field mean densities, field-wide broadcast treatments would have been recommended in six of the nine years. However, based on the median weed density, no treatment would have been recommended in any of the years. Population densities and spatial distribution were variable enough so that site-specific management would have been recommended in every year.

Site-specific management must be balanced with risk management. Uncontrolled weeds produce seeds that can affect future crops and perpetuate the problem. Growers report that the risk of future problems outweighs today's economics, and therefore, many agronomists have a zero threshold for weeds (Stephens, 1982; Wilkerson et al., 2002; Wilson et al., 2008). The goals and objectives of any weed management programme should be well thought out and agreed upon by all interested parties to clarify and achieve the desired outcome.

6.2 Deciding if the variability is great enough to establish multiple management approaches (zones) for the site

The process of deciding if a one management system fits an entire field or if the field should be split into different management areas is a critical step in site-specific management, but is also an area with the least amount of on-farm research (Rew et al., 1998). Easily implemented practices, such as the change from 'conventional' crops to GMO crops, can occur quickly, as producers were familiar with the herbicides and did not need additional equipment or training (Wilkerson et al., 2002) at the time the practice was introduced. In addition, this change simplified the decision process and was easily integrated into operation practices. However, for site-specific management, additional training, scouting and equipment modification may be required (Wilkerson et al., 2002).

Wilson et al. (2008) examined the complexity of Ohio farmers' thoughts about a single weed management treatment decision. They interviewed farmers on a wide variety of issues concerning weed management and control, and summarized the underlying factors by developing an integrated decision-tree model. The model highlights barriers to adoption of innovative practices and agrees with barriers cited in other reports (Waller et al., 1998; Asmus et al., 2013; Reichardt and Jurgens, 2009). The adoption barriers common to these reports include costs of new management techniques, uncertain risks that may be encountered by changing management, and implementation difficulties (e.g. too complicated, too much management needed, or new machinery needed). Adding new issues for consideration – such as the variation of weed problems from site to site within a field, differences among fields and year-to-year variation in weed species present, spatial coverage and weed densities within a patch, even within one field – further complicates the decision process.

Making site-specific treatment decisions for a field ranges from easy (e.g. a uniform treatment is warranted due to high densities of similar species across the field) to difficult (site-specific treatments make sense, but apply what, where and when?). A zero weed tolerance policy with glyphosate-tolerant crops worked fairly well, until resistant weed biotypes became a problem. Now, even zero tolerance is complicated by resistant weed biotypes.

Many different types of bioeconomic models have been developed and tested to compare economic and ecological outcomes of different management practices with varying inputs (Schweizer et al., 1993; Wilkerson et al., 2002; Forcella et al., 1996; Lybecker et al., 1991; Swinton and King, 1994; Schribbs et al., 1990; Wiles et al., 1996; Buhler et al., 1996; Pannell et al., 2004; Gonzalez-Diaz et al., 2015). However, based on the simplicity of control using GMO crops and a limited spectrum of herbicides used in numerous crops, model development and updating with current information stopped after the mid-1990s. With the development of resistant weeds, revision and use of models are regaining importance.

Most models integrate the biology of single or multiple weed species, in either a single crop or crop rotation, with the expected costs for control and outcomes (weed control efficacy, crop yield loss or gain) for a wide selection of treatments (e.g. Rew et al., 1998; Wilkerson et al., 2002; Pannell et al., 2004; Forcella et al., 1996; Gonzalez-Diaz et al., 2015). Models are based on input prices, expected control efficacy, weed densities and expected yield losses or, alternatively, yield gains, with control if different weeds are present. Some models include cropping conditions (crop row spacing, time of weed emergence vs. crop growth) and environmental information to further define expected weed interference. All of these parameters are dynamic and fluctuate based on market prices or environmental influences. The best decision support models must have biological accuracy, be easy to use, and provide accurate, reliable recommendations (Wilkerson et al., 2002). Modelling efforts are often looked on as educational tools for producers. Models can estimate the economics of treatment options, may allow the producer to visualize the future consequences of this year's weed management efforts on future problems and to compare multiple outcomes with varying inputs compared with outcomes from usual practice. For site-specific management decisions, the results from scouting operations can be used for model input for different portions of the field to determine if different management would be justified.

A way to examine if site-specific weed management would be useful is demonstration trials (Reese et al., 2006). These trials could be in replicated plots located at numerous sites in fields that have different weed species or densities. Multiple treatments can then be compared for weed control efficacy and crop yield. These data can be used to refine model inputs (Forcella et al., 1996). Wiles et al. (1992) reported that a model selected the same treatment even when different weed populations were observed. They reported that the largest differences observed were an overestimation of yield loss if no control was recommended in the regular distribution scenarios. Forcella et al. (1996) reported a wide variation in input costs with fewer less expensive weed control options when a modelling approach was compared with standard grower practices. Even with the lower inputs in the modelled treatments, there were minimal changes to weed populations over the four-year study, and similar, if not higher returns. However, the caveat was that the cost of gathering the data had to be less than the ultimate gross margins to realize net profitability.

Alternatively, results from replicated strip trials across field lengths with best management practices varying by site, based on weed species and densities at the time of treatment, could be compared to normal practices (Blackmer and Kyveryga, 2012). This type of participatory research and demonstration allows producers to compare multiple treatments with 'normal' practices under their own field conditions (Reese et al., 2006). There are several problems with these in-the-field approaches, including applicability to other areas outside the treatment areas (i.e. assuming that similar problems and patterns exist in areas external to the differentially treated strips) and the transferability of the best solutions today to future year's problems (Clay et al., 2006b) (Fig. 2).

Control implementation should not be the last step in the weed management process. After a suitable time period, follow-up scouting should be done to determine the effectiveness of the method, escapes or other problems that may not have been noted earlier. Maps of pest locations and historical records should be maintained on a field-by-field basis to help determine both new and long-standing problems and to prioritize solutions.

7 Summary

Weed scouting, whether it is done only with ground-scouting methods or using remote sensing with some ground verification, is an important tool to aid site-specific crop management. Different weed species should be monitored at different times during the season. Remote sensing has the potential to provide real-time analysis to detect problem areas in fields that can assist in making timely management decisions that affect yield; but ground-truthing should be done to verify the problem in order to make the correct decision. Sensor use and resolution may not be sophisticated enough to provide the fine resolution needed to detect sparse patches or weeds that have just emerged (Scotford and Miller, 2005). Once scouting is complete, a decision to implement management via zones or by a uniform method must be made, weighing the costs, benefits and feasibility of each. This is the most important part of site-specific management, and often the least considered. The overriding factors for maintaining the existing weed management systems, rather than implementing site-specific management, have been, and will continue to be, the favourable economic inputs compared with outcomes and the ease of use of current operations (Wilkerson et al., 2002; Wilson et al., 2008; Asmus et al., 2013; Reichardt and Jurgens, 2009; Dickmann and Batte, 2014).

8 Future trends in research

Site-specific weed management, to be successful, should use remote sensing technologies accompanied with good conventional scouting programmes. The benefits of site-specific management must outweigh the cost of the technology inputs, as well as the additional time spent on management. As weed management becomes more complex, improvements for future weed management can outweigh the costs. Records should be kept to compare end-of-season yield maps with the in-season scouting information to aid in future management refinements.

Sensors and drones are being improved and modified for optimizing weed identification and directing control efforts for traditional and solar-powered machines. Robotics and other instruments may gain importance in site-specific weed management. In addition, better understanding of phenotypic responses by genotype based on environmental cues will improve agricultural decisions for inputs. We have just started exploring weed-crop interactions at the gene response and modification levels to get a more holistic response of the crop to weed presence (Moriles et al., 2012; Horvath et al., 2006, 2015). Researchers are using new methods to understand how crops can mitigate weed interference through molecular and physiological changes (Rathore et al., 2014; Choe et al., 2016; Williams II, 2015). This research may lead to selecting crop biotypes that are less sensitive to weed interference or to using techniques to insert or delete genes to make crops blind to weed presence. This information can be paired with improved robust technologies for control (Young et al., 2014) and other scientific discoveries to provide knowledge and opportunities for improved weed management.

9 Where to look for further information

Clay, D. E. (Ed), (2004) *Site-Specific Management Guidelines*. IPNI Publishing, Norcross, Georgia.

Clay, S. A. (Ed) (2011), *GIS applications in Agriculture. Volume 3: Invasive species*. CRC Press.

Pierce, F.J. and D.E. Clay. (Eds) (2007), *GIS applications in Agriculture*. CRC Press.

Srinivasan, A. (Ed) 2006. Handbook of Precision Agriculture. *Principles and Applications*. Haworth Press. Binghamton, NY.

Young, S. L. and Pierce, F. J. (Eds), (2014) *Automation: The Future of Weed Control in Cropping Systems*. Springer.

10 References

Adamchuk, V. I., Ferguson, R. B. and Herget, G. W. (2010), 'Soil heterogeneity and crop growth', pp. 3–16. In Oerke, E.-C., Gerhards, R., Menz, G. and Sikora, R. A. (Eds), *Precision Crop Protection – The Challenge and Use of Heterogeneity*. Springer.

Asmus, A., Clay, S. A. and Ren, C. (2013), 'Summary of certified crop advisors' response to a weed resistance survey', *Agron. J.*, 105(4), 1160–6.

Åstrand, B. and Baerveldt, A.-J. (2005), 'A vision based row-following system for agricultural field machinery', *Mechatronics*, 15(2), 251–69.

Blackmer, T. M. and Kyveryga, P. M. (2012), 'Precision tools to evaluate alternative weed management systems in soybean'. In Khosla R. (Ed.), *Proceedings of the 10th International Conference on Precision Agriculture, Denver, CO, USA*, ASA/SSSA/CSSA CDROM 1187. Available from: http://isafarmnet.com/pdf/Precision_Tools_Evaluate_Alternative_Weed_Management_Systems_Soybean.pdf (accessed November 2016).

Brain, P. and Cousens, R. (1990), 'The effects of weed distribution on predictions of yield loss', *J. Appl. Ecol.*, 27(2), 735–42.

Buhler, D. D., King, R. P., Swinton, S. M., Gunsolus, J. L. and Forcella, F. (1996) 'Field evaluation of a bioeconomic model for weed management in corn (*Zea mays*), *Weed Sci.*, 44(4), 915–23.

Cardina, J., Johnson, G. A. and Sparrow, D. H. (1997), 'The nature and consequences of weed spatial distribution', *Weed Sci.*, 45(3), 364–73.

Chang, F.-L., and Coakley Jr., J. A. (1993), 'Estimating errors in fractional cloud cover obtained with infrared threshold methods', *J. Geophys. Res.*, 98(D5), 8825–39.

Chang, J., Clay, S. A., Clay, D. E. and Dalsted, K. (2004), 'Detecting weed-free and weed-infested areas of a soybean (*Glycine max*) field using near-infrared spectral data', *Weed Sci.*, 52(4), 642–8.

Chang, J., Clay, S. A., Clay, D. E., Aaron, D., Helder, D. and Dalsted, K. (2005), 'Clouds influence precision and accuracy of ground-based spectroradiometers', *Comm. Soil Sci. Plant Anal.*, 36 (13–14), 1799–807.

Choe, E., Drnevich, J. and Williams II, M. M. (2016), 'Identification of crowding stress tolerance co-expression networks involved in sweet corn yield', *PLoS One*, 11(1), e0147418. doi: org/10.1371/journal.pone.0147418.

Christensen, S., Nordbo, E., Heisel, T. and Walter, A. M. (1998), 'Overview of developments in precision weed management, issues of interest and future directions being considered in

Europe', pp. 3–13. In Medd R. W. and Pratley, J. E. (Eds), *Precision Weed Management in Crops and Pastures. Workshop Proceedings.* Charles Sturt Univ., Wagga Wagga, New South Wales, Australia. ISBN: 0 86396 650 0.

Christensen, S., Rasmussen, J., Pedersen, S. M., Dorado, J. and Fernandez-Quintanilla, C. (2014), 'Prospects for site-specific weed management'. RHEA Robotics and associated High Technologies and Equipment for Agriculture and Forestry. 2nd International conference Madrid Spain, May 2014 Conference paper, pp. 541–9.

Clay, D. E., Clay, S. A., DeSutter, T. and Reese, C. (2017), 'From plows, horses, and harnesses to precision technologies in the North American Great Plains'. In *Oxford Research Encyclopedia of Environmental Sci.* http://environmentalscience.oxfordre.com/view/10.1093/acrefore/9780199389414.001.0001/acrefore-9780199389414-e-196.

Clay, S. A. (2006a), 'Developing weed management zones for site specific farming application'. In *Proceedings of the 8th International Conference on Precision Agriculture*, A to Z Session, ASA/CSSA/SSSA, Madison, WI (CD-ROM).

Clay, S. A. (2006b), 'Use of remote sensing for weed management'. In *Proceedings of the 8th International Conference on Precision Agriculture*, A to Z Session, ASA/CSSA/SSSA, Madison, WI (CD-ROM).

Clay, S. A. Ed. (2011), GIS applications in Agriculture. Volume 3: Invasive species. CRC Press.

Clay, S. A., Chang, J., Clay, D. E., Reese, C. L. and Dalsted, K. (2004), 'Using remote sensing to develop weed management zones in soybeans', SSMG #42. In Clay, D. E. (Ed.), *Site-Specific Management Guidelines.* IPNI Publishing, Norcross, Georgia.

Clay, S. A., Clay, D. E., Horvath, D., Pullis, J., Carlson, C. G., Hansen, S. and Reicks, G. (2009), 'Corn (*Zea mays*) responses to competition: Growth alteration vs limiting factor', *Agron. J.*, 101(6), 1522–9.

Clay, S. A. and Johnson, G. (1999), 'Scouting for Weeds', SSMG-15. In Clay, D. E. (Ed.), *Site-Specific Management Guidelines.* IPNI Publishing, Norcross, Georgia.

Clay, S. A. and Kitchen, N. (forthcoming), 'Chapter 2. Understanding and identifying variability'. In Shannon, K. D. (Ed.), *Precision Farming Basics.* ASA.

Clay, S. A., Kleinjan, J., Clay, D. E. and Batchelor, W. (2005), 'Growth and fecundity of several weed species in corn and soybean', *Agron. J.*, 97(1), 294–302.

Clay, S. A., Kleinjan, J. and Clay, D. E. (2006a), 'Weed emergence by landscape position'. In *Proceedings of the 8th International Conference on Precision Agriculture.* ASA/CSSA/SSSA, Madison, WI (CD-ROM).

Clay, S. A., Kreutner, B., Clay, D. E., Reese, C., Kleinjan, J. and Forcella, F. (2006b), 'Spatial distribution, temporal stability, and yield loss estimates for annual grasses and common ragweed (*Ambrosia artimisiifolia*) in a corn/soybean production field over nine years', *Weed Sci.*, 54(2), 380–90.

Clay, S. A., Lems, G. J., Clay, D. E., Forcella, F., Ellsbury, M. M. and Carlson, C. G. (1999), 'Sampling weed spatial variability on a fieldwide scale', *Weed Sci.*, 47(6), 674–81.

Colbach, N. and Forcella, F. (2011), 'Adapting geostatistics to analyze spatial and temporal trends in weed populations', pp. 319–71. In Clay, S. A. (Ed.), *GIS Applications in Agriculture. Volume 3: Invasive species.* CRC Press.

Colbach, N., Forcella, F. and Johnson, G. A. (2000), 'Spatial and temporal stability of weed populations over five years', *Weed Sci.*, 48(May-June), 366–77.

Concurrent Solutions, 2004. An Ultralight Autonomous Robot for Weed Control. Final Report, Phase I Proposal Number 00128, USDA 2003 SBIR Program, USDA CSREES, Washington, D. C.

Dalsted, K. and Queen, L. (1999), 'Interpreting remote sensing data', SSMG-26. In Clay D. E. (Ed.), *Site-Specific Management Guidelines.* IPNI Publishing, Norcross, Georgia.

Darlington, H. T. (1951), 'The seventy-year period for Dr. Beal's seed viability experiment', *Am. J. Bot.*, 38(5), 379–81.

Davis, A. S., Clay, S., Cardina, J., Dille, A., Forcella, F., Lindquist, J. and Sprague, C. (2013), 'Seed burial physical environment explains departures from regional hydrothermal model of giant ragweed (*Ambrosia trifida*) seedling emergence in US Midwest', *Weed Sci.*, 61(3), 415–21.

Dickmann, F. and Batte, M. T. (2014), 'Economics of technology for precision weed control in conventional and organic systems', pp. 203–20. In Young, S. L. and Pierce, F. J. (Eds), *Automation: The Future of Weed Control in Cropping Systems*. Springer.

Dieleman, J. A., Mortensen, D. A., Buhler, D. D., Cambardella, C. A. and Moorman, T. B. (2000), 'Identifying associations among site properties and weed species abundance. I. Multivariate analysis', *Weed Sci.*, 48(5), 567–75.

Dille, A. J., Mortensen, D. A. and Young, L. J. (2002), 'Predicting weed species occurrence based on site properties and previous year's weed presence', *Prec. Ag.*, 3(3), 193–207.

Donald, W. W. (1994), 'Geostatistics for mapping weeds, with Canada thistle (*Cirsium arvense*) patch as a case study', *Weed Sci.*, 42(4), 648–57.

Downey, D., Giles, D. K. and Slaughter, D. C. (2004), 'Weeds accurately mapped using DGPS and ground-based vision identification', *Calif. Agric.*, 58(4), 218–21.

Du, J.-X., Wang, X.-F. and Zhang, G.-J. (2007), 'Leaf shape based plant species recognition', *Appl. Math. Comput.*, 185(2), 883–93.

Egley, G. H. and Williams, R. D. (1990), 'Decline of weed seeds and seedling emergence over five years as affected by soil disturbance', *Weed Sci.*, 38(6), 504–10.

El Aissaoui, A. (2016), A Feasibility Study of Direct Injection Spraying Technology for Small Scale Farming: Modeling and Design of a Process Control System. University of Liege-Gembloux Agro-Bio Tech, Belgium, p. 176.

Forcella, F. (1998), 'Real time assessment of seed dormancy and seedling growth for weed management', *Seed Sci. Res.*, 8(2), 201–10.

Forcella, F., Benech Arnold, R. L., Sanchez, R. and Ghersa, C. M. (2000), 'Modeling seedling emergence', *Field Crops Res.*, 67(2), 123–39.

Forcella, F., King, R. P., Swinton, S. M., Buhler, D. D. and Gunsolus, J. L. (1996), 'Multi-year validation of a decision aid for integrated weed management in row crops', *Weed Sci.*, 44(3), 650–61.

Gao, B.-C., Montes, M. J., Davis, C. O. and Goetz, A. F. H. (2009), 'Atmospheric correction algorithms for hyperspectral remote sensing data of land and ocean', *Remote Sens. Environ.*, 113, 517–24.

Gerhards, R. (2013), 'Site-specific weed control', pp. 273–94. In Heege H. J. (Ed.), *Precision in Crop Farming: Site Specific Concepts and Sensing Methods: Applications and Results*. Springer.

Gehards, R. and Christensen, S. (2006), 'Site-specific weed management', pp. 185–206. In Srinivasan, A. (Ed.), *Handbook of Precision Agriculture: Principles and Applications*. Food Products Press.

Gonzalez-Andujar, J. L., Chantre, G. R., Morvillo, C., Blanco, A. M. and Forcella, F. (2016), 'Predicting field weed emergence with empirical models and soft computing techniques', *Weed Res.*, 56, 415–23.

González-Díaz, L., Blanco-Moreno, J. M. and González-Andújar, J. L. (2015), 'Spatially explicit bioeconomic model for weed management in cereals: Validation and evaluation of management strategies', *J. Appl. Ecol.*, 52(1), 240–9.

Gumz, M. S. and Weller, S. C. (2011), 'Using GIS to map and manage weeds in field crops', pp. 301–17. In Clay, S. A. (Ed.), *GIS applications in Agriculture. Volume 3: Invasive Species*. CRC Press.

Heap, I. (2016), 'The International Survey of Herbicide Resistant Weeds', Online. Internet. Friday, 18 November 2016. Available: www.weedscience.org.

Hemming, J. and Rath, T. (2001), 'Computer-vision-based weed identification under field conditions using controlled lighting', *J. Agr. Eng. Res.*, 78(3), 233–43.

Horvath, D. P., Gulden, R. and Clay, S. A. (2006), 'Microarray analysis of late season velvetleaf (*Abutilon theophrasti*) impact on corn', *Weed Sci.*, 54(6), 983–94.

Horvath, D. P., Hansen, S. A., Moriles-Miller, J. P., Pierik, R., Yan, C., Clay, D. E., Scheffler, B. and Clay, S. A. (2015), 'RNAseq reveals weed-induced PIF3-like as a candidate target to manipulate weed stress response in soybean', *New Phytol.*, 207(1), 196–210.

Hugenholtz, C. H., Whitehead, K., Brown, O. W., Barchyn, T. E., Moorman, B. J., LeClair, A., Riddell, K. and Hamilton, T. (2013), 'Geomorphological mapping with a small unmanned aircraft system (sUAS): Feature detection and accuracy assessment of a photogrammetrically-derived digital terrain model', *Geomorphology*, 194(July), 16–24.

Humburg, D. (1999), 'Variable rate equipment – technology for weed control', SSMG-7. In Clay, D. E. (Ed.), *Site-Specific Management Guidelines*. IPNI Publishing, Norcross, Georgia.

Johnson, G. A., Mortensen, D. A. and Gotway, C. A. (1996), 'Spatial and temporal analysis of weed seedling populations using geostatistics', *Weed Sci.*, 44(3), 704–10.

Johnson G. A., Mortensen, D. A., Young, L. J. and Martin, A. R. (1995), 'The stability of weed seedling population models and parameters in eastern Nebraska corn (*Zea mays*) and soybean (*Glycine max*) fields', *Weed Sci.*, 43(4), 604–11.

Lass, L. W., Prather, T. S., Sharfii, B. and Price, W. J. (2011), 'Tracking invasive weed species in rangeland using probability functions to identify site-specific boundaries: A case study using yellow starthistle (*Centaurea solstitialis* L.)', pp. 277–99. In Clay, S. A. (Ed.), *GIS Applications in Agriculture. Volume 3: Invasive species*. CRC press.

Lewis, J. (1973), 'Longevity of crop and weed seeds: Survival after 20 years in soil', *Weed Res.*, 13(2), 179–91.

Li, J. (2014), *3D machine vision system for robotic weeding and plant phenotyping*, PhD dissertation, Iowa State Univ.

Linsley, C. M. and Bauer, F. C. (1929), 'Test your soil for acidity', Univ. of Illinois. Agric. Exp. Station Circ., p. 246.

Lou, Y., Clay, S. A., Davis, A. S., Dille, A., Felix, J., Ramirez, A. H. M., Sprague, C. L. and Yannarell, A. C. (2014), 'An affinity-effect relationship for microbial communities in plant-soil feedback loops', *Microb. Ecol.*, 67(4), 866–76.

Lybecker, D. W., Schweizer, E. E. and King, R. P. (1991), 'Weed management decisions in corn based on bioeconomic modeling', *Weed Sci.*, 39(1), 124–9.

MacRae, I., Carroll, M. and Zhu, J. (2011), 'Site-specific management of green peach aphid, *Myzus persicae* (Sulzer)', pp. 167–90. In Clay, S. A. (Ed.), *GIS Applications in Agriculture. Volume 3: Invasive species*. CRC Press.

Mandal, D., Baral, K. and Dasgupta, M. K. (2009), 'Developing site-specific appropriate precision agriculture', *J. Plant Prot. Sci.*, 1(1), 44–50.

Marshall, E. J. P. (1988), 'Field-scale estimates of grass weed populations in arable land', *Weed Res.*, 28(3), 191–8.

Masin, R., Loddo, D., Benvenuti, S., Zuin, M. C., Macchia, M. and Zanin, G. (2010), 'Temperature and water potential as parameters for modeling weed emergence in central-northern Italy', *Weed Sci.*, 58(3), 216–22.

Massey, R. E., Myers, D. B., Kitchen, N. R. and Sudduth, K. A. (2008), 'Profitability maps as an input for site-specific management decision making', *Agron. J.*, 100(1), 52–9.

McGowen, I., Frazier, P. and Orchard, P. (2001), 'Remote sensing for broadscale weed mapping- is it possible? Geo-spatial information in agriculture', *Precision Ag Symposium: Commodities and management*. Available from: http://www.regional.org.au/au/gia/13/283mcgowen.htm (accessed November 2016).

Melakeberhan, H. (2002), 'Embracing the emerging precision agriculture technologies for site-specific management of yield-limiting factors', *J. Nematol.*, 34(1), 185–8.

Moriles, J., Hansen, S., Horvath, D. P., Reicks, G., Clay, D. E. and Clay, S. A. (2012), 'Microarray and growth analyses identify differences and similarities of early corn response to weeds, shade, and nitrogen stress', *Weed Sci.*, 60(1), 158–66.

O'Neill, M. and Dalsted, K. (2011), 'Obtaining spatial data', pp. 9–28. In Clay, S. A. (Ed.), *GIS Applications in Agriculture. Vol. 3: Invasive species*. CRC Press.

Pannell, D. J., Stewart, V., Bennett, A., Monjardino, M., Schmidt, C. and Powles, S. B. (2004), 'RIM: A bioeconomic model for integrated weed management of *Lolium rigidum* in Western Australia', *Agr. Sys.*, 79, 305–25.

Pena, J. M., Torres-Sanchez, J., Serrano-Perez, A., de Castro, A. I. and Lopez-Granados, F. (2015), 'Quantifying efficacy and limits of unmanned aerial vehicle (UAV) technology for weed seedling detection as affected by sensor resolution', *Sensors*, 15, 5609–26.

Rajcan, I., Chandler, K. J. and Swanton, C. J. (2004), 'Red-far-red ratio of reflected light: A hypothesis of why early season weed control is important in corn', *Weed Sci.*, 52(Sept-Oct), 774–8.

Rasmussen, J., Nielsen, J., Garcia-Ruiz, F., Christensen, S. and Streibig, J. C. (2013), 'Potential uses of small unmanned aircraft systems (UAS) in weed research', *Weed Res.*, 53(4), 242–8.

Rathore, M., Singh, R., Choudhary, P. P. and Kumar, B. (2014), 'Weed stress in plants', pp. 255–65. In Gaun, R. K. and Sharma, P. (Eds), *Approaches to Plant Stress and Their Management*. Springer.

Reese, C. L., Clay, D. E., Beck, D., Kleinjan, J., Carlson, C. G. and Clay, S. (2006), 'Lessons learned from implementing management zones and participatory research in production fields'. In *Proceedings of the 8th International Conference on Precision Agriculture*, ASA/CSSA/SSSA, Madison, WI (CD-ROM).

Reichardt, M. and Jurgens, C. (2009), 'Adoption and future perspective of precision farming in Germany: Results of several surveys among different agricultural target groups', *Precis. Agr.*, 10(1), 73–94.

Reisinger, P., Lehoczky, E. and Komives, T. (2005), 'Relationships between soil characteristics and weeds', *Comm. Soil Sci. Plant Anal.*, 36(4–6), 623–8.

Reitsma, K. D., Clay, D. E., Clay, S. A., Dunn, B. H. and Reese, C. L. (2016), 'Does the U. S. cropland data layer provide an accurate benchmark for land-use change estimates?', *Agron. J.*, 108(1), 266–72.

Rew, L. J., Medd, R. W. and Farquharson, R. J. (1998), 'The scope for precision weed management in on-farm decision making'. In Medd, R. W. and Pratley, J. E. (Eds), *Precision Weed Management in Crops and Pastures*. Workshop Proceedings. Charles Sturt Univ., Wagga Wagga, New South Wales, Australia. ISBN: 0 86396 650 0.

Robbins, B. (1998), 'Real-time detection and classification via computer vision'. In Medd, R. W. and Pratley, J. E. (Eds), Precision Weed Management in Crops and Pastures. Workshop Proceedings. Charles Sturt Univ., Wagga Wagga, New South Wales, Australia. ISBN: 0 86396 650 0.

Robert, P. C., Rust, R. H. and Larson, W. E. (1996), *Preface to Proceedings of the 3rd International Conference of Precision Agriculture*. ASA/CSSA/SSSA. June 1996. Minneapolis, MN.

Roldan, J. J., Joossen, G., Sanz, D., del Cerro, J. and Barrientos A. (2015), 'Mini-UAV based sensory system for measuring environmental variables in greenhouses', *Sensors*, 15(2), 3334–50.

Schribbs, J. M., Schweizer, E. E., Hergert, L. and Lybecker, D. W. (1990), 'Validation of four bioeconomic weed management models for sugarbeet (*Beta vulgaris*) production', *Weed Sci.*, 38(4/5), 445–51.

Schweizer, E. C., Lybecker, D. W., Wiles, L. J. and Westra, P. (1993), 'Bioeconomic weed management models in crop production', pp. 103–8. In Buxton, D. R., Shibles, R., Forsberg, R. A., Blad, B. L., Asay, K. H., Paulsen, G. M. and Wilson, R. F. (Eds), *International Crop Science I*. CSSA, Ames Iowa, July 1992.

Scotford, I. M. and Miller, P. C. H. (2005), 'Applications of spectral reflectance techniques in northern European cereal production: A review', *Biosys. Eng.*, 90(3), 235–50.

Slaughter, D. C., Giles, D. K., Fennimore, S. A. and Smith, R. F. (2008a), 'Multispectral machine vision identification of lettuce and weed seedlings for automated weed control', *Weed Technol.*, 22(2), 378–84.

Slaughter, D. C., Giles, D. K. and Downey, D. (2008b), 'Autonomous robotic weed control systems: A review', *Comput. Electron. Agr.*, 61(1), 63–78.

Sogaard, H. T. and Lund, I. (2007), 'Application accuracy of a machine vision-controlled robotic micro-dosing system', *Biosys. Eng.*, 96(3), 315–22.

Sokefeld, M., Gerhards, R., Oebel, H. and Therburg, R. D. (2005), 'Development of test bench for measuring of lag time of direct nozzle injection for site-specific herbicide applications', *Agratechnische Forschung*, 11(5), 145–54.

Stephens, R. J. (1982), 'One year's seeds, seven years' weeds', pp. 33–46. In Stephens, R. J. (Ed.), *Theory and Practice of Weed Control*. Springer.

Stevens, O. A. (1932), 'The number and weight of seeds produced by weeds', *Am. J. Bot.*, 19(9), 784–94.

Swinton, S. M. and King, R. P. (1994), 'A bioeconomic model for weed management in corn and soybean', *Agr. Sys.*, 44(3), 313–35.

Toole, E. H. and Brown, E. (1946), 'Final results of the Duvel buried seed experiment', *J. Agr. Res.*, 72, 201–10.

Uscanga-Mortera, E., Clay, S. A., Forcella, F. and Gunsolus, J. (2007), 'Common waterhemp growth and fecundity as influenced by emergence date and competing crop', *Agron. J.*, 99(Sept-Oct), 1265–70.

Vleeshouwers, L. M. and Kropff, M. J. (2000), 'Modelling field emergence patterns in arable weeds', *New Phytol.*, 148(3), 445–57.

Waggoner P. E. and Aylor, D. E. (2000), 'Epidemiology, a science of patterns', *Annu. Rev. Phytopathol.*, 38(1), 1–24.

Waller, B. H., Hoy, C. W., Henderson, J. L., Stinner, B. and Welty, C. (1998), 'Matching innovations with potential users, a case study of potato IPM practices', *Agr. Ecosyst. Environ.*, 70(2–3), 203–15.

Walter, A. M., Christensen, S. and Simmelsgaard, S. E. (2002), 'Spatial correlation between weed species densities and soil properties', *Weed Res.*, 42(1), 26–38.

Watts, A. C., Perry, J. H., Smith, S. E., Burgess, M. A., Wilkinson, B. E., Szantoi, Z., Ifju, P. G. and Percival, H. F. (2010), 'Small unmanned aircraft systems for low-altitude aerial surveys', *J. Wildl. Manag.*, 74(7), 1614–19.

Wiles, L. J., Bobbitt, R. and Westra, P. (2007), 'Site-specific weed management in growers' fields: Predictions from hand-drawn maps', pp. 81–102. In Pierce, F. J. and Clay, D. E. (Eds), *GIS Applications in Agriculture*. CRC Press.

Wiles, L. J., King, R. P., Schweizer, E. E., Lybecker, D. W. and Swinton, S. M. (1996), 'GWM: General weed management model', *Agr. Sys.*, 50, 335–76

Wiles, L. J., Wilkerson, G. G., Gold, H. J. and Coble, H. D. (1992), 'Modeling weed distribution for improved postemergence control decisions', *Weed Sci.*, 40(4), 546–53.

Wilkerson, G. G., Wiles, L. J. and Bennett, A. C. (2002) 'Weed management decision models: Pitfalls, perceptions, and possibilities of the economic threshold approach', *Weed Sci.*, 50(4), 411–24.

Williams II, M. M. (2015), 'Relationships among phenotypic traits of sweet corn and tolerance to crowding stress', *Field Crops Res.*, 185(January), 45–50.

Wilson, R. S., Tucker, M. A., Hooker, N. H., LeJeune, J. T. and Doohan, D. (2008), 'Perceptions and beliefs about weed management: Perspectives of Ohio grain and produce farmers', *Weed Tech.*, 22(2), 339–50.

Young, S. L., Meyer, G. E. and Woldt, W. E. (2014), 'Future directions for automated weed management in precision agriculture', pp. 249–60. In Young, S. L. and Pierce, F. J. (Eds), *Automation: The Future of Weed Control in Cropping Systems*. Springer.

Zhang, C. and Kovacs, J. M. (2012), 'The application of small unmanned aerial systems for precision agriculture', *Precis. Agr.*, 13(6), 693–712.

Zimdahl, R. L. (2004), *Weed-Crop Competition. A Review*, 2nd ed., Blackwell Publishing.

Zwiggelaar, R. (1998), 'A review of spectral properties of plants and their potential use for crop/weed discrimination in row-crops', *Crop Prot.*, 17(3), 189–206.

Chapter 8

Assessing and minimizing the environmental effects of herbicides

Christopher Preston, University of Adelaide, Australia

1 Introduction

Herbicides are chemicals that are placed into the environment to control unwanted plants. Over the past several decades they have been widely used for weed control in both developed and developing countries because they are highly efficacious, easy to use and relatively cheap (Gianessi and Reigner 2007; Gianessi 2013). However, one of the problems that they pose is the effects they may have on the environment other than their intended action of controlling weeds (Freemark and Boutin 1995; McLaughlin and Mineau 1995; Stoate et al. 2001). The extent of adverse effects depends on numerous factors including: the chemistry and environmental behaviour of the herbicide molecule, soil type, rainfall pattern, temperature at the time of application, topography of the site, and the susceptibility, level of exposure and timing of exposure of organisms in the environment.

The type of effects that herbicides can impose on the environment is diverse. These include direct effects on non-target plants and effects on soil microbes aquatic organisms. In addition, they may have indirect effects on these species groups as well as others, such as insects, birds and mammals (Freemark and Boutin 1995; Bünemann et al. 2006); they may also have indirect ecological effects. Which species might be affected and how severe the effect is will depend on the sensitivity of the individual species, the specific herbicide and the size and extent of the herbicide exposure. A number of procedures have been proposed to attempt to assess and/or predict the likely effects of herbicides in the environment (Reus et al. 2002; Hart et al. 2003; Van den Brink et al. 2006; Kniss 2017). However, these are not always successful.

Environmental effects of herbicides may be controlled by avoiding the use of a herbicide, reducing the rate of application, substituting another herbicide or changing the type of

http://dx.doi.org/10.19103/AS.2017.0025.10

application (Hart et al. 2003; Felsot et al. 2010). One must consider the conditions at the time of application, the soil type, rainfall patterns and where in the environment the sensitive species occur (Hart et al. 2003; Van den Brink et al. 2006; Reichenberger et al. 2007). To guarantee that no environmental effects will occur is impossible; however, with a good understanding of herbicide chemistry, behaviour in the environment and likely organisms affected, practices that will minimize the environmental effect of herbicides can be employed.

An essential step in assessing and managing the potential environmental effects is to understand the processes through which herbicides create adverse environmental effects. Therefore, this chapter first discusses such processes. It then discusses the types of effects and the factors that are important in assessing whether the effects are likely to be negative. Finally, it discusses mitigation strategies to limit negative environmental effects. Due to the nature of herbicides and the way they are typically applied, the discussion considers processes and potential effects on the area where the products have been applied, on areas where they have not been applied and in water.

2 Sources and fate of herbicides in the environment

Herbicides usually enter the environment while being applied to plants in order to control weeds (Stoate et al. 2001; Nazarko et al. 2005; Rolando et al. 2013). However, there are other sources, such as accidental and deliberate spillage during manufacture, transport or storage that will not be covered here. Herbicides are usually sprayed when applied to the environment; however, they can also be applied as granules or concentrated liquids through rollers or wipers, although such applications are much less common. The various application methods used can influence movement potential.

Figure 1 illustrates some of the sources and fates of herbicides in the environment. The left side illustrates fates during, or shortly after, the application process and the right side illustrates more long-term fates. The figure also illustrates the most common source of herbicides in the environment, application through a sprayer.

2.1 Spray drift

One of the most common mechanisms under which herbicides spread into areas outside the application area is drift from spray application. Spray drift is the movement of the herbicide spray liquid away from its intended target of deposition in other areas (Gil and Sinfort 2005; Nuyttens et al. 2011; Al Heidary et al. 2014). There are two important types of spray drift. The first is wind-driven drift that occurs as a result of application under windy conditions. The second is inversion-driven drift that occurs as a result of application under inversion conditions (Felsot et al. 1996, 2010). A temperature inversion occurs when instead of the normal decrease in temperature with height from the ground temperature increases with height (see Fig. 2). It prevents normal convective overturning of the atmosphere and can trap drift below the inversion.

The main factors that influence spray drift are the environmental conditions at the time of herbicide application, the topography, spray droplet size, height of boom and travel speed (Gil and Sinfort 2005; Felsot et al. 2010). Typically, spray drift occurs as a result of small droplets remaining longer in the air. These droplets can be trapped by wind

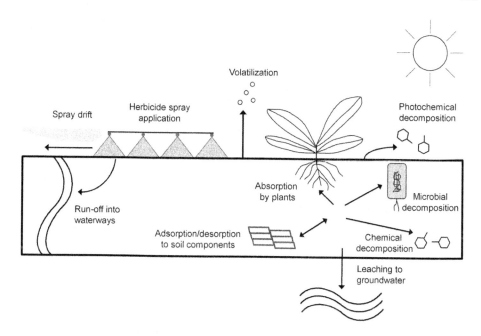

Figure 1 Sources and fate of herbicides in the environment. On the left side are illustrated ways that herbicides can enter the environment at the time or shortly after application: through spray application, spray drift, volatilization or run-off. On the right side is the fate of herbicides after they have been applied to the target: adsorption and desorption to soil, absorption by plants, leaching to groundwater and various methods of degradation.

currents, or in the absence of wind, be trapped below an inversion layer. Droplets smaller than 141 μm have the highest risk of drift (Carlsen et al. 2006; Nuyttens et al. 2007a; 2009). Hence, much of the focus of understanding drift risk is on the behaviour of small droplets. This includes factors that lead to their creation, as well as atmospheric factors that describe how such droplets move (Nuyttens et al. 2007a, 2009; Felsot et al. 2010).

One of the main factors influencing droplet size is the choice of nozzle (Etheridge et al. 2001; Nuyttens et al. 2007a; Al Heidary et al. 2014). Flat fan nozzles are widely used and are preferable for pesticide application. These nozzles are cheap to manufacture and produce droplets in a wide range of sizes that make them ideal for a multipurpose nozzle for contact and systemic pesticides (Knoche 1994; Ramsdale and Messersmith 2001). Unfortunately, these characteristics make these nozzles poor for drift management (Nuyttens et al. 2007a, 2009). While all pesticides drift, it is herbicides that are most destructive due to their potential for damage to sensitive plant species (Marrs et al. 1989, 1991; Kleijn and Snoeijing 1997).

One way to reduce drift is to reduce the number of small droplets produced by selecting different nozzles. Drift reduction and air inclusion nozzles have been developed to reduce the number of fine droplets produced compared to flat fan nozzles (Etheridge et al. 1999; Guler et al. 2007). However, they are not multipurpose nozzles as they may not be suitable for all fungicides and insecticides where good foliar coverage is required (Permin et al. 1992; Lešnik et al. 2005).

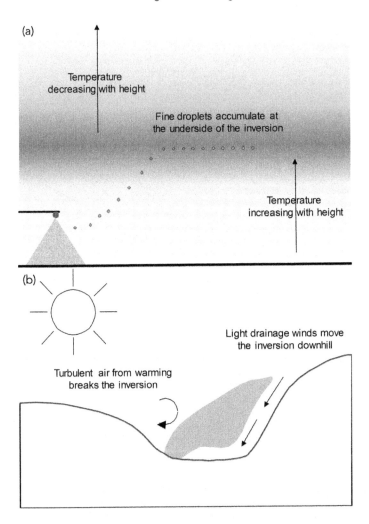

Figure 2 (a) Inversion conditions occur under still air conditions when the atmosphere is warm from the soil surface to the inversion layer and cool from above. Fine-spray droplets collected at the bottom of the inversion layer. (b) The inversion layer is carried away by light drainage winds often getting collected in valleys and other low areas in the landscape. The herbicide falls out when the inversion breaks.

There are numerous other factors that influence droplet size including spray pressure and adjuvants added to the spray mix (Gil and Sinfort 2005; Nuyttens et al. 2011; Creech et al. 2015). Increasing pressure will increase the number of finer droplets produced and increase the risk of spray drift (Nuyttens et al. 2011; Creech et al. 2015). Some adjuvants added to the spray mix can change the size of droplets. Surfactants can result in a greater number of smaller droplets being produced (Dexter and Huddleston 1998; Hilz and Vermeer 2013).

Atmospheric conditions can alter the number of fine driftable droplets. Higher air temperatures and lower humidity will rapidly reduce droplet size, increasing the number of driftable fine droplets (Camp et al. 1989; Gil and Sinfort 2005).

Other management factors can enable droplets to remain suspended in the air and increase their drift potential. Increasing forward speed increases the risk that spray droplets will become entrained behind the vehicle and remain longer in the air (Van de Zande et al. 2005; Nuyttens et al. 2006, 2007b). Raising the boom height will increase the distance the droplets may have to fall to before they reach the target, increasing drift potential (Nuyttens et al. 2006, 2007b).

Wind-driven drift is dependent on wind speed. Typical recommendations are to not spray herbicides in regions where wind speeds are greater than 15 km h^{-1} (Arvidsson et al. 2011; Felsot et al. 2010). Gusty wind can increase wind-driven drift, as changing the wind speed and direction makes management of the process more difficult. Whether wind-driven drift will result in off-target damage depends on the distance the droplets move and whether there are sensitive species down wind. Generally, damage from wind-driven drift is likely to occur within a few metres to 150 m from the release point for ground sprayers (Carlsen et al. 2006). The distance may be greater with aerial sprayers due to the higher release point (Marrs et al. 1992; Robinson et al. 2000).

Several models are available to predict drift under different environmental conditions. The models most widely used by regulatory agencies are AgDRIFT, used mainly in the United States, and AGDISP, used in Canada, Australia and other countries (Duane et al. 1992; Hewitt et al. 2001; Teske et al. 2002; Felsot et al. 2010). These models are designed for near-field applications, rather than for long-distance applications. They also assume flat terrain, constant atmospheric conditions and no barriers to spray drift (Bird et al. 2002; Felsot et al. 2010). These are useful for determining appropriate buffer zones to protect sensitive plants from spray drift damage (Hewitt, 2000; Felsot et al. 2010).

Inversion drift occurs typically under still air conditions (Matthews 2014). Air temperatures are cooler close to the ground and become warm with an increase in height to the inversion layer (Fig. 2). Above the inversion layer, the atmosphere becomes cool with height. Inversions are more likely to occur when the atmosphere is becoming cool and when there is little wind (Fritz et al. 2008). Late evening, night and early morning are the most likely times for inversions to form. With air movement, mixing occurs and inversions do not occur.

When herbicides are applied during inversions, fine droplets that do not reach the ground will rise and become trapped and concentrated under the inversion layer (Fig. 2a). These droplets will remain suspended until the inversion conditions break (Fritz et al. 2008). Light katabatic winds (down slope) can move the inversion layer and associated droplets of pesticide spray to long distances and droplets will typically be collected in low places, such as ditches, hollows and valleys (Fig. 2b).

Inversion layers will break if the wind speed increases above 8 km h^{-1}; however, a stronger breeze is typically required to break up an inversion layer than is required to stop an inversion layer from forming (Fritz 2006). Once an inversion layer has formed, the still air pocket has a tendency to deflect breezes. Commonly, inversion layers break after sunrise when heat from the sun starts warming surfaces. This increases the temperature of the air near the ground, which then starts to rise. The turbulence breaks the inversion layer (Fritz 2006).

When the inversion layer breaks, any herbicide droplets still present in the inversion layer will fall to the ground (Fig. 2b). Should there be susceptible species underneath the inversion, they may be damaged by the herbicide present. Because inversions can move pesticides long distances, unlike wind-driven spray drift, applicators are often unaware that their activities have resulted in damage (Crabbe et al. 1994; Bird et al. 1996; Miller and Staughton 2000). This can make inversion drift much more difficult to predict, recognize and manage.

2.2 Volatilization

Volatilization is evaporation of the herbicide from the surface of the soil or plants after application (Bedos et al. 2002). This causes herbicide vapour to get retained in the air, from where it can move off-site (Egan and Mortensen 2012). The main factors influencing volatilization of herbicides are the chemistry of the herbicide, the nature of the surface, agricultural practices, temperature and surface moisture (Yates 2006a).

All herbicides have some volatility, but for most of them, it is extremely low, indicating that volatilization poses little threat of off-site movement and can be ignored. However, some herbicides have a much higher potential for volatilization. The characteristics that increase volatilization potential include the herbicide vapour pressure and the Henry's Law constant (Spencer et al. 1988; Guth et al. 2004; Yates et al. 2006a). Henry's law is a thermodynamic principle that at a constant temperature the concentration of the gas dissolved in a fluid with which it does not combine chemically is almost directly proportional to the partial pressure of the gas at the surface of the fluid.

Studies on pesticides have shown that there is a good correlation between evaporation rate and vapour pressure of herbicides and other pesticides (Yates et al. 2006b; van Wesenbeeck et al. 2008). Therefore, in terms of risk, herbicide vapour pressure is a significant factor. Table 1 provides data on vapour pressure of several common herbicides. As can be seen, this ranges over many orders of magnitude. It is considered that herbicides with vapour pressure below 0.1 mPa are not volatile (Guth et al. 2004). Some herbicides, most notably 2,4-D and related compounds, are acids that are normally formulated as salts or esters, which can dramatically alter their volatility (Flint et al. 1968; Noble and Hamilton 1990).

Other physicochemical characteristics, such as water solubility, which dictates how the herbicide will interact with soil and plant leaves (Seiber et al. 1986; Yates 2006b), can change the volatility risk of herbicides. Herbicides with lower water solubility will typically have a higher Henry's Law constant, leading to greater volatility (Houbraken et al. 2016). In addition, water on the surface of the soil may reduce binding sites available for the herbicide and increase initial volatility (Prueger et al. 2005; Schneider et al. 2013).

For 2,4-D and other acids, formulation as an ester can increase the vapour pressure and therefore the risk of volatilization (Flint et al. 1968; Hee and Sutherland 1974; Grover 1976). In general, salts will dissociate rapidly to the free acid (Flint et al. 1968). All esters of 2,4-D have higher volatility than the acid forms (Table 2); however, the vapour pressure of low molecular weight esters is much higher than that of higher molecular weight esters [e.g. methyl esters are more volatile than isopropyl esters (Hee et al. 1975; Grover et al. 1985]. The same is the case for aminocyclopyrachlor, where the methyl ester has higher vapour movement than the free acid (Strachan et al. 2010).

Environmental conditions can also have a major influence on the amount and rapidity of volatility. Soil type also influences volatility, which is typically greater from sandy soils than from loam soils (Atienza et al. 2001) due to fewer binding sites and lower absorption. Higher organic matter soils reduce volatilization of trifluralin (Grass et al. 1994) by providing

Table 1 Estimated vapour pressure of various herbicides at 25°C and volatility rating

Herbicide	Estimated vapour pressure (mPa) at 25°C	Volatility rating
Clomazone	19.2	High volatility
Tri-allate	12	High volatility
Trifluralin	9.5	Moderate volatility
S-metolachlor	3.7	Low volatility
Pendimethalin	1.94	Low volatility
Dicamba	1.67	Low volatility
Propyzamide	0.058	Low volatility
Atrazine	0.039	Low volatility
Glyphosate	0.013	Low volatility
2,4-D	0.009	Low volatility
Diflufenican	4.25×10^{-3}	Low volatility
Chlorimuron-ethyl	4.9×10^{-7}	Low volatility

Data from Lewis et al. (2016).

additional binding sites for the herbicide. Plant residue on the soil surface will also reduce volatilization of herbicides (Gish et al. 1995).

Volatility increases at higher air velocities. Higher humidity can increase volatilization for herbicides of low water solubility, such as trifluralin and triallate (Grass et al. 1994; Schneider et al. 2013). The effects of temperature on volatilization are less clear. Increasing temperature should increase volatilization, but this may be affected by changes in other factors, such as humidity and soil water content (Grass et al. 1994; Tabernero et al. 2000; Schneider et al. 2013). At higher temperatures, the surfaces of soil and leaves dry faster, reducing the opportunity for volatilization (Rüdel 1997). For example, vapour movement of dicamba varied by time of application with higher rates of vapour movement from midday and morning applications compared to evening (Mueller et al. 2013).

Lastly, volatilization is often greater from the surface of plant leaves than it is from the soil (Rüdel 1997; Guth et al. 2004). This is likely caused by the longer periods of surface wetness on plant surfaces and the greater number of binding sites available in soil (Houbraken et al. 2016).

Table 2 Effect of ester size on the estimated vapour pressure for 2,4-D esters compared with 2,4-D

2,4-D Ester	Estimated vapour pressure (mPa) at 25°C
Isopropyl	6.13
Butoxyethyl	0.6
Propylene glycol butyl ether	0.4
2-ethylhexyl	0.27
2,4-D acid	0.009

Data for 2,4-D esters from Flint et al. (1968).

2.3 Herbicide run-off

Herbicides may also move from the site of application in water run-off (Baker and Mickelson 1994; Reichenberger et al. 2007; Tang et al. 2012). The main factors that influence run-off are the chemistry of the herbicide, the topography of the application site, soil properties including soil organic matter, water availability, the speed of water movement and the rate of herbicide applied.

Of herbicide properties, the solubility of the herbicide and its ability to bind to soil components are the most important, as run-off is usually the result of rainfall or irrigation shortly after herbicide application. Table 3 summarizes these characteristics for some common herbicides.

Run-off of herbicides is typically only a problem if the herbicide enters an area where there is a risk of environmental damage. The greatest risk is its entry into streams, rivers, lakes and ephemeral water pools. Hence, the focus of this discussion will be on run-off of herbicides into water. Herbicides can move off-site when dissolved in water, but also can attach to soil particles that are moved off-site in water (Fawcett et al. 1994).

Generally, more water-soluble herbicides have a greater risk of run-off (Huber et al. 1998; Carter 2000; Gaynor et al. 2002). However, this is influenced by the binding of the herbicide to soil components, typically soil carbon (Fawcett et al. 1994; Leu et al. 2004) and sorption by montmorillonite clays. Table 3 shows that some herbicides that have lower water solubility and higher binding to soil organic matter are less likely to be affected by run-off. However, as mentioned earlier, if soil is moved off-site, herbicides bound to the soil will also move.

Soil characteristics and topography of the site are important factors in deciding how far herbicides may move (Leu et al. 2004). Vegetation and crop residue left in fields will tend to slow water movement, reducing the likelihood of herbicide run-off (Sauer and Daniel 1987; Fawcett et al. 1994). Cultivation of fields will slow water movement; however, it will increase the likelihood of soil movement and can result in greater herbicide run-off (Sauer and Daniel 1987; Potter et al. 2008).

Table 3 Solubility, binding to organic carbon and typical half-life of some common herbicides

Herbicide	Solubility (mg L^{-1})	K$_{foc}$ (mL g^{-1})	Typical half-life (days)
Atrazine	35	174	75
Simazine	5	750	60
Glyphosate	10500	16331	15
Chlorimuron-ethyl	1200	106	40
Diflufenican	0.05	1996	180
S-metolachlor	480	226	15
Dimethenamid-P	1499	227	11
Diuron	35.6	1067	75.5
Picloram	560	7.2	82.2
Metsulfuron-methyl	2790	12	10
Acetochlor	282	285	14

Data from Lewis et al. (2016).

Field topography influences the amount and rate of run-off through the amount and velocity of water movement (Reichenberger et al. 2007). Sloping fields will have greater risk of herbicide run-off compared with flat fields (Louchart et al. 2001; Leu et al. 2004). In addition, landscape topography will influence the potential damage any herbicide run-off can cause. Landscapes sloping strongly to streams and rivers are likely to collect herbicide run-off from numerous fields and concentrate the herbicide in waterways (Louchart et al. 2001; Leu et al. 2004).

In addition to topography, the amount and timing of rainfall or irrigation after herbicide application is important in enabling run-off from fields (Bowman et al. 1994). Heavy rainfall or irrigation will move more herbicides than light rainfall does. In the initial period after herbicide application, rainfall is more likely to move herbicides before they have a chance to adsorb to soil components or be degraded (Johnson et al. 2000; Kah and Brown 2007). For example, the amount of dissolved metolachlor and terbuthylazine in run-off decreased with an increase in time of irrigation from 3 to 14 days after application (Patdzolt et al. 2007). However, the amount of pendimethalin moving with sediment was dependent on both the amount of sediment being moved and time after application.

The final factor of importance is the amount of herbicide applied. Water-soluble herbicides with low binding capacity to organic carbon which are applied at high rates are more likely to be detected in surface waters (Barbash et al. 2001; Shipitalo and Owens 2006). Surveys of surface waters typically identify pesticides, such as atrazine that are widely used at high rates (Thurman et al. 1991; Barbash et al. 2001). However, changes in frequency of herbicide use can lead to changes in the herbicides detected in surface waters (Scribner et al. 2000; Carabias-Martínez et al. 2003).

2.4 Fates of herbicides in the soil

Once a herbicide has been applied, there are several processes that influence its amount and location in the soil. Herbicide losses occur through photodegradation, chemical degradation, microbial degradation and absorption by plants (Carter 2000). All of these processes reduce the amount of herbicide available. These losses will reduce the risk of environmental effects. The typical half-life of some common herbicides in soil is provided in Table 3. These data are averages of field data, but field half-life can be widely different depending on which of the above factors dominate (Beulke et al. 2000).

Photodegradation is the breakdown of herbicides due to the action of sunlight, either visible or UV, is typically a slow process and usually accounts for only minor losses (Katagi 2004). Photodegradation rates are dependent on the chemistry of the herbicide, the amount of solar energy absorbed and whether the herbicide is bound to soil components (Burrows et al. 2002; Katagi 2004). Binding of herbicides to soil components avoids photodegradation (Curran et al. 1992).

Chemical degradation occurs when chemicals in the environment react with the herbicide with hydrolysis being the major reaction (Sarmah and Sabadie 2002). Chemical degradation can range from being a minor component to a significant component of loss (Kookana et al. 1998; Sarmah and Sabadie 2002). The rate of chemical degradation will be dependent on the chemistry of the herbicide, temperature and moisture availability (Sarmah and Sabadie 2002; Hussain et al. 2015).

Microbial activity is the most common way herbicides are degraded in the environment (Sandmann et al. 1988; Erickson et al. 1989; Sørensen et al. 2003; Holtze et al. 2008). Soils contain a varied population of microorganisms, several of which have the ability to metabolize herbicides (Aislabie and Lloyd-Jones 1995). Many herbicides contain N, P or other elements that can be used for growth by microorganisms (Anderson 1984). The size and activity of the microbial population and its ability to access the herbicide influence the rate of herbicide degradation (Aislabie and Lloyd-Jones 1995; Veeh et al. 1996).

Most of the microbial population occurs in the top few centimetres of soil (Anderson 1984). Herbicides that are leached below this level will be degraded more slowly, as there will be fewer microbes available. Moisture and temperature are required for microbial populations to grow, so microbial degradation is faster in moist, warm soils than in dry or cold soils (Fuesler and Hanafey 1990; Goetz et al. 1990; Flint and Witt 1997). Soils with higher organic matter typically have higher microbial populations and herbicides are degraded more rapidly than soils with low organic matter, provided the herbicide is available to microbes (Walker et al. 1989; Veeh et al. 1996). Lastly, continual use of a herbicide can result in the selection of microbial populations that are able to degrade it more rapidly, leading to loss of efficacy (Roeth 1986).

Herbicides that have low water solubility or are tightly bound to soil components may be unavailable for microbial degradation (Gevao et al. 2000). This may result in greater persistence of the herbicide in the environment. Binding can occur to either organic carbon, clay particles or cations in the soil (Gevao et al. 2000). Trifluralin, for example, has very tight binding to organic matter where it is unavailable for degradation (Wheeler et al. 1979). Paraquat binds tightly to montmorillonite clays (Smith and Mayfield 1978) and glyphosate binds tightly to polyvalent cations, such as Al^{3+} and Fe^{3+} (Sprankle et al. 1975; Moshier and Penner 1978). Because of their complete sorption, paraquat and glyphosate have no soil residual activity.

Soil characteristics can also influence the half-life of the herbicide (Allen and Walker 1987). This is illustrated well by some sulfonylurea herbicides. These herbicides are rapidly degraded by chemical hydrolysis in acid soils, but not in alkaline soils (Fredrickson and Shea 1986; Blair and Martin 1988). In alkaline soils, the herbicides become deprotonated, increasing their solubility and the likelihood they will be moved to soil layers with few microbes (Walker et al. 1989). Together these two factors account for the much greater persistence of these herbicides in alkaline soils compared with acid soils.

Herbicides that are absorbed by plants, either crops or weeds, in the areas where they are applied are then unavailable for environmental effects until the plant dies and breaks down releasing the herbicide. In the plant, the herbicide may be metabolized resulting in less available to return to the environment (Sandermann 1992).

2.5 Leaching of herbicides to groundwater

The leaching of herbicides to groundwater is an area of immense interest due to its potential effects on human health (Ritter et al. 2002). In many parts of the world, groundwater is an important source of water for human consumption, and its contamination with pesticides, including herbicides, carries risks (Stuart et al. 2012). Groundwater flows at many depths below the surface. A significant concern is the contamination of shallow groundwater, as this is more likely to be used for human consumption (Ritter 1990). The major factors that affect leaching to groundwater include: the chemistry of the herbicide, specifically its solubility in water, the soil type, the pattern and intensity of rainfall or irrigation, the rate of herbicide applied, the rate of degradation of the herbicide in the soil and the location of groundwater (Arias-Estévez et al. 2008).

Herbicides that have higher water solubility and lower binding capacity to soil organic carbon are more likely to be leached to groundwater (Kolpin et al. 1998; Ritter 1990). Table 3 lists the water solubility and binding to soil organic matter of a variety of common herbicides. Based on solubility alone, atrazine is more commonly found in groundwater than simazine (Barbash et al. 2001) and metolachlor more commonly found than acetochlor. Surveys of wells in the United States have found atrazine and metolachlor more often in groundwater and at higher concentrations than simazine and acetochlor, respectively (Barbash et al. 2001; Steinheime 1993). However, the amounts present in groundwater are also influenced by the amount applied (see the following section).

Soil type is a major determinant of leaching (Ritter 1990; Morillo et al. 2004; Montoya et al. 2006). Sandy soils tend to have larger soil particles and larger spaces between particles than clay soils. This leads to more rapid water movement through the soil, increasing the likelihood of leaching to groundwater.

The pattern and intensity of rainfall influences leaching. High rainfall or irrigation results in increased movement, particularly after application to dry soil (Sigua et al. 1993; Isensee and Sadeghi 1995; Pot et al. 2005). This is due to the rapid water flow through cracks or macropores where the herbicide has less opportunity to interact with soil components.

Herbicide degradation also influences leaching. Herbicides that are degraded slowly have more opportunity to be moved through the soil profile and into groundwater (Arias-Estévez et al. 2008). The relatively long half-life of atrazine (60 days) makes it a particular risk for being present in groundwater (Barbash et al. 2001). The various factors that affect degradation of herbicides have already been described. Sandy soil types that are more likely to facilitate movement of herbicides also tend to have lower populations of microorganisms and hence lower rates of microbial degradation.

As mentioned earlier, shallow groundwater is typically the focus of concern because it is more likely to be contaminated with herbicides as it is closer to the source and more likely to be used for human consumption (Barbash et al. 2001). Studies in the United States and elsewhere have shown that herbicides are more likely to be detected in shallow wells (Leistra and Boesten 1989; Moorman et al. 1999; Barbash et al. 2001; Kolpin et al. 2002). They are more likely to be detected in groundwater in agricultural regions (Kolpin et al. 2002). However, herbicides are also found in groundwater in urban regions (Van Stempvoort et al. 2014).

Finally, the rate of the herbicide applied has a major effect on herbicide detection in groundwater (Barbash et al. 2001; Kolpin et al. 2002). For example, sulfonylurea herbicides are typically more soluble than atrazine and bind less to organic matter. However, because of their much lower use rates, sulfonylurea herbicides are rarely detected in groundwater (Battaglin et al. 2000).

3 Environmental effects of herbicides

The types of environmental effects of herbicides can be direct or indirect. These can occur off-site in terrestrial environments, in water bodies or on-site.

The degree of the environmental effect a herbicide may impose is dependent on the herbicide concentration, the sensitivity of the species exposed and the possible interaction of a sensitive species with a sufficient concentration of a herbicide. Because the distribution of sensitive species and the concentration of herbicide in space and time are heterogeneous, it is difficult to accurately quantify environmental effects of herbicides. There is a large body of evidence examining the effect of specific herbicides on sensitive organisms that can be drawn on (Freemark and Boutin 1995; McLaughlin and Mineau 1995; Stoate et al. 2001; Cedergreen and Streibig 2005; Bünemann et al. 2006); however, little of it addresses the issue of heterogeneity. A good many of the studies are conducted on mechanisms of damage at concentrations higher than those usually encountered in the environment. In this section, I will address some of the methodologies used to address effects and consider their limitations.

3.1 Off-site terrestrial effects of herbicides

The relevant processes are spray drift and volatilization. Typical losses from drift are hard to estimate, but may be as much as 10% of the product applied (Gil and Sinfort 2005). Most drift occurs within a few metres of the application zone; however, a small percentage may occur at large distances due to inversion (Bird et al. 1996). Attempts to model spray drift under wind conditions to identify the size of buffer zones required have been made and these models can be used to assess likely environmental concentrations (Hewitt 2000; Felsot et al. 2010). Inversion drift has been less studied, because it is difficult to predict the concentration of herbicide that will be present in the environment.

Volatilization losses can be up to 90% of the herbicide applied, depending on the chemistry of the herbicide (Bedos et al. 2002; Guth et al. 2004); however, most of this will be readily dispersed and only a small fraction is likely to have an observable environmental effect. Most studies of volatilization have occurred close to the source and little on long-distance effects (Breeze et al. 1992), which are assumed to be much lower and therefore less important.

Plants are most likely to be affected by off-site movement (Schmitz et al. 2015). Due to the economic effect of spray drift and volatilization on crops, there has been considerable interest in quantifying their effects on crop yield. Much of this work is done by applying a percentage of the field dose over the top of the crop (Table 4). This approach is good for determining damage thresholds and comparing products; however, it rarely represents the true effect of damage due to drift in the environment, as non-crop species are not considered (Egan et al. 2014a).

As would be expected the amount of damage caused by different herbicides varies among crops (Table 4). Broadleaf crops, such as cotton and grapes, are particularly sensitive to herbicides (e.g. 2,4-D) used for broadleaf weed control (Egan et al. 2014a). By contrast, cereal crops are more sensitive to herbicides active on grasses, such as glyphosate. The amount of damage for the same rate of herbicides on crops can be widely variable (Table 4). Plants can have different sensitivity to herbicides at different growth stages, with the early reproductive stage often being the most susceptible. Therefore, depending on when the spray drift occurs, damage to non-target plants will vary (Kurtz and Street 2003; Everitt and Keeling 2009). However, for other herbicide crop combinations, early drift damage can result in greater yield loss (Ellis et al. 2003). Higher concentration of herbicide in the droplets drifting will increase the damage that it causes (Banks and Schroeder 2002; Ellis et al. 2002; Roider et al. 2008).

Much less work has been conducted to examine the effects of herbicide drift on non-crop species (Marrs et al. 1989, 1991). Clearly, herbicide drift into field borders or adjacent woodlands will affect sensitive species (Kleijn and Snoeijing 1997; De Snoo and Van der Poll 1999). Where data are not available, estimates of the potential damage to non-crop species can be derived from related crop species.

Table 4 Examples of the impact of various herbicides at spray drift rates[a] on yield of various crop species

Crop	Herbicide	Yield loss[b] (%)	References
Cotton	2,4-D amine	3–100	Marple et al. (2007), Everitt and Keeling (2009)
	2,4-D ester	97–100	Marple et al. (2007)
	Picloram	27–95	Marple et al. (2007)
	Dicamba	6–19	Marple et al. (2007)
	Fluroxypyr	3–34	Marple et al. (2007)
	Triclopyr	0–9	Marple et al. (2007), Snipes et al. (1991)
	Clopyralid	3–9	Marple et al. (2007)
	Glyphosate	0–6	Ellis and Griffin (2002)
	Glufosinate	0–6	Ellis and Griffin (2002)
Soybean	Dicamba	2–5	Al-Khatib and Peterson (1999)
	Primisulfuron	3–4	Al-Khatib and Peterson (1999)
	Nicosulfuron	0	Al-Khatib and Peterson (1999)
	Glyphosate	0–10	Ellis and Griffin (2002), Al-Khatib and Peterson (1999)
	Glufosinate	0–10	Ellis and Griffin (2002), Al-Khatib and Peterson (1999)
Maize	Glyphosate	0–8	Ellis et al. (2003)
	Glufosinate	2–4	Ellis et al. (2003)
Wheat	Imazamox	0–8	Deeds et al. (2006)
	Glyphosate	0–29	Deeds et al. (2006), Roider et al. (2007)
Rice	Glyphosate	0–14	Ellis et al. (2003)
	Glufosinate	0–8	Ellis et al. (2003)
Potato	Dicamba	6–29	Wall (1994)
	Clopyralid	0–23	Wall (1994)
	Imazamethabenz	0–2	Eberlein and Guttieri (1994)
	Imazethapyr	0–1	Eberlein and Guttieri (1994)
	Imazapyr	30–69	Eberlein and Guttieri (1994)
	Tribenuron	4–5	Wall (1994)
	Glyphosate	0–5	Felix et al. (2011)
Grape	2,4-D	29–41	Al-Khatib et al. (1993)
	Chlorsulfuron	0	Al-Khatib et al. (1993)
	Thifensulfuron	0–6	Al-Khatib et al. (1993)
	Bromoxynil	0	Al-Khatib et al. (1993)
	Glyphosate	2–10	Al-Khatib et al. (1993)

[a] Spray drift rates used ranged from 1 to 5% of the full application rate.
[b] Yield loss measured as reduced grain, product or biomass production compared to nil and expressed as the range of values for different crop stages measured.

Indirect effects of drift may make sensitive species less competitive, leading to reduced seed set and population size (Schmitz et al. 2015). This will inevitably affect the plant species mix through competition (Boutin et al. 2014; Egan et al. 2014b). However, it could also affect the population size of species that rely on those sensitive plant species for food or shelter.

3.2 Effects of herbicides in aquatic ecosystems

Surface run-off of herbicides ranges up to 0.25% of that applied to fields (Carter 2000). While the amount of run-off from a single field may be insufficient to result in a significant effect in water, most water bodies have a catchment and run-off can accumulate over time. Therefore, herbicides can accumulate in water from numerous sources (Leu et al. 2004). The other ways that herbicides can enter surface water bodies is through overspray or through accidental or deliberate contamination.

Many organisms living in water have variable sensitivity to herbicides. There is extensive literature on the effects of individual herbicides on individual species. Typically these experiments examine acute toxicity in static experiments (Cooper 1993; DeLorenzo et al. 2001). Much of the research has been to determine the relative sensitivity of species to identify suitable test species for risk assessment (Fairchild et al. 1997; Schuler and Rand 2008). However, across the range of herbicides of interest no species will be consistently the most sensitive (Fairchild et al. 1998).

An alternative approach is to identify sensitivity distributions for freshwater species (Cedergreen et al. 2004; Van den Brink et al. 2006). Figure 3 shows example responses of some freshwater species to a herbicide. The proportion of the total population composed of species with an EC_{50} to a particular herbicide or lower is plotted against the herbicide EC_{50} (Forbes and Calow 2002). Such an approach can be used to identify the trophic group most likely to be affected by the herbicide. In this example, aquatic plants (macrophytes) are more sensitive than algae, which are more sensitive than invertebrates, and vertebrates are the most tolerant. Responses to herbicides can vary, but often aquatic plants are the most sensitive trophic group (Schmitt-Jansen and Altenburger 2005). For atrazine, algae and aquatic plants showed similar sensitivity, with invertebrates considerably more tolerant followed by vertebrates. A similar pattern occurred with diquat–algae and aquatic plants had similar high sensitivity and vertebrates and invertebrates were more tolerant. The pattern for 2,4-D is quite different, with aquatic plants being the most sensitive, followed by invertebrates, vertebrates and algae (Van den Brink et al. 2006).

Species sensitivity distributions can be used to estimate the potential effect of various concentrations of herbicides in water. Atrazine concentrations in surface waters in the United States were found to rarely exceed 20 µg L^{-1} and then usually only following heavy rainfall (Solomon et al. 1996). Similarly, a survey in Ontario, Canada, identified atrazine at concentrations of up to 3.9 µg L^{-1} and metolachlor at concentrations up to 1.8 µg L^{-1} (Byer et al. 2011). This means that under most circumstances only the most sensitive of primary producers are likely to be affected. However, inhibiting the most sensitive species can result in changes to community structure (Van den Brink et al. 2006).

One of the problems with attempting such risk assessments is that they are typically based on acute exposure over short periods of time (DeLorenzo et al. 2001). In reality, concentrations of pesticides vary widely in water and exposure may occur over extended periods. In addition, studies typically consider one or a small number of species, rather than the whole community.

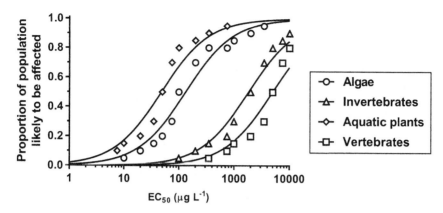

Figure 3 Example responses of freshwater species to herbicides. The figure represents the proportion of the population likely to be affected against EC_{50}. The response of the different groups: algae (O), aquatic plants (◊), invertebrates (Δ) and vertebrates (□).

Herbicides that inhibit photosynthesis are most likely to have the greatest effect on riverine communities due to their effects on primary producers (DeLorenzo et al. 2001). Diuron at 10 µg L^{-1} reduced chlorophyll a production, a measure of the size of the primary producer community, changed algal species richness and reduced bacterial populations (Pesce et al. 2006). This concentration is towards the top end of measured concentrations of diuron in water (Field et al. 2003). Diuron in run-off has also been shown to affect marine communities (Jones et al. 2003).

In contrast, other herbicides may have minimal effects on riverine communities. Glyphosate at 10 µg L^{-1} had virtually no effect on a riverine community (Pesce et al. 2009). Other studies with formulated glyphosate (Schaffer and Sebetich 2004; Pérez et al. 2007; Relyea 2005) found large effects on aquatic communities. However, these studies were conducted at very high glyphosate concentrations. Studies at glyphosate concentrations more typical of those measured in water showed smaller effects (Stachowski-Haberkorn et al. 2008; Relyea 2009), which were likely caused by the surfactant in the formulated herbicide. These surfactants are only likely to enter water through overspray.

3.3 On-site effects of herbicides

Another place that herbicides may affect is the site of application. Much of the research on on-site effects of herbicides is focussed on herbicides in groundwater, because contamination of groundwater is hazardous to human health (Ritter et al. 2002).

As most herbicides end up in the soil after application, there is a likelihood there will be direct effects on soil microbial communities. However, measuring effects on soil microorganisms is complex (Imfield and Vuilleumier 2012). Many soil microorganisms cannot be easily cultured, and measuring them *in situ* can be difficult and time-consuming with the results variable (Schloss and Handelsman 2005). For this reason a lot of measurements are made in artificial microcosms where soil and climate variability can be managed (Jacobsen and Hjelmsø 2014). A large number of such experiments with herbicides have been conducted with 10–100 times the field rate, or the amount used is difficult to calculate back to a field rate (Riah et al. 2014). This makes the results difficult to interpret in relation to normal herbicide use.

A wide range of measurements can be made (Table 5) to determine microbial activity and microbial diversity. All these techniques have drawbacks; however, the best picture is likely to come from a combination of diversity plus activity measurements (Imfield and Vuilleumier 2012). Diversity measurement techniques are increasingly using modern molecular biology methods, including next-generation sequencing, in order to obtain a better distribution of

Table 5 Selected ecological indicators used to determine the effect of herbicides on microbial communities in soil

Type of measurement	Microbial characteristic being measured	Examples of use with herbicides
Biomass	Microbial biomass	Perucci and Scarponi (1994)
Substrate induced respiration	Microbial respiration	Wardle and Parkinson (1990a)
Microcalorimetry	Heat released by microbial activity	Prado and Airoldi (2001)
Soil dehydrogenase	Global measure of enzymatic activity	Beulke and Malkomes (2001)
Fluorescein diacetate hydrolase	Global measure of enzymatic activity	Perucci and Scarponi (1994)
Acid phosphatase	Microbial phosphorus cycle activity	Pozo et al. (1994)
Alkaline phosphatase	Microbial phosphorus cycle activity	Pozo et al. (1994)
Phosphomonoesterase	Microbial phosphorus cycle activity	Sessitsch et al. (2004)
β-Glucosidase	Microbial carbon cycle activity	Sofo et al. (2012)
Cellulase	Microbial carbon cycle activity	Omar and Abdul-Sater (2001)
Urease	Microbial nitrogen cycle activity	Tejada (2009)
Arylsulphatase	Microbial sulfur cycle activity	Sofo et al. (2012)
Plate culture	Culturable microbial population size	Pozo et al. (1994)
Biolog® plates	Culturable microbial diversity	El Fantroussi et al. (1999)
Lipid analysis through PLFA (phospholipid fatty acid analysis) and FAME (fatty acid methyl esters)	Combined estimate of microbial biomass and diversity	Weaver et al. (2007)
DGGE (denaturing gradient gel electrophoresis) of 16S rDNA genes	Microbial taxonomic diversity	El Fantroussi et al. (1999)
SSCP (single-strand conformation polymorphism) of 16S rDNA genes	Microbial taxonomic diversity	Schmalenberger and Tebbe (2002)
T-RFLP (terminal restriction fragment length polymorphism)	Microbial taxonomic diversity	Hart et al. (2009)

all the microbes present (Jacobsen and Hjelmsø 2014). Microbial activity measurements often measure one or a few number of enzymes (Riah et al. 2014). This has the advantage of being relatively easy to do and not requiring radioactivity; however, choice of enzyme is an important factor. Favoured enzymes include dehydrogenase, as it is present in all living cells, or enzymes involved in the phosphorus, nitrogen or carbon cycles (Riah et al. 2014).

Studies on numerous herbicides show that they have varying positive and negative impacts on individual microbial species (Moorman 1989; Wardle and Parkinson 1990a; Johnsen et al. 2001; Sebiomo et al. 2011). However, the environmental effect depends on the rate of herbicide applied, the amount that is free in the soil (not bound to organic matter or other soil components), the rate of degradation and the amount of leaching. In addition, care needs to be taken with assessments of herbicide effects on microbial populations *in situ*, as often other practices are likely to affect microbial communities (Dick 1992; Meriles et al. 2009).

There are several long-term studies of the effects of herbicides on the size or activity of microbial communities. Studies of substituted urea herbicides on soil communities treated for more than ten years showed a reduction in culturable biomass with isoproturon, diuron and diuron plus linuron (El Fantroussi et al. 1999). Studies on repeated applications of glyphosate found a change to the bacterial species present with more gram-negative bacteria and particularly *Burkholdaria* spp. (Lancaster et al. 2010).

By contrast, there have been numerous short-term studies on the effects of glyphosate on soil microbial communities that have found negative (Chakravarty and Chatarpaul 1990; Zobiole et al. 2011), positive (Wardle and Parkinson 1990b; Haney et al. 2002; Mijangos et al. 2009) or no effect (Haney et al. 2000; Lupwayi et al. 2004; Weaver et al. 2007). Differences in application rates, soil types and study length contribute to these different results. Effects of glyphosate are often transient and dependent on the plant community (Haney et al. 2000; Hart et al 2009). There are other factors that need to be considered while assessing effects of the herbicide. For example, glyphosate is a broad-spectrum herbicide that controls most vegetation. This in itself is likely to have an indirect effect on the soil community by depriving it of carbon (Busse et al. 2001). Secondly, glyphosate is often substituted for tillage in agricultural systems. Tillage is known to have substantial and, depending on the intensity, long-lasting effects on soil communities (Lupwayi et al. 1998; Jackson et al. 2003).

Many herbicides are metabolized in the soil by microbes acting alone or in consortia. As such, they provide C, N and other elements for microbial growth (Aislabie and Lloyd-Jones 1995). Continual use of the same herbicide can lead to accelerated breakdown in the soil due to the selection of microbial communities able to use the herbicide for growth. Examples include chlorotoluron (Rouchaud et al. 2000), carbetamide (Hole et al. 2001) and atrazine (Krutz et al. 2007). This will clearly influence the species composition of the microbial community due to competition, although there are few studies exploring this.

There may also be indirect environmental effects from the application of herbicides. Herbicides control unwanted plants, which have implications for species that live in these fields. One example is the reduction in weed seed production in the farm scale evaluation of genetically modified (GM) crops in the United Kingdom when more effective herbicides were used (Heard et al. 2003), which resulted in a reduction in some insect species (Hawes et al. 2003). These reductions in weed seeds and insect populations reduced food available for farmland birds (Gibbons et al. 2006).

A second example is the reduction of milkweed present in corn crops across the US Midwest and the effect of this on monarch butterflies. Even though milkweed is less common in crop fields compared with roadsides (Hartzler and Buhler 2000), the efficacy of glyphosate in corn and soybean fields resulted in a large reduction of milkweed in crops (Hartzler 2010). This reduction in milkweed has been correlated with reduced monarch butterfly populations in the US Midwest (Pleasants and Oberhauser 2013). It is likely that there are other examples where weed control in fields affects other species.

4 Managing environmental effects of herbicides

Having discussed the various ways herbicides can enter the environment and the effects they may have, it is important to consider how to manage these effects. The various strategies that can be used are reduction in herbicide use and its rates, substitution of herbicides or mitigation of herbicide effects. The following discussion will consider some of the approaches that can be taken to minimize these effects.

4.1 Reducing herbicide use or herbicide rates

One way of managing the environmental effects of herbicides is to use herbicides less often. This can be achieved in a couple of ways. For example, other weed control techniques, such as tillage, which can control weeds with no loss of crop yield, can be used (Pimental et al. 1993). There have been some discussion and experimental demonstration of replacement of herbicides with mechanical methods of weed control (Melander et al. 2005; Bajwa et al. 2015). However, the major reason for lack of adoption of mechanical weed control is its cost. Also from an environmental perspective, excessive tillage has negative impacts including increased erosion, reduced water infiltration, greater carbon emissions from tillage operations and detrimental effects on soil organisms (Lal 1993; Pagliai et al. 2004).

An alternative approach is to not apply herbicide to the whole field. Applying herbicide banded over the crop row and using mechanical weed control on the inter-row can reduce the amount applied by up to 80% depending on the row spacing. Band application in maize can be as effective as broadcast herbicide application and provide the same level of crop production for no greater cost (Heydel et al. 1999).

Precision agriculture can reduce the amount applied by applying herbicide only where weeds are present. Depending on the weed distribution, precision application can reduce application up to 90% (Rew et al. 1996; Gerhards and Christensen 2003; Berge et al. 2007). However, this requires an accurate map of the weed infestation or real-time weed identification. Camera-assisted spray technology can be useful in fallow applications (Felton et al. 2002), but in crop applications are currently limited by the availability of weed recognition technology.

Reduction in the rate of application can reduce the amount applied. Several European countries have enacted programmes for reductions in pesticide use (Gianessi et al. 2009). A variety of decision aids to guide farmers on how much of herbicide is needed based on environmental factors, weeds present and weed size have been developed (Kudsk 2008; Sønderskov et al. 2014). This has led to an estimated 60% reduction in herbicide applied in Denmark (Sønderskov et al. 2014).

There are several risks to reducing herbicide rates. The most obvious risk is control failure, particularly if low rates are used under adverse conditions (Zhang et al. 2000;

Blackshaw et al. 2006). Recently, it has been proposed that below-label rates of herbicides may also result in an acceleration of herbicide resistance evolution (Neve and Powles 2005; Renton et al. 2011).

4.2 Substituting for herbicides

An alternative strategy to minimize environmental effects is the substitution of herbicides with lower effect for those of higher effect. There have been several approaches to comparing the environmental effects of pesticides. The most discussed parameter is the environmental impact quotient (EIQ) (Kovach et al. 1992). This is calculated from a range of factors related to the toxicity of the herbicide in addition to its persistence and movement in the environment. The EIQ has three components: an applicator component, a consumer component and an ecological component (Fig. 4).

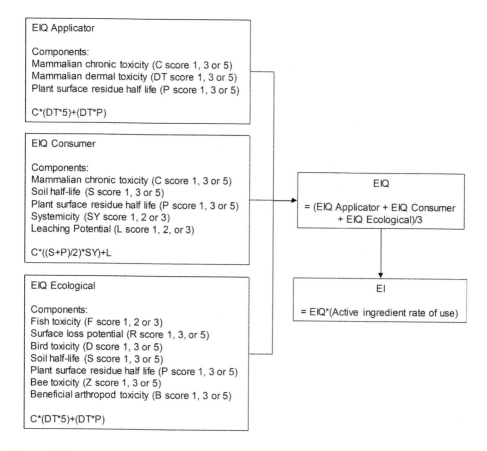

Figure 4 Components and scoring system used to determine EIQ for applicator, consumer and ecological fractions. The final EI is the EIQ multiplied by the active ingredient use rate (after Kovach et al. 1992).

The EIQ can be used to compare individual pesticides for their potential environmental effect (Kovach et al. 1992). However, for practical purposes EIQ is multiplied by the rate used to develop an environmental impact (EI), which can be used to compare weed control programmes for their environmental effect (Soltani et al. 2012). This is done by summing the EI for each pesticide used. There are some shortcomings to using EIQ for herbicides. First, data on some parameters, specifically for beneficial arthropod toxicity, are often lacking (Kovach et al. 1992). In addition, for a variety of reasons, herbicides are treated differently within EIQ compared with other pesticides (Kniss and Coburn 2015). For example, herbicides are all scored low for systemicity. Post-emergent herbicides are all arbitrarily given the highest value for plant surface residual half-life (Kovach et al. 1992).

While EI for herbicide comparisons is likely to be better than simply adding up amounts of herbicide used, it will be imperfect. Kniss and Coburn (2015) conducted simulation analyses for EI of herbicides and found that leaching and soil surface run-off for herbicides explained less than 1% of the variation in EI, despite their obvious importance. Plant surface half-life in contrast explained 26–28% of the variation in EI, despite it being assigned arbitrarily (Kniss and Coburn 2015). Plant surface half-life figures strongly in all three aspects of the EIQ, hence making it important to the final value (Kovach et al. 1992). However, two of these factors, the consumer and the applicator components, focus on human health, rather than environmental effect *per se*. Improvements to EIQ will be necessary for it to be used more effectively for herbicide substitution.

EIQ has not been widely used to compare herbicide programmes, except for the case of GM herbicide-tolerant crops (Brookes and Barfoot 2017). Frequent comparisons have been made comparing herbicide programmes for GM herbicide-tolerant crops with the programmes used in non-GM crops. These comparisons have typically found that herbicide programmes used in GM herbicide-tolerant crops have a lower EI compared to conventional herbicide programmes (Brookes and Barfoot 2017). This is typically the result of herbicide programmes in GM herbicide-tolerant crops being dominated by glyphosate, which has a lower mammalian toxicity and environmental effect than many other herbicides (Kniss 2017).

Another place that EIQ has been used has been to consider the potential effect of glyphosate resistance. Resistance to glyphosate often results in farmers using other herbicides in addition to glyphosate, rather than direct substitution of herbicides. This has led to a higher EI for herbicide programmes in Canada (Beckie et al. 2014). The increase in EI depends on the products added and can range from a marginal increase to doubling the EI.

There are a range of other systems developed to compare the environmental risks of pesticides, mostly in Europe. A number of these have been developed for regulatory purposes (Reus et al. 2002); however, several including the Environmental Yardstick (EY) for pesticides developed in the Netherlands (Reus and Leendertse 2000), SYNOPS developed in Germany (Strassemeyer et al. 2017) and p-EMA (Brown et al. 2003) developed in the United Kingdom are intended for use by pesticide users.

The EY of the Netherlands uses chemical properties, specifically persistence and mobility in the soil, the dose rate, soil organic matter content, time of application, method of application and distance to surface water to determine a risk in Environmental Impact Points (EIP) (Reus 1992). This method provides measures of risk for groundwater, surface water and soil separately. They can then be compared to the regulated maximum permissible concentrations in these environments to determine whether use of the herbicide is likely to exceed the maximum permitted concentration (MPC) (Reus and Leendertse 2000).

SYNOPS is considerably more complex than EY and uses several inputs related to persistence in soil and water, the rate of degradation, vapour pressure, solubility, molecular weight, octanol/water partition coefficient and several toxicological parameters related to earthworms, *Daphnia*, algae, fish, mammals and birds. The highest application rate and most intense application pattern are also used to determine long- and short-term predicted environmental concentrations (Gutsche and Rossberg 1997). An online tool with which the environmental risk assessment can be evaluated on a single field or aggregated level has been developed (Strassemeyer et al. 2017).

The UK p-EMA uses toxicity exposure ratios for birds, mammals, fish, *Daphnia*, algae, *Lemna*, honeybees and earthworms based on the toxicity of the herbicide to the organism adjusted for likely exposure (Hart et al. 2003). This is then coupled with a model of pesticide behaviour in the environment that includes information on pesticide chemistry and behaviour in the environment, the formulation used, the rate of application and the number of repeated applications, crop type and growth stage, spray application, droplet size, quarterly average rainfall, annual average air temperature, soil type, presence of aquifers and surface waters, distance to water depth, and presence of honeybee hives and other wildlife habitats (Brown et al. 2003). An eco-rating is developed that is used to assess whether current practices are good, require review or are poor (Lewis et al. 2003).

4.3 Mitigating herbicide effects

Mitigation programmes to reduce the effect of herbicides on the environment typically consider ways to reduce the movement of herbicides from application sites (Reichenberger et al. 2007). As discussed earlier, the two major ways that herbicides move from the site of application are drift and run-off in water. Mitigation practices can be used in both cases to reduce the amount moving and its likelihood of affecting non-target organisms.

For spray drift, the main mitigation practices include application technology, drift control agents, passive buffer zones and vegetation barriers (Felsot et al. 2010). Spray drift cannot be completely prevented, but it can be minimized. Changes to spray technology including using nozzles that produce fewer fine driftable droplets; changing boom height, forward speed and spray pressure; and using shields are some of the simpler and more practical approaches to reducing spray drift (Nuyttens et al. 2009; Felsot et al. 2010).

Perhaps one of the simplest practices to reduce drift is to change nozzle type. Moving to a drift reduction or air inclusion nozzle type will increase droplet diameter and reduce drift potential (Nuyttens et al. 2009). Reducing spray pressure will likewise reduce production of driftable fine droplets (Nuyttens et al. 2007a). Reducing boom height and forward speed reduces the potential for droplets to become entrained behind the vehicle (Nuyttens et al. 2007b). Spray shields are widely used in row crops, such as vegetables and in other sensitive situations. Correctly designed spray shields will stop the droplets from moving away from the intended target (Felsot et al. 2010). Air-assisted sprayers use air to help drive the droplets onto the target, reducing the potential for drift (Piche et al. 2000).

Drift reduction technologies may affect herbicide efficacy and efficiency of operations. For example, droplets that are too large are likely to bounce off the target or provide insufficient coverage on the target (Creech et al. 2016). Air inclusion in droplets makes them more likely to splatter and spread on the target. In addition, increasing water rates improves coverage (Ferguson et al. 2016).

Buffer zones to sensitive vegetation are widely promoted to reduce the effect of spray drift on sensitive plants or nearby water bodies (Robinson et al. 2000; Brown et al. 2004;

Carlsen et al. 2006). Buffer zones can be bare earth or unsprayed crop (Burn 2003). For buffer zones to be effective, they need to be large enough to intercept sufficient product so that the sensitive areas are not affected. How large the buffer zones need to be will depend on the landscape, wind speeds and the herbicide being applied (Felsot et al. 2010). Buffer zones will need to be larger down wind, on downward slopes, under windy conditions and for herbicides that have effects at low percentages of the spray volume, such as 2,4-D. Buffer zones containing unsprayed crop will be more effective than bare earth, depending on the height of the crop, as they will intercept more of the spray volume (Burn 2003).

Vegetation barriers can be used to intercept spray drift in front of particularly sensitive areas (Ucar and Hall 2001; Brown et al. 2004; Wennecker et al. 2005). However, unless the vegetation barrier is tolerant to all of the herbicides being used, it is likely to suffer damage. Vegetation barriers can also be used as windbreaks to reduce wind speeds at ground level over fields (Ucar and Hall 2001). This can provide more opportunities to spray without it being too windy.

For inversion spray drift, few of the mitigation strategies that can be used for wind-driven spray drift are effective. The most effective minimization strategy is to not apply herbicides when inversions are present or are likely to occur (Fritz et al. 2008). This can be done by monitoring weather and atmospheric conditions. As inversions most often occur between evening and early morning (Matthews 2014), confining pesticide application to the middle of the day will avoid inversions most often. When spraying at other times, wind speeds greater than 3 km h^{-1} are required.

Mitigation strategies for herbicide run-off into water bodies are fewer and harder to implement. The main effective mitigation strategies are vegetated buffer strips and shift in the application date (Riechenberger et al. 2007). There is considerable variation in the efficacy of buffer strips that is unrelated to the width of the buffer strip (Krutz et al. 2005; Otto et al. 2012; Tang et al. 2012). The vegetation in the buffer strip can influence efficacy at reducing run-off, with grassy buffer strips performing better than other types (Krutz et al. 2005).

Mitigation strategies to reduce leaching to groundwater are even fewer. The most effective strategy is substitution by a less leachable herbicide (Businelli et al. 2001; Arias-Estévez et al. 2008). Changes in application date can reduce leaching to groundwater, by applying herbicides when high rainfall events are less likely (Flury 1996). However, this is a less-than-perfect solution, as high rainfall events can be hard to predict.

5 Future trends and conclusion

Herbicides can have numerous effects on the environment, both on-site and off-site. The main causes of herbicides moving off-site are spray drift, volatilization and run-off in surface water. The effects can be extensive including damage to sensitive plants, damage to sensitive species in water and changes to the microbial populations. There can also be indirect effects through control of weeds or reduction in competitiveness, thereby affecting the ecosystem.

Understanding the behaviour of individual herbicides can aid in management of off-site effects. Where practical, substitution of herbicides with a more benign environmental effect or less likely to occur in the environment at damaging concentrations can be an effective way to reduce the effects. There is increasing development of decision support tools, particularly in Europe, to aid users in this regard. Other practices that can be employed include reducing the herbicide rate used, implementing buffer zones, changing spray application practices, times or dates of application or substituting other practices for weed management. Ultimately, the choice of minimization strategy will depend on the situation.

There remains a need for better understanding of herbicide behaviour by users in order to effectively implement appropriate risk minimization. Increasing public concern about herbicide effects on the environment is likely to fuel regulatory approaches to adoption of minimization practices. This can also be seen in Europe with a wide range of approaches that have resulted in reductions in the number of pesticide active ingredients available (Hillocks 2012). There will be a clear need for the development of programmes that will assist farmers to maintain adequate levels of weed control under these circumstances. The increasing evolution of herbicide resistance in weed species (Heap 2014) could exacerbate weed problems, particularly where the availability of effective herbicides is limited.

6 Where to look for further information

A recent comprehensive review on the causes and management of spray drift is by Felsot et al. (2010). See the reference list. The US EPA has recommendations for reducing spray drift at:

https://www.epa.gov/reducing-pesticide-drift

A review of herbicide movement in soils can be found at Arias-Estévez et al. (2008). See the reference list. IUPAC has a pesticide properties database that provides information on the chemical properties, soil behaviour, leaching potential and toxicology of pesticides at:

http://sitem.herts.ac.uk/aeru/iupac/index.htm

A review of factors influencing the appearance and mitigation of herbicides in streams can be found at Riechenberger et al. (2007). See the reference list. The USDA Natural Resources Conservation Service has a National Pesticide Loss Database for the US at:

https://www.nrcs.usda.gov/wps/portal/nrcs/detail/national/technical/nra/ceap/ws/?cid=nrcs143_014167

The US Geological Survey houses the National Water Quality Assessment Project that has information on pesticide risks in groundwater and surface water for the US at:

https://water.usgs.gov/nawqa/pnsp/

7 References

Aislabie, J. and Lloyd-Jones, G. (1995), 'A review of bacterial-degradation of pesticides', *Soil Res.*, 33, 925–42.
Al Heidary, M., Douzals, J. P., Sinfort, C. and Vallet, A. (2014), 'Influence of spray characteristics on potential spray drift of field crop sprayers: A literature review', *Crop Prot.*, 63, 120–30.

Al-Khatib, K. and Peterson, D. (1999), 'Soybean (*Glycine max*) response to simulated drift from selected sulfonylurea herbicides, dicamba, glyphosate, and glufosinate', *Weed Technol.*, 13, 264–70.

Al-Khatib, K., Parker, R. and Fuerst, E. P. (1993), 'Wine grape (*Vitis vinifera* L.) response to simulated herbicide drift', *Weed Technol.*, 7, 97–102.

Allen, R. and Walker, A. (1987), 'The influence of soil properties on the rates of degradation of metamitron, metazachlor and metribuzin', *Pest Manag. Sci.*, 18, 95–111.

Anderson, J. P. E. (1984), 'Herbicide degradation in soil: Influence of microbial biomass', *Soil Biol. Biochem.*, 16, 483–9.

Arias-Estévez, M., López-Periago, E., Martínez-Carballo, E., Simal-Gándara, J., Mejuto, J. C. and García-Río, L. (2008), 'The mobility and degradation of pesticides in soils and the pollution of groundwater resources', *Agric. Ecosyst. Environ.*, 123, 247–60.

Arvidsson, T., Bergström, L. and Kreuger, J. (2011), 'Spray drift as influenced by meteorological and technical factors', *Pest Manag. Sci.*, 67, 586–98.

Atienza, J., Tabernero, M. T., Álvarez-Benedí, J. and Sanz, M. (2001), 'Volatilisation of triallate as affected by soil texture and air velocity', *Chemosphere*, 42, 257–61.

Bajwa, A. A., Mahajan, G. and Chauhan, B. S. (2015), 'Nonconventional weed management strategies for modern agriculture', *Weed Sci.*, 63, 723–47.

Baker, J. L. and Mickelson, S. K. (1994), 'Application technology and best management practices for minimizing herbicide runoff', *Weed Technol.*, 8, 862–9.

Banks, P. A. and Schroeder, J. (2002), 'Carrier volume affects herbicide activity in simulated spray drift studies', *Weed Technol.*, 16, 833–7.

Barbash, J. E., Thelin, G. P., Kolpin, D. W. and Gilliom, R. J. (2001), 'Major herbicides in ground water: Results from the National Water-Quality Assessment', *J. Environ. Qual.*, 30, 831–45.

Battaglin, W. A., Furlong, E. T., Burkhardt, M. R. and Peter, C. J. (2000), 'Occurrence of sulfonylurea, sulfonamide, imidazolinone, and other herbicides in rivers, reservoirs and ground water in the Midwestern United States, 1998', *Sci. Total Environ.*, 248, 123–33.

Beckie, H. J., Sikkema, P. H., Soltani, N., Blackshaw, R. E. and Johnson, E. N. (2014), 'Environmental impact of glyphosate-resistant weeds in Canada', *Weed Sci.*, 62, 385–92.

Bedos, C., Cellier, P., Calvet, R., Barriuso, E. and Gabrielle, B. (2002), 'Mass transfer of pesticides into the atmosphere by volatilization from soils and plants: Overview', *Agronomie*, 22, 21–33.

Berge, T. W., Fykse, H. and Aastveit, A. H. (2007), 'Patch spraying of weeds in spring cereals: Simulated influences of threshold level and spraying resolution on spraying errors and potential herbicide reduction', *Acta Agricult. Scand. Section B - Soil Plant Sci.*, 57, 212–21.

Beulke, S. and Malkomes, H. P. (2001), 'Effects of the herbicides metazachlor and dinoterb on the soil microflora and the degradation and sorption of metazachlor under different environmental conditions', *Biol. Fert. Soil.*, 33, 467–71.

Beulke, S., Dubus, I. G., Brown, C. D. and Gottesbüren, B. (2000), 'Simulation of pesticide persistence in the field on the basis of laboratory data - A review', *J. Environ. Qual.*, 29, 1371–9.

Bird, S. L., Esterly, D. M. and Perry, S. G. (1996), 'Off-target deposition of pesticides from agricultural aerial spray applications', *J. Environ. Qual.*, 25, 1095–104.

Bird, S. L., Perry, S. G., Ray, S. L. and Teske, M. E. (2002), 'Evaluation of the AGDISP aerial spray algorithms in the AgDRIFT model', *Environ. Toxicol. Chem.*, 21, 672–81.

Blackshaw, R. E., O'Donovan, J. T., Harker, K., Clayton, G. W. and Stougaard, R. N. (2006), 'Reduced herbicide doses in field crops: A review', *Weed Biol. Manag.*, 6, 10–17.

Blair, A. M. and Martin, T. D. (1988), 'A review of the activity, fate and mode of action of sulfonylurea herbicides', *Pest Manag. Sci.*, 22, 195–219.

Boutin, C., Strandberg, B., Carpenter, D., Mathiassen, S. K. and Thomas, P. J. (2014), 'Herbicide impact on non-target plant reproduction: What are the toxicological and ecological implications?', *Environ. Pollut.*, 185, 295–306.

Bowman, B. T., Wall, G. J. and King, D. J. (1994), 'Transport of herbicides and nutrients in surface runoff from corn cropland in southern Ontario', *Canad. J. Soil Sci.*, 74, 59–66.

Breeze, V., Thomas, G. and Butler, R. (1992), 'Use of a model and toxicity data to predict the risks to some wild plant species from drift of four herbicides', *Ann. Appl. Biol.*, 121, 669–77.

Brookes, G. and Barfoot, P. (2017), 'Environmental impacts of genetically modified (GM) crop use 1996–2015: Impacts on pesticide use and carbon emissions', *GM Crops Food*, 8, 117–47.

Brown, C., Hart, A., Lewis, K. and Dubus, I. (2003), 'p-EMA (I): Simulating the environmental fate of pesticides for a farm-level risk assessment system', *Agronomie*, 23, 67–74.

Brown, R. B., Carter, M. H. and Stephenson, G. R. (2004), 'Buffer zone and windbreak effects on spray drift deposition in a simulated wetland', *Pest Manag. Sci.*, 60, 1085–90.

Bünemann, E. K., Schwenke, G. D. and Van Zwieten, L. (2006), 'Impact of agricultural inputs on soil organisms - A review', *Soil Res.*, 44, 379–406.

Burn, A. (2003), 'Pesticide buffer zones for the protection of wildlife', *Pest Manag. Sci.*, 59, 583–90.

Burrows, H. D., Santaballa, J. A. and Steenken, S. (2002), 'Reaction pathways and mechanisms of photodegradation of pesticides', *J. Photochem. Photobiol. B: Biol.*, 67, 71–108.

Businelli, D., Tombesi, E. and Trevisan, M. (2001), 'Modelling herbicide treatment impact on groundwater quality in a central Italy area', *Agronomie*, 21, 267–76.

Busse, M. D., Ratcliff, A. W., Shestak, C. J. and Powers, R. F. (2001), 'Glyphosate toxicity and the effects of long-term vegetation control on soil microbial communities', *Soil Biol. Biochem.*, 33, 1777–89.

Byer, J. D., Struger, J., Sverko, E., Klawunn, P. and Todd, A. (2011), 'Spatial and seasonal variations in atrazine and metolachlor surface water concentrations in Ontario (Canada) using ELISA', *Chemosphere* 82, 1155–60.

Camp, C. R., Sadler, E. J. and Busscher, W. J. (1989), 'A water droplet evaporation and temperature model', *Trans. ASAE*, 32, 457–62.

Carabias-Martínez, R., Rodríguez-Gonzalo, E., Fernández-Laespada, M. E., Calvo-Seronero, L. and Sánchez-San Román, F. J. (2003), 'Evolution over time of the agricultural pollution of waters in an area of Salamanca and Zamora (Spain)', *Water Res.*, 37, 928–38.

Carlsen, S. C. K., Spliid, N. H. and Svensmark, B. (2006), 'Drift of 10 herbicides after tractor spray application. 2. Primary drift (droplet drift)', *Chemosphere*, 64, 778–86.

Carter, A. D. (2000), 'Herbicide movement in soils: Principles, pathways and processes', *Weed Res.* 40, 113–22.

Cedergreen, N. and Streibig, J. C. (2005), 'The toxicity of herbicides to non-target aquatic plants and algae: Assessment of predictive factors and hazard', *Pest Manag. Sci.*, 61, 1152–60.

Cedergreen, N., Spliid, N. H. and Streibig, J. C. (2004), 'Species-specific sensitivity of aquatic macrophytes towards two herbicide', *Ecotoxicol. Environ. Safety*, 58, 314–23.

Chakravarty, P. and Chatarpaul, L. (1990), 'Non-target effect of herbicides: I. Effect of glyphosate and hexazinone on soil microbial activity, microbial population, and in-vitro growth of ectomycorrhizal fungi', *Pest Manag. Sci.*, 28, 233–41.

Cooper, C. M. (1993), 'Biological effects of agriculturally derived surface water pollutants on aquatic systems - A review', *J. Environ. Qual.*, 22, 402–8.

Crabbe, R. S., McCooeye, M. and Mickle, R. E. (1994), 'The influence of atmospheric stability on wind drift from ultra-low-volume aerial forest spray applications', *J. Appl. Meteorol.*, 33, 500–7.

Creech, C. F., Henry, R. S., Fritz, B. K. and Kruger, G. R. (2015), 'Influence of herbicide active ingredient, nozzle type, orifice size, spray pressure, and carrier volume rate on spray droplet size characteristics', *Weed Technol.*, 29, 298–310.

Creech, C. F., Moraes, J. G., Henry, R. S., Luck, J. D. and Kruger, G. R. (2016), 'The impact of spray droplet size on the efficacy of 2, 4-D, atrazine, chlorimuron-methyl, dicamba, glufosinate, and saflufenacil', *Weed Technol.*, 30, 573–86.

Curran, W. S., Loux, M. M., Liebl, R. A. and Simmons, F. W. (1992), 'Photolysis of imidazolinone herbicides in aqueous solution and on soil', *Weed Sci.*, 40, 143–8.

De Snoo, G. R. and Van der Poll, R. J. (1999), 'Effect of herbicide drift on adjacent boundary vegetation', *Agricult. Ecosyst. Environ.*, 73, 1–6.

Deeds, Z. A., Al-Khatib, K., Peterson, D. E. and Stahlman, P. W. (2006), 'Wheat response to simulated drift of glyphosate and imazamox applied at two growth stages', *Weed Technol.*, 20, 23–31.

DeLorenzo, M. E., Scott, G. I. and Ross, P. E. (2001), 'Toxicity of pesticides to aquatic microorganisms: A review', *Environmen. Toxicol. Chem.*, 20, 84–98.

Dexter, W. R. and Huddleston, E. W., 1998. Effects of adjuvants and dynamic surface tension on spray properties under simulated aerial conditions. In: *Pesticide Formulations and Application Systems: Eighteenth Volume* (Nalewaja, J. D., Goss, G. R. and Tann, R. S. (Eds)), ASTM International, West Conshohoken, PA, USA, pp. 95–106.

Dick, R. P. (1992), 'A review: Long-term effects of agricultural systems on soil biochemical and microbial parameters', *Agric. Ecosyst. Environ.*, 40, 25–36.

Duan, B., Yendol, W. G., Mierzejewski, K. and Reardon, R. (1992), 'Validation of the AGDISP aerial spray deposition prediction model', *Pest Manag. Sci.*, 36, 19–26.

Eberlein, C. V. and Guttieri, M. J. (1994), 'Potato (*Solanum tuberosum*) response to simulated drift of imidazolinone herbicides', *Weed Sci.*, 42, 70–5.

Egan, J. F. and Mortensen, D. A. (2012), 'Quantifying vapor drift of dicamba herbicides applied to soybean', *Environment. Toxicol. Chem.*, 31, 1023–31.

Egan, J. F., Barlow, K. M. and Mortensen, D. A. (2014a), 'A meta-analysis on the effects of 2, 4-D and dicamba drift on soybean and cotton', *Weed Sci.*, 62, 193–206.

Egan, J. F., Bohnenblust, E., Goslee, S., Mortensen, D. and Tooker, J. (2014b), 'Herbicide drift can affect plant and arthropod communities', *Agric. Ecosyst. Environ.*, 185, 77–87.

El Fantroussi, S., Verschuere, L., Verstraete, W. and Top, E. M. (1999), 'Effect of phenylurea herbicides on soil microbial communities estimated by analysis of 16S rRNA gene fingerprints and community-level physiological profiles', *Appl. Environ. Microbiol.*, 65, 982–8.

Ellis, J. M. and Griffin, J. L. (2002), 'Soybean (*Glycine max*) and cotton (*Gossypium hirsutum*) response to simulated drift of glyphosate and glufosinate', *Weed Technol.*, 16, 580–6.

Ellis, J. M., Griffin, J. L. and Jones, C. A. (2002), 'Effect of carrier volume on corn (*Zea mays*) and soybean (*Glycine max*) response to simulated drift of glyphosate and glufosinate', *Weed Technol.*, 16, 587–92.

Ellis, J. M., Griffin, J. L., Linscombe, S. D. and Webster, E. P. (2003), 'Rice (*Oryza sativa*) and corn (*Zea mays*) response to simulated drift of glyphosate and glufosinate', *Weed Technol.*, 17, 452–60.

Erickson, L. E., Lee, K. H. and Sumner, D. D. (1989), 'Degradation of atrazine and related s-triazines', *Crit. Rev. Environ. Sci. Technol.*, 19, 1–14.

Etheridge, R. E., Womac, A. R. and Mueller, T. C. (1999), 'Characterization of the spray droplet spectra and patterns of four venturi-type drift reduction nozzles', *Weed Technol.*, 13, 765–70.

Etheridge, R. E., Hart, W. E., Hayes, R. M. and Mueller, T. C. (2001), 'Effect of venturi-type nozzles and application volume on postemergence herbicide efficacy', *Weed Technol.*, 15, 75–80.

Everitt, J. D. and Keeling, J. W. (2009), 'Cotton growth and yield response to simulated 2,4-D and dicamba drift', *Weed Technol.*, 23, 503–6.

Fairchild, J. F., Ruessler, D. S., Haverland, P. S. and Carlson, A. R. (1997), 'Comparative sensitivity of *Selenastrum capricornutum* and *Lemna minor* to sixteen herbicides', *Arch. Environ. Contam. Toxicol.*, 32, 353–7.

Fairchild, J. F., Ruessler, D. S. and Carlson, A. R. (1998), 'Comparative sensitivity of five species of macrophytes and six species of algae to atrazine, metribuzin, alachlor, and metolachlor', *Environ. Toxicol. Chem.*, 17, 1830–4.

Fawcett, R. S., Christensen, B. R. and Tierney, D. P. (1994), 'The impact of conservation tillage on pesticide runoff into surface water: A review and analysis', *J. Soil Water Conserv.*, 49, 126–35.

Felix, J., Boydston, R. and Burke, I. C. (2011), 'Potato response to simulated glyphosate drift', *Weed Technol.*, 25, 637–44.

Felsot, A. S., Bhatti, M. A., Mink, G. I. and Reisenauer, G. (1996), 'Biomonitoring with sentinel plants to assess exposure of nontarget crops to atmospheric deposition of herbicide residues', *Environ. Toxicol. Chem.*, 15, 452–9.

Felsot, A. S., Unsworth, J. B., Linders, J. B. H. J., Roberts, G., Rautman, D., Harris, C. and Carazo, E. (2010), 'Agrochemical spray drift; assessment and mitigation - A review', *J. Environ. Sci. Health B*, 46, 1–23.

Felton, W. L., Alston, C. L., Haigh, B. M., Nash, P. G., Wicks, G. A. and Hanson, G. E. (2002), 'Using reflectance sensors in agronomy and weed science', *Weed Technol.*, 16, 520–7.

Ferguson, J. C., Hewitt, A. J. and O'Donnell, C. C. (2016), 'Pressure, droplet size classification, and nozzle arrangement effects on coverage and droplet number density using air-inclusion dual fan nozzles for pesticide applications', *Crop Prot.*, 89, 231–8.

Field, J. A., Reed, R. L., Sawyer, T. E., Griffith, S. M. and Wigington, P. J. (2003), 'Diuron occurrence and distribution in soil and surface and ground water associated with grass seed production', *J. Environ. Qual.*, 32, 171–9.

Flint, J. L. and Witt, W. W. (1997), 'Microbial degradation of imazaquin and imazethapyr', *Weed Sci.*, 45, 586–91.

Flint, G. W., Alexander, J. J. and Funderburk, O. P. (1968), 'Vapor pressures of low-volatile esters of 2,4-D', *Weed Sci.*, 16, 541–4.

Flury, M. (1996), 'Experimental evidence of transport of pesticides through field soils - A review', *J. Environ. Qual.*, 25, 25–45.

Forbes, V. E. and Calow, P. (2002), 'Species sensitivity distributions revisited: A critical appraisal', *Hum. Ecol. Risk Assess.*, 8, 473–92.

Fredrickson, D. R. and Shea, P. J. (1986), 'Effect of soil pH on degradation, movement, and plant uptake of chlorsulfuron', *Weed Sci.*, 34, 328–32.

Freemark, K. and Boutin, C. (1995), 'Impacts of agricultural herbicide use on terrestrial wildlife in temperate landscapes: A review with special reference to North America', *Agric. Ecosyst. Environ.*, 52, 67–91.

Fritz, B. K. (2006), 'Meteorological effects on deposition and drift of aerially applied sprays', *Trans. ASABE*, 49, 1295–301.

Fritz, B. K., Hoffmann, W. C., Lan, Y., Thompson, S. J. and Huang, Y. (2008), 'Low-level atmospheric temperature inversions and atmospheric stability: Characteristics and impacts on agricultural applications', *Agr. Engg. Int.: CIGR eJ.*, X, 10.

Fuesler, T. P. and Hanafey, M. K. (1990), 'Effect of moisture on chlorimuron degradation in soil', *Weed Sci.*, 38, 256–61.

Gaynor, J. D., Tan, C. S., Drury, C. F., Welacky, T. W., Ng, H. Y. F. and Reynolds, W. D. (2002), 'Runoff and drainage losses of atrazine, metribuzin, and metolachlor in three water management systems', *J. Environ. Qual.*, 31, 300–8.

Gerhards, R. and Christensen, S. (2003), 'Real-time weed detection, decision making and patch spraying in maize, sugarbeet, winter wheat and winter barley', *Weed Res.*, 43, 385–92.

Gevao, B., Semple, K. T. and Jones, K. C. (2000), 'Bound pesticide residues in soils: A review', *Environ. Pollut.*, 108, 3–14.

Gianessi, L. P. (2013), 'The increasing importance of herbicides in worldwide crop production', *Pest Manag. Sci.*, 69, 1099–105.

Gianessi, L. P. and Reigner, N. P. (2007), 'The value of herbicides in US crop production', *Weed Technol.*, 21, 559–66.

Gianessi, L., Rury, K. and Rinkus, A. (2009), 'An evaluation of pesticide use reduction policies in Scandinavia', *Outlooks Pest Manag.*, 20, 268–74.

Gibbons, D. W., Bohan, D. A., Rothery, P., Stuart, R. C., Haughton, A. J., Scott, R. J., Wilson, J. D., Perry, J. N., Clark, S. J., Dawson, R. J. and Firbank, L. G. (2006), 'Weed seed resources for birds in fields with contrasting conventional and genetically modified herbicide-tolerant crops', *Proc. Roy. Soc. Lond. B: Biolog. Sci.*, 273, 1921–8.

Gil, Y. and Sinfort, C. (2005), 'Emission of pesticides to the air during sprayer application: A bibliographic review', *Atmospheric. Environ.*, 39, 5183–93.

Gish, T. J., Sadeghi, A. and Wienhold, B. J. (1995), 'Volatilization of alachlor and atrazine as influenced by surface litter', *Chemosphere*, 31, 2971–82.

Goetz, A. J., Lavy, T. L. and Gbur, E. E. (1990), 'Degradation and field persistence of imazethapyr', *Weed Sci.*, 38, 421–8.

Grass, B., Wenclawiak, B. W. and Rüdel, H. (1994), 'Influence of air velocity, air temperature, and air humidity on the volatilisation of trifluralin from soil', *Chemosphere*, 28, 491–9.

Grover, R. (1976), 'Relative volatilities of ester and amine forms of 2,4-D', *Weed Sci.*, 24, 26–8.

Grover, R., Shewchuk, S. R., Cessna, A. J., Smith, A. E. and Hunter, J. H. (1985), 'Fate of 2,4-D iso-octyl ester after application to a wheat field', *J. Environ. Qual.*, 14, 203–10.

Guler, H., Zhu, H., Ozkan, H. E., Derksen, R. C., Yu, Y. and Krause, C. R. (2007), 'Spray characteristics and drift reduction potential with air induction and conventional flat-fan nozzles', *Trans. ASABE*, 50, 745–54.

Guth, J. A., Reischmann, F. J., Allen, R., Arnold, D., Hassink, J., Leake, C. R., Skidmore, M. W. and Reeves, G. L. (2004), 'Volatilisation of crop protection chemicals from crop and soil surfaces under controlled conditions - prediction of volatile losses from physico-chemical properties', *Chemosphere*, 57, 871–87.

Gutsche, V. and Rossberg, D. (1997), 'SYNOPS 1.1: A model to assess and to compare the environmental risk potential of active ingredients in plant protection products', *Agric. Ecosyst. Environ.*, 64, 181–8.

Haney, R. L., Senseman, S. A., Hons, F. M. and Zuberer, D. A. (2000), 'Effect of glyphosate on soil microbial activity and biomass', *Weed Sci.*, 48, 89–93.

Haney, R. L., Senseman, S. A. and Hons, F. M. (2002), 'Effect of Roundup Ultra on microbial activity and biomass from selected soils', *J. Environ. Qual.*, 31, 730–5.

Hart, A., Brown, C., Lewis, K. and Tzilivakis, J. (2003), 'p-EMA (II): Evaluating ecological risks of pesticides for a farm-level risk assessment system', *Agronomie*, 23, 75–84.

Hart, M. M., Powell, J. R., Gulden, R. H., Dunfield, K. E., Pauls, K. P., Swanton, C. J., Klironomos, J. N., Antunes, P. M., Koch, A. M. and Trevors, J. T. (2009), 'Separating the effect of crop from herbicide on soil microbial communities in glyphosate-resistant corn', *Pedobiologia*, 52, 253–62.

Hartzler, R. G. (2010), 'Reduction in common milkweed (*Asclepias syriaca*) occurrence in Iowa cropland from 1999 to 2009', *Crop Prot.*, 29, 1542–4.

Hartzler, R. G. and Buhler, D. D. (2000), 'Occurrence of common milkweed (*Asclepias syriaca*) in cropland and adjacent areas', *Crop Prot.*, 19, 363–6.

Hawes, C., Haughton, A. J., Osborne, J. L., Roy, D. B., Clark, S. J., Perry, J. N., Rothery, P., Bohan, D. A., Brooks, D. R., Champion, G. T. and Dewar, A. M. (2003), 'Responses of plants and invertebrate trophic groups to contrasting herbicide regimes in the Farm Scale Evaluations of genetically modified herbicide–tolerant crops', *Phil. Trans. Roy. Soc. Lond. B: Biol. Sci.*, 358, 1899–913.

Heap, I. (2014), 'Global perspective of herbicide-resistant weeds', *Pest Manag. Sci.*, 70, 1306–15.

Heard, M. S., Hawes, C., Champion, G. T., Clark, S. J., Firbank, L. G., Haughton, A. J., Parish, A. M., Perry, J. N., Rothery, P., Scott, R. J. and Skellern, M. P. (2003), 'Weeds in fields with contrasting conventional and genetically modified herbicide–tolerant crops. I. Effects on abundance and diversity', *Phil. Trans. Roy. Soc. Lond. B: Biol. Sci.*, 358, 1819–32.

Hee, S. Q. and Sutherland, R. G. (1974), 'Volatilization of various esters and salts of 2,4-D', *Weed Sci.*, 22, 313–18.

Hee, S. S. Q., Sutherland, R. G., McKinlay, K. S. and Saha, J. G. (1975), 'Factors affecting the volatility of DDT, dieldrin, and dimethylamine salt of (2, 4-dichlorophenoxy) acetic acid (2, 4-D) from leaf and glass surfaces', *Bull. Environ. Contam. Toxicol.*, 13, 284–90.

Hewitt, A. J. (2000), 'Spray drift: Impact of requirements to protect the environment', *Crop Prot.*, 19, 623–7.

Hewitt, A. J., Teske, M. E. and Thistle, H. E. (2001), 'The development of the AgDRIFT® model for aerial application from helicopters and fixed-wing aircraft', *Aust. J. Ecotoxicol.*, 8, 3–6.

Heydel, L., Benoit, M. and Schiavon, M. (1999), 'Reducing atrazine leaching by integrating reduced herbicide use with mechanical weeding in corn (*Zea mays*)', *Eur. J. Agron.*, 11, 217–25.

Hillocks, R. J. (2012), 'Farming with fewer pesticides: EU pesticide review and resulting challenges for UK agriculture', *Crop Prot.*, 31, 85–93.

Hilz, E. and Vermeer, A. W. (2013), 'Spray drift review: The extent to which a formulation can contribute to spray drift reduction', *Crop Prot.*, 44, 75–83.

Hole, S. J., McClure, N. C. and Powles, S. B. (2001), 'Rapid degradation of carbetamide upon repeated application to Australian soils', *Soil Biol. Biochem.*, 33, 739–45.

Holtze, M. S., Sørensen, S. R., Sørensen, J. and Aamand, J. (2008), 'Microbial degradation of the benzonitrile herbicides dichlobenil, bromoxynil and ioxynil in soil and subsurface environments–insights into degradation pathways, persistent metabolites and involved degrader organisms', *Environ. Pollut.*, 154, 155–68.

Houbraken, M., van den Berg, F., Butler Ellis, C. M., Dekeyser, D., Nuyttens, D., De Schampheleire, M. and Spanoghe, P. (2016), 'Volatilisation of pesticides under field conditions: Inverse modelling and pesticide fate models', *Pest Manag. Sci.*, 72, 1309–21.

Huber, A., Bach, M. and Frede, H. G. (1998), 'Modeling pesticide losses with surface runoff in Germany', *Sci. Total Environ.*, 223, 177–91.

Hussain, S., Arshad, M., Springael, D., SøRensen, S. R., Bending, G. D., Devers-Lamrani, M., Maqbool, Z. and Martin-Laurent, F. (2015), 'Abiotic and biotic processes governing the fate of phenylurea herbicides in soils: A review', *Crit. Rev. Environ. Sci. Technol.*, 45, 1947–98.

Imfeld, G. and Vuilleumier, S. (2012), 'Measuring the effects of pesticides on bacterial communities in soil: A critical review', *Eur. J. Soil Biol.*, 49, 22–30.

Isensee, A. R. and Sadeghi, A. M. (1995), 'Long-term effect of tillage and rainfall on herbicide leaching to shallow groundwater', *Chemosphere*, 30, 671–85.

Jackson, L. E., Calderon, F. J., Steenwerth, K. L., Scow, K. M. and Rolston, D. E. (2003), 'Responses of soil microbial processes and community structure to tillage events and implications for soil quality', *Geoderma*, 114, 305–17.

Jacobsen, C. S. and Hjelmsø, M. H. (2014), 'Agricultural soils, pesticides and microbial diversity', *Curr. Opin. Biotechnol.*, 27, 15–20.

Johnsen, K., Jacobsen, C. S., Torsvik, V. and Sørensen, J. (2001), 'Pesticide effects on bacterial diversity in agricultural soils – a review', *Biol. Fert. Soils*, 33, 443–53.

Johnson, D. H., Shaner, D. L., Deane, J., Mackersie, L. A. and Tuxhorn, G. (2000), 'Time-dependent adsorption of imazethapyr to soil', *Weed Sci.*, 48, 769–75.

Jones, R. J., Muller, J., Haynes, D. and Schreiber, U. (2003), 'Effects of herbicides diuron and atrazine on corals of the Great Barrier Reef, Australia', *Marine Ecol. Progr. Ser.*, 251, 153–67.

Kah, M. and Brown, C. D. (2007), 'Changes in pesticide adsorption with time at high soil to solution ratios', *Chemosphere*, 68, 1335–43.

Katagi, T. (2004), 'Photodegradation of pesticides on plant and soil surfaces', *Rev. Environ. Contam. Toxicol.*, 182, 1–78.

Kleijn, D. and Snoeijing, G. I. J. (1997), 'Field boundary vegetation and the effects of agrochemical drift: Botanical change caused by low levels of herbicide and fertilizer', *J. Appl. Ecol.*, 1413–25.

Kniss, A. R. and Coburn, C. W. (2015), 'Quantitative evaluation of the environmental impact quotient (EIQ) for comparing herbicides', *PloS One*, 10, e0131200.

Kniss, A. R. (2017), 'Long-term trends in the intensity and relative toxicity of herbicide use', *Nat. Commun.*, 8, doi:10.1038/ncomms14865.

Knoche, M. (1994), 'Effect of droplet size and carrier volume on performance of foliage-applied herbicides', *Crop Prot.*, 13, 163–78.

Kolpin, D. W., Barbash, J. E. and Gilliom, R. J. (1998), 'Occurrence of pesticides in shallow groundwater of the United States: Initial results from the National Water-Quality Assessment Program', *Environ. Sci. Technol.*, 32, 558–66.

Kolpin, D. W., Barbash, J. E. and Gilliom, R. J. (2002), 'Atrazine and metolachlor occurrence in shallow ground water of the United States, 1993 to 1995: Relations to explanatory factors', *J. Am. Wat. Res Assoc.* 38, 301–11.

Kookana, R. S., Baskaran, S. and Naidu, R. (1998), 'Pesticide fate and behaviour in Australian soils in relation to contamination and management of soil and water: A review', *Soil Res.*, 36, 715–64.

Kovach J., Petzoldt C., Degni J. and Tette J. (1992), 'A method to measure the environmental impact of pesticides', New York Food and Life Sciences Bulletin Number 139, 8p.

Krutz, L. J., Zablotowicz, R. M., Reddy, K. N., Koger III, C. H. and Weaver, M. A. (2007), 'Enhanced degradation of atrazine under field conditions correlates with a loss of weed control in the glasshouse', *Pest Manag. Sci.*, 63, 23–31.

Krutz, L. J., Senseman, S. A., Zablotowicz, R. M. and Matocha, M. A. (2005), 'Reducing herbicide runoff from agricultural fields with vegetative filter strips: A review', Weed Sci., 53, 353–67.

Kudsk, P. (2008), 'Optimising herbicide dose: A straightforward approach to reduce the risk of side effects of herbicides', Environmentalist, 28, 49–55.

Kurtz, M. E. and Street, J. E. (2003), 'Response of rice (Oryza sativa) to glyphosate applied to simulate drift', Weed Technol., 17, 234–8.

Lal, R. (1993), 'Tillage effects on soil degradation, soil resilience, soil quality, and sustainability', Soil Tillage Res., 27, 1–8.

Lancaster, S. H., Hollister, E. B., Senseman, S. A. and Gentry, T. J. (2010), 'Effects of repeated glyphosate applications on soil microbial community composition and the mineralization of glyphosate', Pest Manag. Sci., 66, 59–64.

Leistra, M. and Boesten, J. J. T. I. (1989), 'Pesticide contamination of groundwater in western Europe', Agric. Ecosyst. Environ., 26, 369–89.

Lešnik, M., Pintar, C., Lobnik, A. and Kolar, M. (2005), 'Comparison of the effectiveness of standard and drift-reducing nozzles for control of some pests of apple', Crop Prot., 24, 93–100.

Leu, C., Singer, H., Stamm, C., Müller, S. R. and Schwarzenbach, R. P. (2004), 'Variability of herbicide losses from 13 fields to surface water within a small catchment after a controlled herbicide application', Environ. Sci. Technol., 38, 3835–41.

Lewis, K., Brown, C., Hart, A. and Tzilivakis, J. (2003), 'p-EMA (III): Overview and application of a software system designed to assess the environmental risk of agricultural pesticides', Agronomie, 23, 85–96.

Lewis, K. A., Tzilivakis, J., Warner, D. and Green, A. (2016), 'An international database for pesticide risk assessments and management', Hum. Ecol. Risk Assess, 22, 1050–64.

Louchart, X., Voltz, M., Andrieux, P. and Moussa, R. (2001), 'Herbicide transport to surface waters at field and watershed scales in a Mediterranean vineyard area', J. Environ. Qual., 30, 982–91.

Lupwayi, N. Z., Rice, W. A. and Clayton, G. W. (1998), 'Soil microbial diversity and community structure under wheat as influenced by tillage and crop rotation', Soil Biol. Biochem., 30, 1733–41.

Lupwayi, N. Z., Harker, K. N., Clayton, G. W., Turkington, T. K., Rice, W. A. and O'Donovan, J. T. (2004), 'Soil microbial biomass and diversity after herbicide application', Can. J. Plant Sci., 84, 677–85.

Marple, M. E., Al-Khatib, K., Shoup, D., Peterson, D. E. and Claassen, M. (2007), 'Cotton response to simulated drift of seven hormonal-type herbicides', Weed Technol., 21, 987–92.

Marrs, R. H., Frost, A. J., Plant, R. A. and Lunnis, P. (1992), 'Aerial applications of asulam: A bioassay technique for assessing buffer zones to protect sensitive sites in upland Britain', Biol. Conserv., 59, 19–23.

Marrs, R. H., Frost, A. J. and Plant, R. A. (1991), 'Effects of herbicide spray drift on selected species of nature conservation interest: The effects of plant age and surrounding vegetation structure', Environ. Pollut., 69, 223–35.

Marrs, R. H., Williams, C. T., Frost, A. J. and Plant, R. A. (1989), 'Assessment of the effects of herbicide spray drift on a range of plant species of conservation interest', Environ. Pollut., 59, 71–86.

Matthews, G. (2014), 'Aerial spray drift - consequences of spraying small droplets of herbicide. Outlooks Pest Manag., 25, 279–83.

McLaughlin, A. and Mineau, P. (1995), 'The impact of agricultural practices on biodiversity', Agric. Ecosyst. Environ., 55, 201–12.

Melander, B., Rasmussen, I. A. and Bàrberi, P. (2005), 'Integrating physical and cultural methods of weed control - examples from European research', Weed Sci., 53, 369–81.

Meriles, J. M., Gil, S. V., Conforto, C., Figoni, G., Lovera, E., March, G. J. and Guzmán, C. A. (2009), 'Soil microbial communities under different soybean cropping systems: Characterization of microbial population dynamics, soil microbial activity, microbial biomass, and fatty acid profiles', Soil Tillage Res., 103, 271–81.

Mijangos, I., Becerril, J. M., Albizu, I., Epelde, L. and Garbisu, C. (2009), 'Effects of glyphosate on rhizosphere soil microbial communities under two different plant compositions by cultivation-dependent and-independent methodologies', Soil Biol. Biochem., 41, 505–13.

Miller, D. R. and Stoughton, T. E. (2000), 'Response of spray drift from aerial applications at a forest edge to atmospheric stability', *Agric. Forest Meteorol.*, 100, 49–58.

Montoya, J. C., Costa, J. L., Liedl, R., Bedmar, F. and Daniel, P. (2006), 'Effects of soil type and tillage practice on atrazine transport through intact soil cores', *Geoderma*, 137, 161–73.

Moorman, T. B. (1989), 'A review of pesticide effects on microorganisms and microbial processes related to soil fertility', *J. Product. Agric.*, 2, 14–23.

Moorman, T. B., Jaynes, D. B., Cambardella, C. A., Hatfield, J. L., Pfeiffer, R. L. and Morrow, A. J. (1999), 'Water quality in Walnut Creek watershed: Herbicides in soils, subsurface drainage, and groundwater', *J. Environ. Qual.*, 28, 35–45.

Morillo, E., Undabeytia, T., Cabrera, A., Villaverde, J. and Maqueda, C. (2004), 'Effect of soil type on adsorption-desorption, mobility, and activity of the herbicide norflurazon', *J. Agric. Food Chem.*, 52, 884–90.

Moshier, L. J. and Penner, D. (1978), 'Factors influencing microbial degradation of ^{14}C-glyphosate to ^{14}CO$_2$ in soil', *Weed Sci.*, 26, 686–91.

Mueller, T. C., Wright, D. R. and Remund, K. M. (2013), 'Effect of formulation and application time of day on detecting dicamba in the air under field conditions', *Weed Sci.*, 61, 586–93.

Nazarko, O. M., Van Acker, R. C. and Entz, M. H. (2005), 'Strategies and tactics for herbicide use reduction in field crops in Canada: A review', *Canad. J. Plant Sci.*, 85, 457–79.

Neve, P. and Powles, S. (2005), 'Recurrent selection with reduced herbicide rates results in the rapid evolution of herbicide resistance in *Lolium rigidum*', *Theor. Appl. Genet.*, 110, 1154–66.

Noble, A. and Hamilton, D. J. (1990), 'Relation between volatility rating and composition of phenoxy herbicide ester formulations', *Pest Manag. Sci.*, 28, 203–14.

Nuyttens, D., De Schampheleire, M., Steurbaut, W., Baetens, K., Verboven, P., Nicolai, B., Ramon, H. and Sonck, B. (2006), 'Experimental study of factors influencing the risk of drift from field sprayers Part 2: Spray application technique', *Asp. Appl. Biol.*, 77, 331–9.

Nuyttens, D., Baetens, K., De Schampheleire, M. and Sonck, B. (2007a), 'Effect of nozzle type, size and pressure on spray droplet characteristics', *Biosyst. Eng.*, 97, 333–45.

Nuyttens, D., De Schampheleire, M., Baetens, K. and Sonck, B. (2007b), 'The influence of operator-controlled variables on spray drift from field crop sprayers', *Trans. ASABE*, 50, 1129–40.

Nuyttens, D., De Schampheleire, M., Verboven, P., Brusselman, E. and Dekeyser, D. (2009), 'Droplet size and velocity characteristics of agricultural sprays', *Trans. ASABE*, 52, 1471–80.

Nuyttens, D., De Schampheleire, M., Baetens, K., Brusselman, E., Dekeyser, D. and Verboven, P. (2011), 'Drift from field crop sprayers using an integrated approach: Results of a five-year study', *Trans. ASABE*, 54, 403–8.

Omar, S. A. and Abdel-Sater, M. A. (2001), 'Microbial populations and enzyme activities in soil treated with pesticides', *Wat. Air Soil Pollut.*, 127, 49–63.

Otto, S., Cardinali, A., Marotta, E., Paradisi, C. and Zanin, G. (2012), 'Effect of vegetative filter strips on herbicide runoff under various types of rainfall', *Chemosphere*, 88, 113–19.

Pagliai, M., Vignozzi, N. and Pellegrini, S. (2004), 'Soil structure and the effect of management practices', *Soil Tillage Res.* 79, 131–43.

Pätzold, S., Klein, C. and Brümmer, G. W. (2007), 'Run-off transport of herbicides during natural and simulated rainfall and its reduction by vegetated filter strips', *Soil Use Manag.*, 23, 294–305.

Pérez, G. L., Torremorell, A., Mugni, H., Rodriguez, P., Vera, M. S., Nascimento, M. D., Allende, L., Bustingorry, J., Escaray, R., Ferraro, M. and Izaguirre, I. (2007), 'Effects of the herbicide Roundup on freshwater microbial communities: A mesocosm study', *Ecol. Applicat.*, 17, 2310–22.

Permin, O., Jørgensen, L. N. and Persson, K. (1992), 'Deposition characteristics and biological effectiveness of fungicides applied to winter wheat and the hazards of drift when using different types of hydraulic nozzles', *Crop Prot.*, 11, 541–6.

Perucci, P. and Scarponi, L. (1994), 'Effects of the herbicide imazethapyr on soil microbial biomass and various soil enzyme activities', *Biol. Fert. Soil.*, 17, 237–40.

Pesce, S., Fajon, C., Bardot, C., Bonnemoy, F., Portelli, C. and Bohatier, J. (2006), 'Effects of the phenylurea herbicide diuron on natural riverine microbial communities in an experimental study', *Aquat. Toxicol.* 78, 303–14.

Pesce, S., Batisson, I., Bardot, C., Fajon, C., Portelli, C., Montuelle, B. and Bohatier, J. (2009), 'Response of spring and summer riverine microbial communities following glyphosate exposure', *Ecotox. Environ. Safety*, 72, 1905–12.

Piche, M., Panneton, B. and Theriault, R. (2000), 'Reduced drift from air-assisted spraying', *Canad. Agric. Eng.*, 42, 117–22.

Pimentel, D., McLaughlin, L., Zepp, A., Lakitan, B., Kraus, T., Kleinman, P., Vancini, F., Roach, W. J., Graap, E., Keeton, W. S. and Selig, G. (1993), 'Environmental and economic effects of reducing pesticide use in agriculture', *Agric. Ecosyst. Environ.*, 46, 273–88.

Pleasants, J. M. and Oberhauser, K. S. (2013), 'Milkweed loss in agricultural fields because of herbicide use: Effect on the monarch butterfly population', *Insect Conserv. Diver.*, 6, 135–44.

Pot, V., Šimůnek, J., Benoit, P., Coquet, Y., Yra, A. and Martínez-Cordón, M. J. (2005), 'Impact of rainfall intensity on the transport of two herbicides in undisturbed grassed filter strip soil cores', *J. Contam. Hydrol.*, 81, 63–88.

Potter, T. L., Truman, C. C., Strickland, T. C., Bosch, D. D. and Webster, T. M. (2008), 'Herbicide incorporation by irrigation and tillage impact on runoff loss', *J. Environ. Qual.*, 37, 839–47.

Pozo, C., Salmeron, V., Rodelas, B., Martinez-Toledo, M. V. and Gonzalez-Lopez, J. (1994), 'Effects of the herbicide alachlor on soil microbial activities', *Ecotoxicology*, 3, 4–10.

Prado, A. G. S. and Airoldi, C. (2001), 'The effect of the herbicide diuron on soil microbial activity', *Pest Manag. Sci.*, 57, 640–5.

Prueger, J. H., Gish, T. J., McConnell, L. L., Mckee, L. G., Hatfield, J. L. and Kustas, W. P. (2005), 'Solar radiation, relative humidity, and soil water effects on metolachlor volatilization', *Environ. Sci. Technol.*, 39, 5219–26.

Ramsdale, B. K. and Messersmith, C. G. (2001), 'Drift-reducing nozzle effects on herbicide performance', *Weed Technol.*, 15, 453–60.

Reichenberger, S., Bach, M., Skitschak, A. and Frede, H. G. (2007), 'Mitigation strategies to reduce pesticide inputs into ground-and surface water and their effectiveness; a review', *Sci. Total Environ.*, 384, 1–35.

Relyea, R. A. (2005), 'The lethal impact of Roundup on aquatic and terrestrial amphibians', *Ecol. Appl.*, 15, 1118–24.

Relyea, R. A. (2009), 'A cocktail of contaminants: How mixtures of pesticides at low concentrations affect aquatic communities', *Oecologia*, 159, 363–76.

Renton, M., Diggle, A., Manalil, S. and Powles, S. (2011), 'Does cutting herbicide rates threaten the sustainability of weed management in cropping systems?', *J. Theor. Biol.*, 283, 14–27.

Reus, J. A. W. A. (1992), 'An environmental yardstick for pesticides: An instrument to measure the environmental impact of pesticides', *Acta Hort.*, 347, 215–24.

Reus, J. A. and Leendertse, P. C. (2000), 'The environmental yardstick for pesticides: A practical indicator used in the Netherlands', *Crop Prot.*, 19, 637–41.

Reus, J. A. W. A., Leendertse, P., Bockstaller, C., Fomsgaard, I., Gutsche, V., Lewis, K., Nilsson, C., Pussemier, L., Trevisan, M., Van der Werf, H. and Alfarroba, F. (2002), 'Comparison and evaluation of eight pesticide environmental risk indicators developed in Europe and recommendations for future use', *Agric. Ecosyst. Environ.*, 90, 177–87.

Rew, L. J., Cussans, G. W., Mugglestone, M. A. and Miller, P. C. H. (1996), 'A technique for mapping the spatial distribution of *Elymus repens*, with estimates of the potential reduction in herbicide usage from patch spraying', *Weed Res.* 36, 283–92.

Riah, W., Laval, K., Laroche-Ajzenberg, E., Mougin, C., Latour, X. and Trinsoutrot-Gattin, I. (2014), 'Effects of pesticides on soil enzymes: A review', *Environ. Chem. Lett.*, 12, 257–73.

Ritter, W. F. (1990), 'Pesticide contamination of ground water in the United States - A review. *J. Environ. Sci. Health B*, 25, 1–29.

Ritter, L., Solomon, K., Sibley, P., Hall, K., Keen, P., Mattu, G. and Linton, B. (2002), 'Sources, pathways, and relative risks of contaminants in surface water and groundwater: A perspective prepared for the Walkerton inquiry', *J. Toxicol. Environ. Health A*, 65, 1–142.

Robinson, R. C., Parsons, R. G., Barbe, G., Patel, P. T. and Murphy, S. (2000), 'Drift control and buffer zones for helicopter spraying of bracken (*Pteridium aquilinum*)', *Agric. Ecosyst. Environ.*, 79, 215–31.

Roeth, F. W. (1986), 'Enhanced herbicide degradation in soil with repeat application', *Rev. Weed Sci*, 2, 45–65.

Roider, C. A., Griffin, J. L., Harrison, S. A. and Jones, C. A. (2007), Wheat response to simulated glyphosate drift', *Weed Technol.*, 21, 1010–15.

Roider, C. A., Griffin, J. L., Harrison, S. A. and Jones, C. A. (2008), 'Carrier volume affects wheat response to simulated glyphosate drift', *Weed Technol*, 22, 453–8.

Rolando, C. A., Garrett, L. G., Baillie, B. R. and Watt, M. S. (2013), 'A survey of herbicide use and a review of environmental fate in New Zealand planted forests', *New Zeal. J. For. Sci.*, 43, doi:10.1186/1179–5395-43–17.

Rouchaud, J., Neus, O., Blucke, R., Cools, K., Eelen, H. and Dekkers, T. (2000), 'Soil dissipation of diuron, chlortoluron, simazine and propyzamide, and diflufencian herbicides after repeated application in fruit tree orchards', *Environ. Contam. Toxicol.* 39, 60–5.

Rüdel, H. (1997), 'Volatilisation of pesticides from soil and plant surfaces', *Chemosphere*, 35, 143–52.

Sandermann, H. (1992), 'Plant metabolism of xenobiotics', *Trends Biochem. Sci.*, 17, 82–4.

Sandmann, E. R. I. C., Loos, M. A. and Van Dyk, L. P. (1988) 'The microbial degradation of 2,4-dichlorophenoxyacetic acid in soil', *Rev. Environ. Contam. Toxicol.*, 101, 1–53.

Sarmah, A. K. and Sabadie, J. (2002), 'Hydrolysis of sulfonylurea herbicides in soils and aqueous solutions: A review', *J. Agric. Food Chem.*, 50, 6253–65.

Sauer, T. J. and Daniel, T. C. (1987), 'Effect of tillage system on runoff losses of surface-applied pesticides', *Soil Sci. Soc. Am. J.*, 51, 410–15.

Schaffer, J. D. and Sebetich, M. J. (2004), 'Effects of aquatic herbicides on primary productivity of phytoplankton in the laboratory', *Bull. Environ. Contam. Toxicol.*, 72, 1032–7.

Schloss, P. D. and Handelsman, J. (2005), 'Metagenomics for studying unculturable microorganisms: Cutting the Gordian knot', *Genome Biol.*, 6, 229. doi:10.1186/gb-2005-6-8-229.

Schmalenberger, A. and Tebbe, C. C. (2002), 'Bacterial community composition in the rhizosphere of a transgenic, herbicide-resistant maize (*Zea mays*) and comparison to its non-transgenic cultivar Bosphore', *FEMS Microbiol. Ecol.*, 40, 29–37.

Schmitt-Jansen, M. and Altenburger, R. (2005), 'Predicting and observing responses of algal communities to photosystem ii-herbicide exposure using pollution-induced community tolerance and species-sensitivity distributions', *Environ. Toxicol. Chem.*, 24, 304–12.

Schmitz, J., Stahlschmidt, P. and Brühl, C. A. (2015), 'Assessing the risk of herbicides to terrestrial non-target plants using higher-tier studies', *Hum. Ecol. Risk Assess.*, 21, 2137–54.

Schneider, M., Endo, S. and Goss, K. U. (2013), 'Volatilization of pesticides from the bare soil surface: Evaluation of the humidity effect', *J. Environ. Qual.*, 42, 844–51.

Schuler, L. J. and Rand, G. M. (2008), 'Aquatic risk assessment of herbicides in freshwater ecosystems of south Florida' *Arch. Environ. Contam. Toxicol.*, 54, 571–83.

Scribner, E. A., Battaglin, W. A., Goolsby, D. A. and Thurman, E. M. (2000), 'Changes in herbicide concentrations in Midwestern streams in relation to changes in use, 1989-1998', *Sci. Total Environ.*, 248, 255–63.

Sebiomo, A., Ogundero, V. W. and Bankole, S. A. (2011), 'Effect of four herbicides on microbial population, soil organic matter and dehydrogenase activity', *Afr. J. Biotechnol.*, 10, 770–8.

Seiber, J. N., McChesney, M. M., Sanders, P. F. and Woodrow, J. E. (1986), 'Models for assessing the volatilization of herbicides applied to flooded rice fields', *Chemosphere*, 15, 127–38.

Sessitsch, A., Gyamfi, S., Tscherko, D., Gerzabek, M. H. and Kandeler, E. (2004), 'Activity of microorganisms in the rhizosphere of herbicide treated and untreated transgenic glufosinate-tolerant and wildtype oilseed rape grown in containment', *Plant Soil*, 266, 105–16.

Shipitalo, M. J. and Owens, L. B. (2006), 'Tillage system, application rate, and extreme event effects on herbicide losses in surface runoff', *J. Environ. Qual.*, 35, 2186–94.

Sigua, G. C., Isensee, A. R. and Sadeghi, A. M. (1993), 'Influence of rainfall intensity and crop residue on leaching of atrazine through intact no-till soil cores', *Soil Sci.*, 156, 225–32.

Smith, E. A. and Mayfield, C. I. (1978), 'Paraquat: Determination, degradation, and mobility in soil', *Wat. Air Soil Pollut.*, 9, 439–52.

Snipes, C. E., Street, J. E. and Mueller, T. C. (1991), 'Cotton (*Gossypium hirsutum*) response to simulated triclopyr drift', *Weed Technol.*, 5, 493–8.

Sofo, A., Scopa, A., Dumontet, S., Mazzatura, A. and Pasquale, V. (2012), 'Toxic effects of four sulphonylureas herbicides on soil microbial biomass', *J. Environ Sci. Health B*, 47, 653–9.

Solomon, K. R., Baker, D. B., Richards, R. P., Dixon, K. R., Klaine, S. J., La Point, T. W., Kendall, R. J., Weisskopf, C. P., Giddings, J. M., Giesy, J. P. and Hall, L. W. (1996), 'Ecological risk assessment of atrazine in North American surface waters', *Environ. Toxicol. Chem.*, 15, 31–76.

Soltani, N., Nurse, R. E., Shropshire, C. and Sikkema, P. H. (2012), 'Weed control, environmental impact and profitability of pre-plant incorporated herbicides in white bean', *Am. J. Plant Sci.*, 3, 846–53.

Sønderskov, M., Kudsk, P., Mathiassen, S. K., Bøjer, O. M. and Rydahl, P. (2014), 'Decision support system for optimized herbicide dose in spring barley', *Weed Technol.*, 28, 19–27.

Sørensen, S. R., Bending, G. D., Jacobsen, C. S., Walker, A. and Aamand, J. (2003), 'Microbial degradation of isoproturon and related phenylurea herbicides in and below agricultural fields', *FEMS Microbiol. Ecol.*, 45, 1–11.

Spencer, W. F., Cliath, M. M., Jury, W. A. and Zhang, L. Z. (1988), 'Volatilization of organic chemicals from soil as related to their Henry's law constants', *J. Environ. Qual.*, 17, 504–9.

Sprankle, P., Meggitt, W. F. and Penner, D. (1975), 'Adsorption, mobility, and microbial degradation of glyphosate in the soil', *Weed Sci.*, 23, 229–34.

Stachowski-Haberkorn, S., Becker, B., Marie, D., Haberkorn, H., Coroller, L. and De La Broise, D. (2008), 'Impact of Roundup on the marine microbial community, as shown by an in situ microcosm experiment', *Aquat. Toxicol.*, 89, 232–41.

Steinheimer, T. R. (1993), 'HPLC determination of atrazine and principal degradates in agricultural soils and associated surface and ground water', *J. Agric. Food Chem.*, 41, 588–95.

Stoate, C., Boatman, N. D., Borralho, R. J., Carvalho, C. R., De Snoo, G. R. and Eden, P. (2001), 'Ecological impacts of arable intensification in Europe', *J. Environ. Manag.*, 63, 337–65.

Strachan, S. D., Casini, M. S., Heldreth, K. M., Scocas, J. A., Nissen, S. J., Bukun, B., Lindenmayer, R. B., Shaner, D. L., Westra, P. and Brunk, G. (2010), 'Vapor movement of synthetic auxin herbicides: Aminocyclopyrachlor, aminocyclopyrachlor-methyl ester, dicamba, and aminopyralid', *Weed Sci.*, 58, 103–8.

Strassemeyer, J., Daehmlow, D., Dominic, A. R., Lorenz, S. and Golla, B. (2017), 'SYNOPS-WEB, an online tool for environmental risk assessment to evaluate pesticide strategies on field level', *Crop Prot.*, 97, 28–44.

Stuart, M., Lapworth, D., Crane, E. and Hart, A. (2012), 'Review of risk from potential emerging contaminants in UK groundwater', *Sci. Total Environ.* 416, 1–21.

Tabernero, M. T., Álvarez-Benedí, J., Atienza, J. and Herguedas, A. (2000), 'Influence of temperature on the volatilization of triallate and terbutryn from two soils', *Pest Manag. Sci.*, 56, 175–80.

Tang, X., Zhu, B. and Katou, H. (2012), 'A review of rapid transport of pesticides from sloping farmland to surface waters: Processes and mitigation strategies', *J. Environ. Sci.*, 24, 351–61.

Tejada, M. (2009), 'Evolution of soil biological properties after addition of glyphosate, diflufenican and glyphosate + diflufenican herbicides', *Chemosphere*, 76, 365–73.

Teske, M. E., Bird, S. L., Esterly, D. M., Curbishley, T. B., Ray, S. L. and Perry, S. G. (2002), 'AgDrift®: A model for estimating near-field spray drift from aerial applications', *Environ. Toxicol. Chem.*, 21, 659–71.

Thurman, E. M., Goolsby, D. A., Meyer, M. T. and Kolpin, D. W. (1991), 'Herbicides in surface waters of the midwestern United States: The effect of spring flush', *Environ. Sci. Technol.*, 25, 1794–6.

Ucar, T. and Hall, F. R. (2001), 'Windbreaks as a pesticide drift mitigation strategy: A review', *Pest Manag. Sci.*, 57, 663–75.

Van de Zande, J. C., Stallinga, H., Michielsen, J. M. G. P. and Van Velde, P. (2005) 'Effect of sprayer speed on spray drift', *Annu. Rev. Agric. Eng.*, 4, 129–42.

Van den Brink, P. J., Blake, N., Brock, T. C. and Maltby, L. (2006), 'Predictive value of species sensitivity distributions for effects of herbicides in freshwater ecosystems', *Hum. Ecol. Risk Assess.*, 12, 645–74.

Van Stempvoort, D. R., Roy, J. W., Brown, S. J. and Bickerton, G. (2014), 'Residues of the herbicide glyphosate in riparian groundwater in urban catchments', *Chemosphere*, 95, 455–63.

van Wesenbeeck, I., Driver, J. and Ross, J. (2008), 'Relationship between the evaporation rate and vapor pressure of moderately and highly volatile chemicals', *Bull. Environ. Contamin. Toxicol.*, 80, 315–18.

Veeh, R. H., Inskeep, W. P. and Camper, A. K. (1996), 'Soil depth and temperature effects on microbial degradation of 2,4-D', *J. Environ. Qual.*, 25, 5–12.

Walker, A., Cotterill, E. G. and Welch, S. J. (1989), 'Adsorption and degradation of chlorsulfuron and metsulfuron-methyl in soils from different depths', *Weed Res.*, 29, 281–7.

Wall, D. A. (1994), 'Potato (*Solanum tuberosum*) response to simulated drift of dicamba, clopyralid, and tribenuron', *Weed Sci.*, 42, 110–14.

Wardle, D. A. and Parkinson, D. (1990a), 'Effects of three herbicides on soil microbial biomass and activity', *Plant Soil*, 122, 21–8.

Wardle, D. A. and Parkinson, D. (1990b), 'Influence of the herbicide glyphosate on soil microbial community structure', *Plant Soil*, 122, 29–37.

Weaver, M. A., Krutz, L. J., Zablotowicz, R. M. and Reddy, K. N. (2007), 'Effects of glyphosate on soil microbial communities and its mineralization in a Mississippi soil', *Pest Manag. Sci.*, 63, 388–93.

Wenneker, M., Heijne, B. and Van de Zande, J. C. (2005), 'Effect of natural windbreaks on drift reduction in orchard spraying', *Commun. Agric Appl. Biol. Sci.*, 70, 961–9.

Wheeler, W. B., Stratton, G. D., Twilley, R. R., Ou, L. T., Carlson, D. A. and Davidson, J. M. (1979), 'Trifluralin degradation and binding in soil', *J. Agric. Food Chem.*, 27, 702–6.

Yates, S. R. (2006a), 'Simulating herbicide volatilization from bare soil affected by atmospheric conditions and limited solubility in water', *Environ. Sci. Technol.*, 40, 6963–8.

Yates, S. R. (2006b), 'Measuring herbicide volatilization from bare soil', *Environ. Sci. Technol.*, 40, 3223–8.

Zhang, J., Weaver, S. E. and Hamill, A. S. (2000), 'Risks and reliability of using herbicides at below-labeled rates', *Weed Technol.*, 14, 106–15.

Zobiole, L. H. S., Kremer, R. J., Oliveira, R. S. and Constantin, J. (2011), 'Glyphosate affects micro-organisms in rhizospheres of glyphosate-resistant soybeans', *J. Appl. Microbiol.* 110, 118–27.

Chapter 9

Trends in the development of herbicide-resistant weeds

Ian Heap, International Survey of Herbicide-Resistant Weeds, USA

1 Introduction

Since their introduction in the mid-1940s modern synthetic herbicides have revolutionized weed control and quickly became the most cost-effective and efficient means of controlling weeds. Herbicides are still the primary weed control method employed in crop production; however, they face several major challenges. One challenge is the real and perceived risks to people and the environment, which has led to ever-increasing stringent regulations necessary to register a new product. Regulations that have increased the cost of bringing new herbicides to market, in combination with the domination of Roundup Ready crops in key market segments, have led to the cessation of registration of herbicides with novel sites of action (Fig. 1). This is a major problem for the future of weed control, as the use of herbicides with new sites of action was the primary way in which growers dealt with the appearance of herbicide-resistant weeds. It is clear that weed scientists have chosen to emphasize finding ways to maintain the utility of existing herbicides until new cost-effective alternatives eventually replace them rather than searching for new sites of action and developing new, novel ways to control weeds (e.g. robotic weeders).

Like insecticide, fungicide and antibiotic resistance, the phenomena of herbicide resistance can be understood as an evolutionary process, resulting from a strong selection

http://dx.doi.org/10.19103/AS.2017.0025.11

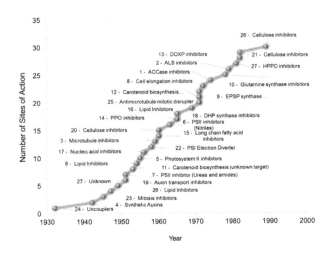

Figure 1 Introduction time of new herbicide sites of action (Weed Science Society of America codes).

pressure on a population that enriches the proportion of rare resistant individuals, which eventually leads to resistance. Populations of weeds differ in many characteristics, one of them being their sensitivity to herbicides. Before companies register a herbicide the US Environmental Protection Agency requires extensive trials, which results in a recommended rate that has been shown to kill the majority of a weed species over a range of environmental and geographic conditions. Within weed populations there are very rare individuals (usually between 1 in 10 000 and 1 in 10 000 000) that possess genetic traits that allow them to survive this recommended rate, multiply and eventually dominate the population resulting in weed control failure. Once resistance appears on one site, subsequent cases in the same area are often due to the spread of small numbers of resistant individuals (by movement of seed, pollen or plant parts) from the original site to neighbouring fields that are then selected through herbicide use and create a resistant population.

2 Herbicide resistance definitions

2.1 Target site resistance

Target site resistance does not affect the amount of herbicide that reaches the target site. Once the herbicide reaches the target site it has less of an effect on the resistant biotype. There are several variations of target site resistance.

Single-base pair alteration

The most common type of resistance is a modification of the herbicide binding site on the target (often an enzyme) that prevents binding, yet still enables the function of the target. Resistance to ALS, ACCase and PSII inhibitors is often due to a single-base pair alteration of a codon that changes one amino acid in the target protein and changes the binding site.

Multiple-base pair alteration

Less frequently two base pair changes in the same codon can result in amino acid changes that confer resistance through changes in the binding site (Dayan et al. 2014). For example, a higher level of resistance to glyphosate was conferred by a change in two base pairs which changed two amino acids in the 5-enolpyruvylshikimate-3-phosphate synthase (EPSPS) enzyme (Yu et al. 2015).

Codon deletion

Resistance can also result from a deletion of an entire codon, which occurs with PPO inhibitors (Patzoldt et al. 2006).

Gene amplification

Gene amplification is another form of target site resistance and, although it has only recently been identified in weeds, it is a common resistance mechanism in insects (Bass and Field 2011). Gene amplification results in an increase in gene copy numbers and consequently an increase in the production of the target enzyme, effectively diluting the herbicide in relation to the target site (Gaines et al. 2010).

2.2 Non-target site resistance

Non-target site resistance refers to a resistance mechanism that prevents a lethal dose of herbicide reaching the target site. These mechanisms usually result in a low level of resistance and are often combined in individuals (through outcrossing) to eventually result in higher levels of resistance.

Enhanced metabolism

Enhanced metabolism refers to the ability of the weed to degrade/metabolize the herbicide into an inactive molecule before it can seriously affect the plant. Metabolic resistance to ALS, ACCase and PSII inhibitors is most common in grasses (e.g. *Alopecurus myosuroides*, *Echinochloa phyllopogon* and *Lolium rigidum*) and is often a result of an increase in cytochrome P450 enzymes that can metabolize herbicides (Preston et al. 1996; Fisher et al. 2000, Hana et al. 2014).

Decreased absorption and/or translocation

Resistance can occur when there is a decrease in the absorption or translocation of an herbicide to such an extent that it does not reach the site of action in sufficient concentration to cause death. Sometimes a combination of decreased translocation and increased sequestration work together to increase resistance (Sammons and Gaines 2014).

Sequestration

Resistance can occur through the sequestration of herbicides into parts of the plant (vacuoles, or sorption onto cell walls), thus keeping the herbicide from the site of action. Glyphosate resistance in *Conyza canadensis* and *Lolium rigidum* has been shown to be due to sequestration of glyphosate into the vacuole (Ge et al. 2010, Sammons and Gaines 2014).

From a resistance management perspective, it is important to note that weeds can exhibit cross-resistance and multiple resistance.

Cross-resistance

Cross-resistance occurs where a single resistance mechanism confers resistance to several herbicides. The most common type of cross-resistance is target site cross-resistance, where an altered target site (enzyme) confers resistance to many or all the herbicides that inhibit the same enzyme.

Multiple resistance

Multiple resistance occurs when two or more resistance mechanisms occur within the same plant, often due to sequential selection by different herbicide sites of action. A diagnosis of multiple resistance requires knowledge of the resistance mechanisms (Heap and LeBaron 2001).

3 Resistant weeds by site of action

The International Survey of Herbicide-Resistant Weeds can be found online at http://www.weedscience.org and is the source of all the data used in graphs and tables of this chapter (Heap 2016). Although not well publicized at the time, the first cases of herbicide resistance in weeds were synthetic auxin (2,4-D) resistance in *Daucus carota* (Switzer 1957) and *Commelina diffusa* (Hilton 1957). The first well-documented case of herbicide resistance was triazine-resistant *Senecio vulgaris* reported by Ryan (1970). Research soon followed that identified numerous triazine-resistant weeds in corn in the mid-1970s, and subsequently to many other herbicide sites of action. Since 1975 we have seen the appearance of approximately 11 unique (species by site of action) cases of herbicide-resistant weeds being identified each year (Fig. 2). There are 26 known herbicide sites of action and Table 1 illustrates that different herbicide sites of action differ in their propensity to select herbicide resistance.

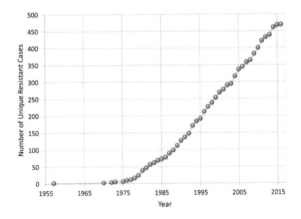

Figure 2 Global increase in unique (species x site of action) herbicide-resistant weed cases.

Table 1 The occurrence of herbicide-resistant weed species to herbicide groups

#	Herbicide group	WSSA group	Example herbicide	Dicots	Monocots	Total
1	ALS inhibitors	1	Chlorsulfuron	97	62	159
2	Photosystem II inhibitors	2	Atrazine	50	23	73
3	ACCase inhibitors	3	Sethoxydim	0	48	48
4	EPSP synthase inhibitors	9	Glyphosate	18	17	35
5	Synthetic Auxins	4	2,4-D	24	8	32
6	PSI Electron Diverter	22	Paraquat	22	9	31
7	PSII inhibitor (Ureas and amides)	7	Chlorotoluron	10	18	28
8	Microtubule inhibitors	3	Trifluralin	2	10	12
9	PPO inhibitors	14	Oxyfluorfen	9	1	10
10	Lipid inhibitors	16	Triallate	0	10	10
11	Long chain fatty acid inhibitors	15	Butachlor	0	5	5
12	PSII inhibitors (Nitriles)	6	Bromoxynil	3	1	4
13	Carotenoid biosynthesis inhibitors	12	Diflufenican	3	1	4
14	Carotenoid biosynthesis (unknown target)	11	Amitrole	1	3	4
15	Cellulose inhibitors	26	Quinclorac	0	3	3
16	Antimicrotubule mitotic disrupter	25	Flamprop-methyl	0	3	3
17	HPPD inhibitors	27	Isoxaflutole	2	0	2
18	DOXP inhibitors	13	Clomazone	0	2	2
19	Glutamine synthase inhibitors	10	Glufosinate	0	2	2
20	Mitosis inhibitors	23	Propham	0	1	1
21	Unknown	27	Endothall	0	1	1
22	Cell elongation inhibitors	8	Difenzoquat	0	1	1
23	Nucleic acid inhibitors	17	MSMA	1	0	1
				242	229	471

3.1 Weed Science Society of America group 2: ALS inhibitors

ALS inhibitors prevent the biosynthesis of branch chain amino acids (valine, leucine and isoleucine) by binding to and inhibiting the acetolactate synthase enzyme (Duggleby and Pang 2000). One hundred and fifty-nine weed species have evolved resistance to ALS inhibitors since they were first introduced in the 1980s. They account for one-third of all herbicide resistance cases and are the most prone herbicide site of action for selection of herbicide-resistant weeds. There are several reasons why more weed species have evolved resistance to ALS inhibitors than to any other herbicide group. The first is that there are more ALS inhibitor herbicides (56 herbicides belonging to 5 chemical classes, which is twice as many as any other herbicide group), and they have been used on a greater area than for any other herbicide group. Also, many of the ALS inhibitor herbicides have soil residual activity, which can increase the selection pressure for resistance. But most importantly there is a very high initial resistance gene frequency in weed populations because of the many mutations in the ALS enzyme that confer ALS inhibitor resistance but allow the ALS enzyme to function. Eight amino acid resistance conferring substitutions (Ala 122, Pro 197, Ala 205, Asp 376, Arg 377, Trp 574, Ser 653 and Gly 654) have been identified on the ALS gene. It should be noted that each of these substitutions confer different patterns of cross-resistance to the 5 classes of ALS inhibitors.

3.2 Weed Science Society of America group 5: PSII inhibitors (triazines)

PSII inhibitors were first registered in the 1950s and there are now 26 commercial PSII inhibitors that belong to six chemical classes (Fig. 1). PSII inhibitors block the binding site of plastoquinone by binding to the D1 protein in the photosystem II complex in chloroplasts (Gronwald 1994). The resistance of *Senecio vulgaris* to the PSII inhibitor simazine (Ryan 1970) was of little economic significance (isolated case of simazine resistance around potted plants in a nursery) but was of major scientific importance because it alerted scientists to the risk of herbicide-resistant weeds in corn, which at that time relied heavily on the PSII inhibitor atrazine for weed control in the United States and Europe. Weed scientists conducted intensive research on triazine-resistant weeds in the mid-1970s and within ten years they had identified 40 triazine-resistant species in the United States and Europe. Today 73 weed species have been identified with triazine resistance (Fig. 3), after which research attention was turned to resistance to other herbicide sites of action. Whilst some cases of triazine resistance have been attributed to enhanced metabolism the majority are due to a mutation (Ser264 to Gly) in the *psbA* gene, which results in an altered D1 protein and a reduction in herbicide binding to the thylakoid membrane in chloroplasts (Gronwald 1997). Triazine herbicides are still used extensively in corn production despite widespread triazine resistance in *Amaranthus, Chenopodium, Solanum* spp. and many other weed species in corn. They are used because they are effective on many weeds and growers learnt to deal with triazine-resistant weeds by adding herbicides that have a different site of action.

3.3 Weed Science Society of America group 1: ACCase inhibitors

ACCase inhibitors were introduced in the mid-1970s (Fig. 1). The 21 commercial inhibitors target grass weeds and belong to three chemical classes:

1 Blockage of fatty acid biosynthesis by inhibiting the enzyme acetyl-co-enzyme A carboxylase (Buchanan et al. 2000). Weeds have evolved several mechanisms to resist ACCase enzyme, the most common mechanism is target site mutations of acetyl-co-enzyme A carboxylase which prevents the herbicide from binding to the enzyme.
2 Enhanced metabolism of ACCase inhibitor has been identified (relatively common in *Avena* and *Lolium* spp.).
3 The least common mechanism is the over-expression of the ACCase enzyme (Brown et al. 2002; De'lye et al. 2005; Hochberg et al. 2009; Liu et al. 2007).

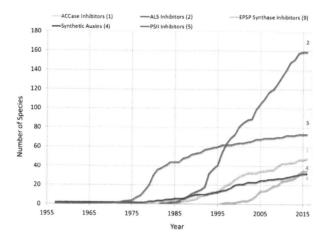

Figure 3 Number of resistant weed species for several herbicide sites of action (WSSA codes).

They are used extensively for selective control of grass weeds in cereal crops, and that is where the greatest resistance problems have occurred. The most important ACCase inhibitor-resistant grasses in cereals belong to *Avena*, *Lolium*, *Setaria*, *Phalaris* and *Alopecurus* spp. in North America, Europe, Australia and India. ACCase inhibitor resistance in *Echinochloa* spp., *Ischaemum rugosum* and *Leptochloa chinensis* is of major economic significance in rice.

3.4 Weed Science Society of America group 9: glyphosate

Glyphosate, a non-selective herbicide, was first used commercially in 1974 and has become the most successful and widely used herbicide ever discovered, because it delivers excellent broad-spectrum weed control with minimal environmental impacts (Duke and Powles 2008). Glyphosate disrupts the shikimate pathway by inhibiting the chloroplast enzyme EPSPS which results in the inhibition of aromatic amino acid production (Baylis 2000; Woodburn 2000). Initially glyphosate was expensive and its use was restricted to high-value applications in orchards and non-crop sites. As its price declined it was used in fallow and pre-plant applications particularly by farmers practising zero tillage. Glyphosate's use increased dramatically with the introduction of Roundup Ready crops, the first being Roundup Ready soybean in 1996. Coincidentally the first glyphosate-resistant weed, *Lolium rigidum*, was identified in 1996, which had evolved resistance in an apple orchard after 15 years of multiple glyphosate treatments per year (Pratley et al. 1999). There has been a steady increase in glyphosate-resistant weeds since 1996 (Fig. 4). A total of 35 glyphosate-resistant weeds have been identified, 17 of them monocots and 18 dicots (Table 1). About half of the glyphosate-resistant weeds have evolved resistance in Roundup Ready cropping systems, with the other half in orchards, non-crop and fallow situations. Glyphosate-resistant weeds that have evolved in Roundup Ready crops are of much greater economic importance than those in orchards, non-crop and fallow situations. *C. canadensis* was the first glyphosate-resistant weed to be identified in a Roundup Ready cropping system (VanGessel 2001). It had evolved resistance in a Roundup Ready soybean field in Delaware in 2000,

Figure 4 Increase in the number of glyphosate-resistant weed species worldwide.

and is now widespread throughout the United States, not only in the Roundup Ready soybean, corn and cotton crops but also in orchards and non-crop situations. Whilst *C. canadensis* is the most widespread glyphosate-resistant weed in the world (found in 11 countries and 25 US states) it is not the worst glyphosate-resistant weed, as there are sufficient herbicidal alternatives to manage it. *Amaranthus palmeri* is the most serious glyphosate-resistant weed, first identified in cotton crops in Georgia in 2005, but has since become widespread throughout the Southern and Midwestern states (27 US states in total). In addition to cotton it is commonly found in corn and soybean and less frequently in orchards, pecans and even horseradish. Similarly, glyphosate-resistant *A. tuberculatus* is a major threat to the corn/soybean rotation in the Midwest (now found in 18 US states), in large part because it has become resistant to most of the other major herbicide groups in addition to glyphosate. More recently *Kochia scoparia* has become a widespread resistant weed. It was first identified in 2007 (Waite et al. 2012) and has rapidly spread across 10 US states and parts of Canada because of its very effective tumbleweed dispersal mechanism. Glyphosate-resistant *Ambrosia artemisiifolia* has been identified in 15 US states. Resistant *Ambrosia trifida* has been identified in 12 US states. It can devastate yield in corn and soybean. Because of their larger seed size and lack of a highly effective dispersal mechanism the cases of glyphosate-resistant *Ambrosia* spp. are not as widespread as that of glyphosate-resistant *Conyza*, *Amaranthus* spp. or *Kochia*. Glyphosate-resistant *Sorghum halepense* has been identified in three US states but has not become a widespread problem, unlike in Argentina where it now infests many thousands of soybean acres. Another glyphosate-resistant grass weed, *Digitaria insularis* has become widespread in soybean crops in Paraguay and Brazil.

Lolium spp., *Conyza* spp. and *Eleusine indica* are the most common and damaging glyphosate-resistant weeds that have evolved in non-Roundup Ready situations. Glyphosate-resistant weeds have been found in 27 countries, with the United States having 16 species, Australia 12, Argentina 9, Brazil 8, Canada 5 and Spain 5 species.

Although there are some target site mutations that confer glyphosate resistance (Kaundun 2008), most of the cases have been identified with decreased translocation/sequestration (Feng et al. 2004) or gene amplification (Gaines et al. 2010). Once resistant weeds appear in Roundup Ready crops growers often continue to use glyphosate, because it is a broad-spectrum herbicide and will control many other weeds. They add other herbicide chemistries to glyphosate to control glyphosate-resistant weeds. The real problems arise when the glyphosate-resistant weeds are also resistant to other herbicides. The recent introduction of synthetic auxin-resistant crops is aimed at management of glyphosate-resistant broadleaf weeds; however, this may be a short-term solution if these crops are not managed to avoid over reliance on synthetic auxins alone.

3.5 Weed Science Society of America group 4: synthetic auxins

Synthetic auxin herbicides (2,4-D and MCPA) were introduced in 1944. There are now 24 commercial synthetic auxin herbicides that belong to six chemical classes. Synthetic auxins act by mimicking the plant hormone indole-3-acetic acid (IAA) which affects cell division, elongation and differentiation. Symptoms of synthetic auxins are twisting, faciation of the crown and leaf petioles, hypertrophy and premature abscission of the leaves (Stirling and Hall 1997). Even though synthetic auxins have been used over large areas for more than 70 years they are still, on the whole, effective. The majority of the 32 synthetic auxin-resistant weeds (5 grasses and 27 broadleaves) that have been identified so far are isolated due to scientific curiosities, and have not caused much economic damage in comparison to other major herbicide groups (Table 1). Quinclorac,a synthetic auxin, is unusual in that it controls grasses through a novel mechanism. The five grass species that have evolved resistance to quinclorac, *Digitaria ischaemum* and four *Echinochloa* spp. (*E. crus-galli*, *E. crus-pavonis*, *E. zelayensis* and *E. colona*), should be considered separately to the synthetic auxin-resistant broadleaves due to the novel mechanism. Eight synthetic auxin-resistant weed species have been identified in the United States, the most important being dicamba- and fluroxypyr-resistant *K. scoparia* common in cereal crops; 2,4-D resistant *Daucus carota* along roadsides; and *Lactuca serriola* with resistance to 2,4-D, dicamba and MCPA. The isolated appearance of 2,4-D-resistant *A. tuberculatus* in Nebraska is particularly concerning, as it shows that this common species, which has already evolved resistance to glyphosate, ALS inhibitors, PSII inhibitors, PPO inhibitors and 4-HPPD inhibitors will also become resistant to the synthetic auxins being used in the recently released synthetic auxin-resistant crops. Other synthetic auxin-resistant broadleaves that infest more than a few thousand hectares are 2,4-D-resistant *Raphanus raphanistrum* from Australia, 2,4-D-resistant *Papaver rhoeas* in Europe, *Gallium* sp. in Canada and *Carduus* sp. in New Zealand.

3.6 Other herbicides

Other important herbicide-resistant weed problems occur in Group 22 (e.g. paraquat) with 31 species, Group 7 (e.g. chlorotoluron) with 28 species, Group 3 (e.g. trifluralin) with 12 species and Group 14 (e.g. fomesafen). The Group 14 PPO inhibitors are of high concern as this site of action is very useful for controlling glyphosate-resistant *Amaranthus* spp. in the United States.

4 Resistant weeds by crop, region and weed family

4.1 Resistant weeds by crop

Wheat

More weeds have evolved herbicide resistance in wheat than in any other crop. There are 135 unique types (species x site of action) of herbicide-resistant weeds identified in wheat (Fig. 5), belonging to 74 weed species (Fig. 6). *Lolium rigidum* and *Alopecurus myosuroides* were the first weeds to evolve herbicide resistance in wheat in 1982 (Heap 1982, 1986). Indeed, these two species are the most widespread and economically damaging

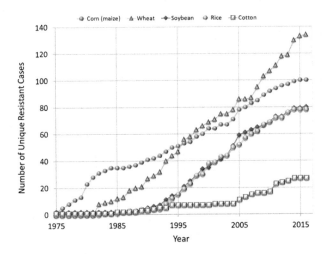

Figure 5 Increase in unique herbicide-resistant weed cases for selected crops.

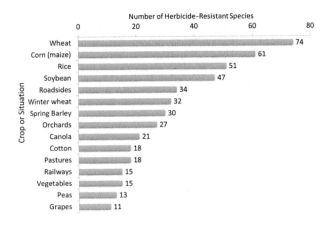

Figure 6 Number of herbicide-resistant weed species by crop.

herbicide-resistant weeds in wheat today, with *Lolium rigidum* being the most common resistant weed of wheat in Australia, and *Alopecurus myosuroides* in wheat in Europe. Both species are particularly troublesome because they have evolved resistance to all of the major wheat-selective herbicides. *Avena fatua* is also a major herbicide-resistant weed of wheat, having evolved resistance to 7 herbicide sites of action and being of greatest economic importance in Canada, the United States, Europe and Australia. Although only 22 of the 74 herbicide-resistant weeds in wheat are grasses, they account for about 90% of the economic damage, as there are sufficient site of action alternatives for control of broadleaf weeds in wheat. *Phalaris* spp. in India and Mexico are major herbicide-resistant weeds of wheat, and *Bromus* and *Setaria* spp. in the United States. Of the broadleaf species, *K. scoparia* has evolved widespread resistance to ALS inhibitors in addition to synthetic auxin, triazine and glyphosate resistance and will continue to be problematic in wheat in the United States and Canada. In Australia the broadleaf weed *R. raphanistrum* has evolved resistance to a wide range of herbicides and is among the worst herbicide-resistant weeds of wheat, after *Lolium rigidum*. In European wheat crops *Papaver rhoeas* has evolved widespread resistance to ALS inhibitors and several other sites of action and *Galium* spp. are also of major concern.

Corn

There are 100 unique types of herbicide-resistant weeds in corn (Fig. 5), belonging to 61 species (Fig. 6). From the mid-1970s to the early 1990s more herbicide-resistant weeds were found in corn than any other crop. The widespread use of atrazine in corn production in the United States and Europe led to the widespread occurrence of triazine-resistant weeds. Common triazine-resistant weeds are from the *Amaranthus*, *Chenopodium*, *Polygonum* and *Setaria*. In all there are 45 triazine-resistant weeds in corn. With increasing triazine resistance problems, corn growers in the United States rapidly adopted ALS inhibitors in the 1990s and target site resistance soon followed. Twenty-six weed species have evolved ALS inhibitor resistance in corn with the most problematic being *Amaranthus*, *Ambrosia*, *Solanum*, *Setaria*, *Digitaria*, *Sorghum* and *Conyza* spp. Fortunately, there are numerous herbicide chemistries available for weed control in corn and although resistance was of some economic importance there were often pre-emergence and post-emergence herbicides that would control triazine and ALS inhibitor-resistant weeds. Growers rapidly adopted Roundup Ready soybean (1996) and corn (1998) in part to control herbicide-resistant weeds, but unfortunately this has led to the selection of 17 glyphosate-resistant weeds in these crops. Europeans did not adopt ALS inhibitors extensively, nor did they grow Roundup Ready corn, therefore they do not have similar problems with ALS inhibitors or glyphosate-resistant weeds. *A. tuberculatus* is the biggest threat to corn production in the United States. Weed populations with multiple resistance to glyphosate, ALS inhibitors, PSII inhibitors, PPO inhibitors, 4-HPPD inhibitors and the synthetic auxins are likely to become common in the near future.

Soybean

There are 80 unique types of herbicide-resistant weeds that have been identified in soybean production (Fig. 5), belonging to 47 weed species (20 monocots and 27 dicots) (Fig. 6). The early 1990s saw the evolution of many ALS and ACCase inhibitor-resistant weeds evolving in soybean, which was partially solved by rapid adoption of Roundup Ready soybeans in 1996, but resulted in the selection of glyphosate-resistant weeds by 2000. In soybean there are currently 30 species that have evolved resistance to ALS inhibitors, 14 grass species with ACCase inhibitor resistance, 17 species with resistance to glyphosate and 7 species with resistance to PPO inhibitors, with the remainder of cases belonging to other herbicide sites of action. The worst herbicide-resistant broadleaf weeds in soybean are *Amaranthus* spp. (*A. tuberculatus*, *A. palmeri*, *A. hybridus*, *A. retroflexus* and *A. powellii*), *C. canadensis*, *Ambrosia* spp. (*A. artemisiifolia* and *A. trifida*), *Xanthium strumarium*, *Chenopodium album* and *K. scoparia*. The worst herbicide-resistant grass weeds in soybean are *Sorghum halepense*, *Setaria* spp. (*S. faberi* and *S. pumila*), *Digitaria insularis*, *Eleusine indica* and *Echinochloa* spp. (*E. colona*, *E. crus-galli*). The greatest threat for soybean production in the United States is multiple resistance in *A. tuberculatus* in the mid-western states and *A. palmeri* in the southern states.

Rice

Herbicide-resistant weeds in rice developed in the early 1990s and have steadily increased by about 3 cases per year. Globally there are 79 unique types of herbicide-resistant weeds (Fig. 5) that have evolved resistance in rice production, belonging to 51 species (Fig. 6). The greatest problems of herbicide-resistant weeds of rice occur in the United States (26 unique types), Japan (24 unique types) and South Korea (16 unique types). ALS inhibitor-resistant weeds account for half of the resistance cases in rice (40 of the 79 unique types). The main weed species that have evolved ALS inhibitor resistance in rice are monocots: *Cyperaceae* (10), *Poaceae* (7) and *Alismataceae* (7), and the dicot *Scrophulariaceae* (7) and *Lythraceae* (4). Nine grasses have evolved resistance to ACCase inhibitors. The most serious are *Echinochloa* spp. (*E. colona*, *E. crus-galli*, *E. oryzoides* and *E. phyllopogon*), *Leptochloa chinensis* and *L. panicoides*, and *I. rugosum*. The worst herbicide-resistant weeds in rice belong to *Echinochloa* spp. because they have evolved resistance to the majority of available rice-selective herbicides. In rice production *Echinochloa* spp. have evolved resistance to ACCase inhibitors (fenoxaprop + others), ALS inhibitors (bensulfuron + others), PSII inhibitors (propanil), EPSP synthase inhibitors (glyphosate), DOXP inhibitors (clomazone), synthetic auxins (quinclorac), chloracetamide long-chain fatty acid inhibitors (butachlor) and thiocarbamate lipid inhibitors (thiobencarb and molinate). In China *E. crus-galli* with resistance to butachlor and thiobencarb now infests more than 2 million hectares (Huang and Gressel 1997).

Cotton

The majority of herbicide-resistant weeds have been found in Southern United States; however, some cases have been reported from Brazil, Paraguay, Israel and Greece. The first herbicide-resistant weed found was *Eleusine indica* which was identified in 1973 with resistance to trifluralin (Group 3, dinitroaniline) and eventually was found in seven Southern US states. *X. strumarium* with resistance to MSMA and DSMA (Group 17, nucleic acid inhibitor) appeared in 1985 in Alabama and is now found in six states. Whilst *Sorghum halepense* and *C. canadensis* present serious herbicide resistance challenges in cotton, *A. palmeri* has been devastating. Resistance first appeared to trifluralin in 1989, then to ALS inhibitors (Group 2, imazaquin and pyrithiobac) in 1994, and to glyphosate (Group 9) in Roundup Ready in 2005. In corn and soybean *A. palmeri* has also been found with PSII inhibitor resistance (Group 5), PPO inhibitor resistance (Group 14) and HPPD inhibitor resistance (Group 27). Glyphosate-resistant *A. palmeri* is found in all cotton-growing regions of the United States and has the greatest economic effect on cotton production.

Perennial crops

Seventy-two weed species have evolved resistance to one or more herbicide sites of action in perennial crops (orchards, forestry, plantations and pastures). The most common herbicide-resistant weeds in perennial crops are *Poa annua*, *Lolium rigidum*, *C. canadensis*, *Senecio vulgaris* and *Eleusine indica*. Orchards (apples, olives, pears, citrus, almonds, grapes, etc.) and plantations (palm oil, coffee, rubber, etc.) often rely upon frequent applications of broad-spectrum herbicides and 15 weeds have evolved resistance to glyphosate, two to glufosinate and 22 to paraquat. Glufosinate-resistant *Eleusine indica* (found in palm oil plantations) and glufosinate-resistant *Lolium multiflorum* (found in hazelnuts and grapes) are the only known cases of glufosinate resistance worldwide.

Non-crop

Herbicides have long been used on roadsides, railways and industrial sites for weed control. Repeat applications of highly residual herbicides are used in non-crop weed control to maintain bare ground, which is a recipe for the rapid evolution of herbicide-resistant weeds. There are 67 unique types of herbicide-resistant weeds that have evolved resistance to herbicides under non-crop situations, belonging to 45 weed species (25 dicots and 20 monocots). The majority of these cases are on roadsides (47 unique cases), railways (23 unique cases) and industrial sites (8 cases). Three sites of action account for most non-crop resistance: PSII inhibitors (21), glyphosate (16) and ALS inhibitors (12). Because PSII and ALS inhibitors have been used for many years in the United States there is widespread multiple resistance in *K. scoparia*, *Salsola tragus* and *C. canadensis*. These weeds have such good dispersal mechanisms (tumble weed for the first two and parachute seeds for *Conyza*) that they are surely spreading resistance to farmer's fields.

4.2 Resistant weeds by region

One hundred and thirty-six herbicide-resistant weeds have been identified in the United States, more than in any other country (Fig. 7). Herbicide-resistant weeds were documented

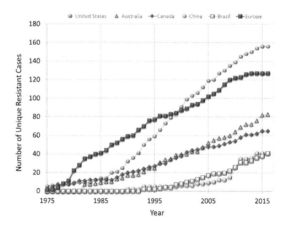

Figure 7 Increase in unique resistant cases for selected countries and Europe.

Table 2 The occurrence of herbicide-resistant weeds in countries

Rank	Country	# Resistant weeds	Rank	Country	# Resistant weeds
1	United States	156	34	Mexico	7
2	Australia	82	35	Colombia	6
3	Canada	64	36	Austria	5
4	France	46	37	Portugal	5
5	Brazil	41	38	Thailand	5
6	China	41	39	Bulgaria	4
7	Spain	37	40	India	4
8	Israel	36	41	Egypt	3
9	Japan	36	42	Indonesia	3
10	Germany	32	43	Philippines	3
11	Italy	30	44	Cyprus	2
12	United Kingdom	27	45	Finland	2
13	Belgium	22	46	Paraguay	2
14	Poland	22	47	Sri Lanka	2
15	Malaysia	21	48	Uruguay	2
16	Czech Republic	18	49	Ecuador	1
17	Turkey	17	50	El Salvador	1
18	Chile	16	51	Ethiopia	1
19	South Korea	16	52	Fiji	1
20	Argentina	15	53	Guatemala	1
21	New Zealand	15	54	Honduras	1
22	Switzerland	15	55	Hungary	1
23	South Africa	14	56	Ireland	1
24	Venezuela	13	57	Jordan	1
25	Iran	12	58	Kenya	1
26	Greece	10	59	Lithuania	1
27	Denmark	9	60	Nicaragua	1
28	Norway	9	61	Panama	1
29	Sweden	9	62	Russia	1
30	Bolivia	8	63	Saudi Arabia	1
31	Netherlands	8	64	Slovenia	1
32	Serbia	8	65	Taiwan	1
33	Costa Rica	7	66	Tunisia	1

Table 3 The number and percentage of herbicide-resistant species by family, and the percentage of species considered principal weeds by Holm et al. (1991, 1997) for each of these families

Family	Number of resistant species in family	Resistant species (% of total)	Weed species (% of world's principal weeds)
Poaceae	80	34	25
Asteraceae	39	17	16
Brassicaceae	22	8	4
Amaranthaceae	10	5	3
Chenopodiaceae	8	4	2
Polygonaceae	7	3	5
Scrophulariaceae	7	3	1
Cyperaceae	6	3	5
Caryophyllaceae	5	2	1
Alismataceae	5	2	1
Solanaceae	4	2	2
Lythraceae	4	2	1
23 other families pooled	30	14	18
Total	**210**	**100**	**84**

early in many European countries. Those with the most reported weeds are France (46), Spain (37), Germany (32), Italy (30) and the United Kingdom (27) (Table 2). Europe now reports 125 unique cases. The first was triazine resistance in corn, but greater problems were found with multiple resistance in grasses, particularly *Alopecurus myosuroides*, *Lolium* spp. and *Avena* spp. Australia and Canada have remarkably similar profiles in the report of herbicide-resistant weeds up until 2005, with multiple resistance in grass weeds in wheat driving the numbers to around 50 cases. From 2005 onwards the occurrence of glyphosate-resistant weeds in fallow and non-crop areas in Australia has contributed to Australia now having 82 unique cases of herbicide resistance, whilst Canada has 64 cases. China and Brazil did not rely upon herbicides for weed control in crops as early as the developed countries, and hence resistant weeds did not present a problem until the mid-1990s. However, herbicides have been used extensively in these countries over the last 30 years, and as a result the number of cases has significantly increased in the last 20 years, with both having 41 resistant cases. Whilst extensive use of ALS inhibitors has led to the rapid increase in herbicide resistance in China, the introduction of Roundup Ready soybeans has led to numerous glyphosate and ALS resistance in Brazil. It should be noted that some countries do not have good systems for documenting herbicide resistance, hence it is severely under reported; for instance, only one herbicide-resistant weed has been reported from Russia (ALS inhibitor-resistant *Picris hieracioides* in 2000). Israel and Japan have 36 cases of herbicide-resistant weeds each (Table 2).

4.3 Resistant weeds by weed family

Grasses are particularly prone to evolving herbicide resistance, more than would be predicted by their abundance in agronomic crops (Table 3). They account for 34% of all herbicide resistance cases, yet they only constitute 25% of the world's principal weeds (Table 3). In second place is the large family of Asteraceae which account for 17% of herbicide-resistant weeds in line with being 16% of the world's principal weeds. The plant families of Brassicaceae and Chenopodiaceae are also over represented in their percentage of herbicide-resistant weeds in comparison to their general abundance as weed pests.

5 Management of herbicide-resistant weeds

The best management strategy to delay herbicide resistance is to rely upon destabilizing/redirecting the selection pressure. No matter what weed control strategy is employed (chemical, mechanical, biological, etc.), if it is used for long enough on its own weeds will evolve and render it useless. Thus the use of as many weed control strategies in combination and rotation as practical is the best way to minimize the selection for herbicide-resistant weeds. This starts with using herbicide sites of action in rotation and combination, as the probability of an individual processing of two rare resistance genes is exceedingly small. But rotation of chemistry alone is not enough, integration of all economically viable weed control techniques are required to minimize resistance. Employing crop rotations, crop management, tillage systems, mowing, burning and even allelopathy may all be useful methods of reducing/redirecting the selection pressure.

Confirmation of herbicide-resistant weeds

In order to track and manage herbicide resistance there must be good systems in place to accurately and cheaply confirm its presence. The best way to confirm resistance is a classic dose–response experiment, where weed seed is collected and seedlings grown (with a known susceptible population of the same species) in the greenhouse and treated at the appropriate growth stage with a range of herbicide rates (usually 6 or more) that encompass survival through to complete mortality of both the susceptible and resistant populations. For large-scale screening where resistance has already been characterized in dose–response tests, there are several options that permit cost-effective screening. Whole-plant tests are carried out using one or two herbicide rates (chosen on the basis of previous dose–response assays) and are the best method for large-scale screening, because they are not mechanism specific. Other methods of herbicide resistance detection should first be compared and calibrated to whole-plant dose–response experiments. Petri dish assays that measure seedling growth on an herbicide-impregnated substrate of filter paper, agar or silica sand can be used for cheap and rapid screening. Other techniques, such as enzyme-based tests, DNA sequencing and leaf disc assays can be useful but are very specific to the mechanism of resistance.

Methods of spread of herbicide-resistant weeds

After the initial selection of rare genetic mutations that confer herbicide resistance, the spread of resistance becomes the primary source of new infestations, particularly for herbicides such as glyphosate and the synthetic auxins, where the initial gene frequencies for resistance are exceedingly rare. A knowledge of weed biology (dispersal mechanisms) is necessary to accurately predict the potential rate of spread of resistance and identify ways to limit it. Some weeds, for example, C. canadensis, have wind-borne seeds that can travel for many miles and disperse very quickly. *K. scoparia* and *Salsola tragus* disperse their seeds by breaking off at the base and tumbling for many miles, releasing seed as they go. Others, for example, pigweeds, have millions of small seeds that are easily transferred from field to field embedded in mud on shoes, pickup trucks and farm equipment.

Integrated weed management

All weed control practices apply a selection pressure on weed populations, and when used alone and given enough time weeds will evolve and survive the practice. The solution is to avoid using a single weed control practice for an extended period. Integrated weed management involves designing a comprehensive weed control plan that considers the use of all economically available weed control practices in such a way that no one practice is relied upon exclusively for weed control over an extended period. Such measures include preventing the spread of weeds, monitoring weed populations, tillage, crop rotations, nutrition, burning, biological weed control, crop competition, rotating herbicide sites of action and use of herbicide-resistant crops. A large source of new infestations of herbicide-resistant weeds comes from the spread of a few seeds from neighbouring fields. Farmers can reduce the spread of weeds by using certified seed, cleaning farm equipment between fields, covering grain trucks, only allowing weed-free livestock and hay onto the property and controlling weed seed nurseries along roads, fences, irrigation ditches and stockyards. Monitoring and creating an inventory of serious weed problems is key in developing an effective integrated weed management programme. Crops can be managed to maximize their competitiveness through use of competitive varieties (particularly those that have allelopathy), stale seed beds, cover crops, reduced row spacing, early seeding, high seeding rates, shallow seeding and good crop nutrition. Whilst many growers have gone towards zero tillage there is a case to be made for inclusion of sporadic tillage in any cropping system. In the United Kingdom some farmers mold board plough once every five years, burying weed seeds to a depth from which they cannot emerge. Other types of tillage include strip tillage, inter-row tillage, harrowing, spring and fall tillage, rod weeders and rotary hoeing. In Australia a campaign to manage herbicide-resistant *Lolium rigidum* includes targeting weed seeds through burning, baling chaff, burning chaff in rows and seed destruction as the seed comes out of the combine (Harrington seed destructor, Walsh et al. 2012). Some of these seed destruction techniques are being evaluated and appear to be successful for control of glyphosate-resistant *A. palmeri* in Southern United States. A combination of non-chemical weed controls can significantly reduce weed populations but is often not sufficient to avoid some crop losses from weeds without the use of herbicides. Herbicides are also part of the integrated weed management plan, and by reducing weed populations through non-chemical methods before using herbicides there is a much lower probability of selecting herbicide-resistant weeds. To minimize the risk of resistance, different herbicide sites of action should be used in rotation, sequences and mixtures. Beckie and Reboud (2009) have shown through field trials that mixtures of herbicides with different sites of action can be more effective to delay resistance than herbicide rotations. For herbicide mixtures to be effective the active ingredients should have different target sites and applied at rates that either herbicide alone would control the target weed problems. The reason this strategy works is that it is exceedingly rare that two mutations for different herbicide sites of action will occur spontaneously in the same individual. Be aware that herbicide mixtures are often used in a way that does not minimize resistance. Examples would be when the two herbicides in the mixture target different species, or where a weed has already evolved resistance to one of the herbicides in the mixture. In order to use a wider selection of herbicide sites of action growers are returning to pre-emergence herbicides to reduce the selection on the post-emergence herbicides. In addition, the use of herbicide-resistant crops can provide a wider selection of herbicide sites of action and will play an increasingly important role in weed control. Integrated weed management is successful because it aims to reduce the number of individuals that are controlled by any one weed control method. Evolution is a numbers game – the larger the population the higher the risk of finding rare resistant individuals.

Multiple resistance and non-target site resistance

In recent years scientists, growers, industry and the media have paid a lot of attention to the increase in glyphosate-resistant weeds, particularly in Roundup Ready cropping systems. However, glyphosate resistance alone is not a major issue, it is the combination of glyphosate with resistance to other herbicide sites of action that poses the biggest threat to sustained weed control. Glyphosate was rapidly adopted in part to solve the problems of weeds with resistance to PSII, ALS and ACCase herbicides. Now many of the weeds that have evolved glyphosate resistance in Roundup Ready crops also have resistance to one or more other herbicide sites of action. Globally there are nearly 100 weed species that are resistant to more than one herbicide site of action and 25 species that have resistance to four herbicide sites of action (Fig. 8). Grasses and *Amaranth* spp. have the greatest propensity to evolve resistance to multiple herbicide sites of action (Fig. 9). In addition, non-target site resistance, particularly in grasses (*Lolium*, *Alopecurus* and *Echinochloa* spp.) presents added difficulty because it often leads to unpredictable multiple resistance and cannot be managed as easily by rotating herbicide modes of action.

Impact of herbicide-resistant crops on herbicide-resistant weeds

Herbicide-resistant crops were rapidly adopted in North America, Brazil, Argentina and some other countries because they simplified weed control, provided better weed control at lower cost and with less crop injury. In addition, they allowed farmers to move

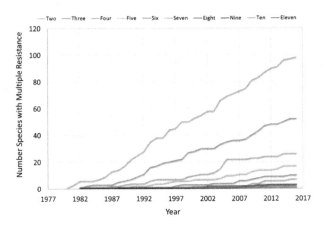

Figure 8 Weed species with resistance to more than one site of action.

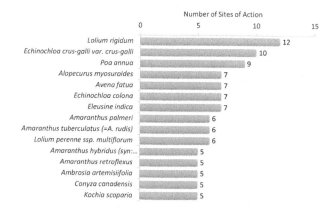

Figure 9 Top 15 weed species with resistance to multiple herbicide sites of action.

towards reduced or zero-tillage systems that reduce fuel use, conserve moisture and most importantly reduce soil erosion. While Roundup Ready crops have also been beneficial in combatting existing cases of herbicide-resistant weeds, over reliance on glyphosate has led to the rapid appearance of glyphosate-resistant weeds (Fig. 4), which often already have resistance to older herbicide sites of action. Industry is trying to cope with this challenge by introducing more herbicide-resistant crops (e.g. to synthetic auxins) that can be used in conjunction with the Roundup Ready trait. This is a reasonable resistance management strategy if it is used prior to weeds evolving resistance to glyphosate, but not wise if growers intend to rely upon a single site of action (synthetic auxin) to control glyphosate-resistant weeds.

6 Future outlook on herbicide resistance

Despite a lack of discovery of a new herbicide site of action in over 30 years, and the ever-increasing number of herbicide-resistant weeds, farmers are likely to rely on herbicides for 20 or 30 years as the backbone of their weed control programmes. For this to happen herbicides must be utilized in a more sustainable way, which will involve smart herbicide mixtures, rotations and more intensive integrated weed management programmes. It is clear that glyphosate-resistant weeds will continue to diminish the utility of this great herbicide, and other herbicides will also continue to fail over time. There are ways to save crop production from herbicide-resistant weeds.

1 Herbicide discovery stalled, but given the reality of herbicide-resistant weeds there are reasonable incentives for some companies to embark on new herbicide discovery programmes.
2 Herbicide-resistant crops will continue to help by allowing the use of old herbicides in new ways. Crops with stacked herbicide-resistant genes will facilitate smart herbicide mixtures to delay and avoid resistance.
3 Integrated weed management will decrease the selection pressure on herbicides and delay herbicide resistance.

4 Advances in RNAi technology may assist by providing weed control. However, there are large technological and economic issues that mean this technology will be at least 15 years before it is economically viable.

5 New technologies will slowly replace herbicides. For example, advances in computers that dramatically increase the speed and accuracy of weed identification in conjunction with robotics are likely to provide economical weed control in row crops, orchards, plantations and non-crop situations within 20 years.

7 Conclusion

The widespread occurrence of herbicide resistance and the increase in multiple resistance should be regarded as a warning that chemical control of weeds will be a temporary weed management solution. Certainly, there are a number of practices, encompassing integrated weed management, that can delay the appearance of herbicide resistance, but in the long run herbicides should only be seen as stopgap measures until better weed control solutions can be developed. No matter what weed control strategies are employed, there will always be selection pressure directed towards survival of weeds, and that is why a combination of weed control strategies is required to destabilize the evolution of resistance to any weed control practice.

8 Where to look for further information

Key organisations/groups:
The International Survey of Herbicide-Resistant Weeds: http://www.weedscience.org
The Herbicide Resistance Action Committee: http://www.hracglobal.com
The Weed Science Society of America: http://www.wssa.net
The European Weed Research Society: http://www.ewrs.org/
The Western Australian Herbicide-Resistance Initiative: http://ahri.uwa.edu.au/

Key journals publishing articles on herbicide-resistant weeds:
Weed Science: http://www.wssajournals.org/loi/wees
Weed Technology: http://www.wssajournals.org/loi/wete
Pest Management Science: http://www.wiley.com/WileyCDA/WileyTitle/productCd-PS.html
Weed Research: http://www.ewrs.org/weedresearch.asp
Pesticide Biochemistry and Physiology: http://www.journals.elsevier.com/pesticide-biochemistry-and-physiology
Plant Physiology: www.plantphysiol.org/

9 References

Bass, C. and Field, L. M. 2011. Gene amplification and insecticide resistance. *Pest Manag. Sci.*, 67: 886–90.

Baylis, A. D. 2000. Why glyphosate is a global herbicide: Strengths, weaknesses and prospects. *Pest Manag. Sci.*, 56: 299–308.

Beckie, H. J. and Reboud, X. 2009. Selecting for weed resistance: Herbicide rotation and mixture. *Weed Technol.*, 23, 363–70.

Brown, A. C., Moss, S. R., Wilson, Z. A. and Field, L. M. 2002. An isoleucine to leucine substitution in the ACCase of *Alopecurus myosuroides* (black-grass) is associated with resistance to the herbicide sethoxydim. *Pest. Biochem. Physiol.*, 72: 160–8.

Buchanan, B. B., Gruissem, W. and Jones, R. L. 2000. *Biochemistry and Molecular Biology of Plants*. American Society of Plant Physiology: Rockville, MD, Courier Companies, p. 1280.

Dayan, F. E., Rimando, A. M., Pan, Z., Baerson, S. R., Gimsing, A. L. and Duke, S. O. 2010 Molecules of interest: Sorgoleone. *Phytochemistry*, 71: 1032–9.

Délye, C., Zhang, X. Q., Michel, S., Matejicek, A. and Powles, S. B. 2005. Molecular bases for sensitivity to acetyl-coenzyme-A-carboxylase inhibitors in black-grass. *Plant Physiol.*, 137: 794–806.

Duggleby, R. G. and Pang, S. S. 2000. Acetohydroxyacid synthase. *J. Biochem. Mol. Biol.*, 33:1–36.

Duke, S. O. and Powles, S. B. (2008) Glyphosate: A once in a century herbicide. *Pest Manag. Sci.*, 64:319–25.

Feng, P. C. C., Tran, M., Sammons, R. D., Heck, G. R. and Cajacop, C. A. 2004. Investigations into glyphosate-resistant horseweed (*Conyza canadensis*): Retention, uptake, translocation, and metabolism. *Weed Sci.*, 52: 498–505.

Fisher, A. J., Bayer, D. E., Carriere, M. D., Ateh, C. M. and Yim, K. O. 2000. Mechanisms of resistance to bispyribac-sodium in an *Echinochloa phyllopogon* accession. *Pest. Biochem. Physiol.*, 28: 156–65.

Gaines, T. A., Preston, C., Leach, J. E., Chisholm, S. T. and Shaner, D. L. 2010. Gene amplification is a mechanism for glyphosate resistance evolution. *Proc. Natl. Acad. Sci. U. S. A.*, 107: 1029–34.

Gaines, T. A., Zhang, W., Wang, D., Bukun, B., Chisholm, S. T., Shaner, D. L., Nissen, S. J., Patzoldt, W. L., Tranel, P. J., Culpepper, A. S., Grey, T. L., Webster, T. M., Vencill, W. K., Sammons, R. D., Jiang, J, Preston, C., Leach, J. E. and Westra, P. 2010. Gene amplification confers glyphosate resistance in *Amaranthus palmerii*. *Proc. Natl. Acad. Sci. U. S. A.*, 107: 1029–34.

Ge, X., d'Avignon, D. A., Ackerman, J. J. H. and Sammons, R. D. 2010. Rapid vacuolar sequestration: The horseweed glyphosate resistance mechanism. *Pest Manag. Sci.*, 66: 345–8.

Gronwald, J. W. 1997. Resistance to PSII inhibitor herbicides. In De Prado, R., Jorrin, J. and Garcia-Torres, L. (Eds), *Weed and Crop Resistance to Herbicides*. Kluwer Academic Publishers: Dordrecht, the Netherlands, p. 340.

Hana, H., Yua, Q., Vila-Aiubb, M. and Powles, S. B. 2014. Genetic inheritance of cytochrome P450-mediated metabolic resistance to chlorsulfuron in a multiple herbicide resistant *Lolium rigidum* population. *Crop Prot.*, 65: 57–63.

Heap, I. and Knight, R. 1982. A population of ryegrass tolerant to the herbicide diclofop-methyl. *J. Aust. Inst. Agr. Sci.*, 48: 156–7.

Heap, I. and Knight, R. 1986. The occurrence of herbicide cross-resistance in a population of annual ryegrass, *Lolium rigidum*, resistant to diclofop-methyl. *Aust. J. Agr. Res.*, 37, 149–56.

Heap, I. M. and LeBaron, H. 2001. Introduction and overview of resistance. In Powles, S. B. and Shaner, D. L. (Eds), *Herbicide Resistance and World Grains* (pp. 1–22). Boca Raton, Florida, USA: CRC Press, p. 308.

Heap, I. 2016. The International Survey of Herbicide Resistant Weeds Online. Internet. 12 September 2016. Available www.weedscience.com.

Hilton, H. W. 1957. Herbicide tolerant strains of weeds. Hawaiian Sugar Plant Association Annual Report 69.

Hochberg, O., Sibony, M. and Rubin, R. 2009. The response of ACCase-resistant *Phalaris paradoxa* populations involves two different target site mutations. *Weed Res.*, 49: 37–46.

Holm, L. J., Plucknett, D. L., Pancho, J. V. and Herberger, J. 1991. *The World's Worst Weeds: Distribution and Biology*. Malabar, Florida, USA: Krieger.

Holm, L., Doll, J., Holm, E., Pancho, J. and J. Herberger. 1997. *The World's Worst Weeds: Natural Histories and Distribution*. New York: Wiley, p. 1152.

Huang, B. Q. and Gressel, J. 1997. Barnyardgrass (*Echinochloa crus-galli*) resistance to both butachlor and thiobencarb in China. *Resistant Pest Manag.*, 9: 5.

Kaundun, S. S., Zelaya, I. A., Dale, R. P., Lycett, A. J. and Carter, P. 2008. Importance of the P106S target-site mutation in conferring resistance to glyphosate in a goosegrass (*Eleusine indica*) population from the Philippines. *Weed Sci.*, 56: 637–46.

Liu, W. J., Harrison, D. K. and Chalupska, D. 2007. Single-site mutations in the carboxyl transferase domain of plastid acetyl-CoA carboxylase confer resistance to grass-specific herbicides. *Proc. Natl. Acad. Sci. U. S. A.*, 104(9): 3627–32.

Patzoldt W. L., Hager A. G., McCormick J. S. and P. J. Tranel. 2006. A codon deletion confers resistance to herbicides inhibiting protoporphyrinogen oxidase. *Proc. Natl. Acad. Sci. U. S. A.*, 103: 12329–34.

Pratley, J., Urwin, N., Stanton, R., Baines, P., Broster, J., Cullis, K., Schafer, D., Bohn, J. and Krueger, R. 1999. Resistance to Glyphosate in Lolium rigidum. I. Bioevaluation. *Weed Sci.*, 47(4): 405–11.

Preston, C., Tardif, F. J., Christopher, J. T. and Powles, S. B. 1996. Multiple resistance to dis-similar herbicide chemistries in a biotype of *Lolium rigidum* due to enhanced activity of several herbicide degrading enzymes. *Pest. Biochem. Physiol.*, 54: 123–34.

Ryan, G. F. 1970. Resistance of common groundsel to simazine and atrazine. *Weed Sci.*, 18: 614–16.

Sammons R. D. and Gaines, T. A. 2014. Glyphosate resistance: State of knowledge. *Pest Manag. Sci.* 70: 1367–77.

Sterling, T. M. and Hall, J. C. 1997. Mechanism of action of natural auxins and the auxinic herbicides. In Roe, R. M., Burton, J. D. and Kuhr, R. J. (Eds), *Toxicology, Biochemistry and Molecular Biology of Herbicide Activity* (pp. 111–41). Amsterdam: IOS Press.

Switzer, C. M. 1957. The existence of 2,4-D resistant strains of wild carrot. *Proc. Northeast Weed Control Conf.*, 11: 315–18.

VanGessel, M. J. 2001. Glyphosate-resistant horseweed from Delaware. *Weed Sci.*, 49: 703–5.

Waite, J., Thompson, C. R., Peterson, D. E., Currie, R. S., Olson, B. L. S., Stahlman, P. W. and Al-Khatib, K. 2013. Differential kochia (*Kochia scoparia*) populations response to glyphosate. *Weed Sci.*, 61: 193–200.

Walsh, M. J., Harrington, R. B. and Powles, S. B. 2012. Harrington seed destructor: A new nonchemical weed control tool for global grain crops. *Crop Sci.*, 52: 1343–7.

Woodburn, A. T. 2000. Glyphosate: production, pricing and use world-wide. *Pest Manag. Sci.*, 56: 309–12.

Yu, Q., Jalaludin, A., Han, H., Chen, M., Sammons, R. D. and Powles, S. B. 2015. Evolution of a double amino acid substitution in the EPSP synthase in *Eleusine indica* conferring high level glyphosate resistance. *Plant Physiol.* 167(4): 1440–7. doi:http://dx.doi.org/10.1104/pp.15.00146.

Part 4

Cultural and physical methods for weed control

Chapter 10

The role of herbicide-resistant crops in integrated weed management

Prashant Jha, Montana State University, USA; and Krishna N. Reddy, USDA-ARS, USA

1 Introduction

Chemical weed control began with the use of 2,4-D in the mid-1940s. Since then, a wide array of herbicides has been commercialized and that has greatly contributed to increased crop yields. Herbicide use in 21 major crops in the United States increased over 13-fold from 16 million kg in 1960 to 217 million kg in 1981. By 1980, over 90% of the corn, cotton and soybean areas were treated with herbicides compared to less than 10% of these crops planted in 1952 (Fernandez-Cornejo et al. 2014). With the introduction of several new, more specific and more effective herbicides, the cost of weed control with herbicides decreased relative to other control practices (labour, fuel and machinery). These benefits of lower production costs, higher crop yields and quality, and increased profit margins for farmers resulted in over-dependence on herbicides for weed management. Use of the same herbicide year after year has led to evolution of herbicide-resistant (HR) weeds. Development of herbicide resistance in weeds is widely recognized as a result of adaptive evolution of weed populations to repetitive use of same herbicide or class of herbicides (Jasieniuk et al. 1996). In response to selection pressure exerted by herbicides, weed populations change in genetic composition by selection of genes already present or arisen newly through mutation resulting in evolution of resistance (Délye et al. 2013; Jasieniuk et al. 1996). The first case of resistance to triazines was reported in 1968 (Ryan 1970). Since then, there has been an alarming increase in evolution and spread of HR weeds. As of 2017, globally, 252 weed species (147 dicots and 105 monocots) have evolved resistance

http://dx.doi.org/10.19103/AS.2017.0025.12

to 161 different herbicides representing 23 of the 26 known herbicide sites of action (SOA) in 91 crops in 68 countries (Heap 2017).

HR crops, both transgenic (created through integration of transgene) and non-transgenic (created through traditional plant breeding or mutagenesis) as shown in Tables 1 and 2, have been widely grown in several countries since their commercialization in the early 1980s to mid-1990s (Green 2012; Powles 2008; Reddy and Jha 2016). HR crop technology was a blessing for growers as it provided simple, flexible, effective and economical weed management options. Each specific HR crop (viz., glyphosate-resistant, glufosinate-resistant, imidazolinone-tolerant) provided a unique opportunity to manage specific weeds. Furthermore, HR crops offered simplicity and flexibility to manage a broad spectrum of weeds and weeds resistant to other herbicides. For example, use of glyphosate in glyphosate-resistant (GR) crops offered a tremendous advantage to manage weeds resistant to other herbicides such as ALS inhibitors, acetyl CoA carboxylase (ACCase) inhibitors, dinitroanilines and organo-arsenicals. Among all HR crops, GR crops offered farmers more simplicity and flexibility to manage weeds. The rapid adoption of GR crops by growers was mainly because of weed-free fields, increased yields with less input, and increased profit per unit area (Castle et al. 2006). The high rate of adoption of GR soybean, cotton and corn in North America resulted in unprecedented impact because glyphosate was often the sole herbicide used over large production areas. Its use was accompanied by a drastic decline in mechanical and cultural methods to manage weed seed banks (Green 2011; Jha et al. 2017; Owen and Zelaya 2005; Shaw et al. 2009). Ultimately, over-reliance on glyphosate, especially in conservation tillage systems, resulted in evolution of GR weeds. There are now 37 GR weed species globally (Heap 2017).

Table 1 Commercially available transgenic herbicide-resistant (HR) crops (Adapted from Green and Castle 2010; Green 2012)

Crop	Resistance trait	Trait gene(s)	Year available
Canola	Glufosinate	*pat*	1995
Canola	Glyphosate	*cp4 epsps, gox v247*	1996
Corn	Glufosinate	*pat*	1996
Corn	Glyphosate	Multiple *zm-2mepsps*	1998
		Two *cp4 epsps* cassettes	2001
Soybean	Glyphosate	*cp4 epsps*	1996
Soybean	Glufosinate	*pat*	2009
Cotton	Glyphosate	*cp4 epsps*	1997
		Two *cp4 epsps*	2006
		zm-2mepsps	2009
Cotton	Glufosinate	*bar*	2005
Rice	Glufosinate	*bar*	2006
Sugar beet	Glyphosate	*cp4 epsps*	2007
Alfalfa	Glyphosate	Two *cp4 epsps*	2011*

* Glyphosate-resistant alfalfa was first released in 2006, but got legal clearance for sale in 2011.

Table 2 Commercially available non-transgenic herbicide-resistant (HR) crops (Adapted from Green and Castle 2010; Green 2012)

Crop	Resistance trait	Selection method	Year available
Soybean	Triazine	Tissue culture	1981
	Sulfonylureas	Seed mutagenesis	1994
Canola	Triazine	Whole plant	1984
	Imidazolinone	Microspore selection	1997
Corn	Imidazolinone	Pollen mutagenesis/tissue culture	1993
	Cyclohexanediones (sethoxydim)	Tissue culture	1996
Wheat	Imidazolinone	Seed mutagenesis	2002
Rice	Imidazolinone	Seed mutagenesis	2002
Sunflower	Imidazolinone	Transfer from weedy relative	2003
	Sulfonylureas	Transfer from weedy relative	2006
Sorghum	Sulfonylureas	Transfer from weedy relative	2013

The increasing number of HR weeds led to development and commercialization of several multiple HR (stacked-trait) crops as tools to manage weeds that had become difficult-to-control or resistant to glyphosate and other herbicides (Duke 2005; Owen 2008; Reddy and Jha 2016). However, diversification of weed control methods is critical to future use of HR technology, otherwise, shifts in weed populations related to ecological adaptation, natural tolerance or evolved resistance (Owen and Zelaya 2005), will continue to pose an economic threat to production agriculture. Lessons need to be learnt and integrated weed management (IWM) programmes need to be implemented to maintain sustainability of GR and other HR crop technologies (Powles 2008). This chapter provides an outlook on major HR crops (commercialized or under development), their benefits and pitfalls, and outlines a direction forward for growers to manage weeds, regardless of herbicide resistance.

2 Glyphosate-resistant crops

Commercialization of HR crops, particularly GR crops, has created a paradigm change in weed management tactics adopted by growers on their farms. GR soybean, cotton and canola were introduced in 1996 and corn in 1998. By 2016, 94% of soybean, 89% of cotton and 89% of corn areas were planted with GR cultivars in the United States (USDA 2016). Globally, 83% of soybean, 75% of cotton, 29% of corn and 24% of canola areas were planted with GR cultivars in 2015 (James 2015). The rapid adoption of GR crop technology was attributed to the effective, easy-to-use, economical and safe use of glyphosate for broad-spectrum weed control. Agronomic advantages such as early planting and conservation tillage also facilitated rapid adoption and commercial success of GR crops to enhance global food security (Green 2012; Powles 2008). Conservation tillage (particularly no tillage) in GR crop systems is considered more environmentally sustainable, compared with the conventional tillage systems, with regard to soil erosion and water quality (Cerdeira and Duke 2006; Price et al. 2011). Anecdotal evidence suggests that corn, soybean and cotton growers valued consistency in weed control and protection against yield loss as important reasons for adopting GR crop technology. At the outset, the GR crops (viz. corn, cotton and soybean) offered a tremendous opportunity to manage weeds resistant to other herbicides (ALS inhibitors, ACCase inhibitors, dinitroanilines and organo-arsenicals) (Green 2012). Prudent use of GR crops could have increased herbicide diversity for weed control by enabling use of herbicide tank mixtures, herbicide rotations or sequential herbicide programmes. Instead, the simplicity and convenience of glyphosate-based GR cropping systems has been over-exploited, with growers often relying on glyphosate only for weed control in GR corn, soybean and cotton (Bayliss 2000; Duke 2005; Gianessi 2005; Green 2011). This situation could partially be attributed to the common perception that GR weeds would never evolve, since no weeds developed resistance to glyphosate even after more than two decades (prior to 1996) of non-selective glyphosate use in non-crop situations (Bradshaw et al. 1997).

One of the major consequences of this unprecedented change following the rapid adoption of GR crops has been a greater selection pressure on the weed community (Duke 2005). There has been a decline in number of herbicides used to manage weeds. 'The number of herbicide active ingredients used on at least 10% of the US soybean area declined from 11 in 1995 to only 1, glyphosate, in 2002' (Green and Owen 2011). This lack of diversity in weed control tactics resulted in weed population shifts to species that have natural tolerance to or have evolved resistance to glyphosate (Duke 2005; Owen 2008). With an increase in land area under GR soybean, corn and cotton production in the United States, weed species such as pigweeds (*Amaranthus* spp.), horseweed (*Conyza canadensis* (L.) Cronq.), common lambsquarters (*Chenopodium album* L.), velvetleaf (*Abutilon theophrasti* Medik.), Asiatic dayflower (*Commelina communis* L.) and tropical spiderwort (*Commelina benghalensis* L.) well adapted to no-till systems and/or difficult to control with glyphosate, became dominant in the weed community (Culpepper 2006; Hilgenfeld et al. 2001; Owen 2008; Scursoni et al. 2007).

With the first discovery of GR rigid ryegrass (*Lolium rigidum* Gaudin) in Australia in 1996 (Powles et al. 1998), by 2017, 37 weed species were resistant to glyphosate globally (Heap 2017). In the United States, 17 weed species evolved resistance to glyphosate mostly in GR cropping systems (Heap 2017). Of particular significance is GR Palmer amaranth (*Amaranthus palmeri* S. Watson) that first appeared in GR cotton in Georgia in 2008, and has now become a threat to the conservation tillage system in corn, soybean and cotton crops across south-eastern, Midsouth and Midwestern USA (Price et al. 2011). Other economically significant weed species that evolved glyphosate resistance with the massive adoption of GR crops over large areas in the United States include common ragweed (*Ambrosia artemisiifolia* L.), giant ragweed (*Ambrosia trifida* L.) and various *Conyza* and *Lolium* spp. Likewise, the rapid adoption of GR soybean in Argentina and Brazil resulted in field-evolved GR biotypes of johnsongrass (*Sorghum halepense* L. Pers) and wild poinsettia (*Euphorbia heterophylla* L.), respectively (Vila-Aiub et al. 2007; Vidal et al. 2007).

Because of rapid reproduction potential and spread of these GR weeds, growers have to face drastic crop yield reductions and have to change their crop production and weed control practices, which in most cases, are cost prohibitive (Shaw et al. 2011). For instance, herbicide input costs to manage GR Palmer amaranth in cotton in Georgia, USA, have more than doubled due to complex and expensive weed control programmes required for successful management (Sosnoskie and Culpepper 2014). A recent survey suggests that nearly 50% of US growers are now dealing with GR weeds in their fields (Fraser 2013). Therefore, weed management practices must integrate other herbicide SOAs, if this novel, once-in-a-century herbicide (glyphosate), and GR crop technology, are to be sustained for future use.

3 Glufosinate-resistant crops

Glufosinate-resistant corn, cotton and soybean were commercialized in 1997, 2004 and 2009, respectively, a similar time frame as their GR counterparts. Glufosinate resistance trait has provided US cotton and soybean growers a valuable tool to manage GR weeds, such as Palmer amaranth (Norsworthy et al. 2008). Stacked-trait cultivars of soybean, corn and cotton that confer resistance to both glufosinate and glyphosate are now commercially available and allow growers to diversify their weed management programmes. Greater cost, narrow spectrum of weeds and more restrictive timing of application (effective mostly on smaller weeds) are the major factors contributing to the slower adoption of glufosinate versus glyphosate (Green and Owen 2011). Furthermore, glufosinate is not very effective on grasses and perennial weeds. Three weed species goosegrass (*Eleusine indica* L.), perennial ryegrass (*Lolium perenne* L. ssp. perenne) and Italian ryegrass (*Lolium perenne* L. ssp. *multiflorum*) have already evolved resistance to glufosinate (Heap 2017), which may impede the long-term utility of this HR crop technology if not used as a component of IWM.

4 Imidazolinone and sulphonylurea-tolerant crops

Imidazolinone (IMI) herbicides including imazapyr, imazapic, imazethapyr, imazamox, imazamethabenz and imazaquin control weeds by inhibiting the acetohydroxyacid synthase (AHAS) or acetolactate synthase (ALS) enzyme, thereby disrupting the biosynthesis of branched chain amino acids in plants (Tan et al. 2005). These herbicides are used for broad-spectrum grass and broadleaf control in IMI-tolerant crops. The IMI-tolerance trait, also referred as the Clearfield™ trait, was commercialized in corn in 1993, followed by canola (1997), wheat (2002), rice (2002) and sunflower (2003). IMI herbicides are effective for control of certain difficult-to-control weeds such as shattercane [*Sorghum bicolor* (L) Moench] and johnsongrass [*Sorghum halepense* (L) Pers] in IMI-tolerant corn, red rice (*Oryza sativa* var. *sylvatica*) in IMI-tolerant rice, wild mustard [*Brassica kaber* (DC) LC Wheeler] and stinkweed [*Pluchea camphorata* (L) DC] in IMI-tolerant oilseed rape, and downy brome (*Bromus tectorum* L.), jointed goatgrass (*Aegilops cylindrica* Host) and Italian ryegrass in IMI-tolerant wheat (Tan et al. 2005).

Similarly, the sulphonylurea-tolerant (ST) trait in crops provides increased tolerance to chlorimuron and other compounds in the sulphonylurea family of ALS inhibitors applied post-emergence for weed control (Reddy and Whiting 2000). The ST soybean offers additional flexibility to growers in double crop situations (soybean after wheat) by mitigating herbicide carryover injury concerns in soybean from soil residual sulphonylurea herbicides applied in wheat.

Because these non-transgenic, HR traits, are incorporated in crops using traditional breeding techniques (mutagenesis and selection), regulatory barriers related to commercialization are significantly less than the transgenic HR traits. These IMI/ST traits are often stacked with other HR trait(s) to allow use of herbicide mixtures because of the widespread distribution of ALS-resistant weeds (Green and Owen 2011; Heap 2017). Reports of gene flow from IMI-tolerant crops to closely related weed species such as red rice, wild sunflower (*Helianthus annuus* L.) and jointed goatgrass (Tan et al. 2005) are other classical examples of why diversity and stewardship programmes are needed for using these HR crops in future.

5 New HR crop technologies

Unfortunately, over-reliance on HR crop technology over the past two decades has led to rapid evolution of HR weeds because of massive selection pressure (Duke and Powles 2009). Evolution of weed resistance to glyphosate has diminished the utility of glyphosate considerably. As a solution to this problem, the development of a new generation of multiple HR crops has been pursued vigorously by several agrochemical industries. Currently, Monsanto, Dow, Bayer, Syngenta and BASF are developing new stacked-trait crops in combination with the GR trait. They are glyphosate-glufosinate (soybean, corn, cotton), glyphosate-ALS inhibitors (soybean, corn, canola), glyphosate-glufosinate-2,4-D (soybean, cotton), glyphosate-glufosinate-dicamba (soybean, corn, cotton), glyphosate-glufosinate-HPPD inhibitors (soybean and cotton), glyphosate-glufosinate-2,4-D-ACCase inhibitors (corn) and glufosinate-dicamba (wheat) (Green 2014). Transgenic, protoporphyrinogen oxidase (PPO)-resistant corn has also been developed (Green and Owen 2011). The relatively new HR traits (Table 3) when used in stacked-trait crops will provide new options with existing herbicides and can potentially be used to control GR- and ALS inhibitor-resistant weeds. However, these stacked-trait crops will not be a total weed management solution because several weeds have already evolved resistance to these herbicides. For example, 87 weeds resistant to various ALS inhibitors, 48 weeds resistant to ACCase inhibitors and 34 weeds resistant to synthetic auxins have already evolved (Heap 2017). Furthermore, PPO-resistant *Amaranthus* species have been documented in Midwestern and Southern US states, including multiple-resistant biotypes of Palmer amaranth (resistant to glyphosate and PPO inhibitors) and common waterhemp (resistant to glyphosate/2, 4-D, ALS, PS II, HPPD and PPO inhibitors) (Heap 2017). Therefore, the utility of stacked-trait crops depends on the specific weed problem to be addressed and requires knowledge of the herbicide SOA to match the specific HR weed problem (Green and Owen 2011; Shaner and Beckie 2014). Improved formulation and application technologies for using 2,4-D (2,4-D choline) and dicamba (DGA salt) will provide growers much-needed tools to manage GR broadleaf weeds in stacked-trait crops, with reduced off-target herbicide drift and injury to sensitive broadleaf plants (Green and Owen 2011). Multiple HR crops will continue to evolve, thereby allowing growers to use new herbicide mixtures with multiple SOAs, but there is an urgent need to use this technology more pragmatically and judiciously to maintain the long-term sustainability of existing herbicides.

Table 3 New transgenic herbicide-resistant (HR) traits stacked with glyphosate- and/or glufosinate resistance trait(s) (Adapted from Green and Castle 2010; Green 2012)

Resistance trait	Trait characteristics	Crop(s)*
2,4-D	Microbial degradation enzyme	Corn, cotton, soybean
Dicamba	*Pseudomonas maltophilia, O*-demethylase	Corn, cotton, soybean
HPPD inhibitor	Over-expression, alternate pathway, and increased pathway flux	Soybean, cotton
PPO inhibitors	Resistant microbial and *Arabidopsis thaliana* PPO	Corn
AOPP, ACCase inhibitor and synthetic auxin	Microbial, aryloxyalkanoate dioxygenase	Corn
Multiple herbicides	Glutathione *S*-transferase, *Escherichia coli* P450, *Zea mays*	TBD

* Some of these publicly disclosed HR traits will be commercialized in the near future.
HPPD, 4-hydroxyphenylpyruvate dioxygenase; PPO, protoporphyrinogen oxidase; AOPP, aryloxyphenoxypropionate; ACCase, acetyl CoA carboxylase; TBD, to be determined.

6 HR crops as part of an IWM programme

Herbicides (with or without HR crops) are still essential for weed management in modern cropping systems. HR crop technology alone cannot provide total weed control. HR crops must be integrated with other weed control tactics. It is best regarded as supplementary to other weed control methods that increase the diversity of weed control tactics. There is a greater need for IWM, a holistic approach that integrates different methods of weed control to manage weeds and maintain crop yields (Harker and O'Donovan 2013; Swanton and Murphy 1996). The IWM approach must include use of combinations of mechanical (tillage before planting, in-crop cultivation, hand hoeing, post-harvest tillage), cultural (competitive cultivars, plant densities, row spacing, crop rotation, winter crops in rotation, cover crops), chemical (residual herbicides, herbicide full-labelled rate, tank mixtures at the label rate, sequences, application timing, herbicide rotation with different modes of action), biological tactics where and when available, as well as preventive (weed seed bank management, clean equipment) techniques. Also, use of combinations of different herbicide application methods: post-harvest (fallow seedbed), pre-plant foliar (burndown), pre-plant incorporated, pre-emergence, post-emergence over-the-top, directed-post-emergence and spot treatment is critical to manage weeds. Due to high short-term costs associated with the use of an array of weed control tactics, growers often are reluctant to diversify management tactics. Sustainable weed management requires a longer-term strategy than that of a single-season approach. Herbicide dependence has failed as evident from the severity of the evolution of weeds resistant to 23 of the 26 herbicide SOA. Growers have no choice. They must diversify to achieve sustainable weed management.

7 Summary

The HR (single or stacked-trait) crops represent a revolutionary breakthrough in weed control technology, but they are only one of several weed control tactics. The HR weed management strategies must be diversified in order to curtail or disrupt HR weeds from evolving and spreading, with an ultimate goal of not allowing any weed to survive and set seed. Integration of HR crop technology with cultural, mechanical and chemical (along with biological where available) tactics is critical in the management of herbicide resistance and to ensure sustained food and fibre production. The future weed management tactics look a lot more like the ones used in the past, that is, the pre-HR crop era. HR crops will still not eliminate the need for discovery of new SOA herbicides and other new technologies (robotics and site-specific weed management tools) to manage the 'wicked' nature (Shaw 2016) of the problem of HR weeds.

8 Where to find further information

Additional information on HR (single or stacked-trait) crops and IWM approaches is readily available in the literature and at various websites maintained by state cooperative extension services of land-grant universities and agrochemical companies. Several research articles, reviews and book chapters have been published on various aspects of HR crops and IWM systems. Some of them have been listed (by no means exhaustive) in the following references section and others can be found by diligent search of literature.

9 References

Bayliss, A. D. (2000), 'Why glyphosate is a global herbicide: strengths, weaknesses and prospects', *Pest Manag. Sci.*, 56, 299–308.

Bradshaw, L. D., Padgette, S. R., Kimbal, S. L. and Wells, B. H. (1997), 'Perspectives on glyphosate resistance', *Weed Technol.*, 11, 189–98.

Castle, L. A., Wu. G. and McElroy, D. (2006), 'Agricultural input traits: Past, present and future', *Curr. Opin. Biotechnol.*, 17, 105–12.

Cerdeira, A. L. and Duke, S. O. (2006), 'The current status and environmental impacts of glyphosate-resistant crops: A review', *J. Environ. Qual.*, 35, 1633–58.

Culpepper, A. S. (2006), 'Glyphosate-induced weed shifts', *Weed Technol.*, 20, 277–81.

Délye, C., Jasieniuk, M. and Le Corre, V. (2013), 'Deciphering the evolution of herbicide resistance in weeds', *Trends Genet.*, 29, 649–58.

Duke, S. O. (2005), 'Taking stock of herbicide-resistant crops ten years after introduction', *Pest Manag. Sci.*, 61, 211–18.

Duke, S. O. and Powles, S. B. (2009), 'Glyphosate-resistant crops and weeds: Now and in the future', *AgBioForum*, 12, 346–57.

Fernandez-Cornejo, J., Nehring, R., Osteen, C., Wechsler, S., Martin, A. and Vialou, A. (2014), 'Pesticide use in U.S. agriculture: 21 selected crops, 1960–2008', United States Department of Agriculture, Economic Research Service, Economic Information Bulletin Number 124, pp. 80. May 2014.

Fraser, K. (2013), 'Glyphosate Resistant Weeds – Intensifying', *Stratus Agri-Marketing*, http://www.stratusresearch.com/blog07.htm (accessed 10 February 2017).

Gianessi, L. P. (2005), 'Economic and herbicide use impacts of glyphosate-resistant crops', *Pest Manag. Sci.*, 61, 241–5.

Green, J. M. and Castle L. A. (2010), 'Transitioning from single to multiple herbicide resistant crops', In *Glyphosate Resistance in Crops and Weeds: History, Development, and Management* (Nandula, V. K. (Ed.)). Wiley: Hoboken, NJ, pp. 67–91.

Green, J. M. (2011), 'Outlook on weed management in herbicide-resistant crops: Need for diversification', *Outlooks Pest Manag.*, 22, 100–4.

Green, J. M. and Owen, M. D. K. (2011), 'Herbicide-resistant crops: Utilities and limitations for herbicide-resistant weed management', *J. Agric. Food Chem.*, 59, 5819–29.

Green, J. M. (2012), 'The benefits of herbicide-resistant crops', *Pest Manag. Sci.*, 68, 1323–31.

Green, J. M. (2014), 'Current state of herbicides in herbicide-resistant crops', *Pest Manag. Sci.*, 70, 1351–7.

Harker, K. N. and O'Donovan, J. T. (2013), 'Recent weed control, weed management, and integrated weed management', *Weed Technol.*, 27, 1–11.

Heap, I. (2017), 'The international survey of herbicide-resistant weeds', www.weedscience.org. (Accessed26 February 2017).

Hilgenfeld, K. L., Martin, A. R., Mason, S. C. and Mortensen, D. A. (2001), 'Mechanisms involved in weed species shifts in a glyphosate-tolerant system', *Proc. North Cent. Weed Sci. Soc. Abstr.*, 56.

James, C. (2015), 'Global Status of Commercialized Biotech/GM Crops', ISAAA Brief No. 51. ISAAA: Ithaca, NY, http://www.isaaa.org/resources/publications/briefs/51/executivesummary/default. asp, (assessed 1 March 2017).

Jasieniuk, M., Brûlé-Babel, A. and Morrison, J. N. (1996), 'The evolution and genetics of herbicide resistance in weeds', *Weed Sci.*, 44, 176–93.

Jha, P., Kumar, V., Godara, R. K. and Chauhan, B. S. (2017), 'Weed management using crop competition in the United States: A review', *Crop Prot.*, 95, 31–7.

Norsworthy, J. K., Griffith, G. M., Scott, R. C., Smith, K. L. and Oliver, L. R. (2008), 'Confirmation and control of glyphosate-resistant Palmer amaranth (*Amaranthus palmeri*) in Arkansas', *Weed Technol.*, 22, 108–13.

Owen, M. D. K. (2008), 'Weed species shifts in glyphosate-resistant crops', *Pest Manag. Sci.*, 64, 377–87.

Owen, M. D. and Zelaya, I. A. (2005), 'Herbicide-resistant crops and weed resistance to herbicides', *Pest Manag. Sci.*, 61, 301–11.

Powles, S. B. (2008), 'Evolved glyphosate-resistant weeds around the world: Lessons to be learnt', *Pest Manag. Sci.*, 64, 360–5.

Powles, S. B., Lorraine-Colwill, D. F., Dellow, J. J. and Preston, C. (1998), 'Evolved resistance to glyphosate in rigid ryegrass (*Lolium rigidum*) in Australia', *Weed Sci.*, 46, 604–7.

Price, A. J., Balkcom, K. S., Culpepper, S. A., Kelton, J. A., Nichols, R. L. and Schomberg, H. (2011), 'Glyphosate-resistant Palmer amaranth: A threat to conservation tillage', *J. Soil Water Conserv.*, 66, 265–75.

Reddy, K. N. and Whiting, K. (2000), 'Weed control and economic comparisons of glyphosate-resistant, sulfonylurea-tolerant, and conventional soybean (*Glycine max*) systems', *Weed Technol.*, 14, 204–11.

Reddy, K. N. and Jha, P. (2016), 'Herbicide-resistant weeds: Management strategies and upcoming technologies', *Indian J. Weed Sci.*, 48, 108–11.

Ryan, G. F. (1970), 'Resistance of common groundsel to simazine and atrazine', *Weed Sci.*, 18, 614–16.

Scursoni, J. A., Forcella, F. and Gunsolus, J. (2007), 'Weed escapes and delayed emergence in glyphosate-resistant soybean', *Crop Prot.*, 26, 212–18.

Shaner, D. L. and Beckie, H. J., (2014), 'The future for weed control and technology', *Pest Manag. Sci.*, 70, 1329–39.

Shaw, D. R. (2016), 'The 'wicked' nature of the herbicide resistance problem', *Weed Sci.*, 64(Special Issue), 552–8.

Shaw, D. R., Givens, W. A., Farno, L. A., Gerard, P. D., Jordan, D., Johnson, W. G., Weller, S. C., Young, B. G., Wilson, R. G. and Owen, M. D. K. (2009), 'Using a grower survey to assess the benefits and challenges of glyphosate-resistant cropping systems for weed management in U.S. corn, cotton, and soybean', *Weed Technol.*, 23, 134–49.

Shaw, D. R., Owen, M. D., Dixon, P. M., Weller, S. C., Young, B. G., Wilson, R. G. and Jordan, D. L. (2011), 'Benchmark study on glyphosate-resistant cropping systems in the United States. Part 1: Introduction to 2006–2008', *Pest Manag. Sci.*, 67, 741–6.

Sosnoskie, L. M. and Culpepper, A. S. (2014), 'Glyphosate-resistant Palmer amaranth (*Amaranthus palmeri*) increases herbicide use, tillage, and hand-weeding in Georgia cotton', *Weed Sci.*, 62, 393–402.

Swanton, C. J. and Murphy, S. D. (1996), 'Weed science beyond the weeds: the role of integrated weed management (IWM) in agroecosystems health', *Weed Sci.*, 44, 437–45.

Tan, S., Evans, R. R., Dahmer, M. L., Singh, B. K. and Shaner, D. L. (2005), 'Imidazolinone-tolerant crops: History, current status, and future', *Pest Manag. Sci.*, 61, 246–57.

USDA, National Agricultural Statistics Service NASS, (2016), 'Economics, Statistics and Market Information System', *Acreage*, http://usda.mannlib.cornell.edu/MannUsda/viewDocumentInfo. do.documentID=1000 (Accessed 30 January 2016).

Vidal, R. A., Trezzi, M. M., De Prado, R., Ruiz-Santaella, J. P. and Vila-Aiub, M. (2007), 'Glyphosate-resistant biotypes of wild poinsettia (*Euphorbia heterophylla* L.) and its risk analysis on glyphosate-tolerant soybeans', *J. Food Agric. Environ.*, 5, 265–9.

Vila-Aiub, M. M., Balbi, M. C., Gundel, P. E., Ghersa, C. M. and Powles, S. B. (2007), 'Evolution of glyphosate-resistant Johnsongrass (*Sorghum halepense*) in glyphosate-resistant soybean', *Weed Sci.*, 55, 566–71.

Chapter 11

Cultural techniques to manage weeds

Matt Liebman, Iowa State University, USA

1 Introduction

Over the last half-century, herbicides have become the dominant tool for weed management in agricultural systems of industrialized nations (Gianessi 2013), as well as the focus of much of the research in weed science (Harker and O'Donovan 2013). Use of herbicides has, in many cases, increased farm profitability, facilitated the adoption of reduced tillage practices that contribute to soil and water conservation, increased farm labour efficiency, and improved farmers' quality of life (Gianessi and Reigner 2007; Gianessi 2013; Zimdahl 2013). Concomitantly, heavy reliance on herbicides has also resulted in cases of environmental contamination and widespread problems with herbicide resistance in weed populations (Liebman et al. 2016).

Integrated weed management (IWM) is seen by many analysts as a useful approach for improving the long-term effectiveness and reliability of weed suppression, while decreasing environmental contamination (Swanton and Wiese 1991; Harker et al. 2012; Shaner 2014; Liebman et al. 2016). Integrated weed management is characterized by the use of sets of farming practices that as a group suppress weed emergence, survival, growth, resource use, and competition against crops. Because IWM spreads the burden of weed suppression and crop protection across multiple tactics, risks of failure can be

http://dx.doi.org/10.19103/AS.2017.0025.13

reduced relative to approaches that rely heavily on only one type of control tactic (Liebman and Gallandt 1997). Additionally, by minimizing the exposure of weed populations to any single control tactic, for example, particular groups of herbicides, rates of weed adaptation and resistance evolution are expected to be lower than for strategies that rely heavily on single tactics (Bottrell and Weil 1995; Owen 2016). While IWM does not exclude the use of herbicides, the development and implementation of IWM systems is contingent on a better understanding of the effects and coordinated use of non-chemical as well as chemical tactics (Harker et al. 2012; Harker and O'Donovan 2013).

Elsewhere in this volume, the authors examine a wide range of farming practices and weed control tactics that can be included in IWM strategies, including cover cropping, intercropping, rotation sequencing, release or conservation of biological control agents, cultivation with specialized machinery, and site-specific herbicide application. The focus of this chapter is on cultural techniques that can also contribute to effective weed management strategies, including choice of crop density, crop arrangement, and crop genotype, and manipulation of initial crop size, soil fertility, and soil moisture conditions. Weed management strategies that make use of cultural factors seek to reduce weed density, resource consumption, biomass production and competition with crops. They also seek to prevent colonization of fields by weed species not previously present. Additionally, by altering the availability of light, water and nutrients in space and time, and by challenging weeds with allelochemicals, cultural tactics are intended to improve crop performance (Liebman and Mohler 2001; Mohler 2001).

As noted earlier, a key feature of IWM strategies is that by combining complementary control tactics, practices that may be individually weak can collectively provide much greater levels of weed suppression (Anderson 2007, 2009; Liebman and Davis 2009). The effects of combinations of tactics may be additive or synergistic. Examples of the consequences of multi-tactic weed management strategies are examined at the end of this chapter.

2 Crop population density

Planting crops at higher densities can increase crop competitiveness against weeds, thereby reducing weed growth, lowering weed seed production and increasing crop yield under weed-infested conditions (Mohler 2001a; Lemerle et al. 2004). This approach is especially well suited to low-input and organic farming systems or conventional systems in which herbicide resistance in weeds is problematic. Ecological theory predicts that as the density of a crop population increases, the proportion of available light, water and nutrients captured by the crop and usurped from associated weeds should increase (Mohler 2001a; Lemerle et al. 2004). These shifts in interspecific competition for resources are also accompanied by shifts in intraspecific competition. To understand these relationships, it is important to distinguish between individual- and population-level responses.

Evaluated at the level of an individual plant, as crop density increases, intraspecific competition among crop plants increases, leading to reductions in individual plant size and reproductive output. However, evaluated at a population level, interspecific competition by the crop against associated weeds also increases with increased crop density, leading to reduced weed growth and, in many cases, higher crop yields per unit area under weedy conditions (Lemerle et al. 2004; Kristensen et al. 2008; Place et al. 2009; Olsen et al.

2012; Lutman et al. 2013; Marín and Weiner 2014). Mohler (2001a) reviewed 91 cases in the literature and found only six failed to show decreasing weediness with increasing crop density; neutral or positive responses for crop yield under weedy conditions were noted for virtually all test cases. Crops for which increased density reduced weed biomass or weed density included barley, bean, cabbage, cotton, cowpea, flax, lentil, maize, oat, pea, peanut, perennial ryegrass, rapeseed, rice, safflower, soybean, sweet potato, timothy and wheat.

Several caveats for the high crop density approach for weed management should be noted. First, though it works well with cereals and pulse crops for which seed size is relatively constant despite variation in numbers of seeds per plant, the high-density approach is not appropriate for all crops, especially vegetables for which crowding-related decreases in size of harvestable units reduce market value (Mohler 2001a). Second, large increases in crop density are likely to result in lodging, disease, and other problems affecting crop yield and quality (Håkansson 2003). Third, the high crop density approach used alone can be insufficient for weed suppression under commercial production conditions. For example, Williams and Boydston (2013) reported that in experiment plots examining interactions between wild proso millet (*Panicum miliaceum*) and sweet corn grown across a wide range of densities, increased crop density led to a taller and thicker crop canopy, less weed biomass and lower weed seed production. However, in comparisons between a crop population currently used by commercial growers and a higher crop population known to optimize yield of certain sweet corn hybrids, there were only small reductions in growth and seed production by the weed. Thus, a combination of increased crop density with other weed suppression tactics is desirable (Lemerle 2004).

3 Crop spatial arrangement

Crop spatial arrangement has been shown to affect weed–crop interactions in some situations. For a given crop population density, narrowing the distance between crop rows increases a crop's uniformity in space. In many experimental studies, narrower row spacing is confounded with higher crop densities, but a considerable number of studies have been conducted in which crop arrangement and density are manipulated independently, allowing for examination of row spacing effects separately.

Mohler (2001a) assessed 48 studies testing the effects of narrower rows at constant crop population density (i.e., more equidistant crop spacing) on weed density or weed biomass production and found weed suppression in 48% of the cases, a neutral effect in 17% of the cases, a positive effect in 2% of the cases, and variable responses in 35% of the cases. Crops included in the review were barley, bean, cotton, flax, lupin, corn, oat, peanut, pearl millet, pigeonpea, rapeseed, safflower, sorghum, soybean, sunflower and wheat. In cases where narrower rows do result in fewer and smaller weeds, the effect is often coincident with increased crop ground cover, leaf area index, dry matter production, and light interception, especially early in the growing season (Mohler 2001a; Harder et al. 2007; Drews et al. 2009). In a study that investigated the effects of variable row spacing for wheat grown with and without nitrogen fertilizer, narrower rows increased weed suppression regardless of fertility level (Kristensen et al. 2008). In contrast, in another study with wheat, narrower rows led to weed suppression in one of three years when rainfall was adequate, but not in two years that were exceptionally dry (Olsen et al., 2012).

For both corn (Marín and Weiner 2014) and wheat (Kristensen et al. 2008; Weiner et al. 2010), increasing crop density while narrowing row spacing can have complementary

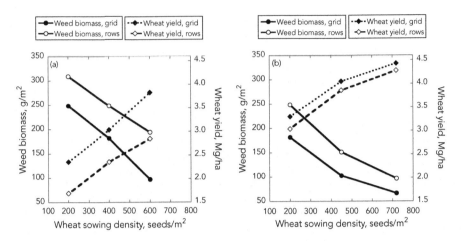

Figure 1 The effects of wheat sowing density and planting pattern on weed biomass production and wheat grain yield under weed-infested conditions in 1998 (a) and 1999 (b). Individual wheat seeds sown in the grid pattern were spaced nearly equidistantly, whereas wheat seeds in the row pattern were sown in rows spaced 12.8 cm apart. The experiment was conducted by Weiner et al. (2001).

effects, with the greatest weed suppression occurring with increased density and greater spatial uniformity. High-density, narrow row arrangements for corn resulted in an average of 65% less weed biomass and 46% greater yield relative to a standard lower-density, wider row arrangement (Marín and Weiner 2014). Sowing wheat at a high density in narrow rows rather than at lower density and in wider rows resulted in large reductions in weed biomass and substantial increases in wheat yield under weed-infested conditions relative to a standard lower-density, wider-row sowing pattern (Weiner et al. 2001, Fig. 1).

The utility of narrow row spacing may be limited for systems in which interrow cultivation is required and row distances interfere with the passage of cultivation equipment. Alternatively, in systems that use herbicides, computer-assisted guidance systems for precision cultivation, or full-width harrows that pass over a crop without damaging it significantly, narrow row spacing may be a useful addition to a grower's portfolio of weed management tactics.

4 Sowing time and transplanting

Weed species have distinctive patterns of release from dormancy, germination and seedling emergence that are driven by interactions between seed physiology and environmental conditions, especially soil temperature and moisture (Forcella et al. 1997; Hartzler et al. 1999; Myers et al. 2004; Werle et al. 2014). Through knowledge of the timing of species-specific pulses of weed emergence, and adjustments in crop planting date and preplanting weed control practices, weed emergence patterns can be exploited to reduce densities of weeds infesting crops.

For warm season crops such as corn and soybean, delaying planting for several weeks can allow large numbers of weed seedlings to emerge and be killed by tillage or herbicides before the crop is sown (Gunsolus 1990; Forcella et al. 1993). For example, in an experiment conducted over a three-year period in Minnesota, mean density of giant foxtail (*Setaria faberi*) in soybean grown with no supplementary weed control was 213 plants m^{-2} for an early-planted treatment, sown on 12–16 May, compared with 40 plants m^{-2} in a late-planted treatment, sown on 2–7 June (Buhler and Gunsolus 1996). It should be noted that while delayed planting may offer opportunities to reduce weed density, it can also lead to reductions in corn and soybean yield potential (Gunsolus 1990).

Delayed planting can also be used to reduce weed densities in cool season crops such as wheat. Lutman et al. (2013) conducted a meta-analysis of 19 studies investigating the effect of wheat planting date on densities of blackgrass (*Alopecurus myosuroides*) and found that delaying planting from September until late October reduced blackgrass densities about 50%. This effect was attributed to the destruction of a greater proportion of weed seedlings prior to sowing the crop, since the weed typically begins to emerge in late summer. In a subset of the studies that compared blackgrass densities in September-sown wheat (i.e. winter wheat) and wheat sown the following spring, densities of the weed were 88% lower for the spring-sown crop. Lutman et al. (2013) noted that while delayed seeding could offer weed management advantages, farmers in the United Kingdom might be reluctant to adopt the practice due to the risk of being prevented from sowing the crop at all due to progressively wetter conditions after September.

Weed emergence prior to planting a crop can be enhanced by using preplanting tillage to expose weed seeds to light and other environmental cues for germination. The false and stale seedbed approaches exploit this phenomenon and have been used in the production of sugar beet, carrot, onion, and other crops. Soil is disturbed by tillage several weeks ahead of crop sowing to promote weed emergence; weeds are then killed by flaming, shallow cultivation or herbicide application, and the crop is sown with minimal further soil disturbance (Mohler 2001b; Rasmussen 2003). Rasmussen et al. (2011) reported that planting 29 days after seedbed preparation and flaming eight days after the crop was sown reduced weed density in sugar beet 89% relative to a treatment that was sown one day after seedbed preparation and flamed 10 days later.

Transplanting crops allows crops to grow under relatively weed-free conditions in seedling beds or greenhouse trays during the early portion of the growing season before being placed in fields where weeds have been removed by tillage, flaming, flooding or herbicides. This 'head-start' in crop size relative to emerging weeds can convey large weed management benefits including reductions in the length of the critical period for weed control, weed biomass production and crop yield loss to weed competition (Weaver et al. 1992). In field experiments conducted in Ontario, Weaver (1984) noted that to prevent yield loss, the minimum weed-free period for transplanted tomato was five weeks as compared with nine weeks for direct-seeded tomato. While not suitable for all crops due to labour costs and other factors, transplanting is a valuable weed management technique for many horticultural crops and for rice in some regions.

5 Choice of crop genotype and breeding for competitive and allelopathic abilities

Within a given crop species, genotypes can differ substantially in their ability to suppress weed growth and sustain yield in the presence of weeds. Collectively, these two phenomena are often called 'crop competitive ability', though allelopathic interactions as well as resource competition can be involved (Bertholdsson 2011; Worthington and Reberg-Horton 2013). Mohler (2001a) noted that crop genotypes with superior weed-suppressive ability or weed tolerance have been reported for at least 25 crop species, including cereals, pulses, forages and vegetables.

As might be expected from studies of other ecological phenomena, a crop genotype's competitive ability can vary substantially among different environments and experiments (Lutman et al. 2013; Andrew et al. 2015; Jacob et al. 2016). An additional complicating factor is that while weed suppression and weed tolerance can be positively correlated, negative correlations or a lack of relation between the two phenomena can also exist (Lemerle et al. 2001; Watson et al. 2006; Colquhoun et al. 2009; Bertholdsson 2010). These complications notwithstanding, interest in developing crop genotypes with improved weed competitive ability is increasing in response to the evolution of herbicide resistance in weeds, environmental concerns associated with herbicide use, and the needs of organic producers and smallholder farmers who eschew or lack access to herbicides (Worthington and Reberg-Horton 2013; Andrew et al. 2015). Christensen et al. (1994), Williams et al. (2008a) and Gealy et al. (2014) have shown that competitive cultivars of wheat, barley, rye, sweet corn and rice can better reduce weed growth and maintain yield when treated with reduced rather than full doses of herbicides.

Crop competitive ability against weeds can be conferred by a number of heritable traits including rapid emergence, rapid early growth, greater numbers of tillers and branches, tall shoots, and greater canopy area and light interception, as well as differences in the size and depth of root systems that affect access to water and nutrients (Lemerle et al. 1996; Watson et al. 2006; Zhao et al. 2006; Williams et al. 2008b; Beckie et al. 2008; Colquhoun et al. 2009; Drews et al. 2009; Andrew et al. 2015). These traits might serve as selection targets in breeding programmes, rather than selecting for competitive ability directly. A complementary approach to breeding for increased resource capture involves breeding for increased allelopathic ability, that is, an enhanced capacity to interfere with weed germination, growth and development via chemicals exuded from crop roots or shoots. Genotypes of rice, wheat, barley, oat, rye and sorghum have been identified with high allelopathic activity against weeds (Olofsdotter et al. 2002; Belz 2007; Seal and Pratley 2010; Bertholdsson 2011; Gealy et al. 2014), and efforts have been initiated to breed high-yielding, weed-suppressive cultivars through traditional techniques, quantitative trait loci mapping and marker-assisted selection (Belz 2007).

Selection of crop genotypes for both enhanced ability to compete with weeds for light, water and nutrients, and increased allelopathic ability has been pursued in research programmes for a number of cereal crops. Weed-suppressive rice cultivars are now commercially available in the United States and China, and weed-suppressive wheat and barley cultivars are being bred for commercial release in Sweden (Worthington and Reberg-Horton 2013).

6 Mulching

Mulch materials applied to the soil surface can suppress the emergence of weed seedlings and consequently reduce weed plant densities, biomass production and competition with crops. Mulch materials can also promote retention of soil moisture and alter soil temperature regimes (Liebman and Mohler 2001). As noted by Grundy and Bond (2007), mulch materials can be sheeted or particulate in form. Examples of the former include black polyethylene; geotextiles; needle-punched fabrics made from natural fibres; various types of paper products; and carpeting. Examples of particulate mulches include shredded and chipped bark or wood; sawdust; crushed rock or gravel; hay, grass clippings, straw, and other crop residues; and various industrial waste materials, such as shredded tyres.

Choice of mulch materials for weed suppression is determined by local availability, cost and management considerations. In general, mulching is used for weed suppression in high value crops, such as tomato (Anzalone et al. 2010), apple (Arentoft et al. 2013) and medicinal herbs (Duppong et al. 2004), for which the investment in materials and labour for application and removal is justified by the savings in total weed control costs and increases in crop quality or yield. Mulching can strongly suppress weed growth (Fig. 2). However, managing weeds that emerge at the edges of sheeted mulches can be problematic, while particulate mulches are generally ineffective against established perennial weeds (Grundy and Bond 2007).

Sheeted mulches can restrict access to light needed to cue weed seed germination, as well as physically obstruct the emergence of weed seedlings (Grundy and Bond 2007). Additionally, polyethylene sheet mulches can be used for soil 'solarization' whereby tarped soil is heated by sunlight well above ambient conditions, and weed seeds and newly germinated seedlings are killed thermally (Liebman and Mohler 2001). Weed seed death due to solarization is also related to changes in soil ethylene and carbon dioxide concentrations, and can be promoted by increasing soil moisture levels and incorporating phytotoxic crop residues prior to tarping (Liebman and Mohler 2001). Unlike other sheet mulching techniques used when crops are growing, solarization is used before or after crop production.

Weed suppression by particulate mulches increases as mulch depth increases (Ozores-Hampton et al. 2001a) and generally requires a mulch depth greater than 7 cm for satisfactory levels of control (Marble 2015). Teasdale and Mohler (2000) investigated seedling emergence of four weed species from beneath a variety of particulate mulch

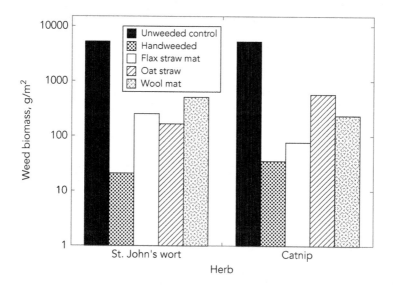

Figure 2 Weed growth in response to different weeding and mulching treatments used for St. John's wort and catnip production in an experiment conducted by Duppong et al. (2004). Means of two consecutive years of observations (2001–2002) on the same plots are shown.

materials and attributed successful emergence to the capacity of seedlings to grow around obstructing mulch elements under limiting light conditions. Large-seeded species were more likely to be able to emerge through mulch layers than small-seeded species. Two-parameter models that included terms for mulch area index (mulch surface area per unit of ground area), and either light extinction by mulch or the fraction of mulch volume that was solid, explained 53–75% of the variation in suppression of seedling emergence, depending on weed species. Particulate mulches made from organic materials such as municipal solid waste, yard trimmings and livestock manure can also suppress weed emergence and growth chemically, through emissions of organic acids, phenolic compounds, ammonia and other materials (Ozores-Hampton et al. 2001b). Care must obviously be exercised to minimize threats to crop growth and development by such materials. Consequently, mulch materials with phytotoxic properties have been proposed for use between, rather than in, crop rows (Ozores-Hampton et al. 2001b).

7 Soil fertility management

The maintenance of soil fertility through the application of mineral fertilizers and/or organic materials, such as plant residues, manure and compost, is a critical component of sustainable cropping systems. Nitrogen, phosphorus and potassium are typically the nutrients that most limit crop productivity in the absence of fertilization. Weeds can also be highly responsive to fertilizer application, though differences exist among species in the degree of response (Blackshaw et al. 2003, 2004a; Blackshaw and Brandt 2008; Storkey et al. 2010; Moreau et al. 2014). When crop and associated weed species differ in their height growth and canopy production responses to soil fertility conditions, large shifts in competitive relations can occur (Liebman and Mohler 2001). Variations in the timing, placement and form of fertility amendments have been shown to be capable of affecting weed population dynamics and crop-weed competition, and thus might serve as components of IWM strategies (Liebman and Mohler 2001).

Blackshaw and colleagues investigated the effects of fall versus spring fertilizer application on crops and weeds in a series of experiments conducted in Alberta and Saskatchewan (Blackshaw et al. 2004b, 2005a, 2005b). Cropping systems included continuous spring wheat, a spring wheat-canola rotation and a barley-field pea rotation. Weed infestations were created intentionally by sowing weed seeds as single species or species mixtures at the inception of the experiments, each of which ran for four years. In general, application of fertilizers (N, P and/or S, depending on the experiment) when crops were sown in the spring (April or May) maintained or increased crop yields, and had a neutral or negative effect on weed biomass production, relative to treatments in which fertilizer was applied the previous fall. After four years, spring rather than fall fertilization lowered weed seed densities in the soil seed bank by 21–24% for the wheat-canola rotation (Blackshaw et al. 2005a), but had no effect on weed seed densities for the barley-field pea rotation (Blackshaw et al. 2005b). For the continuous wheat experiment, in which individual weed species were sown separately, spring fertilization reduced soil seed bank density of wild oat (*Avena fatua*) and common lambsquarters (*Chenopodium album*), but had no effect on seed density of green foxtail (*Setaria viridis*) and wild mustard (*Brassica kaber*) (Blackshaw et al. 2004b). With the observed desirable or neutral, but not negative, effects on crops and weeds, spring fertilization would appear to offer advantages over fall fertilization for spring-sown crops on the Canadian prairies.

Johnson et al. (2007) investigated the effects of delayed fertilizer application on corn and giant ragweed (*Ambrosia trifida*) performance in Indiana, comparing a full dose of N fertilizer (200 kg N ha⁻¹) at planting with a late fertilizer treatment, in which all N was applied at the five- or eight-leaf stage of corn development, and a split fertilizer treatment, in which a half-dose of N was applied at corn planting and another half-dose was applied at the five- or eight-leaf stage. The latter two treatments were investigated as possible strategies to increase corn N use efficiency by better matching crop N demand with N supply. The crop and weed were grown in mixture at fixed densities, and a weed-free corn treatment was also included in the experiment. Compared with at-planting N application, the late and split fertilization treatments increased giant ragweed late season biomass 83% and 42%, respectively. In contrast, corn grain yield was unaffected by N fertilizer timing, and giant ragweed reduced corn yield 19% regardless of N fertilizer timing. Thus, from the perspective of crop yield protection and weed suppression, the at-planting N fertilization strategy was superior. These results and those of Blackshaw et al. (2004b, 2005a, 2005b) indicate that the impacts of fertilization practices may be site-, crop- and weed-specific.

Placement of fertilizer into the soil in bands near crop rows rather than broadcasting it on the soil surface can improve crop performance and constrain weed growth. This has been shown for crops that include bean, soybean, peanut, alfalfa and rice (DiTomaso 1995), and presumably reflects improved access to nutrients by crops and reduced access by weeds growing between crop rows. Rasmussen et al. (1996) compared the effects of fertilizer placement in three years of field trials in Denmark and reported that band application of N into soil 5 cm below rows of barley rather than surface broadcasting increased barley yield an average of 28% and decreased weed biomass an average of 55%. Similarly, in a four-year experiment conducted with spring wheat in Alberta, incorporating N fertilizer into the soil in bands and especially by point injection, rather than surface broadcasting, tended to reduce weed growth (Fig. 3) and weed seed density in soil and increase wheat yields (Blackshaw et al. 2004b).

Nutrients can be supplied to crops through manure applications, but manure may also contain weed seeds (Cudney et al. 1992), leading to concerns over infesting fields

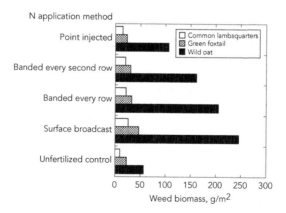

Figure 3 The effects of different methods for nitrogen fertilizer on biomass produced by common lambsquarters (*Chenopodium album*), green foxtail (*Setaria viridis*) and wild oat (*Avena fatua*) grown with spring wheat in an experiment conducted by Blackshaw et al. (2004b). Means of four consecutive years of observations (1998–2001) on the same plots are shown.

with species not previously present, or increasing the densities of resident weed species. However, results of an on-farm study conducted in Wisconsin indicate such threats may be minimized where weed control practices are effective. Cook et al. (2007) measured weed densities on 11 cash grain farms that received manure from neighbouring dairy farms and found that manuring did not introduce new weed species, nor did it increase weed densities; these results were attributed to the high levels of weed control achieved with existing practices used by the farmers. Additionally, threats of introducing or augmenting weed populations through seeds applied in manure can be greatly reduced by composting the manure, which kills seeds thermally and perhaps through their exposure to phytotoxins such as organic acids that are generated during the composting process (Ozores-Hampton et al. 1999; Eghball and Lesoing 2000; Larney and Blackshaw 2003).

Much remains to be learned concerning the effects of composted manure on crop-weed interactions and weed dynamics. In an experiment investigating corn grown in mixture with each of three weed species, application of composted swine manure increased seed production by common waterhemp (Amaranthus rudis) and velvetleaf (Abutilon theophrasti), but had no effect on seed production by giant foxtail and little or no effect on corn grain yield (Liebman et al. 2004). Composted swine manure also increased the competitive effect of common waterhemp on soybean (Menalled et al. 2004). Blackshaw (2005) conducted a four-year fertility regime experiment with wheat and a mixed species weed community, and found that after fertility treatments had been in place for a year, weed N uptake and growth with fresh and composted cattle manure was similar to or greater than that with broadcast N fertilizer. Manure and compost tended to have a greater positive effect on weeds than spring wheat, and at the conclusion of the study, the ranking of weed seed densities in soil was composted manure = fresh manure ≥ broadcast N fertilizer > banded N fertilizer. In contrast, Lindsey et al. (2013) examined composted dairy manure effects on potato grown with three weed species (common lambsquarters, giant foxtail and hairy nightshade (Solanum physalifolium)), and reported that compost did not increase biomass or seed production of any of the weed species, while increasing potato yield 5–15%. In a study of the effects of different fertility amendments on weed seed banks in a fodder beet-winter wheat-cabbage-perennial ryegrass-silage corn rotation sequence, De Cauwer et al. (2011) found that total weed seed bank density was lowest in plots amended with compost and highest in plots amended with liquid cattle manure. Reductions in weed seed densities in soil, especially of hard-coated species such as Chenopodium spp., were correlated with increases in total microbial biomass and soil organic carbon content. Taken together, results of these studies indicate that while manure and compost can have beneficial effects on soil fertility and crop production, effective weed control practices are needed to limit the establishment, growth, and reproduction of species that are stimulated by amendments. More needs to be understood about the effects of different organic amendments on weed seed decay and mortality.

After decomposition in and on the soil, residues of certain legume crops can be important sources of N for succeeding crops, while also influencing weed dynamics and crop-weed interactions (Liebman and Ohno 1998). In field experiments, residues of crimson clover and red clover reduced common lambsquarters and wild mustard density, emergence rate, relative growth rate, biomass production and competitive ability, while enhancing sweet corn growth and yield (Dyck and Liebman 1994; Dyck et al. 1995; Davis and Liebman 2001). Aqueous extracts of crimson clover, hairy vetch and red clover have been shown to be allelopathic under laboratory conditions (White et al. 1989; Liebman and Sundberg 2006); for the latter species, phenolic compounds are believed to be responsible (Ohno et al. 2000). Allelopathic responses can differ among target species, creating the possibility of selective control. Liebman and Sundberg (2006) found that red clover extracts had little or no effect on large-seeded crop species, such as corn, but strongly suppressed the germination and growth of small-seeded weeds, such as common lambsquarters and wild mustard. Weed-suppressive effects of phenolic acids in red clover residues can be enhanced by soil-borne pathogens such as *Pythium* spp., which can attack small-seeded weeds, such as wild mustard, to a greater degree than corn (Conklin et al. 2002).

In low-external-input and organic farming systems, farmers often combine the use of manure, compost and crop residues for soil improvement and enhanced fertility over the long term (Liebman and Davis 2000). Consequently, effects on weed and crop performance in such systems may reflect accumulated changes in soil properties, such as temporal patterns of nutrient release, water-holding capacity, bulk density, and microbial community composition and activity (Gallandt et al. 1999). In a study of weed and potato performance in plots amended with soil-improving crop residues ('green manure'), cattle manure, and cull potato compost versus barley residues and high rates of synthetic fertilizers, Gallandt et al. (1998) found that after the treatments had been in place four years, organic matter amendments had enhanced soil physical properties and fertility, and increased potato canopy production and tuber yield. When herbicides were not applied and cultivation comprised the only direct form of weed control, weed growth in the treatment receiving organic amendments was 75% lower than in the treatment that relied heavily on fertilizers, despite similar weed densities. Gallandt et al. (1998) attributed the latter effect to improvements in soil quality that promoted a more vigorous potato crop that was better able to compete with weeds.

Similar results were obtained by Ryan et al. (2009, 2010), who measured the competitive effects of mixed-species stands of weeds on corn in two contrasting systems that had been in place for 27 years: a diversified organic rotation that received residues of legume green manures and manure versus a simpler, conventionally managed rotation without legume green manures and manure. The investigators found that a given density of weeds and a given amount of weed biomass caused more yield loss for corn in the conventional than the organic system. Greater crop tolerance of weeds in the organic system was attributed to improved soil quality, diversification of nutrient sources, and niche differentiation and resource partitioning between crop and weed species (Ryan et al. 2009, 2010; Smith et al. 2010).

8 Irrigation and flooding: depth, timing and placement

Globally, soil moisture is the main factor limiting crop production in much of the world where rainfall is insufficient to meet crop demand (Steduto et al. 2012). Consequently, various forms of irrigation are widely used to enhance crop production. Water management can also be used to suppress weeds, though the responses of weed individuals and species are affected by the magnitude, timing and location of changes in soil moisture conditions.

Water management is especially important for rice production, the staple crop for about half the world's population and one of the few major crops that is adapted to flooded soil conditions (Rao et al. 2007). Rice can be grown under rain-fed conditions, but most rice production occurs with inundation for at least part of the crop cycle. Weed species in the Poaceae and Cyperaceae dominate the weed floras of rice crops, and many of the weed species present in rice fields are adapted to flooded conditions (Rao et al. 2007).

Due to rising costs of production, in many rice-producing regions there has been a shift in crop establishment practices from manual transplanting of seedlings to direct-seeding. Whereas transplanting creates a size differential between rice seedlings and newly emerging weeds that creates a competitive advantage for the crop, by removing this size difference, direct seeding increases the potential effects of weed competition on the crop. Water management, in concert with other weed management techniques, thus plays an important role in controlling weeds in direct-seeded rice (Rao et al. 2007; Chauhan and Johnson 2011).

Gealy et al. (2014) found that weed competition against rice was much greater with non-flooded furrow-irrigated conditions than with full-field flooding; rice yields were 76% lower in the former conditions than the latter. The investigators identified rice cultivars with high levels of competitive ability and allelopathic activity against weeds as desirable components of IWM strategies for rice. Chauhan and Johnson (2011) compared the effects of times of water application and flooding depth on *Echinochloa crus-galli* in direct-seeded rice and found that maximum reduction in the height and biomass of the weed was achieved when soil was flooded to a depth of 10 cm within two days of sowing the crop; decreasing water depth and delaying flooding increased the weed's growth. Similar results were reported by Williams et al. (1990). Other factors identified by Chauhan and Johnson (2011) as contributing to suppression of *E. crus-galli* included deep burial of seeds (>8 cm) by tillage prior to planting the crop, or alternatively, maintenance of a thick mulch of crop residues from the previous crop. Continual use of flooded conditions for rice production often creates shifts in weed floras towards water-tolerant species (Rao et al. 2007). Consequently, rotations of different crops with rice that create large differences in soil moisture regimes is likely necessary to disturb weed community dynamics and prevent increases in densities of adapted weed species (Williams et al. 1990).

The timing of irrigation water application can be used as a weed-suppression tactic not only in rice but also in non-flooded crops grown in arid conditions. In an experiment evaluating weed management strategies for lettuce production in California, Shem-Tov et al. (2006) compared the use of pre-plant irrigation of raised beds followed 7 or 14 days later by shallow tillage with a no pre-plant irrigation control treatment. Pre-irrigating and then cultivating resulted in the emergence and removal of up to 127 weeds m^{-2} before the crop was sown. The pre-irrigation and pre-plant cultivation treatments also reduced in-crop weed densities (Fig. 4a) and hand-weeding time up to 77% and 50%, respectively. The greatest gains in weed control were obtained by waiting 14 days between irrigation and cultivation (Fig. 4b); this treatment maximized the number of weed seedlings that emerged prior to their removal.

Discrete placement of irrigation water can strongly suppress weed growth while enhancing crop performance and resource use efficiency. Grattan et al. (1988) compared three water management systems for tomato production in California: sprinkler irrigation, which spread water uniformly over a plot; furrow irrigation, which concentrated water between crop rows; and buried drip irrigation, which concentrated water directly beneath crop rows. For each irrigation treatment, weed growth and crop yields were compared in plots not treated with herbicides and in those treated with napropamide and pebulate. All treatments were cultivated and hand weeded for seven weeks after planting the crop. In the absence of herbicides, weed biomass in the sprinkler and furrow irrigation treatments was >17-fold greater than in the buried drip irrigation treatment, with most of the weed growth occurring between crop rows. For the buried drip treatment, weed growth was similar with or without herbicide application and tomato fruit yield was higher than in the

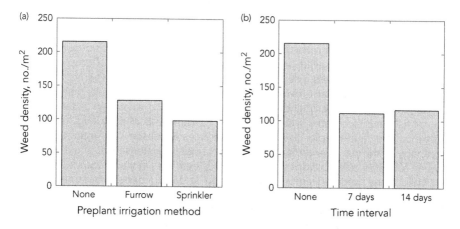

Figure 4 The effects on weed density of preplant irrigation methods (a) and time interval (b) between irrigation and tillage for seedbed preparation in a field experiment with lettuce conducted by Shem Tov et al. (2006). Means of four site-year combinations are shown.

other water management treatments, regardless of herbicide use, reflecting increased water use efficiency.

9 Effects of combining multiple practices: examples of 'many little hammers' at work

Liebman and Gallandt (1997) used the term 'many little hammers' (MLH) to characterize multi-tactic weed management strategies in which individual tactics may be insufficient to provide effective weed suppression, but whose cumulative effects prevent weed population growth and competition against crops. An MLH approach may also be expected to spread the burden of crop protection across a diverse range of stress and mortality factors acting on weeds, thereby reducing selection of weed populations for resistance, limiting shifts in weed community composition towards especially problematic species, and minimizing the risks of weed control failures due to mitigating factors such as weather conditions. Indirectly, by removing some of the burden of crop protection from 'large hammers,' that is, herbicides and soil disturbance through cultivation, MLH approaches may contribute to reductions in herbicide emissions to the environment and to better protection of soil quality. Other analysts of weed management strategies have expressed similar opinions (Anderson 2003; Blackshaw et al. 2008; Chauhan 2012; Mortensen et al. 2012; Norsworthy et al. 2012).

Over the past two decades, a considerable amount of empirical evidence has accumulated to support the MLH hypothesis with regard to the effects of combining cultural tactics for weed suppression. For example, Malik et al. (1993) evaluated the use of different bean cultivars, seeding rates and row spacings, and noted significant effects of all factors on weed biomass production. Two indeterminate, semi-vining bean cultivars suppressed weed biomass production 17–28% relative to a determinate, bush cultivar and use of a narrow-row, high-density planting pattern rather than a traditional wider-row, lower-density pattern reduced weed biomass 19–22%. Under weed-infested conditions, bean seed yields were higher for the two more competitive cultivars grown at higher-than-normal density in narrower rows than for the determinate bush cultivar grown with normal row spacing and density. In an experiment examining the effects of crop row spacing, population density, and herbicide rates, Teasdale (1995) found that when corn was grown with narrow row spacing at a twice-normal population density, reduced herbicide rates (¼ X) provided weed control and corn yield equivalent to what was obtained for corn grown at full herbicide rates with normal row spacing and density (Fig. 5). This effect was attributed to more rapid closure and greater early season light interception by the corn canopy in the narrow-row/high-density treatment. Kristensen et al. (2008) reported that when full rates of herbicides were used, no differences in weed competitiveness and wheat yield were detected among row spacing and crop density treatments, whereas when herbicides were not applied, increasing crop density consistently reduced weed biomass and increased crop yield; the effects of narrowing rows were less consistent, but generally reduced weed biomass and increased wheat yield.

Interactions between cultural practices, cropping system diversity and herbicide use can have important effects on weed dynamics and crop performance. O'Donovan et al. (2013) compared combinations of seeding rates (conventional versus twice-normal) and rotation sequence (barley–canola–barley–field pea–barley vs. continuous barley) grown with three herbicide rates at two sites in Alberta. The diverse rotation combined with the higher barley seeding rate resulted in higher barley yields and reduced wild oat biomass compared to continuous barley grown at a lower seeding rate. Wild oat (*Avena fatua*) seed population density in the soil tended to decline in a manner that paralleled aboveground biomass production of the weed, with up to 40-fold reductions in weed seed numbers observed for barley grown at high density in the diverse rotation with full herbicide rates relative to continuous culture of barley at normal density with ¼ X herbicide rate. At one site, the ¼ X herbicide rate in combination with the diverse rotation and the higher barley seeding rate resulted in less wild oat seed in the soil than the ½ X rate used with

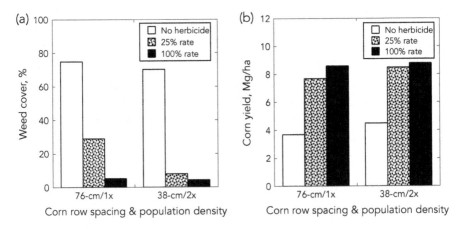

Figure 5 The effects of contrasting corn row spacing and population density combinations and different herbicide rates on weed cover (a) and corn grain yield (b) in a field experiment conducted by Teasdale (1995). The 25% and 100% herbicide rate treatments received a mixture of atrazine, metolachlor and paraquat. Means of four years of observations (1989–1992) are shown.

continuous normal-density barley, suggesting that cultural practices may compensate for suboptimal herbicidal effects with regard to reducing wild oat in the soil seed bank. Similar results were reported from studies of barley-field pea and wheat-canola rotations grown with sets of cultural weed suppression tactics: higher crop density and spring rather than fall fertilizer application tended to reduce weed biomass, increase crop yields and permit reductions in herbicide application while maintaining effective weed control (Blackshaw et al. 2005a, 2005b).

Synergistic effects of combinations of multiple cultural tactics for weed suppression were reported by Anderson (2005), who examined the use of increased crop population density, narrower row spacing and delayed planting for sunflower production, and the use of increased crop density, narrower rows and banded rather than broadcast fertilizer placement for corn production. Relative to conventional crop production practices, the use of any one tactic reduced weed biomass 5–10%, combinations of two tactics reduced weed biomass 20–25%, but the use of three practices reduced weed biomass 60–85%. Synergistic effects of cultural practices were also noted by Ryan et al. (2011), who investigated the effects of combining mulching with higher than normal seeding rates for soybean production. Weed biomass decreased with increasing amounts of mulch composed of rye residue, and also decreased with increasing soybean density in two of four site-years. Also in two of the four site-years, combining rye mulch with higher soybean seeding rates resulted in greater weed suppression than would be predicted by the efficacy of each tactic alone. In practical terms, increasing soybean planting rate was able to compensate for lower rye mulch levels when the tactics were combined.

10 Future trends in research

Evaluating the combined effects of multiple weed suppression tactics in field experiments can be expensive, large in spatial extent, and challenging to manage if each individual tactic and each combination of tactics are included in a factorial design with multiple replications of each treatment. While the empirical data so gained are valuable, models of weed population dynamics in response to various farming practices can be complementary tools with which to design weed management strategies (Holst et al. 2007; Colbach and Mézière 2013). Such models can be especially useful for identifying key points in weed life cycles for interventions (Davis 2006). As computing speeds and power increase, it is likely that future research concerning cultural techniques for weed management will link empirical field studies ('in vivo') with modelling analyses ('in silico'[1]) to examine the main and interactive effects of multiple factors driving weed dynamics (Colbach and Mézière 2013).

Colbach and Mézière (2013) and Colbach et al. (2013) constructed a simulation model to examine the effects of various farming practices and soil and climate conditions on the winter annual weed blackgrass (*Alopecurus myosuroides*), one of the most challenging species to manage in autumn-sown crops of Atlantic European countries. In addition to assessing the effects of tillage practices, herbicides, mechanical weeding operations and crop rotation systems, the investigators evaluated the effects of cultural practices such as crop sowing date, crop density, nitrogen fertilization, manure application, and straw

---

1. An expression used to mean performed on computer or via computer simulation.

burial or removal. In general, soil and weather conditions, initial weed seedbank density, and direct weed control tactics (i.e. herbicides and mechanical weeding) had much larger effects on the number of *A. myosuroides* plants present at crop maturity than did cultural techniques for weed suppression. Nonetheless, delayed planting, increased crop density, straw removal and reduction of viable weed seeds in manure through composting were found to have beneficial effects on weed suppression. Colbach and Mézière (2013) concluded that the major advantage of the modelling approach they used was its ability to reveal the effects of different farming practices across a wide range of weather, soil and weed seedbank conditions. For example, according to the model, the first tillage operations for preparing a seedbed for winter wheat should be delayed until at least 50 mm of rain has fallen since harvest of the preceding crop; this would insure that a high proportion of potential weed recruits germinates and is killed before a winter wheat crop is sown. Colbach et al. (2013) noted that their modelling results highlighted the types of information that should be collected in farmer surveys and field monitoring activities. However, they also noted the need to develop population dynamics models for other weed species with dissimilar life histories.

Liebman and Davis (2009) used a population dynamics model to evaluate multi-tactic strategies for managing the creeping broadleaf perennial weed Canada thistle (*Cirsium arvense*). The model was parameterized with empirical data concerning the weed's life cycle and then used to examine the individual and interactive effects of tillage, mowing, competition from a short-duration cover crop, competition from a multiyear stand of alfalfa and seed predation. Model results indicated that a combination of competition from a cover crop and alfalfa, mowing, and high rates of seed predation resulting from improved seed predator habitat most effectively reduced the weed's rosette survival, seed survival and plant population density. Moreover, model results indicated that when multiple tactics were applied, the need for a high rate of efficacy of any individual tactic to suppress the weed was reduced, supporting the MLH concept. Though this example does not focus exclusively on the use of cultural techniques for weed suppression, it illustrates that modelling may be a fruitful approach for identifying non-chemical management tactics for perennial weeds.

11 Summary

Producing enough food and farm income while protecting environmental quality is one of the critical challenges facing humanity in the twenty-first century. Weeds constitute ubiquitous and recurrent threats in virtually all cropping systems and require careful treatment if farm productivity and profitability are to be sustained.

This chapter has described a diverse set of cultural techniques that can reduce weed population density, biomass production and competition against crops. Compared with the 'large hammers' that modern cropping systems rely upon for weed suppression—herbicides and mechanical cultivation—cultural techniques are 'little hammers' with generally weaker effects on weeds. Nonetheless, when used in particular combinations, the cumulative effects of cultural tactics may be substantial and can lessen the burden of crop protection placed on chemical and mechanical controls. When this occurs, selection pressure for herbicide resistance may be decreased, chemical pollution of air and water by herbicides may be minimized, and soil degradation due to cultivation may be ameliorated. Cultural techniques for weed management are not panaceas for the shortcomings of current strategies, but they could play an increasingly important role in future approaches.

Like many other approaches for improving agricultural sustainability, the use of cultural techniques for weed suppression is based on improved knowledge and decision-making rather than commercial products. Consequently, much of the innovation and refinement for the use of cultural practices is likely to come from public sector researchers and farmers, rather than from industry. It is worth considering that public sector funding for agricultural research in the U.S. has declined over the past quarter century (Wang et al. 2013). If this pattern continues and is representative of the situation globally, further development and provision of information about cultural techniques for weed management could become considerably more difficult in the future.

12 Where to look for further information

The subject of cultural techniques for weed management is not one for which abundant amounts of information are available. Insight into cultural techniques for weed management requires a reasonably well-developed appreciation of weed ecology, including the processes of germination, seedling establishment, resource use, interspecific competition, seed production, vegetative propagation and seed mortality. The more prominent role of weed ecology in informing weed management practices distinguishes cultural techniques from chemical and physical practices. The following list of information sources is by no means exhaustive, but is offered with the intention of pointing the reader in useful directions.

Foundational ecological literature for developing and implementing cultural weed management strategies includes Harper's (2010) *Population Biology of Plants*, first published in 1977; Grime's (2006) *Plant Strategies, Vegetation Processes, and Ecosystem Properties*, first published in 1981; and Grubb's (1977) review of the concept of the 'regeneration niche' in maintaining species richness in plant communities.

Cousens and Mortimer's (1995) *Dynamics of Weed Populations* provides a highly lucid analysis of the intersection between plant ecology, population genetics, agricultural practices and weed management. Håkansson's (2003) *Weeds and Weed Management on Arable Land: An Ecological Approach* offers a large amount of information on weed ecology in the context of agroecosystem management. Radosevich et al.'s (2007) text, *Ecology of Weeds and Invasive Plants: Relationship to Agriculture and Natural Resource Management*, covers similar topics, but includes consideration of weed management from social perspectives. Zimdahl's (2004) book *Weed-Crop Competition: A Review*, originally published in 1980, constitutes a thorough review of competitive interactions in agroecosystems and the different factors mitigating them. Texts with a particular emphasis on cultural weed management practices include Liebman et al.'s (2001) *Ecological Management of Agricultural Weeds*, and Upadhyaya and Blackshaw's (2007) *Non-Chemical Weed Management: Principles, Concepts and Technology*.

Over the past decade, an increasing number of articles concerning cultural weed management has appeared in the journals *Weed Research, Weed Science, Weed Technology*. The European Weed Research Society maintains a working group focused on physical and cultural methods for weed control.

13 References

Anderson, R. L. 2003. An ecological approach to strengthen weed management in the semiarid Great Plains. *Adv. Agron.* 80: 33–62.

Anderson, R. L. 2005. A multi-tactic approach to managing weed population dynamics in crop rotations. *Agron. J.* 97: 1579–83.

Anderson, R. L. 2007. Managing weeds with a dualistic approach of prevention and control. A review. *Agron. Sustain. Dev.* 27: 13–18.

Anderson, R. L. 2009. Impact of preceding crop and cultural practices on rye growth in winter wheat. *Weed Technol.* 23: 564–8.

Andrew, I. K. S., Storkey, J. and Sparkes, D. L. 2015. A review of the potential for competitive cereal cultivars as a tool in integrated weed management. *Weed Res.* 55: 239–48.

Anzalone, A., Cirujeda, A., Aibar, J., Pardo, G. and Zaragoza, C. 2010. Effect of biodegradable mulch materials on weed control in processing tomatoes. *Weed Technol.* 24: 369–77.

Arentoft, B. W., Ali, A., Streibig, J. C. and Andreasen, C. 2013. A new method to evaluate the weed-suppressing effect of mulches: A comparison between spruce bark and cocoa husk mulches. *Weed Res.* 53: 169–75.

Beckie, H. J., Johnson, E. N., Blackshaw, R. E. and Gan, Y. 2008. Weed suppression by canola and mustard cultivars. *Weed Technol.* 22: 182–5.

Belz, R. G. 2007. Allelopathy in crop/weed interactions–an update. *Pest Manag. Sci.* 63: 308–26.

Bertholdsson, N.-O. 2010. Breeding spring wheat for improved allelopathic potential. *Weed Res.* 50: 49–57.

Bertholdsson, N.-O. 2011. Use of multivariate statistics to separate allelopathic and competitive factors influencing weed suppression ability in winter wheat. *Weed Res.* 51: 273–83.

Blackshaw, R. E. 2005. Nitrogen fertilizer, manure, and compost effects on weed growth and competition with spring wheat. *Agron. J.* 97: 1612–21.

Blackshaw, R. E. and Brandt, R. N. 2008. Nitrogen fertilizer rate effects on weed competitiveness are species dependent. *Weed Sci.* 56: 743–7.

Blackshaw, R. E., Brandt, R. N., Janzen, H. H. and Entz, T. 2004a. Weed species response to phosphorus fertilization. *Weed Sci.* 52: 406–12.

Blackshaw, R. E., Brandt, R. N., Janzen, H. H., Entz, T., Grant, C. A. and Derksen, D. A. 2003. Differential response of weed species to added nitrogen. *Weed Sci.* 51: 532–9.

Blackshaw, R. E., Harker, K. N., O'Donovan, J. T., Beckie, H. J. and Smith, E. G. 2008. Ongoing development of integrated weed management systems on the Canadian prairies. *Weed Sci.* 56: 146–50.

Blackshaw, R. E., Molnar, L. J. and Janzen, H. H. 2004b. Nitrogen fertilizer timing and application method affect weed growth and competition with spring wheat. *Weed Sci.* 52: 614–22.

Blackshaw, R. E., Beckie, H. J., Molnar, L. J., Entz, T. and Moyer, J. R. 2005a. Combining agronomic practices and herbicides improves weed management in wheat-canola rotations within zero tillage production systems. *Weed Sci.* 53: 528–35.

Blackshaw, R. E., Moyer, J. R., Harker, K. N. and Clayton, G. W. 2005b. Integration of agronomic practices and herbicides for sustainable weed management in a zero-till barley-field pea rotation. *Weed Technol.* 19: 190–6.

Bottrell, D. G. and Weil, R. R. 1995. Protecting crops and the environment: Striving for durability, pp. 55–73. In Juo, A. S. and Freed, R. D. (Eds), *Agriculture and Environment: Bridging Food Production and Environmental Protection in Developing Countries.* American Society of Agronomy: Madison, WI.

Buhler, D. D. and Gunsolus, J. L. 1996. Effect of date of preplant tillage and planting on weed populations and mechanical weed control in soybean (*Glycine max*). *Weed Sci.* 44: 373–9.

Chauhan, B. S. 2012. Weed ecology and weed management strategies for dry-seeded rice in Asia. *Weed Technol.* 26: 1–13.

Chauhan, B. S. and Johnson, D. E. 2011. Ecological studies on *Echinochloa crus-galli* and the implications for weed management in direct-seeded rice. *Crop Prot.* 30: 1385–91.

Christensen, S. 1994. Crop-weed competition and herbicide performance in cereal varieties and species. *Weed Res.* 34: 29–37.

Colbach, N., Granger, S. and Mézière, D. 2013. Using a sensitivity analysis of a weed dynamics model to develop sustainable cropping systems. II. Long-term effect of past crops and management techniques on weed infestation. *J. Agric. Sci.* 151: 229–45.

Colbach, N. and Mézière, D. 2013. Using a sensitivity analysis of a weed dynamics model to develop sustainable cropping systems. I. Annual interactions between crop management techniques and biophysical field state variables. *J. Agric. Sci.* 151: 229–45.

Colquhoun, J. B., Konieczka, C. M. and Rittmeyer, R. A. 2009. Ability of potato cultivars to tolerate and suppress weeds. *Weed Technol.* 23: 287–91.

Conklin, A. E., Erich, M. S., Liebman, M., Lambert, D., Gallandt, E. R. and Halteman, W. A. 2002. Effects of red clover (*Trifolium pratense*) green manure and compost soil amendments on wild mustard (*Brassica kaber*) growth and incidence of disease. *Plant Soil* 238: 245–56.

Cook, A. R., Posner, J. L. and Baldock, J. O. 2007. Effects of dairy manure and weed management on weed communities in corn on Wisconsin cash-grain farms. *Weed Technol.* 21: 389–95.

Cousens, R. and Mortimer, M. 1995. *Dynamics of Weed Populations.* Cambridge University Press: Cambridge, UK.

Cudney, D. W., Wright, S. D., Schultz, T. A. and Reints, J. S. 1992. Weed seed in dairy manure depends on collection site. *Calif. Agric.* 46(3): 31–2.

Davis, A. S. 2006. When does it make sense to target the weed seed bank? *Weed Sci.* 54: 558–65.

Davis, A. S. and Liebman, M. 2001. Nitrogen source influences wild mustard growth and competitive effect on sweet corn. *Weed Sci.* 49: 558–66.

De Cauwer, B., D'Hose, T., Cougnon, M., Leroy, B., Bulcke, R. and Reheul, D. 2011. Impact of the quality of organic amendments on size and composition of the weed seed bank. *Weed Res.* 51: 250–60.

DiTomaso, J. M. 1995. Approaches for improving crop competitiveness through the manipulation of fertilization strategies. *Weed Sci.* 43: 491–7.

Drews, S., Neuhoff, D. and Köpke, U. 2009. Weed suppression ability of three winter wheat varieties at different row spacings under organic farming conditions. *Weed Res.* 49: 526–33.

Duppong, L. M., Delate, K., Liebman, M., Horton, R., Romero, F., Kraus, G., Petrich, J. and Chowdbury, P. K. 2004. The effect of natural mulches on crop performance, weed suppression and biochemical constituents of catnip and St. John's wort. *Crop Sci.* 44: 861–9.

Dyck, E. and Liebman, M. 1994. Soil fertility management as a factor in weed control: The effect of crimson clover residue, synthetic N fertilizer, and their interaction on emergence and early growth of lambsquarters and sweet corn. *Plant Soil* 167: 227–37.

Dyck, E., Liebman, M. and Erich, M. S. 1995. Crop-weed interference as influenced by a leguminous or synthetic fertilizer nitrogen source: 1. Double-cropping experiments with crimson clover, sweet corn, and lambsquarters. *Agric. Ecosyst. Environ.* 56: 93–108.

Eghball, B. and Lesoing, G. W. 2000. Viability of weed seeds following manure windrow composting. *Compost Sci. Util.* 8: 46–53.

Forcella, F., Eradat-Oskoui, K. and Wagner, S. W. 1993. Application of weed seedbank ecology to low-input weed management. *Ecol. Appl.* 3: 74–88.

Forcella, F., Wilson, R. G., Dekker, J., Kremer, R. J., Cardina, J., Anderson, R. L., Alm, D., Renner, K. A., Harvey, R. G., Clay, S. and Buhler, D. D. 1997. Weed seed bank emergence across the Corn Belt. *Weed Sci.* 45: 67–76.

Gallandt, E. R., Liebman, M., Corson, S., Porter, G. A. and Ullrich, S. D. 1998. Effects of pest and soil management systems on weed dynamics in potato. *Weed Sci.* 46: 238–48.

Gallandt, E. R., Liebman, M. and Huggins, D. R. 1999. Improving soil quality: implications for weed management. *J. Crop Prod.* 2: 95–121.

Gealy, D. R., Anders, M., Watkins, B. and Duke, S. 2014. Crop performance and weed suppression by weed-suppressive rice cultivars in furrow and flood-irrigated systems under reduced herbicide inputs. *Weed Sci.* 62: 303–20.

Gianessi, L. P. 2013. The increasing importance of herbicides in worldwide crop production. *Pest Manag. Sci.* 69: 1099–105.

Gianessi, L. P. and Reigner, N. P. 2007. The value of herbicides in U.S. crop production. *Weed Technol.* 21: 559–66.

Grattan, S. R., Schwankl, L. J. and Lanini, W. T. 1988. Weed control by subsurface drip irrigation. *Calif. Agric.* 42: 22–4.

Grime, J. P. 2006. *Plant Strategies, Vegetation Processes, and Ecosystem Properties*, 2nd edition. Wiley: New York, NY.

Grubb, P. J. 1977. The maintenance of species richness in plant communities: The importance of the regeneration niche. *Biol. Rev.* 52: 107–45.

Grundy, A. C. and Bond, B. 2007. Use of non-living mulches for weed control, pp. 135–53. In: Upadhyaya, M. K. and Blackshaw, R. E. (Eds), *Non-Chemical Weed Management: Principles, Concepts and Technology*. CAB International: Wallingford, Oxfordshire, UK.

Gunsolus, J. L. 1990. Mechanical and cultural weed control in corn and soybeans. *Am. J. Altern. Agric.* 5: 114–19.

Håkansson, S. 2003. *Weeds and Weed Management on Arable Land: An Ecological Approach*. CABI Publishing: Wallingford, UK.

Harder, D. B., Sprague, C. L. and Renner, K. A. 2007. Effect of soybean row width and population on weeds, crop yield, and economic return. *Weed Technol.* 21: 744–52.

Harker, K. N. and O'Donovan, J. T. 2013. Recent weed control, weed management, and integrated weed management. *Weed Technol.* 27: 1–11.

Harker, K. N., O'Donovan, J. T., Blackshaw, R. E., Beckie, H. J., Mallory-Smith, C. and Maxwell, B. D. 2012. Our view. *Weed Sci.* 60: 143–4.

Harper, J. L. 2010. *Population Biology of Plants*. Blackburn Press: Caldwell, NJ.

Hartzler, R. G., Buhler, D. D. and Stoltenberg, D. E. 1999. Emergence characteristics of four annual weed species. *Weed Sci.* 47: 578–84.

Holst, N., Rasmussen, I. A. and Bastiaans, L. 2007. Field weed population dynamics: A review of model approaches and applications. *Weed Res.* 47: 1–14.

Jacob, C. E., Johnson, E. N., Dyck, M. F. and Willenborg, C. J. 2016. Evaluating the competitive ability of semileafless field pea cultivars. *Weed Sci.* 64: 137–45.

Johnson, W. G., Ott, E. J., Gibson, K. D., Nielsen, R. L. and Bauman, T. T. 2007. Influence of nitrogen application timing on low density giant ragweed (*Ambrosia trifida*) interference in corn. *Weed Technol.* 21: 763–7.

Kristensen, L., Olsen, J. and Weiner, J. 2008. Crop density, sowing pattern, and nitrogen fertilization effects on weed suppression and yield in spring wheat. *Weed Sci.* 56: 97–102.

Larney, F. J. and Blackshaw, R. E. 2003. Weed seed viability in composted beef cattle feedlot manure. *J. Environ. Qual.* 32: 1105–13.

Lemerle, D., Cousens, R. D., Gill, G. S., Peltzer, S. J., Moerkerk, M., Murphy, C. E., Collins, D. and Cullis, B. R. 2004. Reliability of higher seeding rates of wheat for increased competitiveness with weeds in low rainfall environments. *J. Agric. Sci.* 142: 395–409.

Lemerle, D., Verbeek, B., Cousens, R. D. and Coombes, N. E. 1996. The potential for selecting spring wheat varieties strongly competitive against weeds. *Weed Res.* 36: 505–13.

Lemerle, D., Verbeek, B. and Orchard, B. 2001. Ranking the ability of wheat varieties to compete with *Lolium rigidum*. *Weed Res.* 41: 197–209.

Liebman, M., Baraibar, B., Buckley, Y., Childs, D., Christensen, S., Cousens, R., Eizenberg, H., Heijting, S., Loddo, D., Merotto Jr., A., Renton, M. and Riemens, M. 2016. Ecologically sustainable weed management: How do we get from proof-of-concept to adoption? *Ecol. Appl.* 26: 1352–69. doi:10.1002/15-0995.

Liebman, M. and Davis, A. S. 2000. Integration of soil, crop, and weed management in low-external-input farming systems. *Weed Res.* 40: 27–47.

Liebman, M. and Davis, A. S. 2009. Managing weeds in organic farming systems: an ecological approach, pp. 173–96. In Francis, C. A. (Ed.), *Organic Farming: The Ecological System*. American Society of Agronomy: Madison, WI.

Liebman, M. and Gallandt, E. R. 1997. Many little hammers: Ecological management of crop-weed interactions, pp. 287–339. In Jackson, L. E. (Ed.), *Ecology in Agriculture*. Academic Press: Orlando, FL.

Liebman, M., Menalled, F. D., Buhler, D. D., Richard, T. L., Sundberg, D. N., Cambardella, C. A. and Kohler, K. A. 2004. Impacts of composted swine manure on weed and corn nutrient uptake, growth, and seed production. *Weed Sci*. 52: 365–75.

Liebman, M. and Mohler, C. L. 2001. Weeds and the soil environment, pp. 210–68. In Liebman, M., Mohler, C. L. and Staver, C. P. (Eds), *Ecological Management of Agricultural Weeds*. Cambridge University Press: Cambridge, UK.

Liebman, M., Mohler, C. L. and Staver, C. P. 2001. *Ecological Management of Agricultural Weeds*. Cambridge University Press: Cambridge, UK.

Liebman, M. and Ohno, T. 1998. Crop rotation and legume residue effects on weed emergence and growth: Applications for weed management, pp. 181–221. In Hatfield, J. L., Buhler, D. D. and Stewart, B. A. (Eds.), *Integrated Weed and Soil Management*. Ann Arbor Press: Chelsea, MI.

Liebman, M. and Sundberg, D. N. 2006. Seed mass affects the susceptibility of weed and crop species to phytotoxins extracted from red clover shoots. *Weed Sci*. 54: 340–5.

Lindsey, A. J., Renner, K. A. and Everman, W. J. 2013. Cured dairy compost influence on weed competition and on 'Snowden' potato yield. *Weed Technol*. 27: 378–88.

Lutman, P. J. W., Moss, S. R., Cook, S. and Welham, S. J. 2013. A review of the effects of crop agronomy on the management of *Alopecurus myosuroides*. *Weed Res*. 53: 299–313.

Malik, V. S., Swanton, C. J. and Michaels, T. E. 1993. Interaction of white bean (*Phaseolus vulgaris* L.) cultivars, row spacing, and seed density with annual weeds. *Weed Sci*. 41: 62–8.

Marble, S. C. 2015. Herbicide and mulch interactions: A review of the literature and implications for the landscape maintenance industry. *Weed Technol*. 29: 341–9.

Marín, C. and Weiner, J. 2014. Effects of density and sowing pattern on weed suppression and grain yield in three varieties of maize under high weed pressure. *Weed Res*. 54: 467–74.

Menalled, F. D., Liebman, M. and Buhler, D. D. 2004. Impact of composted swine manure and tillage on common waterhemp (*Amaranthus rudis*) competition with soybean. *Weed Sci*. 52: 605–13.

Mohler, C. L. 2001a. Enhancing the competitive ability of crops, pp. 269–321. In Liebman, M., Mohler, C. L. and Staver, C. P. (Eds), *Ecological Management of Agricultural Weeds*. Cambridge University Press: Cambridge, UK.

Mohler, C. L. 2001b. Mechanical management of weeds, pp. 139–209. In Liebman, M., Mohler, C. L. and Staver, C. P. (Eds), *Ecological Management of Agricultural Weeds*. Cambridge University Press: Cambridge, UK.

Moreau, D., Busset, H., Matejicek, A. and Munier-Jolain, N. 2014. The ecophysiological determinants of nitrophily in annual weed species. *Weed Res*. 54: 335–46.

Mortensen, D. A., Egan, J. F., Maxwell, B. D., Ryan, M. R. and Smith, R. G. 2012. Navigating a critical juncture for sustainable weed management. *Bioscience* 62: 75–84.

Myers, M. W., Curran, W. S., VanGessel, M. J., Calvin, D. D., Mortensen, D. A., Majek, B. A., Karsten, H.D. and Roth, G. W. 2004. Predicting weed emergence for eight annual species in the northeastern United States. *Weed Sci*. 52: 913–19.

Norsworthy, J. K., Ward, S. M., Shaw, D. R., Llewellyn, R. S., Nichols, R. L., Webster, T. M., Bradley, K. W., Friswold, G., Powles, S. B., Burgos, N. R., Witt, W. W. and Barrett, M. 2012. Reducing the risks of herbicide resistance: best management practices and recommendations. *Weed Sci*. 60 (special issue): 31–62.

O'Donovan, J. T., Harker, K. N., Turkington, T. K. and Clayton, G. W. 2013. Combining cultural practices with herbicides reduces wild oat (*Avena fatua*) seed in the soil seed bank and improves barley yield. *Weed Sci*. 61: 328–33.

Ohno, T., Doolan, K., Zibilske, L. M., Liebman, M., Gallandt, E. R. and Berube, C. 2000. Phytotoxic effects of red clover amended soils on wild mustard seedling growth. *Agric. Ecosys. Environ.* 78: 187–92.

Olsen, J. M., Griepentrog, H.-W., Nielsen, J. and Wiener, J. 2012. How important are crop spatial pattern and density for weed suppression by spring wheat? *Weed Sci.* 60: 501–9.

Olofsdotter, M., Jensen, L. B. and Courtois, B. 2002. Improving crop competitive ability using allelopathy–an example from rice. *Plant Breed.* 121: 1–9.

Owen, M. D. K. 2016. Diverse approaches to herbicide-resistant weed management. *Weed Sci.* 64 (special issue): 570–84.

Ozores-Hampton, M., Obreza, T. A. and Stoffella, P. J. 2001a. Mulching with composted municipal solid waste (MSW) for biological control of weeds in vegetable crops. *Compost Sci. Util.* 9: 352–60.

Ozores-Hampton, M., Obreza, T. A. and Stoffella, P. J. 2001b. Weed control in vegetable crops with composted organic mulches, pp. 275–86. In Stoffella, P. J. and Kahn, B. A. (Eds), *Compost Utilization in Horticultural Cropping Systems*. CRC Press: Boca Raton, FL.

Ozores-Hampton, M., Stoffella, P. J., Bewick, T. A., Cantliffe, D. J. and Obreza, T. A. 1999. Effect of age of co-composted MSW and biosolids on weed seed germination. *Compost Sci. Util.* 7: 51–7.

Place, G. T., Reberg-Horton, S. C., Dunphy, J. E. and Smith, A. N. 2009. Seed rate effects on weed control and yield for organic soybean production. *Weed Technol.* 23: 497–502.

Radosevich, S. R., Holt, J. S. and Ghersa, C. 2007. *Ecology of Weeds and Invasive Plants: Relationship to Agriculture and Natural Resource Management*. Wiley: Hoboken, NJ.

Rao, A. N., Johnson, D. E., Sivaprasad, B., Ladha, J. K. and Mortimer, A. M. 2007. Weed management in direct-seeded rice. *Adv. Agron.* 93: 153–255.

Rasmussen, J. 2003. Punch planting flame weeding and stale seedbed for weed control in row crops. *Weed Res.* 43: 393–403.

Rasmussen, J., Henriksen, C. B., Gripentrog, H. W. and Nielsen, J. 2011. Punch planting, flame weeding and delayed sowing to reduce intra-row weeds in row crops. *Weed Res.* 51: 489–98.

Rasmussen, K., Rasmussen, J. and Petersen, J. 1996. Effects of fertiliser placement on weeds in weed harrowed spring barley. *Acta Agric. Scand. Sect. B S. P.* 46: 192–6.

Ryan, M. R., Mirsky, S. B., Mortensen, D. A., Teasdale, J. R. and Curran, W. S. 2011. Potential synergistic effects of cereal rye biomass and soybean planting density on weed suppression. *Weed Sci.* 59: 238–46.

Ryan, M. R., Mortensen, D. A., Bastiaans, L., Teasdale, J. R., Mirsky, S. B., Curran, W. S., Seidel, R., Wilson, D. O. and Hepperly, P. R. 2010. Elucidating the apparent maize tolerance to weed competition in long-term organically managed systems. *Weed Res.* 50: 25–36.

Ryan, M. R., Smith, R. G., Mortensen, D. A., Teasdale, J. R., Curran, W. S., Seidel, R. and Shumway, D. L. 2009. Weed-crop competition relationships differ between organic and conventional cropping systems. *Weed Res.* 49: 572–80.

Seal, A. N. and Pratley, J. E. 2010. The specificity of allelopathy in rice (*Oryza sativa*). *Weed Res.* 50: 303–11.

Shaner, D. L. 2014. Lessons learned from the history of herbicide resistance. *Weed Sci.* 62: 427–31.

Shem-Tov, S., Fennimore, S. A. and Lanini, W. T. 2006. Weed management in lettuce (*Lactuca sativa*) with preplant irrigation. *Weed Technol.* 20: 1058–65.

Smith, R. G., Mortensen, D. A. and Ryan, M. R. 2010. A new hypothesis for the functional role of diversity in mediating resource pools and weed-crop competition in agroecosystems. *Weed Res.* 50: 37–48.

Steduto, P., Hsiao, T. C., Fereres, E. and Raes, D. 2012. *Crop Yield Response to Water*. FAO Irrigation and Drainage Paper 66. Food and Agriculture Organization of the United Nations: Rome, Italy.

Storkey, J., Moss, S. R. and Cussans, J. W. 2010. Using assembly theory to explain changes in a weed flora in response to agricultural intensification. *Weed Sci.* 58: 39–46.

Swanton, C. J. and Wiese, S. F. 1991. Integrated weed management: the rationale and approach. *Weed Technol.* 5: 657–63.

Teasdale, J. R. 1995. Influence of narrow row/high population density corn (*Zea mays*) on weed control and light transmittance. *Weed Technol.* 9: 113–18.

Teasdale, J. R. and Mohler, C. L. 2000. The quantitative relationship between weed emergence and the physical properties of mulches. *Weed Sci.* 48: 385–92.

Upadhyaya, M. K. and Blackshaw, R. E. (Eds), 2007. *Non-Chemical Weed Management: Principles, Concepts and Technology*. CABI Publishing: Wallingford, UK.

Wang, S. L., Heisey, P. W., Huffman, W. E. and Fuglie, K. E. 2013. Public R & D, private R & D, and U.S. agricultural productivity growth: Dynamic and long-run relationships. *Am. J. Agric. Econ.* 95, 1287–93. doi:10.1093/ajae/aat032.

Watson, P. R., Derksen, D. A. and Van Acker, R. C. 2006. The ability of 29 barley cultivars to compete and withstand competition. *Weed Sci.* 54: 783–92.

Weaver, S. E. 1984. Critical period of weed competition in three vegetable crops in relation to management practices. *Weed Res.* 24: 317–25.

Weaver, S. E, Kropff, M. J. and Groeneveld, R. M. W. 1992. Use of ecophysiological models for crop-weed interference: The critical period of weed interference. *Weed Sci.* 40: 302–7.

Weiner, J., Andersen, S. B., Wille, W. K.-M., Griepentrog, H. W. and Olsen, J. M. 2010. Evolutionary agroecology: The potential for cooperative, high density, weed-suppressing cereals. *Evol. Appl.* 3: 473–9.

Weiner, J., Griepentrog, H.-W. and Kristensen, L. 2001. Suppression of weeds by spring wheat *Triticum aestivum* increases with crop density and spatial uniformity. *J. Appl. Ecol.* 38: 784–90.

Werle, R., Sandell, L. D., Buhler, D. D., Hartzler, R. G. and Lindquist, J. L. 2014. Predicting emergence of 23 summer annual weed species. *Weed Sci.* 62: 267–79.

White, R. H., Worsham, A. D. and Blum, U. 1989. Allelopathic potential of legume debris and aqueous extracts. *Weed Sci.* 37: 674–9.

Williams, J. F., Roberts, S. S., Hill, J. E., Scardaci, S. C. and Tibbits, G. 1990. Managing water for weed control in rice. *Calif. Agric.* 44(5): 7–10.

Williams II, M. M. and Boydston, R. A. 2013. Crop seeding level: Implications for weed management in sweet corn. *Weed Sci.* 61: 437–42.

Williams II, M. M., Boydston, R. A. and Davis, A. S. 2008a. Crop competitive ability contributes to herbicide performance in sweet corn. *Weed Res.* 48: 58–67.

Williams II, M. M., Boydston, R. A. and Davis, A. S. 2008b. Differential tolerance in sweet corn to wild-proso millet (*Panicum miliaceum*) interference. *Weed Sci.* 56: 58–67.

Worthington, M. and Reberg-Horton, S. C. 2013. Breeding cereal crops for enhanced weed suppression: Optimizing allelopathy and competitive ability. *J. Chem. Ecol.* 39: 213–31.

Zhao, D. L., Atlin, G. N., Bastiaans, L. and Spiertz, J. H. J. 2006. Comparing rice germplasm groups for growth, grain yield and weed-suppressive ability under aerobic soil conditions. *Weed Res.* 46: 444–52.

Zimdahl, R. L. 2004. *Weed-Crop Competition: A Review*, 2nd edition. Wiley: Oxford, UK.

Zimdahl, R. L. 2013. *Fundamentals of Weed Science*, 4th edition. Elsevier/Academic Press: Amsterdam, The Netherlands.

Chapter 12

The use of rotations and cover crops to manage weeds

John R. Teasdale, USDA-ARS, USA

1 Introduction

Crop rotation defines the sequence of crops grown on a given site. Agriculturalists have understood the positive benefits of rotating crops for centuries and, even in this age of intensive crop production, farmers usually recognize the need for more than one crop species in their cropping sequences. With the recognition that agricultural production must achieve a sustainable balance of high production, soil stewardship and environmental protection, rotational optimization has become a high priority. Long-term sustainability can be achieved only by understanding the cumulative effects of soil and crop management practices and how they are best managed to maintain the productivity and health of agricultural land.

Weedy plants are colonizing species that have the capacity to multiply and persist in the perpetually disturbed habitats that define agricultural soils. Much of the history of weed control involves the trade-off between creating optimum conditions for crops and performing practices that are unfavourable for weeds, and that also may be unfavourable to crops. In addition, practices used to manage weeds in a specific crop will usually be selective for other weedy species with traits adapted to avoid or tolerate those practices. The use of crop rotations is, therefore, an important tool to prevent the continued adaptation of weed communities to practices associated with any given crop (Mohler 2001).

Comprehensive reviews of crop rotation strategies for weed management have been written by Liebman and Dyck (1993) and Liebman and Staver (2001). As described in these reviews, the weed management effects of crop rotations do not simply result from the biological traits of the alternating crop species, but are the result of the suite of production practices associated with each crop including tillage, the timing of planting and harvesting, fertility sources and specific weed management practices. Consequently, an understanding of crop rotation effects on weed populations necessarily involves an understanding of the performance of individual production and weed management practices employed with each crop. This chapter will focus primarily on the literature that has been published since the reviews cited above.

Many design approaches have been used for the study of crop rotations. From a broad perspective, the experimental design of rotation research has mirrored the changing interests of weed scientists from a chemically based focus to inclusion of ecologically based management approaches. Interest in rotation often occurs because of inadequate control by current practices as practitioners search for alternative measures to reinforce management practices. Introduction of reduced tillage practices led to an interest in rotation as a means of controlling weeds that were poorly or simply not controlled by herbicides in the absence of tillage for seedbed preparation. Interest in the interactions between rotations and reduced herbicide rates was a response to the need to mitigate environmental contamination of surface- and groundwater by soil-active herbicides. Current interest in controlling the spread of herbicide-resistant traits in weed populations has led to a resurgence of interest in rotations, as well as other integrated weed management options. And finally, interest in rotations as part of ecologically based cropping systems, such as organic farming, has focused on the integration of soil, fertility and pest management for sustainable crop production without the benefit of herbicides. Success of these systems relies on potential synergies among complex biological interactions, however, they can be more difficult to study experimentally because of the confounded effects of multiple contributing factors and the inadequacies of traditional factorial designs to address these interactions.

The selection of metrics for measuring weed responses and analytical approaches for evaluating weed management have also changed. Earlier research focused on the density and biomass of the emerged above-ground weed community within each crop or selected test crops. During the past two decades, interest has grown in analysing the weed seedbank as potentially representative of the cumulative legacy of cropping system effects on weed populations over time. There also has been a shift towards multivariate analysis and use of various diversity indices as a means for comprehensively understanding changes in the structure of weed communities in response to cropping and management influences. Focus has shifted from a simple description of the dynamics of dominant problematic species to a comprehensive analysis of weed traits that define the assembly of weed communities within cropping systems. Thus, the current literature describing the effects of crop rotation on weed communities includes a diverse array of designs and metrics depending on the interests and objectives of the research.

2 Crop rotation in weed management

2.1 Weed suppression by crop rotation

Several reports have compared the effect of rotation as a sole factor or in a factorial design with other management factors including tillage and herbicide intensity. The standard rotation is usually a continuous monoculture of an important grain or oilseed crop in the region. The alternative rotations usually include a more diverse set of two or more crops. Table 1 summarizes reports where rotations affected a significant comprehensive change in weed abundance and/or community structure in comparison with the standard rotation. This table focuses on abundance and structure of the overall weed community, whereas effects on individual weed species and traits will be addressed in Sections 2.5 and 2.6.

In most instances in Table 1, weed abundance was reduced, weed community structure was changed and weed diversity was increased by longer, more complex, rotations. The alternative rotations often included a forage crop in the rotation which would form a sod of vegetation and be harvested by intermittent mowing, thus creating a substantially different competitive and disruptive environment for weeds compared to annual row crops. This would be expected to reduce the most abundant weed species in the monoculture crops, which in turn would provide for the proliferation of other weed species that are better adapted to the forage crop environment. In this case, the structure of the weed community would be shifted and the overall diversity and evenness of the community increased. This is the result most commonly reported in Table 1. For example, forage species introduced into cereal rotations in Quebec (Légère and Stevenson 2002), Manitoba (Gulden et al. 2011) and Alberta (Blackshaw et al. 2015), as well as meadow and forage species introduced into vegetable rotations in England (Eyre et al. 2011) and into corn rotations in Italy (Tomasoni et al. 2003) and Nigeria (Chikoye et al. 2008), reduced at least one measure of weed abundance and affected at least one measure of community structure. One exception was observed in Ohio, where a corn–oat–hay rotation increased the weed seedbank because weeds proliferated in the oat–hay phase of the rotation where no herbicides were used (Sosnoski et al. 2006). This emphasizes the point that alternative crops with poor weed control attributes can reduce, rather than enhance, overall weed control.

Alternative crop rotations that introduced cereal crops with enhanced capacity to suppress weeds were also effective in reducing overall weed abundance and changing community structure. A competitive grain crop such as barley added to vegetable rotations in Quebec (Benoit et al. 2003) and England (Eyre et al. 2011) reduced the weed seedbank and weed cover, respectively (Table 1). Wheat alternated every year or every second year reduced weed density and increased weed diversity in canola (Cathcart et al. 2006). The presence of a cereal crop in rotation with corn and/or soya bean reduced weed density in a corn uniformity test crop at the conclusion of a rotation experiment in Ontario (Doucet et al. 1999). In central South Dakota, alternating warm season crops with cool season crops reduced weed density compared to a winter wheat-fallow standard rotation (Anderson and Beck 2007). However, in western South Dakota, moisture limitations precluded the addition of crops that would deplete soil moisture and thereby reduce the competitiveness of the succeeding winter wheat crop with weeds (Anderson et al. 2007).

Several experiments have reported where crop rotation had no or minimal overall influence on weed communities. These were because of little difference between crops included in the rotation (Barberi and Lo Cascio 2001; Stevenson and Johnston 1999; Shrestha et al. 2002), the short duration of the experiment (Gulden et al. 2010) or other treatment factors such as herbicide (Brainard et al. 2008; Doucet et al. 1999; Gulden et al. 2009) or tillage (Barberi and Lo Cascio 2001; Murphy et al. 2006; Thorne et al. 2007) that had a relatively stronger influence on weeds. It is interesting that a comparison of three highly diverse six-year rotations differing in grassland components showed little difference in overall rotational effect on weed communities after more than 25 years in Sweden, despite there being clear associations between weed species and individual crops

Table 1 Crop rotational impact on various weed metrics. Weed metrics included above-ground total weed density (D), total weed biomass or cover (B), total soil seedbank (S), community multivariate structure (C), species richness (R), species diversity (Di) and species evenness (E). Symbol indicates whether the metric change relative to the standard rotation was an increase (↑), no change (-) or a decrease (↓). A significant multivariate influence on the weed community structure is designated by 'C+'. Where data from multiple years were presented in a reference, an estimate of the most consistent trend is reported in this table

Location	Conditions	Standard rotation	Rotations with similar weed metrics	Rotations with changed weed metrics	Weed metrics	References
Ohio	Three tillage treatments	continuous Corn (Co)	Co-Soya bean (Sb)	Co-Oat-Hay	S↑,C+,R↑,Di↑,E↑	Snosnoski et al. 2006 Snosnoski et al. 2009
South Dakota (Central)	Standard	Winter wheat (WW)-Fallow (Fa)	WW-Co-Cowpea (Cp)	WW-Co-Fa Pea-WW-Co Pea-WW-Co-Sb	D↓ D↓ D↓	Anderson and Beck 2007
South Dakota (Western)	Standard	WW-Fa	Spring wheat (SW) -WW-Co-Sunflower	WW-Millet (M) WW-Safflower (Sa)-M WW-WW-Sa-Fa	D↑ D↑ D↑	Anderson et al. 2007
Quebec	No-till, Low input	continuous Barley (Ba)	-	Ba-Red clover	B↓,C+ S-,C+,R-,Di-,E-	Légère and Stevenson 2002 Légère et al. 2005
Quebec	No-till	continuous Ba		Ba-Canola-SW-Sb	S↓,C+,R-,Di-,E-	Légère et al. 2011
Quebec	Standard	continuous Carrot	Carrot-Onion-Carrot	Ba-Carrot-Carrot Ba-Onion-Carrot	D-,S↓ D-,S↓	Benoit et al. 2003
Ontario	Plus & minus herbicide	Co and/or Sb	-	Co and/or Sb plus cereal	D↓	Doucet et al. 1999
Manitoba	Standard	Spring cereal after cereal crop	-	Spring cereal after Alfalfa	C+	Ominski et al. 1999
Manitoba	No herbicide	SW-Pea	Spring triticale-Pea Alfalfa-Pea Fallow-Pea	Winter triticale-Pea Sorghum/sudangrass-Pea	C+ C+	Schoofs and Entz 2000
Manitoba	No-tillage	Flax-Oat -Canola-SW	-	Flax-Oat-Alfalfa-Alfalfa	S↓,C+,R-,Di-	Gulden et al. 2011

Location	System	Monoculture	Rotation	Crop sequence	Response	Reference
Alberta	Standard	continuous WW	-	WW-Canola WW-Flax WW-Fallow	D↓,C+ D↓,C+ D↓,C+	Blackshaw et al. 2001
Alberta	Conservation tillage & inputs	continuous SW	-	Bean (Bn)-SW-Potato (Po) Bn-Po-SW-Sugar beet (Be) Bn-Po-SW-Be-SW Bn-Po-Oat-Hay-Hay-Be	D↓,S↓,D↑ D↓,S↓,D↑ D↓,S↓,D↑ D↓,S↓,D↑	Blackshaw et al. 2015
Alberta	Minimum tillage	continuous Canola (Ca)	-	Ca-SW-Ca-SW-Ca Ca-Ca-SW-SW-Ca Ca-SW-SW-SW-Ca	D↓,D↑,E↑ D↓,D↑,E↑ D-,D↑,E↑	Cathcart et al. 2006
England	Conventional and Organic	Vegetables as current or preceding crop	-	Grass/clover or Winter barley as current or preceding crop	B↓,C+ B↓,C+	Eyre et al. 2011
France	Standard	Annual crops preceding wheat	-	Alfalfa Alfalfa preceding wheat	C+,R-,Di-	Meiss et al. 2010
Italy	High and Low inputs	continuous Co	Silage (Si)-Si-Co	Si-Si-Si Si-Si-Si-Meadow/3 yr Meadow/6 yr	S↓,R- S↓,R↑ S↓,R-	Tomasoni et al. 2003
Italy	Plough and No-tillage	continuous WW	-	WW-Faba bean WW-Berseem clover	S↓,C+,R-,Di↑,E↑ S↓,C+,R-,Di↑,E↑	Ruisi et al. 2015
Nigeria	Standard	continuous Co	Fa-Fa-Co	Sb-Sb-Co Sb/Cp-Sb/Cp-Cg Green manure/2 yr - Co Forage legume/2 yr - Co	D-,B↓,C+ D-,B↓,C+ D↓,B↓,C+ D↓,B↓,C+	Chikoye et al. 2008

within the rotations (Andersson and Milberg 1998). Given that no overall weed problems developed in these rotations, the authors suggest that 'a varied rotation seems to be of importance, but the exact type of rotation seems to be less important.' This implies that a specifically programmed sequence of species may be less important than the simple maintenance of a highly diverse rotation.

2.2 Soil quality and weed suppression

Improvement in soil quality and fertility is one of the prime objectives in the design of crop rotations. Often legumes are introduced to fix nitrogen and winter cereals or perennial hay/pasture species are introduced to prevent soil erosion and nutrient losses. Since these rotational crops are also influential in reducing weed populations, opportunities for simultaneously improving soil quality and suppressing weed populations have been implemented in the design of crop rotations, particularly for sustainable and organic systems (Liebman and Staver 2001). Several experiments described in Table 2 contain rotations designed to simultaneously enhance soil fertility and weed suppression by using diversified rotations with the objective of maintaining similar crop performance with reduced external inputs compared to a less diverse conventionally managed cropping system. Although external inputs are confounded with rotational diversity and many other management details (as opposed to including all factor levels in a traditional factorial design), these system experiments are focused on demonstrating the capacity of optimally managed integrated systems to obtain similar yields and weed populations as conventional systems.

Diversification of a corn–soya bean rotation with a winter small grain and alfalfa was investigated in Iowa for promotion of ecosystem services that would supplement, and eventually displace, synthetic external inputs used to maintain crop productivity (Davis et al. 2012). Grain yields and profit in the more diverse systems were similar to, or greater than, those in the conventional system, despite substantial reductions of fertilizer and herbicide inputs (Fig. 1). Weeds were suppressed effectively in all systems, while the toxicity potential to freshwater organisms of the more diverse systems was two orders of magnitude lower than that of the conventional system. Reduced emergence of selected weed species in these low input systems may have resulted from reduced fertilizer inputs at planting and from increased weed seed mortality from seed predation during an extended period without tillage in the diverse rotation (Heggenstaller and Liebman 2006; Liebman et al. 2014; Westerman et al. 2005). A diverse four-year rotation in North Dakota with a low-cost mulch tillage system and reduced herbicide rates had similar emerged weed populations and rates of seedbank decline as did a shorter two-year wheat–soya bean rotation with full tillage and herbicide inputs over an eight-year period (Ramsdale et al. 2006). Likewise, a diversified six-year rotation with conservation practices (including reduced tillage, cover crops and manure) had fewer emerged weeds and lower seedbank populations than continuous conventional spring wheat after 12 years in Alberta (Blackshaw et al. 2015). In a comprehensive analysis of reduced herbicide cropping systems across Canada for controlling *Avena fatua*, Harker et al. (2016) concluded that, 'Management systems that effectively combine diverse and optimal cultural practices against weeds, and limit herbicide use, reduce selection pressure for weed resistance to herbicides and prolong the utility of threatened herbicide tools' (p. 170).

In the complete absence of herbicides (Table 2), diverse rotational and fertility systems were not sufficient to control weeds as well as in conventional systems (Gallandt et al. 1998;

Table 2 Impact of alternative rotational systems on various weed metrics, where alternative rotations were confounded with reduced chemical inputs and enhanced conservation and biologically based services relative to a standard system. Weed metrics included above-ground total weed density (D), total weed biomass or cover (B), total soil seedbank (S), community multivariate structure (C), species richness (R), species diversity (Di) and crop yield loss to weeds (YL). Where data from multiple years were presented in a reference, an estimate of the most consistent trend is reported in this table

Location	Standard rotation and management	Alternative rotation and management	Weed metrics	Reference
Alternative systems have some herbicide inputs				
Iowa	Corn (C)-Soya bean (S) 1X fertilizer 1X herbicide	Corn-Soya bean-Triticale-Alfalfa 0.14X fertilizer 0.11X herbicide	B: similar in C & S S: similar decline in both	Davis et al. 2012
North Dakota	Wheat-soya bean Conventional tillage 1X herbicide	Wheat-Sweet clover-Rye-Soya bean Mulch tillage Reduced herbicide	D: similar S: similar decline in both	Ramsdale et al. 2006
Alberta	3–4-year rotation Conventional tillage No cover crops Fertilizer Herbicide	6-year rotation Reduced tillage Plus cover crops Composted manure + fertilizer Herbicide	D: similar S: similar Di: higher	Blackshaw et al. 2015
Canada	Canola (Ca)-Wheat No tillage 1X herbicide, 1X seeding rate	Ca-Barley-Winter wheat-Barley-Ca or Ca-Alfalfa-Alfalfa-Alfalfa-Ca Reduced herbicide, 2X seeding rate	D: similar B: similar S: similar	Harker et al. 2016
Alternative systems have no herbicide inputs				
Maine	Potato-Barley/Red clover 1X fertilizer No amendments Herbicide	Potato-Pea/Oat/Vetch 0.5X fertilizer Manure/compost Cultivation	D: higher B: higher YL: lower	Gallandt et al. 1998
Nebraska	Corn/Sorghum-Soya bean Conventional fertilizer Herbicide	Alfalfa-Alfalfa-Corn/Sorghum-Wheat Organic fertilizer Cultivation & Rotary hoe	B: higher S: higher R: higher Di: higher	Wortman et al. 2010
Germany	Winter cereal & oilseed rape Conventional fertilizers Herbicides	Winter and spring crops, grass/clover Organic fertilizers Cultivation only	B: higher R: higher C: different YL: lower	Ulber et al. 2009

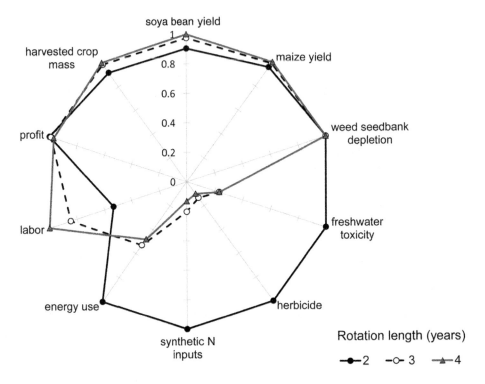

Figure 1 Multiple indicators of cropping system performance of corn–soya bean (2 yr), corn–soya bean–small grain/red clover (3 yr) and corn–soya bean–small grain–alfalfa (4 yr) cropping systems in Boone, IA, averaged over the 2003–2011 study period. Variable means are normalized on a 0–1 scale, with 1 representing the cropping system with the largest absolute value for that variable (N = 36 per cropping system). From Davis et al. (2012), doi:10.1371/journ.

Ulber et al. 2009; Wortman et al. 2010). But several experiments showed that enhancing soil fertility through diversified crop rotations and organic amendments may aid in reducing crop yield loss rate per unit weed biomass (Anderson 2011; Gallandt et al. 1998; Ulber et al. 2009). Indeed, determination of crop yield loss either by direct measurement within a single systems experiment (Ryan et al. 2010) or by meta-analysis of several long-term experiments (Smith et al. 2009b) has shown that the yield loss rate to weeds can be reduced in more diverse rotations with organic amendments compared to simpler rotations with external inputs. Regardless, enhancement of crop competitiveness in these systems is usually not sufficient to eliminate the requirement for at least a reduced level of herbicide at critical points within the rotation to maximize production.

2.3 Simulations

Research on crop rotations is time-consuming and expensive. The experimental area needs to be large enough to accommodate farm-scale operations so as to realistically represent farm conditions and to include all phases of each rotation in every year. In addition, rotation experiments ideally need to be conducted long enough to include more than one cycle of the longest rotation in order to adequately understand the long-term trajectory of rotations on soil conditions and crop productivity as well as on weed populations. Because of these limitations, only a small number of alternative rotations and potential crop sequences can be assessed by experimentation. By comparison, simulation models allow for a much broader exploration of potential sequential arrangements of crops and management practices than can be evaluated experimentally (Colbach et al. 2014a,b).

Several simulations have been published that identify optimal rotational sequences for reducing weed populations. Mertens et al. (2002) systematically examined all rotational sequences between two and six years long including two crops, A and B, and determined that weed population growth rate would be 25% lower with an AABB sequence than an ABAB sequence. Garrison et al. (2014) defined the number of years the same crop is grown in a row as the 'stacking number' (e.g. CSWCSW and CCSSWW have stacking numbers of one and two, respectively). Increasing the stacking number from one to four in a corn–soya bean–wheat rotation potentially could reduce total density of a model weed community by 15%. Anderson (2004) found that arranging four different crops in sequences of two cool season crops followed by two warm season crops was most beneficial for managing simulated populations of both cool season and warm season weeds. This 'cycle of four design' allowed for a two-year period to obtain substantial seed mortality of weeds with different life histories than the crops (e.g. cool season weeds during warm season crops), but this two-year interval was short enough to avoid exponential population growth when crops and weeds had similar life histories. Research in South Dakota (Anderson and Beck 2007) and Alberta (Cathcart et al. 2006) provide experimental support for this theory. However, in a model that explored the effect of weed seed dispersal on a landscape level, a crop sequence with a stacking number of one was superior to a sequence with a stacking number of two when two consecutive years of a particular crop would permit a high dispersion of seeds across the landscape (Gonzalez-Diaz et al. 2012).

Simulations of cropping sequences can be a valuable tool when coupled with bioeconomic models to identify rotations that not only optimize weed suppression but also are economically viable. For example, a resistance and integrated management model was used to determine the efficacy of including alfalfa and pasture in annual grain rotations to control herbicide-resistant *Lolium rigidum* Gaud. (rigid ryegrass) in Australia (Doole and Pannell 2008). These alfalfa and pasture rotational options were found to be economical only in the presence of high populations of herbicide-resistant weeds and high sheep prices.

Accuracy of the weed life history parameters is an essential component of simulation models; however, sound data is not always available, particularly with respect to defining how alternative crops or their management will influence the behaviour of model parameters (Jordan et al. 1995). There is a need to collect life history data on important weed species as they respond to management within actual rotational experiments (Liebman et al. 2014). There also is a need to define important weed traits that underlie the response of weed communities to crop rotations and management as has been simulated with a recent multi-specific weed dynamics model (Colbach et al. 2014b). One area of particular importance is defining weed responses to cropping systems with stochastic weather perturbations included in the model (Colbach et al. 2014a). Research to be discussed in the next section demonstrates that annual variability in weather conditions can greatly obscure the more subtle effects of rotations on weed populations.

2.4 Variability across years and crops

One of the most significant factors present in long-term rotation experiments is the year-to-year variability that affects the response of target weed populations to rotational treatments. Partitioning variance of a long-term experiment in Ontario demonstrated that weed management (ranging from standard to no herbicide) accounted for 38% of the variance, while rotation accounted for only 5.5% of the variance (Doucet et al. 1999). However, their data also show that rotation by year interactions accounted for 29% of the variance, suggesting that annual variability in weed response to rotation accounted for approximately 5 times as much variance as the rotation main effect itself. In an eight-year experiment, Blackshaw et al. (2001) used the canonical correlation coefficient in a canonical discriminant analysis to demonstrate that weed community composition was affected more by yearly variation than by the rotation or tillage management factors in the experiment. They suggest that weeds in the Canadian prairies may be affected more by climatic conditions in a particular year than by agronomic practices. Brainard et al. (2008) describe an experiment in which rotational sequences were initiated in two successive years. Large year-to-year variability and lack of consistency between years led the authors to conclude that 'continued efforts to characterize interactive effects of management and environmental factors on weed population dynamics' were needed. Many others have also reported significant rotation by year interactions in their experiments (Davis et al. 2005; Heggenstaller and Liebman 2006; Swanton et al. 2006; Stevenson and Johnston 1999; Teasdale et al. 2004; Thorne et al. 2007). One analytical approach to negating annual weather variability is grouping data over years, such as pooling data from a set of years early in an experiment in comparison with a set of years at the end of the experiment (Swanton et al. 2006).

Weed seedbank density has become a commonly used metric for analysis of long-term experiments. Since the seedbank represents the cumulative seed inputs over preceding years, it is thought to buffer against variability in annual above-ground weed populations and integrate weed management effects across years. For example, the legacy effects of weed management treatments discontinued 12 years earlier were still detectable in the seedbank of a long-term experiment in Quebec (Légère et al. 2011). However, the magnitude of weed populations within the seedbank can also exhibit extreme fluctuations when samples are taken annually for several successive years (Légère et al. 2011; Smith and Gross 2006; Teasdale et al. 2004). Despite the capacity for selected seeds of most weed species to persist in the seedbank, seed mortality rates of many species can exceed 50% annually, resulting in the potential for one or two years of good weed control to quickly reduce the seedbank following a year with high-seed inputs (Buhler et al. 2001; Smith et al. 2009a; Sjursen 2001). This dynamic nature of seedbanks in response to annual conditions that affect weed establishment and fecundity suggests that they are as subject to year by management interactions as are above-ground weed populations.

There is often a systematic variation in weed response to specific crops within rotations, but little response to the overall rotation. Diverse six-year rotations including winter, spring and meadow crops in Sweden had distinctive weed community associations with each crop, but there was little difference in community response to the overall rotations (Andersson and Milberg 1998). Weed community response to the summer annual crops corn and soya bean were shown to be distinct from that of a winter annual wheat crop (Smith and Gross

2007; Swanton et al. 2006), but there was little difference in weed community density and structure between a corn–soya bean and a corn–soya bean–wheat rotation (Gulden et al. 2009; Gulden et al. 2010; Murphy et al. 2006). Crop characteristics, including season for planting (e.g. spring for corn and soya bean vs fall for winter wheat) and the crop life cycle (annual vs perennial), are important factors influencing weed communities and will be discussed more in Section 2.6.

The experimental design of rotation experiments usually includes all cropping phases of a rotation in each year to avoid confounding crop and year. This results in the same number of sequences of each rotation as the number of crops in the rotation (e.g. a three-year corn(C)–soya bean(S)–wheat(W) rotation would have three sequences, CSWCSW..., SWCSWC... and WCSWCS...). Despite being the same overall rotation, these sequences often display a high degree of variability because there is an inherent confounding of crop and annual weather affecting each sequence. Sequences have been shown to differ as much within a rotation as between rotations (Liebman et al. 2014; Sosnoski et al. 2009; Teasdale et al. 2004). Consequently, variability in the performance of rotational sequences also contributes to the overall variability in weed response in crop rotation experiments and can complicate the interpretation of results.

A final source of variability in rotation performance is the variability in execution of routine farm operations across crop sequences. It is rare that operations are performed equally effectively in all years; instead, it is common to find discussion of unplanned and unexpected anomalies in protocol in most research reports. This can take the form of poor crop stands, poor growing conditions, variable planting dates, inability to cultivate or apply herbicides as scheduled and other unexpected opportunities for weed growth. These lapses in the execution of operations are, in fact, a common occurrence in crop production that often requires flexibility to counteract, but that also can complicate the interpretation of results.

2.5 Weed species associated with rotational crops

Most research results demonstrate that only a few weed species dominate the weed community and its response to management. To understand weed responses to rotations, it is important to understand how the dominant species in weed communities respond to crops and their associated management practices, and some of the key traits driving their responses. A compilation of several weed species that have been dominant in recent research is listed in Table 3. Weed phenology, defined as the seasonal pattern of emergence and flowering of a species, and its association with seasonal cropping patterns becomes an important consideration for evaluating weed responses to crop rotations. Generally, weeds are most compatible with crops that have a phenology, form and function similar to their own (Liebman and Staver 2001). Weeds are more likely to thrive in crops which mimic their emergence requirements, growth form and seasonal duration (Navas 2012). For example, optimum A. emergence occurs during a similar season and under similar conditions as those of spring cereal crops, but the capacity to shed viable seeds before grain harvest ensures survival of the next generation. *Bromus tectorum* is a winter annual weed that thrives in winter wheat, specifically under the soil moisture-conserving crop residue that is characteristic of no-tillage conditions. *Taraxacum officinale* is a prostrate perennial species that tolerates mowing and is, therefore, highly adapted to perennial hay and sod crops.

Table 3 Favourable and unfavourable rotation crops for selected weed species. Only species were included which were listed as a major species in at least two references. Abbreviations are NT=no-tillage, C=corn, S=soya bean, and W=wheat

Weed species	Weed traits	Favourable rotation crops	Unfavourable rotation crops	References
Avena fatua L. (wild oat)	Annual grass, spring emergence, sheds seed before cereal harvest	Spring cereal crops Spring cereal crops Summer annual crops	Alfalfa Winter forage cereals Alfalfa, Cereals@2X seed rate	Ominski et al. 1999 Schoofs and Entz 2000 Harker et al. 2016
Bromus tectorum L. (downy brome)	Winter annual grass, fall emergence, promoted by surface residue	NT winter wheat (WW) NT WW 1 in 2 yrs NT WW 2 straight years	Summer annual crops Summer annuals 2 of 4 yr WW once every 3 or 4 yr	Blackshaw et al. 2001 Anderson and Beck 2007 Anderson et al. 2007
Digitaria sanguinalis (L.) Scop. (large crabgrass)	Annual grass, spring emergence, heat and drought tolerant, decumbent growth form, tolerates mowing	Corn-Oat-Alfalfa Organic C-S-W-Hay Organic Alfalfa+grain crops Peanut-Cotton-Bahiagrass	Corn and Corn-Soya bean Organic Corn-Soya bean Organic annual grain crops Peanut-Cotton	Cardina et al. 2002 Teasdale et al. 2004 Wortman et al. 2010 Leon et al. 2015
Panicum dichotomiflorum Michx. (fall panicum)	Annual grass, late spring emergence, tolerant of residue, triazine tolerant	NT Corn Corn	Corn-Oat-Alfalfa Permanent meadow	Cardina et al. 2002 Tomasoni et al. 2003
Setaria faberi Herrm. (giant foxtail)	Annual grass, spring emergence, light-insensitive germination, survivability within residue and crop canopy	NT Corn NT Corn but, Oat C-S-Triticale-Alfalfa Organic Alfalfa + grain crops	Corn-Soya bean-Wheat Corn-Oat-Alfalfa	Schreiber 1992 Cardina et al. 2002 Buhler et al. 2001 Heggenstaller 2006 Wortman et al. 2010
Setaria pumila (Poir.) Roemer & J.A. Schultes (yellow foxtail)	Annual grass, spring emergence, decumbent growth form, tolerates mowing	Oat Corn-Oat-Alfalfa Organic C-S-W-Hay	Corn, Soya bean Corn, Corn-Soya bean Organic Corn-Soya bean	Buhler et al. 2001 Cardina et al. 2002 Teasdale et al. 2004
Setaria viridis (L.) Beauv. (green foxtail)	Annual grass, spring emergence, tolerates mowing, late summer seeding	Spring cereals, Alfalfa NT Barley monoculture Warm season crops	Sorghum-sudangrass Cereal-Oilseed crop rotation Cool season crops 2 of 4 yr	Schoofs and Entz 2000 Légère et al. 2011 Anderson and Beck 2007

Weed species	Characteristics	Crops/rotations favouring	Crops/rotations suppressing	References
Amaranthus sp. (pigweed species)	Annual broadleaf, late spring emergence, adapted to tilled high N soils and long-season row crops, upright habit, high seed production	Organic Corn-Soya bean Corn, Soya bean Cereal-Oilseed rotation Warm season crops but, Oat Corn-Oat-Alfalfa	Organic C-S-W-Hay Winter wheat Barley monoculture Cool season crops 2 of 4 yr Corn, Soya bean Corn-Soya bean	Teasdale et al. 2004 Swanton et al. 2006 Légère et al. 2011 Anderson and Beck 2007 Buhler et al. 2001 Cardina et al. 2002
Chenopodium album L. (common lambsquarters)	Annual broadleaf, similar to Amaranthus except early spring emergence and optimum growth at cooler temperatures	Organic C-S Organic spring crops Corn, Soya bean Cereal-Oilseed rotation Following annual crops	Organic C-S-W-Hay Grass/clover, winter barley Winter wheat Barley monoculture Established alfalfa	Teasdale et al. 2004 Eyre et al. 2011 Swanton et al. 2006 Légère et al. 2011 Meiss et al. 2010
Polygonum sp. (smartweed species)	Annual broadleaf, spring emergence, capacity to regrow after cutting	Spring cereals Corn-Oat-Alfalfa Winter wheat	Alfalfa Corn, Corn-Soya bean Corn, Soya bean	Ominski et al. 1999 Cardina et al. 2002 Swanton et al. 2006
Polygonum aviculare L. (prostrate knotweed)	Annual broadleaf, spring emergence, opportunistic growth	Following annual crops Organic spring crops WW monoculture	Established alfalfa Grass/clover, winter barley WW-spring legumes	Meiss et al. 2010 Eyre et al. 2011 Ruisi et al. 2015
Taraxacum officinale G.H. Weber ex Wiggers (dandelion)	Perennial broadleaf, prostrate rosette growth habit, tolerates mowing	Alfalfa Established alfalfa	Spring cereals Following annual crops	Ominski et al. 1999 Meiss et al. 2010

Many spring emerging, upright summer annual grass weeds (*Panicum dichotomiflorum*, *Setaria faberi* and *S. viridis*) are adapted to spring-planted, summer crops (Table 3). These C_4 weeds are well adapted to growth and competition within monocultures of corn or spring-planted grain crops where they dominate. In addition, they are highly adapted to germination near the soil surface in high-residue, no-tillage conditions (Cardina et al. 2002; Légère et al. 2011; Schreiber 1992; Tomasoni et al. 2003). But they are less adapted to more diverse rotations, particularly those including a hay or meadow crop (Cardina et al. 2002; Schreiber 1992; Tomasoni et al. 2003) where an established sod prevents establishment during the typical spring window when these species emerge. Results that appear to contradict this generalization show greater foxtail abundance in selected small grain and alfalfa crops (Buhler et al. 2001; Heggenstaller and Liebman 2006; Schoofs and Entz 2000; Wortman et al. 2010). However, in all of these cases, growth and seed production were attributed to opportunities created by lack of weed control after midsummer small grain harvest or suboptimal forage stands. Foxtail species are adapted to emergence within a high-residue environment or an established crop because they lack a light requirement for germination (Teasdale and Mirsky 2015), and emerged seedlings, when released from competition after a grain harvest or in a thin hay stand, can flourish and readily reseed the population (Heggenstaller and Liebman 2006). Schoofs and Entz (2000) emphasize the importance of employing crops such as sorghum-sudangrass that can maintain competition and suppression of annual grass weeds for the full duration of the summer.

Decumbent annual grasses such as *Digitaria sanguinalis* and *Setaria pumila* are selected by hay crops primarily because of their prostrate growth form and tolerance to mowing (Cardina et al. 2002; Teasdale et al. 2004; Wortman et al. 2010). These species also have no light requirement for germination and can emerge within established sod. When combined with tolerance of hot, dry conditions when hay growth slows in mid-summer, these species are well adapted to grow and set seed in late summer. In one case, spring-seeded bahiagrass provided an opportunity for large crabgrass to establish and compete during the establishment year, but bahiagrass became highly competitive during the subsequent season (Leon et al. 2015).

Amaranthus species and *Chenopodium album* are upright broadleaf weeds with prodigious seed production and consequently often dominate the weed seedbank of cropping system experiments. They have a positive germination requirement for light and nitrates and a high relative growth rate once established, thus making them adapted to disturbed agricultural environments, particularly in organic or reduced herbicide systems. Resistance to several herbicide modes of action also has enhanced adaptation to herbicide-intensive conventional systems. These species are, therefore, adapted to spring-planted summer crops that lack effective management tools, particularly in corn–soya bean or mixed summer vegetable rotations that lack the phenological diversity to reduce population growth. Because they establish from small seeds, they can be easily suppressed by established vegetation during the spring germination period, and because they have an upright growth habit, they also can be suppressed by mowing. Thus, an established winter grain or hay crop effectively reduces populations of these weeds (Fig. 2) by creating more phenological and disturbance diversity in the rotation (Anderson and Beck 2007; Eyre et al. 2001; Légère et al. 2011; Meiss et al. 2010; Swanton et al. 2006; Teasdale et al. 2004). However, exceptions can occur when opportunities for establishment are created after midsummer harvest of winter grain crops or in gaps in hay (Buhler et al. 2001; Cardina et al. 2002).

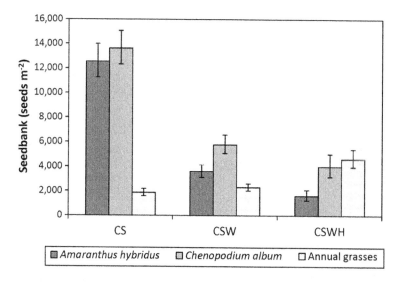

Figure 2 The early spring soil seedbank of *Amaranthus hybridus*, *Chenopodium album* and annual grass weeds (primarily *Digitaria sanguinalis* and *Setaria glauca*) in response to corn–soya bean (CS), corn–soya bean–wheat (CSW) and corn–soyabean–wheat–hay (CSWH) organic rotations at the Farming Systems Project, Beltsville, Maryland. Data are means (with standard error bars) of the five-year period following the first full year of organic management. Means were derived from log-transformed data and back-transformed for presentation. More detailed data by year and cropping sequence within rotation are presented in Teasdale et al. (2004).

2.6 Community assembly of weed traits

Communities of weed species can be conceptualized as the assembled outcome of an ongoing process whereby species are filtered by environmental influences and management operations (Booth and Swanton 2002). This analytical approach focuses on traits that allow selected groups of species to thrive and others to diminish within the environmental and management constraints of cropping systems. In the preceding section, the focus was on the dominant species that were selected (filtered) by rotational crops, along with a discussion of some of the traits likely responsible for their selection. A community approach involves consideration of salient traits across all species within the community and identification of those traits or trait groupings that are selected for or against by management or environmental factors (Navas 2012). For example, Gunton et al. (2011) analysed the association between crop characteristics and weed traits, and found that crop planting date was an important characteristic that was strongly related to many weed traits, including time of germination, time of flowering and length of flowering period, whereas crop type was not a strong predictor of any weed trait. Fried et al. (2009) used a trait-based approach to explain changes in weed communities associated with the rapid increase in sunflower culture in France over 30 years; this change favoured a sunflower mimicking functional group that was nitrophilous, upright and heliophilous, insensitive to sunflower herbicides and shared a rapid summer life cycle. A similar trait-based analysis of the change in weed communities in winter wheat in France showed that, along with the increase in summer annual crops such as sunflower and corn in rotation with winter wheat during this period, there was a corresponding increase in generalist weeds characterized by a broader range of germination and flowering phenology (Fried et al. 2012). Traits conferring a high degree of phenological variability would allow adaptation to more diverse crop rotations than species whose phenology was highly synchronous with winter wheat but poorly adapted to other rotational crops. Andersson and Milberg (1998) also found that 'germination generalists' that were capable of emergence in both fall and spring were most ubiquitous in the long, diverse rotations which they describe. Meiss et al. (2010) demonstrated that rotational hay crops reduced the presence of functional groups with traits including upright and climbing growth form and increased functional groups including rosette growth forms and perennial life history. All of these reports provide additional understanding of how selected traits permit weed communities to modulate and coexist within the varied disturbance and cropping environments of diverse agroecosystems.

3 Cover crops in weed management

Cover crops are plants that do not generate a saleable product directly, but are included within rotations to provide a variety of services to enhance the sustainability of agroecosystems (Blanco-Canqui et al. 2015). There is a wealth of information on the potential uses for cover crops, the range of available species and the best management practices for realizing their benefits (see resource section at the end of this chapter). A survey of cover crop use in the United States (Anonymous 2016) found that there were three main perceived benefits of using cover crops: 'increases overall soil health' (86% of respondents), 'reduces soil erosion' (83%) and 'increases soil organic matter' (82%). Weed control was listed in the middle of the list of benefits by 45% of respondents. Thus, the majority of cover crop users are primarily interested in long-term soil improvement benefits, but weed control is an important additional benefit to almost half of the users. As discussed below, these weed control benefits are, indeed, of primary importance for producers confronted with weed control challenges ranging from conventional growers addressing the control of herbicide-resistant weeds to organic growers attempting to reduce tillage.

3.1 Weed suppression by cover crops

Several generalizations concerning weed suppression by cover crops have been established by early research and were reviewed by Teasdale et al. (2007). These generalizations are diagrammed in Fig. 3 and can be summarized as follows. Weed suppression by cover crops most consistently occurs during the cover crop growth phase, although the residue remaining after cover crop termination can also significantly affect weed populations. A growing cover crop occupies a fallow or off-season period in crop rotations that would otherwise provide an opportunity for weed growth. Once established, the presence of the cover crop inhibits weed growth through competition for resources as well as by absorbing sufficient radiation in the red spectrum to inhibit phytochrome-mediated weed seed germination. Generally, cover crops that provide rapid leaf canopy establishment and high biomass are most effective for weed suppression. After cover crop termination, the remaining residue is less reliable for weed suppression and can become a detriment to

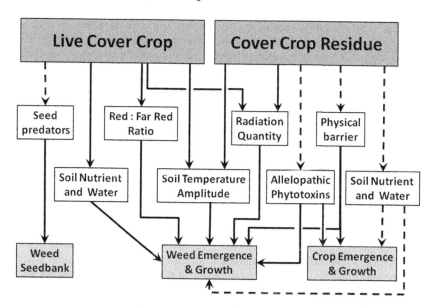

Figure 3 Influences of live cover crops and of cover crop residue at the soil surface on environmental parameters and soil resources and chemistry, and their subsequent effects on weeds and crops. Solid arrows indicate a reduction and dashed arrows indicate an increase in the receiving parameter. Processes depicted here are modified by cover crop biomass and carbon nitrogen ratio as described in the text.

establishment of succeeding crops unless properly managed. Residue incorporated into soil will release phytotoxic compounds for one to two weeks, thus providing a short period of weed suppression, but also potential inhibition of crops if planted too soon. Residue on the soil surface can be highly suppressive of weed establishment, particularly at high residue levels (to be defined in more detail below), but high amounts of surface residue can also interfere with crop establishment.

Turning to more recent research, there has been an expanding body of literature published on cover crops, as their use has become increasingly popular. It is not possible to comprehensively summarize this work here. Instead, this chapter will highlight several recent themes, particularly within the framework of crop rotations and long-term sustainable weed management.

3.2 Cover crop mixtures versus monocultures

According to the CITC/SARE survey (Anonymous 2016), rye was the most commonly used cover crop, but mixtures of cover crops were used more often than any single cover crop species other than rye. On average, 3.6 species were included in these mixtures. There is a clear perception that mixtures can provide a greater diversity of services and greater consistency than individual species. Ecological research has shown that increased species diversity in a system can increase system stability and resilience. Tilman et al. (2012) showed that an increase in plant diversity from 4 to 16 species caused as large an increase in productivity as the addition of 54 kg·ha^{-1}·y^{-1} of fertilizer nitrogen, and was as influential in maintaining productivity as eliminating a dominant herbivore or a major natural drought. However, the authors caution that this represents an average response, and that, if low-diversity communities contained the most productive species, 'their productivity would not differ as much from the productivity of high-diversity communities as biodiversity experiments would predict'. Recent research with cover crop mixtures confirms this assessment.

Several reports demonstrated that the biomass of the most productive cover crop in monoculture was similar or higher than the most productive multi-species mixtures. In Pennsylvania, an analysis of 12 species demonstrated that mixtures with complementary functional groups, either for winter hardiness or for nitrogen, did not result in higher biomass of mixtures compared to the highest yielding monocultures (Finney et al. 2016). Although increasing species richness to eight species in a mixture did increase biomass and reduce variability against the average performance of all monocultures, it did not produce more than that of the most productive canola and cereal rye monocultures. Similarly, in Nebraska, biomass of mixtures with up to eight cover crop species was not higher than that of the highest yielding mustard cover crops in monoculture; however, a 6-species mixture did have a low coefficient of variation, suggesting a capacity to buffer against low productivity (Wortman et al. 2012). A mixture of five species in New Hampshire exhibited evidence of over-yielding, but did not produce more biomass or greater ecological services than the highest-performing monoculture species (Smith et al. 2014). In comparisons of two-species mixtures versus monocultures, there have been variable results, with some reporting mixtures producing no more biomass than the best monoculture (Brainard et al. 2011; Halde et al. 2014; Hodgdon et al. 2016; Lawson et al. 2015; Rühlemann and Schmidtke 2015) and others showing increased biomass in mixtures (Hayden et al. 2014; Wiggins et al. 2016). The choice of seeding rates in these experiments was highly variable and often included higher seeding densities in mixtures than monocultures. In some cases, a replacement series design was followed wherein seeding rates were proportional to designated species proportions in the mixture (Hayden et al. 2014; Smith et al. 2014; Wortman et al. 2012). More research is needed to determine optimum proportions and seeding densities for mixtures and monocultures of cover crops.

Recent research confirms that biomass is the cover crop attribute most highly correlated with weed suppression. Increasing cover crop biomass progressively reduced biomass of weeds growing in the cover crop up to a plateau at a cover crop biomass of 4625 kg ha^{-1}, above which, weeds were completely suppressed (Finney et al. 2016). Sorghum-sudangrass produced high levels of biomass and, either in monoculture or in mixtures, suppressed annual weed biomass 90–99% in New York (Brainard et al. 2011) and *Cirsium arvense* (L.) Scop. (Canada thistle) biomass 50–87% in Ohio (Wedryk and Cardina 2012). Summer monocultures and mixtures of buckwheat and sorghum-sudangrass in New Hampshire reduced the number of weed species relative to a no-cover-crop control by 25–59%, and both weed biomass and species richness were inversely related with cover crop biomass (Smith et al. 2015). Biomass of fall-planted cover crops in New Hampshire was also highly correlated with weed suppression in late fall (Hodgdon et al. 2016).

3.3 Cover crop carbon-to-nitrogen ratio

In addition to biomass, the relative nitrogen content of cover crop tissue and their residues can be an important determinant of weed response to cover crops. Legume cover crops with a lower carbon-to-nitrogen (C:N) ratio often allowed a higher abundance of weeds than cereal cover crops with a higher C:N ratio. Finney et al. (2016) showed that suppression of winter annual weeds growing with a wide range of 18 cover crop treatments increased as C:N ratio rose to 20.7, with almost complete suppression above this level. They concluded, however, that cover crop biomass probably played a greater role in weed suppression than did C:N ratio because biomass was more highly correlated with weed suppression ($R^2 = 0.54$) than was C:N ratio ($R^2 = 0.27$). Haydon et al. (2014) found that abundance of winter annual weeds growing with the cover crop increased progressively as the proportion of hairy vetch in a rye–vetch mixture increased, despite similar biomass of both cover crops and higher light penetration within the rye than the vetch canopy. Radicetti et al. (2013) showed that biomass of winter annual weeds growing with the cover crop was a function of cover crop biomass overall, but that hairy vetch allowed more weed biomass per unit of cover crop biomass than did an oat cover crop. Generally, higher suppression of weeds by cereal cover crops with a high C:N ratio may be related to 1) development of a more extensive and competitive root system than that of legumes and 2) extraction of inorganic nitrogen in the soil profile that could otherwise stimulate germination of nitrophilous weeds.

Weed suppression by residue after cover crop termination is also influenced by residue C:N ratio. Higher weed density has been reported in residue from legumes such as hairy vetch compared to that from cereal cover crops such as barley or oats (Hodgdon et al. 2016; Radicetti et al. 2013; Wayman et al. 2015). This result is probably the consequence of more rapid degradation of legume residue and stimulation of nitrophilous weed germination by inorganic N released during degradation of legumes. A structural equation modelling analysis of red clover and rye cover crops showed that decreasing cover crop C:N ratio increased weed biomass during the reproductive development of a subsequent dry bean crop, and that this weed biomass increase was associated with an increase in cover crop N content which in turn increased soil N following cover crop termination (Hill et al. 2016). On the other hand, rye residue leads to immobilization of soil inorganic nitrogen and demonstrable nitrogen deprivation in nitrophilous weeds such as *Amaranthus retroflexus* (Wells et al. 2013). Soil chemistry is also affected by the release of allelopathic phytotoxins after termination of these cover crops, but these are generally short-lived during the first couple of weeks after incorporation of red clover (Lou et al. 2016) and rye (Teasdale et al. 2012b). So cover crop C:N ratio may have a more persistent effect on final weed abundance throughout the season than allelopathy.

3.4 Organic no-tillage production with high cover crop biomass

Organic no-tillage production is one situation where attaining high biomass of cover crop residue is critical for weed suppression. Because the reliance on tillage for seedbed preparation and control of weeds in organic farming usually is destructive of soil tilth and increases fuel expenses, organic farmers have sought means for reducing tillage. Unlike conventional systems where herbicide technology has been foundational to successful no-tillage crop production, creation of a smothering barrier of cover crop residue is the primary practice available for eliminating weeds in organic no-till systems. Research has shown that approximately 8000 kg ha^{-1} of residue with a thickness of 10 cm and almost complete soil coverage is needed to adequately suppress weeds (Teasdale and Mohler 2000). Many recent evaluations of no-till systems have confirmed this biomass level as a reasonable threshold for controlling weeds (Bernstein et al. 2014; Blackshaw et al. 2010; Canali et al. 2013; Davis 2010; Halde et al. 2014; Smith et al. 2011; Wortman et al. 2013). Additional synergisms can be obtained between cover crop residue and crop canopy closure where early season weed suppression from cover crop residue can complement mid-season suppression from an early-closing crop leaf canopy (Ryan et al. 2011; Wells et al. 2014).

Much research has focused on the optimum management of cover crops to achieve consistently high biomass production for organic production. Optimizing cover crop growth becomes as high a priority as optimizing cash crop growth in these systems. Generally, early planting and late termination dates extend the cover crop growing period, while a high seeding rate and suitable fertility (particularly for cereal cover crops) ensures optimum growth (Mirsky et al. 2013). However, fitting cover crops into existing crop rotations to realize optimum biomass without jeopardizing crop planting requirements can be challenging. In addition, the need to mechanically terminate cover crop growth usually requires that termination be delayed until the cover crop advances to the flowering stage of growth (Mirsky et al. 2012; Reberg-Horton et al. 2012). Together, these constraints require a sufficiently long period following harvesting of one cash crop to grow a cover crop, and a sufficiently long growing season after cover crops are terminated at flowering to plant and produce a second cash crop. The growing seasons in the mid-Atlantic states and south-eastern US are sufficiently long to provide for developing high biomass of fall-planted winter annual cover crops without jeopardizing crop production (Mirsky et al. 2012; Reberg-Horton et al. 2012). But in the US Northern Great Plains or Pacific Northwest, there is often an insufficiently long period for growing a winter annual cover crop nor is there a sufficiently long growing season following cover crop flowering and termination to grow a cash crop (Carr et al. 2012; Luna et al. 2012). This constraint has been overcome in Western Canada where a full growing season has been devoted to producing a green fallow, in preparation for an organic no-till flax system in the following year (Halde and Entz 2014).

Much research and engineering effort has been expended on developing approaches and equipment for mechanically terminating cover crops and planting cash crops into the heavy residue layer (Mirsky et al. 2013; Reberg-Horton et al. 2012). Mowing which cuts and shreds, rolling which crimps and flattens, and undercutting which severs roots can successfully kill the cover crop if performed when the cover crop has reached the flowering stage. Crops planted into a dense mulch of surface residue can suffer from reduced stands, inadequate fertility and lower yields unless care is taken to overcome the interference from residue to seed placement in soil, destruction of seedlings by soil pests and residue influences on nutrient availability. Comparisons of organic no-tillage versus tillage-based production have given mixed results (Fig. 4), with yields and profitability varying considerably depending on location and crop (Bernstein et al. 2014; Canali et al. 2013; Delate et al. 2012; Forcella 2013; Halde and Entz 2014; Luna et al. 2012; Teasdale et al. 2012a; Wortman et al. 2013).

The relative success of organic crops grown with primary tillage suggests that rotations may need intermittent tillage to manage weeds and initiate thorough decomposition and nutrient release from residues. Several reports describe transitions wherein weeds established in a preceding crop or cover crop can survive mechanical mowing or rolling operations and persist into the succeeding crop in the absence of soil disturbance or herbicides (Dorn et al. 2013; Mirsky et al. 2010; Nord et al. 2012; Rühlemann and Schmidtke 2015; Saunders Bulan et al. 2015; Teasdale and Mirsky 2015). In addition, in

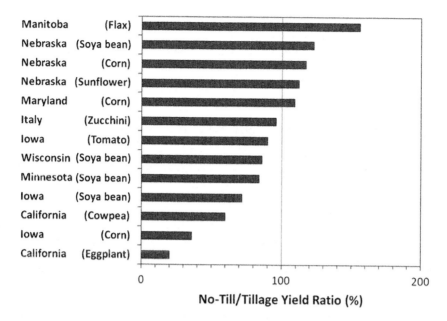

Figure 4 Crop yield ratio of an organic no-tillage system to a tillage-based system for selected locations and crops. Percentages > 100 indicate that yield in the no-till system was higher than that in the tillage system, and percentages < 100 indicate the opposite. Data are averages over years and sites presented in each reference. References are Manitoba-flax, Halde and Entz (2014); Nebraska-soya bean/corn/sunflower, Wortmann et al. (2013); Maryland-corn, Teasdale et al. (2012a); Italy-zucchini, Canali et al. (2013); Iowa-tomato/soya bean/corn, Delate et al. (2012); Wisconsin-soya bean, Bernstein et al. (2014); Minnesota-soya bean, Forcella (2013); California-cowpea/eggplant, Luna et al. (2012).

some of the few long-term evaluations of reduced-tillage organic rotations with cover crops, development of perennial weeds suggests that continuous cover cropping with reduced tillage may be impractical in organic production (Halde et al. 2015; Melander et al. 2016). Consequently, opportunities to introduce tillage into rotations need to be carefully evaluated on the basis of the developing composition of weed communities and their phenology with respect to cover crop and crop rotational sequences. Tillage can play an important role in providing positive disturbance diversity in rotational systems, especially if it is timed to coincide with emergence and establishment of problematic weeds (Mirsky et al. 2010).

Post-planting cultivation offers an option for controlling escaped weeds in no-tillage organic systems. High-residue cultivators have been developed that employ a coulter in front of the cultivator shank to cut through mulch and a relatively flat cultivator sweep that severs weed shoots from their roots beneath the soil surface with minimum disturbance to the residue (Mirsky et al. 2013). However, this arrangement often does not dislodge weed seedlings from soil particles and permit desiccation as effectively as cultivation in a tilled seedbed. A comparison of post-plant cultivation of corn in a tilled versus rolled hairy vetch cover crop demonstrated that weed biomass was reduced by a three-year average of 75% in the tilled vetch system compared to 51% with a high-residue cultivator in the rolled vetch (Teasdale et al. 2012a). This level of weed control with a high-residue cultivator is similar to that found by other investigators who have reported weed biomass reductions ranging from 38% to 63% for high-residue cultivation of soya beans in a rye mulch (Nord et al. 2011) and 52% for data averaged over several experiments summarized by Mirsky et al. (2013). More research is needed to refine equipment for cultivating weeds in a dense layer of residue and to provide a viable alternative to organic growers for managing escaped weeds in no-tillage production.

3.5 Control of herbicide-resistant weeds

Many weed species have developed herbicide-resistant biotypes in the last two to three decades (Heap 2016; Owen et al. 2014), with glyphosate-resistant species becoming particularly problematic for current no-tillage weed management (Duke 2015). This has resulted in the exploration of integrated weed management programmes for controlling herbicide-resistant weeds, including the use of cover crops for early season weed suppression (Norseworthy et al. 2012; Owen et al. 2014). This section will discuss several recent reports on *Amaranthus palmeri* S. Wats. (Palmer amaranth), which has developed resistance to glyphosate and other herbicide modes of action in the south-eastern United States. High biomass levels of rye residue, similar to those for controlling weeds in organic no-tillage as described above, can control *A. palmeri* (Price et al. 2016; Webster et al. 2013; Webster et al. 2016). Most experiments that have tested cover crops for control of herbicide-resistant weeds, however, have not produced high levels of biomass, and consequently have achieved only early season weed suppression that was insufficient to prevent crop yield loss or weed seed production (DeVore et al. 2013; Wiggins et al. 2015; Wiggens et al. 2016). Addition of a pre-emergence herbicide to cover crops did not provide satisfactory control (Wiggins et al. 2016), but addition of post-emergence herbicide mixes did achieve season-long control, suggesting that the integration of cover crops with appropriate post-emergence herbicides could aid in preventing further selection for herbicide resistance in *A. palmeri* (Wiggins et al. 2015). The critical period for control of *A. palmeri* and other annual weeds was minimally affected by a rye cover crop, but rye reduced the biomass and size of weeds that did emerge (Korres and Norsworthy 2015), thus offering potentially more options for post-emergence weed control. Tillage can also offer an option for controlling *A. palmeri* following a pulse of seeding, wherein one deep tillage event to bury surface seeds, followed by two years of a rye cover crop preceding full-season soya beans controlled *A. palmeri* emergence by 92% compared with 69% by deep tillage alone (DeVore et al. 2013).

3.6 Multifunctional role of cover crops

Since cover crops are included within rotations to perform many functions, there is interest in developing analytical approaches to assess the multifunctionality of cover crops (Blanco-Canqui et al. 2015). One approach for quantifying multiple ecosystem services is use of functional traits to predict the ecosystem function of different cover crop communities (Storkey et al. 2015). Models were used to design legume-based cover crop communities that delivered the optimal balance of six ecosystem services: early productivity, regrowth following mowing, weed suppression, support of invertebrates, soil fertility building (measured by yield of the following crop) and conservation of nutrients in the soil (measured by residue lignin + polyphenol:N ratio). An experimental species pool of 12 cultivated legume species was screened for a range of functional traits and ecosystem services at five sites across a geographical gradient in the United Kingdom. Low-to-intermediate levels of species richness (one to four species) that exploited functional contrasts in growth habit and phenology were identified as providing optimal services. It is interesting that traits that conferred early-season legume productivity, regrowth potential and improved yield of subsequent crops were also advantageous for suppressing weeds.

Research in Pennsylvania explored the role of cover crop mixtures of up to 8 species with diverse functional traits relating to phenological diversity and nitrogen utilization diversity in the provision of five ecosystem services (Finney et al. 2016). Cover crop biomass was more highly correlated than C:N ratio to weed suppression, nitrogen retention and cover crop biomass nitrogen content. But, cover crop C:N ratio had greater explanatory power for soil nitrogen supply and subsequent corn yield than did cover crop biomass. Simultaneous measurement of five diverse ecosystem services in this study identified expected limitations of cover crop monocultures with regard to their ability to support multiple ecosystem services; high-yielding non-legumes provided weed suppression and N retention, while legume monocultures supplied inorganic N and promoted corn yields. Mixtures that incorporated diversity in key functional traits may have the potential to simultaneously provide a higher number of services.

Schipanski et al. (2014) developed an assessment framework for a corn–soya bean–wheat rotation to estimate the multifunctional ecological services provided by cover crops. Cumulatively, across all years of the rotation, cover crops increased the provisioning of 8 of 11 ecosystem services relative to the rotation without cover crops. Cover crops increased almost all supporting and regulating services, including biomass production, nitrogen supply, soil carbon storage, nitrate retention, erosion control, weed suppression, mycorrhizal colonization and beneficial insect conservation, while nitrous oxide emissions, pest suppression and yield of all cash crops were equivalent between systems with and without cover crops. In addition, cover crops often provided important ecosystem services at critical points within the crop rotation such as rye providing nitrogen retention following corn harvest or red clover providing weed suppression after wheat harvest. Ideally, cover crop selection and management should focus on provision of services at the most critical and vulnerable points within each rotational cycle.

Insertion of cover crops into crop rotations is constrained by the environmental requirements for production of cash crops and the time and resources growers are willing to invest in growing cover crops for their ecological benefits. This discussion of cover crops has focused on the weed suppressive benefits that cover crops can provide and some of the trade-offs to crop yield potential that may result depending on cover crop selection and management. Weed dynamics, however, is only one component of a complex of physical, chemical and biological processes that make up a sustainably functioning agroecosystem. Cover crops must be assessed within a multifunctional framework to determine the most desirable traits for delivery of overall services needed throughout the crop rotation. Most cover crop research has focused on performance during a single season, often in conjunction with a single subsequent cash crop. As important as these experiments are for determining optimum species, management and agronomic practices, it may be more important to determine the multi-year temporal dynamics of cover crop services. Long-term experiments along with simulations with integrated cash crop and cover crop rotations (Colbach et al. 2014a) can provide valuable information on delivery of multiple ecosystem services and their sustainability.

4 Opportunities for weed establishment within rotations

This chapter has established that an adapted, vigorous crop or cover crop can suppress weeds during its growth phase and a dense mat of surface residue can suppress weeds after cover crop termination. However, throughout the sequence of operations from harvesting of a cash crop to growing a cover crop followed by a second cash crop, there are many opportunities for weed establishment. Several research reports have documented these windows of opportunity either by design or through accounts of their inadvertent occurrence. Table 4 catalogues several potential opportunities within a sequence running from a winter annual cover crop to spring cash crop, but the general principles can be adapted to any sequence of two crops in rotation. The most important characteristic of the transition periods described in Table 4 is the absence of crop leaf cover and the penetration of radiation to the soil surface. For example, light and temperature were identified as important variables controlling emergence of weeds in an alfalfa canopy (Huarte and Benech Arnold 2003), whereby an increase in red-to-far-red ratio and soil temperature amplitude were cues promoting weed germination and growth. Periods with a temporary absence of crop canopy cover will be sensed from the weed perspective as an increase in radiation regardless of the specific crop or management operation that caused it. All effects of crops along with their associated management practices can be described by how they alter the environment of each individual weed (Navas 2012). From this weed-centric perspective, crop rotations are made up of a series of disturbances and periods with and without plant cover that can be characterized by the radiation, temperature, soil moisture and nutrient environments that accompany them. Weed communities will modulate in response to the presence and timing of those environments that either promote or deter selected traits.

The success of crop rotations depends on their capacity to interrupt and prevent conditions that provide signals and resources needed for weed establishment during the windows of opportunity described in Table 4. From a broad perspective, the optimum sequencing of transitions between crops should occur at different seasons and include different types of disturbances throughout a rotation (Liebman and Staver 2001). In this way, no specific species of weed or functional group of weed community traits can predominate. As an example, Anderson (2010) proposed a nine-year rotation for organic production in the Northern Great Plains that includes both phenological and operational diversity. Warm season crops and cool season crops are alternated in two-year sequences for six years followed by three years of alfalfa. Tillage to incorporate legumes before grain crops is alternated with no tillage during forage and cool season crop sequences. This approach to rotational planning could be adapted to include several cover crops in regions with longer seasons, including high-residue-producing cover crop mixes before full-season, summer row crops or late-summer planted, winter-killed cover crops preceding early-spring-planted crops. Sequences without tillage can be useful for reducing weed seed populations when no seed production occurs, but interspersed tillage can target specific transitions that coincide with the emergence periodicity of problematic weed species. More detailed approaches to rotation planning and appropriate integrated management options within rotations are listed in the resources section.

Table 4 Periods of opportunity for weed establishment and growth within a winter annual cover crop followed by summer cash crop sequence

Crop	Period of opportunity relative to crop operations	Weed traits advantaged	References
Cover crop	Between cash crop harvest and cover crop planting	Late summer/fall germination, short-day flowering response	Rühlemann and Schmidtke 2015
	After cover crop planting and before leaf canopy establishment	Fall germination, cold tolerance	Brust et al. 2014 Dorn et al. 2013
	Gaps in cover crop canopy	Winter annual or early spring germination, light responsive	Gieske et al. 2016 Mirsky et al. 2010 Nord et al. 2012 Hill et al. 2016
	Within the canopy	Spring germination, no light requirement, capacity to survive termination operation	Teasdale and Mirsky 2015 Saunders Bulan et al. 2015
Cash crop	After cover crop termination/crop planting and before crop leaf canopy establishment	Germination periodicity matches crop, responsive to tillage and nitrogen or capable of penetrating no-till surface mulch	Anderson and Beck 2007 Bernstein et al. 2016 Hill et al. 2016 Radicetti et al. 2013 Teasdale et al. 2012a Wiggins et al. 2015
	In gaps or poor crop stands	Late spring or summer germination, upright growth or vining habit	Mirsky et al. 2010 Teasdale et al. 2012a Wells et al. 2016
	In short or uncompetitive crops	Late spring or summer germination, upright growth habit	Benoit et al. 2003 Eyre et al. 2011
	After mowing a hay crop	Prostrate growth habit, tolerant of heat and drought	Cardina et al. 2002 Mirsky et al. 2010 Ramsdale et al. 2006 Schoofs and Entz 2000
	Crop leaf senescence before crop harvest	Late summer/fall germination, short-day flowering response	Buhler et al. 2001 Rühlemann and Schmidtke 2015 Smith and Gross 2007

5 Conclusion

Review of the recent literature confirms that the most important features of crop rotations for suppressing weeds are crop phenological diversity and management disturbance diversity. Alternating crops with different planting dates and growth periods and with summer annual, winter annual and perennial life forms will limit the proliferation of groups of weed species with traits that mimic specific crops. Similarly, alternating weed management disturbances in association with rotated crops will not permit weed species adapted to one disturbance regime to proliferate. A specifically programmed sequence of crop species seems to be less important than the simple maintenance of a highly diverse rotation. Within diverse rotations, weed communities often fluctuate more amongst the crops within a sequence than between overall rotations.

The prevalence of large weather-by-rotation interactions highlights the important role that environmental conditions can play in modifying weed behaviour within crop rotations. If weeds respond primarily to the alterations in their physical and chemical environment related to crop production practices, then it is not surprising that weed responses would fluctuate as a result of interacting crop and environmental conditions. This stresses the importance of not only designing crop sequences to optimize weed suppression, but also executing and adapting these plans to close windows of opportunity for weed establishment as they develop.

Cover crops can play a significant role in filling gaps where adapted weed species could develop in crop rotations. Although cover crops require additional management, they can provide multifunctional ecosystem services beyond their role in weed suppression. Enhanced soil quality and fertility are often the most important benefits perceived by farmers, but there are specific situations when weed suppression becomes the primary rationale for their use. Control of weeds in no-tillage organic crop production requires a dense mat of cover crop residue to suppress weeds in the absence of herbicides. An integrated programme of tactics including cover crops can contribute to the control of herbicide-resistant weeds that have proliferated in recent decades and to the reduction of herbicide selection pressure, thereby maintaining the utility of current herbicide modes of action.

Management of the resident weed community is an important function of sustainably managed cropping systems. Agroecosystems are not static, but are created and controlled by the dynamic processes associated with their constituent biological systems. Therefore, sustainable management should not be focused on static routines, but should provide a dynamic framework that integrates biological components to provide the desired productivity and ecological services. Likewise, rotations should not be considered as rigid programmes, but rather as guidelines to be adjusted as required by specific environmental conditions and weed species shifts. Flexible management will be required to address not only short-term modulations of weed communities, but also long-term evolutionary adaptations that will inevitably occur. Sustainable weed management must, therefore, be a dynamic process that requires short-term adjustments and long-term perspectives to manage ever-adapting weed communities.

6 Future trends

This chapter has identified several objectives for future research that will contribute towards sustainable weed management systems.

- Determine and quantify the abiotic and biotic characteristics of specific crops that most influence the response and population trajectory of problematic weed species
- Define the functional traits of weed communities that can be managed by crop rotational design
- Gain a clearer understanding of rotation by environment interactions and how to design rotational sequences that dampen the influence of fluctuating weather conditions

- Create comprehensive datasets from broad-based farm surveys to complement controlled research station rotation experiments
- Develop a comprehensive database of weed life history responses to management tactics, which supports more robust rotation simulation modelling
- Determine optimum densities and proportions for maximizing agroecosystem services from cover crop mixtures
- Develop flexible systems for integrating tillage, no tillage and cover crops to optimize weed management and soil quality in organic production
- Develop integrated weed management programmes including rotations and cover crops to control and prevent development of herbicide-resistant weeds
- Conduct more long-term (multi-year) assessments of cover crop function within crop rotations and the durability of the ecosystems services provided
- Create collaborative research, tactics and management of cover crop and rotational systems on a regional and national scale
- Develop improved tools and frameworks for assessing the multifunctional properties and overall sustainability of cropping system rotations.

7 Where to look for further information

Mohler, C. L. and Johnson, S. E., editors. (2009), *Crop rotation on organic farms: a planning manual.* NRAES-177. http://www.sare.org/Learning-Center/Books/Crop-Rotation-on-Organic-Farms (accessed 5 November 2016). This is a very detailed planning guide describing principles and practical advice for planning rotations. It is based on the acquired knowledge of expert farmers and agricultural specialists.

Clark, A., editor. (2007), *Managing cover crops profitably*, Third edition. Sustainable Agriculture Network. http://www.sare.org/Learning-Center/Books/Managing-Cover-Crops-Profitably-3rd-Edition (accessed 5 November 2016). This book contains much practical information on a wide range of cover crop species and their management.

Midwest Cover Crop Council website has a wealth of information, including cover crop selection tools, catalogues of numerous extension publications, videos and much more. http://www.mccc.msu.edu/ (accessed 5 November 2016).

8 References

Anderson, R. L. (2004), Sequencing crops to minimize selection pressure for weeds in the Central Great Plains, *Weed Technol.* 18(1), 157–64.

Anderson, R. L. (2010), A rotation design to reduce weed density in organic farming, Renew. *Agric. Food Syst.* 25(3), 189–95.

Anderson, R. L. (2011), Corn tolerance to weed interference varies with preceding crop, *Weed Technol.* 25(3), 486–91.

Anderson, R. L. and Beck, D. L. (2007), Characterizing weed communities among various rotations in Central South Dakota, *Weed Technol.* 21(1), 76–9.

Anderson, R. L., Stymiest, C. E., Swan, B. A. and Rickertsen, J. R. (2007), Weed community response to crop rotations in Western South Dakota, *Weed Technol.* 21(1), 131–5.

Andersson, T. N. and Milberg, P. (1998), Weed flora and the relative importance of site, crop, crop rotation, and nitrogen, *Weed Sci.* 46(1), 30–8.

Anonymous. (2016), Annual Report 2015/2016 Cover Crop Survey, http://www.ctic.org/media/CoverCrops/2016CoverCropSurvey_Final.pdf (accessed 5 November 2016).

Barberi, P. and Lo Cascio, B. (2001), Long-term tillage and crop rotation effects on weed seedbank size and composition, *Weed Res.* 41, 325–40.

Benoit, D. L., Leroux, G. and Banville, S. (2003), Influence of carrot/onion/barley cropping sequence on the weed seed bank and field flora of an organic soil in Quebec, Canada, *Aspects Appl. Biol.* 69, 69–75.

Bernstein, E. R., Stoltenberg, D. E., Posner, J. L. and Hedtcke, J. L. (2014), Weed community dynamics and suppression in tilled and no-tillage transitional organic winter rye–soybean systems, *Weed Sci.* 62(1), 125–37.

Blackshaw, R. E., Larney, F. J., Lindwall, C. W., Watson, P. R. and Derksen, D. A. (2001), Tillage intensity and crop rotation affect weed community dynamics in a winter wheat cropping system, *Can. J. Plant Sci.* 81(4), 805–13.

Blackshaw, R. E., Molnar, L. J. and Moyer, J. R. (2010), Sweet clover termination effects on weeds, soil water, soil nitrogen, and succeeding wheat yield, *Agron. J.* 102(2), 634–41.

Blackshaw, R. E., Pearson, D. C., Larney, F. J., Regitnig, P. J., Nitschelm, J. J. and Lupwayi, N. Z. (2015), Conservation management and crop rotation effects on weed populations in a 12-year irrigated study, *Weed Technol.* 29(4), 835–43.

Blanco-Canqui, H., Shaver, T. M., Lindquist, J. L., Shapiro, C. A., Elmore, R. W., Francis, C. A. and Hergert, G. W. (2015), Cover crops and ecosystem services: Insights from studies in temperate soils, *Agron. J.* 107(6), 2449–74.

Booth B. D. and Swanton, C. J. (2002), Assembly theory applied to weed communities, *Weed Sci.* 50(1), 2–13.

Brainard, D. C., Bellinder, R. R., Hahn, R. R. and Shah, D. A. (2008), Crop rotation, cover crop, and weed management effects on weed seedbanks and yields in snap bean, sweet corn, and cabbage, *Weed Sci.* 56(3), 434–41.

Brainard, D. C., Bellinder, R. R. and Kumar, V. (2011), Grass-legume mixtures and soil fertility affect cover crop performance and weed seed production, *Weed Technol.* 25(3), 473–9.

Brust, J., Claupein, W. and Gerhards, R. (2014), Growth and weed suppression ability of common and new cover crops in Germany, *Crop Protect.* 63, 1–8.

Buhler, D. D., Kohler, K. A. and Thompson, R. L. (2001), Weed seed bank dynamics during a five-year crop rotation, *Weed Technol.* 15(1), 170–6.

Canali, S., Campanelli, G., Ciaccia, C., Leteo, F., Testani, E. and Montemurro, F. (2013), Conservation tillage strategy based on the roller crimper technology for weed control in Mediterranean vegetable organic cropping systems, *Europ. J. Agron.* 50, 11–18.

Cardina, J., Herms, C. P. and Doohan, D. J. (2002), Crop rotation and tillage system effects on weed seedbanks, Weed Sci. 50(4), 448–60.

Carr, P. M., Anderson, R. L., Lawley, Y. E., Miller, P. R. and Zwinger, S. F. (2012), Organic zero-till in the northern US Great Plains Region: Opportunities and obstacles, *Renew. Agric. Food Syst.* 27(1), 12–20.

Cathcart, R. J., Topinka, A. K., Kharbanda, P., Lange, R., Yang, R. C. and Hall, L. M. (2006), Rotation length, canola variety and herbicide resistance system affect weed populations and yield, *Weed Sci.* 54(4), 726–34.

Chikoye, D., Ekeleme, F., Lum, A. F. and Schulz, S. (2008), Legume-maize rotation and nitrogen effects on weed performance in the humid and subhumid tropics of West Africa, *Crop Protect.* 27(3–5), 638–47.

Colbach, N., Collard, A., Guyot, S. H. M., Mézière, D. and Munier-Jolain, N. (2014a), Assessing innovative sowing patterns for integrated weed management with a 3D crop:weed competition model, *Europ. J. Agron.* 53, 74–89.

Colbach, N., Granger, S., Guyota, S. H. M. and Mézière, D. (2014b), A trait-based approach to explain weed species response to agricultural practices in a simulation study with a cropping system model, *Agric. Ecosyst. Environ.* 183, 197–204.

Davis, A. S. (2010), Cover-crop roller–crimper contributes to weed management in no-till soybean, *Weed Sci.* 58(3), 300–9.

Davis A. S., Hill, J. D., Chase, C. A., Johanns, A. M. and Liebman, M. (2012), Increasing cropping system diversity balances productivity, profitability and environmental health, *PLoS ONE* 7(10), e47149. doi:10.1371/journal.pone.0047149.

Davis, A. S., Renner, K. A. and Gross, K. L. (2005), Weed seedbank and community shifts in a long-term cropping systems experiment, *Weed Sci.* 53(3), 296–306.

Delate, K., Cwach, D. and Chase, C. (2012), Organic no-tillage system effects on soybean, corn and irrigated tomato production and economic performance in Iowa, USA. Renew. *Agric. Food Syst.* 27(1), 49–59.

DeVore, J. D., Norsworthy, J. K. and Brye, K. R. (2013), Influence of deep tillage, a rye cover crop, and various soybean production systems on Palmer amaranth emergence in soybean, *Weed Technol.* 27(2), 263–70.

Doole, G. J. and Pannell, D. J. (2008), Role and value of including lucerne (*Medicago sativa* L.) phases in crop rotations for the management of herbicide-resistant *Lolium rigidum* in Western Australia, *Crop Prot.* 27 (3–5), 497–504.

Dorn, B., Stadler, M., van der Heijden, M. and Streit, B. (2013), Regulation of cover crops and weeds using a roll-chopper for herbicide reduction in no-tillage winter wheat, *Soil Tillage Res.* 134, 121–32.

Doucet, C., Weaver, S. E., Hamill, A. S. and Zhang, J. H. (1999), Separating the effects of crop rotation from weed management on weed density and diversity, *Weed Sci.* 47(6), 729–35.

Duke, S. O. (2015), Perspectives on transgenic, herbicide-resistant crops in the United States almost 20 years after introduction, *Pest. Manag. Sci.* 71, 652–7.

Eyre, M. D., Critchley, C. N. R., Leifert, C. and Wilcockson, S. J. (2011), Crop sequence, crop protection and fertility management effects on weed cover in an organic/conventional farm management trial, *Europ. J. Agron.* 34(3), 153–62.

Finney, D. M., White, C. M. and Kaye, J. P. (2016), Biomass production and carbon/nitrogen ratio influence ecosystem services from cover crop mixtures, *Agron. J.* 108(1), 39–52.

Forcella, F. (2013), Short- and full-season soybean in stale seedbeds versus rolled-crimped winter rye mulch, *Renew. Agric. Food Syst.* 29(1), 92–9.

Fried, G., Chauvel, B. and Reboud, X. (2009), A functional analysis of large-scale temporal shifts from 1970 to 2000 in weed assemblages of sunflower crops in France, *J. Veg. Sci.* 20, 49–58.

Fried, G., Kazakou, E. and Gaba, S. (2012), Trajectories of weed communities explained by traits associated with species' response to management practices, *Agric. Ecosyst. Environ.* 158, 147–55.

Gallandt, E. R., Liebman, M., Corson, S., Porter, G. A. and Ullrich, S. D. (1998), Effects of pest and soil management systems on weed dynamics in potato, *Weed Sci.* 46(2), 238–48.

Garrison, A. J., Miller, A. D., Ryan, M. R., Roxburgh, S. H. and Shea, K. (2014), Stacked crop rotations exploit weed-weed competition for sustainable weed management, *Weed Sci.* 62(1), 166–76.

Gieske, M. F., Wyse, D. L. and Durgan, B. R. (2016), Spring- and fall-seeded radish cover-crop effects on weed management in corn, *Weed Technol.* 30(2), 559–72.

González-Diaz, L., van den Berg, F., van den Bosch, F. and González-Andújar, J. L. (2012), Controlling annual weeds in cereals by deploying crop rotation at the landscape scale: *Avena sterilis* as an example, *Ecol. Applic.* 22(3), 982–92.

Gulden, R. H., Lewis, D. W., Froese, J. C., Van Acker, R. C., Martens, G. B., Entz, M. H., Derksen, D. A. and Bell, L. W. (2011), The effect of rotation and in-crop weed management on the germinable weed seedbank after 10 years, *Weed Sci.* 59(4), 553–61.

Gulden, R. H., Sikkema, P. H., Hamill, A. S., Tardif, F. J. and Swanton, C. J. (2009), Conventional vs. glyphosate-resistant cropping systems in Ontario: weed control, diversity, and yield, *Weed Sci.* 57(6), 665–72.

Gulden, R. H., Sikkema, P. H., Hamill, A. S., Tardif, F. J. and Swanton, C. J. (2010), Glyphosate-resistant cropping systems in Ontario: multivariate and nominal trait-based weed community structure, *Weed Sci.* 58(3), 278–88.

Gunton, R. M., Petit, S. and Gaba, S. (2011), Functional traits relating arable weed communities to crop characteristics, *J. Veg. Sci.* 22, 541–50.

Halde, C., Bamford, K. C., and Entz, M. H. (2015), Crop agronomic performance under a six-year continuous organic no-till system and other tilled and conventionally-managed systems in the northern Great Plains of Canada, *Agric. Ecosyst. Environ.* 213, 121–30.

Halde, C. and Entz, M. H. (2014), Flax (*Linum usitatissimum* L.) production system performance under organic rotational no-till and two organic tilled systems in a cool subhumid continental climate, *Soil Till. Res.* 143, 145–54.

Halde, C., Gulden, R. H. and Entz, M. H. (2014), Selecting cover crop mulches for organic rotational no-till systems in Manitoba, Canada, *Agron. J.* 106(4), 1193–204.

Harker, K. N., O'Donovan, J. T., Turkington, T. K., Blackshaw, R. E., Lupwayi, N. Z., Smith, E. G., Johnson, E. N., Pageau, D., Shirtliffe, S. J., Gulden, R. H., Rowsell, J., Hall, L. M. and Willenborg, C. J. (2016), Diverse rotations and optimal cultural practices control wild oat (*Avena fatua*), *Weed Sci.* 64(1), 170–80.

Hayden, Z. D., Ngouajio, M. and Brainard D. C. (2014), Rye–vetch mixture proportion tradeoffs: cover crop productivity, nitrogen accumulation, and weed suppression, *Agron. J.* 106(4), 904–14.

Heap, I. (2016), The International Survey of Herbicide Resistant Weeds, www.weedscience.org. (accessed 5 November 2016).

Heggenstaller, A. H. and Liebman, M. (2006), Demography of *Abutilon theoprasti* and *Setaria faberi* in three crop rotation systems, *Weed Res.* 46(2), 138–51.

Hill, E. C., Renner, K. A., Sprague, C. L. and Davis, A. S. (2016), Cover crop impact on weed dynamics in an organic dry bean system, *Weed Sci.* 64(2), 261–75.

Hodgdon, E. A., Warren, N. D., Smith, R. G. and Sideman, R. G. (2016), In-season and carry-over effects of cover crops on productivity and weed suppression, *Agron. J.* 108(4), 1624–35.

Huarte, H. R. and Benech Arnold, R. L. (2003), Understanding mechanisms of reduced annual weed emergence in alfalfa, *Weed Sci.* 51(6), 876–85.

Jordan, N., Mortensen, D. A., Prenzlow, D. M. and Curtis Cox, K. (1995), Simulation analysis of crop rotation effects on weed seedbanks, *Amer. J. Bot.* 82(3), 390–8.doi:10.2307/2445585

Korres, N. E. and Norsworthy, J. K. (2015), Influence of a rye cover crop on the critical period for weed control in cotton, *Weed Sci.* 63(1), 346–52.

Lawson, A., Cogger, C., Bary, A. and Fortuna, A.-M. (2015), Influence of seeding ratio, planting date, and termination date on rye-hairy vetch cover crop mixture performance under organic management, *PLoS ONE* 10(6): e0129597. doi:10.1371/journal.pone.0129597.

Légère, A. and Stevenson, F. C. (2002), Residual effects of crop rotation and weed management on a wheat test crop and weeds, *Weed Sci.* 50(1), 101–11.

Légère, A., Stevenson, F. C. and Benoit, D. L. (2005), Diversity and assembly of weed communities: contrasting responses across cropping systems, *Weed Res.* 45, 303–15.

Légère, A., Stevenson, F. C. and Benoit, D. L. (2011), The selective memory of weed seedbanks after 18 years of conservation tillage, *Weed Sci.* 59(1), 98–106.

Leon, R. G., Wright, D. L. and Marois, J. J. (2015), Weed seed banks are more dynamic in a sod-based, than in a conventional, peanut-cotton rotation, *Weed Sci.* 63(4), 877–87.

Liebman, M. and Dyck, E. (1993), Crop rotation and intercropping strategies for weed management, *Ecol. Appl.* 3(1), 92–122.

Liebman, M., Miller, Z. J., Williams, C. L., Westerman, P. R., Dixon, P. M., Heggenstaller, A., Davis, A. S., Menalled, F. D. and Sundberg, D. N. (2014), Fates of *Setaria faberi* and *Abutilon theophrasti* seeds in three crop rotation systems, *Weed Res.* 54, 293–306.

Liebman, M. and Staver, C. P. (2001), Crop diversification for weed management. In Liebman, M., Mohler, C. L. and Staver, C. P. (eds), *Ecological Management of Agricultural Weeds*, Cambridge University Press, Cambridge, UK, pp. 322–74.

Lou, Y., Davis, A. S. and Yannarell, A. C. (2016), Interactions between allelochemicals and the microbial community affect weed suppression following cover crop residue incorporation into soil, *Plant Soil* 399, 357–71.

Luna, J. M., Mitchell, J. P. and Shrestha, A. (2012), Conservation tillage for organic agriculture: Evolution toward hybrid systems in the western USA, *Renew. Agric. Food Syst.* 27(1), 21–30.

Meiss, H., Médiène, S., Waldhardt, R., Caneill, J., Bretagnolle, V., Reboud, X. and Munier-Jolain, N. (2010), Perennial lucerne affects weed community trajectories in grain crop rotations, *Weed Res.* 50, 331–40.

Melander, B., Rasmussen, I. A. and Olesen, J. E. (2016), Incompatibility between fertility building measures and the management of perennial weeds in organic cropping systems, *Agric. Ecosyst. Environ.* 220, 184–92.

Mertens, S. K., van den Bosch, F. and Heesterbeek, J. A. P. (2002), Weed populations and crop rotations: exploring dynamics of a structured periodic system, *Ecol. Applic.* 12(4), 1125–41.

Mirsky, S. B., Gallandt, F. R., Mortensen, D. A., Curran, W. S. and Shumway, D. L. (2010), Reducing the germinable weed seedbank with soil disturbance and cover crops, *Weed Res.* 50, 341–52.

Mirsky, S. B., Ryan, M. R., Curran, W. S., Teasdale, J. R., Maul, J., Spargo, J. T. Moyer, J., Grantham, A. M., Weber, D., Way, T. R. and Camargo, G. G. (2012), Conservation tillage issues: Cover crop-based organic rotational no-till grain production in the mid-Atlantic region, USA, *Renew. Agric. Food Syst.* 27(1), 31–40.

Mirsky, S. B., Ryan, M. R., Teasdale, J. R., Curran, W. S., Reberg-Horton, C. S., Spargo, J. T., Wells, M. S., Keene, C. L. and Moyer, J. W. (2013), Overcoming weed management challenges in cover crop–based organic rotational no-till soybean production in the Eastern United States, *Weed Technol.* 27(1), 193–203.

Mohler, C. L. (2001), Weed evolution and community structure. In Liebman, M., Mohler, C. L. and Staver, C. P. (eds), *Ecological Management of Agricultural Weeds*, Cambridge University Press, Cambridge, UK, pp. 444–93.

Murphy, S. D., Clements, D. R., Belaoussoff, S., Kevan, P. G. and Swanton, C. J. (2006), Promotion of weed species diversity and reduction of weed seedbanks with conservation tillage and crop rotation, *Weed Sci.* 54(1), 69–77.

Navas, M. L. (2012), Trait-based approaches to unravelling the assembly of weed communities and their impact on agroecosystem functioning, *Weed Res.* 52, 479–88.

Nord, E. A., Curran, W. S., Mortensen, D. A., Mirsky, S. B. and Jones, B. P. (2011), Integrating multiple tactics for managing weeds in high residue no-till soybean, *Agron. J.* 103(5), 1542–51.

Nord, E. A., Ryan, M. R., Curran, W. S., Mortensen, D. A. and Mirsky, S. B. (2012), Effects of management type and timing on weed suppression in soybean no-till planted into roll-crimped cereal rye, *Weed Sci.* 60(4), 624–33.

Norsworthy, J. K., Ward, S. M., Shaw, D. R., Llewellyn, R. S., Nichols, R. L., Webster, T. M., Bradley, K. W., Frisvold, G., Powles, S. B., Burgos, N. R., Witt, W. W. and Barrett, M. (2012), Reducing the risks of herbicide resistance: best management practices and recommendations, *Weed Sci.* 60(Special Issue), 31–62.

Ominski, P. D., Entz, M. H. and Kenkel, N. (1999), Weed suppression by *Medicago sativa* in subsequent cereal crops: a comparative survey, *Weed Sci.* 47(3), 282–90.

Owen, M. D. K., Beckie, H. J., Leeson, J. Y., Norsworthy, J. K. and Steckel, L. E. (2014), Integrated pest management and weed management in the US and Canada, *Pest. Manag. Sci.* 71, 357–76.

Price, A. J., Duzy, L. M., Monks, C. D., Kelton, J. A., Culpepper, A. S., Sosnoskie, L. M., Marshall, M. W., Steckel, L. E. and Nichols, R. L. (2016), High-residue cover crops alone or with strategic tillage to manage glyphosate resistant Palmer amaranth (*Amaranthus palmeri*) in southeastern cotton (*Gossypium hirsutum*), *J. Soil Water Conserv.* 71(1), 1–11.

Radicetti, E., Mancinelli, R. and Campiglia, E. (2013), Impact of managing cover crop residues on the floristic composition and species diversity of the weed community of pepper crop (*Capsicum annuum* L.), *Crop Prot.* 44, 109–19.

Ramsdale, B. K., Kegode, G. O., Messersmith, C. G., Nalewaja, J. D. and Nord, C. A. (2006), Long-term effects of spring wheat-soybean cropping systems on weed populations, *Field Crops Res.* 97, 197–208.

Reberg-Horton, S. C., Grossman, J. M., Kornecki, T. S., Meijer, A. D., Price, A. J., Place, G. T. and Webster, T. M. (2012), Utilizing cover crop mulches to reduce tillage in organic systems in the southeastern USA, *Renew. Agric. Food Syst.* 27(1), 41–8.

Rühlemann, L. and Schmidtke, K. (2015), Evaluation of monocropped and intercropped grain legumes for cover cropping in no-tillage and reduced tillage organic agriculture, *Europ. J. Agron.* 65, 83–94.

Ruisi, P., Frangipane, B., Amato, G., Badagliacca, G., Di Miceli, G., Plaia, A. and Giambalvo, D. (2015), Weed seedbank size and composition in a long-term tillage and crop sequence experiment, *Weed Res.* 55(3), 320–8.

Ryan, M. R., Mirsky, S. B., Mortensen, D. A., Teasdale, J. R. and Curran, W. S. (2011), Potential synergistic effects of cereal rye biomass and soybean planting density on weed suppression, *Weed Sci.* 59(2), 238–46.

Ryan, M. R., Mortensen, D. A., Bastiaans, L., Teasdale, J. R., Mirsky, S. B., Curran, W. S., Seidel, R., Wilson, D. O. and Hepperly, P. R. (2010), Elucidating the apparent maize tolerance to weed competition in long-term organically managed systems, *Weed Res.* 50, 25–36.

Saunders Bulan, M. T., Stoltenberg, D. E. and Posner, J. L. (2015), Buckwheat species as summer cover crops for weed suppression in no-tillage vegetable cropping systems, *Weed Sci.* 63(3), 690–702.

Schipanski, M. E., Barbercheck, M., Douglas, M. R., Finney, D. M., Haider, K., Kaye, J. P., Kemanian, A. R., Mortensen, D. A., Ryan, M. R., Tooker, J. and White, C. (2014), A framework for evaluating ecosystem services provided by cover crops in agroecosystems, *Agric. Syst.* 125, 12–22.

Schoofs, A. and Entz, M. H. (2000), Influence of forages on weed dynamics in a cropping system, *Can. J. Plant Sci.* 80, 187–98.

Schreiber, M. M. (1992), Influence of tillage, crop rotation, and weed management on giant foxtail population dynamics and corn yield, *Weed Sci.* 40(4), 645–53.

Shrestha, A., Knezevic, S. Z., Roy, R. C., Ball-Coelho, B. R. and Swanton, C. J. (2002), Effect of tillage, cover crop and crop rotation on composition of weed flora in a sandy soil. *Weed Res.* 42, 76–87.

Sjursen, H. (2001), Change of the weed seed bank during the first complete six-course crop rotation after conversion from conventional to organic farming, *Biol. Agric. Hortic.* 19, 71–90.

Smith, A. N., Reberg-Horton, C. S., Place, G. T., Meijer, A. D., Arellano, C. and Mueller, J. P. (2011), Rolled rye mulch for weed suppression in organic no-tillage soybeans, *Weed Sci.* 59(2), 224–31.

Smith, R. G., Atwood, L. W., Pollnac, F. W. and Warren, N. D. (2015), Cover-crop species as distinct biotic filters in weed community assembly, *Weed Sci.* 63(1), 282–95.

Smith, R. G., Atwood, L. W. and Warren, N. D. (2014), Increased productivity of a cover crop mixture is not associated with enhanced agroecosystem services, *PLoS ONE* 9. doi:10.1371/journal. pone.0097351.

Smith, R. G. and Gross, K. L. (2006), Rapid changes in the germinable fraction of the weed seed bank in crop rotations, *Weed Sci.* 54(6), 1094–100.

Smith R. G. and Gross K. L. (2007), Assembly of weed communities along a crop diversity gradient, *J. Appl. Ecol.* 44, 1046–56.

Smith, R. G., Jabbour, R., Hulting, A. G., Barbercheck, M. E. and Mortensen, D. A. (2009a), Effects of initial seed-bank density on weed seedling emergence during the transition to an organic feed-grain crop rotation, *Weed Sci.* 57(5), 533–40.

Smith, R. G., Mortensen, D. A. and Ryan, M. R. (2009b), A new hypothesis for the functional role of diversity in mediating resource pools and weed–crop competition in agroecosystems, *Weed Res.* 50, 37–48.

Sosnoskie, L. M., Herms, C. P. and Cardina, J. (2006), Weed seed community composition in a 35-year-old tillage and rotation experiment, *Weed Sci.* 54(2), 263–73.

Sosnoskie, L. M., Herms, C. P., Cardina, J. and Webster, T. M. (2009), Seedbank and emerged weed communities following adoption of glyphosate-resistant crops in a long-term tillage and rotation study, *Weed Sci.* 57(3), 261–70.

Stevenson, F. C. and Johnston, A. M. (1999), Annual broadleaf crop frequency and residual weed populations in Saskatchewan Parkland, *Weed Sci.* 47(2), 208–14.

Storkey, J., Döring, T., Baddeley, J., Collins, R., Roderick, S., Jones, H. and Watson, C. (2015), Engineering a plant community to deliver multiple ecosystem services, *Ecol. Appl.* 25(4), 1034–43.

Swanton, C. J., Booth, B. D., Chandler, K., Clements, D. R. and Shrestha, A. L. (2006), Management in a modified no-tillage corn-soybean-wheat rotation influences weed population and community dynamics, *Weed Sci.* 54(1), 47–58.

Teasdale, J. R., Brandsæter, L. O., Calegari, A. and Neto, F. S. (2007), Cover crops and weed management. In Upadhyaya, M. K. and Blackshaw, R. E. (eds), *Non-Chemical Weed Management*, CAB Int., Chichester, UK, pp. 49–64.

Teasdale, J. R., Mangum, R. W., Radhakrishnan, J. and Cavigelli, M. A. (2004), Weed seedbank dynamics in three organic farming crop rotations, *Agron. J.* 96(5), 1429–35.

Teasdale, J. R. and Mirsky, S. B. (2015), Tillage and planting date effects on weed dormancy, emergence, and early growth in organic corn, *Weed Sci.* 63(2), 477–90.

Teasdale, J. R., Mirsky, S. B., Spargo, J. T., Cavigelli, M. A. and Maul, J. E. (2012a), Reduced-tillage organic corn production in a hairy vetch cover crop, *Agron. J.* 104(3), 621–8.

Teasdale, J. R. and Mohler, C. L. (2000), The quantitative relationship between weed emergence and the physical properties of mulches, *Weed Sci.* 48(3), 385–92.

Teasdale, J. R., Rice, C. P., Cai, G. and Mangum, R. W. (2012b), Expression of allelopathy in the soil environment: soil concentration and activity of benzoxazinoid compounds released by rye cover crop residue, *Plant Ecol.* 213, 1893–905.

Thorne, M. E., Young, F. L. and Yenish, J. P. (2007), Cropping systems alter weed seed banks in Pacific Northwest semi-arid wheat region, *Crop Prot.* 26(8), 1121–34.

Tilman, D., Reich, P. B. and Isbell, F. (2012), Biodiversity impacts ecosystem productivity as much as resources, disturbance, or herbivory, *Proc. Nat. Acad. Sci.* 109(26), 10394–7.

Tomasoni, C., Borrelli, L. and Pecetti, L. (2003), Influence of fodder crop rotations on the potential weed flora in the irrigated lowlands of Lombardy, Italy, *Eur. J. Agron.* 19(3), 439–51.

Ulber, L., Steinmann, H. H., Klimek, S. and Isselstein, J. (2009), An on-farm approach to investigate the impact of diversified crop rotations on weed species richness and composition in winter wheat, *Weed Res.* 49(5), 534–43.

Wayman, S., Cogger, C., Benedict, C., Collins, D., Burke, I. and Bary, A. (2015), Cover crop effects on light, nitrogen, and weeds in organic reduced tillage, *Agroecol. Sustain. Food Syst.* 39(6), 647–65.

Webster, T. M., Scully, B. T., Grey, T. L. and Culpepper, A. S. (2013), Winter cover crops influence *Amaranthus palmeri* establishment, *Crop Prot.* 52, 130–5.

Webster, T. M., Simmons, D. B., Culpepper, A. S., Grey, T. L., Bridges, D. C. and Scully, B. T. (2016), Factors affecting potential for Palmer amaranth (*Amaranthus palmeri*) suppression by winter rye in Georgia, USA, *Field Crops Res.* 192, 103–9.

Wedryk, S. and Cardina, J. (2012), Smother crop mixtures for Canada thistle (*Cirsium arvense*) suppression in organic transition, *Weed Sci.* 60(4), 618–23.

Wells, M. S., Reberg-Horton, S. C. and Mirsky, S. B. (2014), Cultural strategies for managing weeds and soil moisture in cover crop based no-till soybean production, *Weed Sci.* 62(3), 501–11.

Wells, M. S., Reberg-Horton, S. C. and Mirsky, S. B. (2016), Planting date impacts on soil water management, plant growth, and weeds in cover-crop-based no-till corn production, *Agron. J.* 108(1), 162–70.

Wells, M. S., Reberg-Horton, S. C., Smith, A. N. and Grossman, J. M. (2013), The reduction of plant-available nitrogen by cover crop mulches and subsequent effects on soybean performance and weed interference, *Agron. J.* 105(2), 539–45.

Westerman, P. R., Liebman, M., Menalled, F. D., Heggenstaller, A. H., Hartzler, R. G. and Dixon, P. M. (2005), Are many little hammers effective? Velvetleaf population dynamics in two- and four-year crop rotation systems, *Weed Sci.* 53(3), 382–92.

Wiggins, M. S., Hayes, R. M. and Steckel, L. E. (2016), Evaluating cover crops and herbicides for glyphosate-resistant Palmer amaranth (*Amaranthus palmeri*) control in cotton, *Weed Technol.* 30(2), 415–22.

Wiggins, M. S., McClure, M. A., Hayes, R. M. and Steckel, L. E. (2015), Integrating cover crops and POST herbicides for glyphosate-resistant Palmer amaranth (*Amaranthus palmeri*) control in corn, *Weed Technol.* 29(3), 412–18.

Wortman, S. E., Francis, C. A., Bernards, M. A., Blankenship, E. E. and Lindquist, J. L. (2013), Mechanical termination of diverse cover crop mixtures for improved weed suppression in organic cropping systems, *Weed Sci.* 61(1), 162–70.

Wortman, S. E., Francis, C. A. and Lindquist, J. L. (2012), Cover crop mixtures for the western Corn Belt: Opportunities for increased productivity and stability, *Agron. J.* 104(3), 699–705.

Wortman, S. E., Lindquist, J. L., Haar, M. J. and Francis, C. A. (2010), Increased weed diversity, density and above-ground biomass in long-term organic crop rotations, *Renew. Agric. Food Syst.* 25(4), 281–95.

Chapter 13

Developments in physical weed control

Eric R. Gallandt, University of Maine, USA; Daniel Brainard, Michigan State University, USA; and Bryan Brown, University of Maine, USA

1 Introduction

2 Tillage

3 Physical weed control: overview

4 Tools, weeds and soil conditions

5 Weed–crop selectivity

6 Fundamental problems with cultivation

7 Future research priorities

8 Where to look for further information

9 References

1 Introduction

Physical weed control (PWC) refers to the use of a tool or implement to kill a weed seedling, an operation that some authors call *cultivation* or *mechanical weed control*. PWC is a foundational practice for organic farmers, and for diversified vegetable farmers, who have concerns regarding herbicide drift and carryover. Intractable cases of herbicide resistance, notably multiple resistance in *Amaranthus palmeri* and *Conyza canadensis*, and lack of novel herbicide sites of action (Heap 2014) may inspire a return to PWC by grain and cotton farmers as well.

Several important terms are being used widely to describe PWC. *Tillage* involves soil manipulation to enhance crop production, including both primary and secondary tillage that are used to bury plant residues and prepare a seedbed (ASAE 2005). This aggressive physical disturbance makes it an effective weed management operation. *Cultivation* refers to shallow tillage operations aimed to promote crop growth, including control of pests ASAE 2005), and the term is often used synonymously with physical or mechanical weed control. PWC is executed with a given *implement, tool* or *machine*; examples include tine- and rolling harrows, sweeps, torsion or finger weeders. Photos, drawings and descriptions of many common tools are featured in the farmer-focused publications by Bowman (2002) and Van der Shans et al. (2006).

Efficacy describes the proportion of weed seedlings killed by a given PWC event. Ideally, efficacy values are based on pre- and post-cultivation censuses, with sufficient time between counts that surviving and dead seedlings are easily distinguished, but not

http://dx.doi.org/10.19103/AS.2017.0025.15

so much time that a subsequent cohort of seedlings confounds the measurement (Vanhala et al. 2004). Unfortunately, many PWC studies report weed density, perhaps biomass, at some point late in the growing season (Mohler et al. 2016), thereby confounding PWC efficacy with other sources of mortality and recruitment.

Selectivity refers to killing weeds but not the crop. This has long been based on the size advantage that good management provides the crop, for example, by transplanting or by exploiting the greater seed mass and initial size of many crop species (Mohler 1996). Planting crops in rows makes it relatively easy to disturb soil and control weeds between crop rows, that is, the '*inter-row*' zone. As tools are operated closer to the crop row, chances of crop injury or mortality increase. Thus, weeds emerging within the crop row – that is, the '*intra-row*' zone – are essentially protected by the crop. Control of intra-row weeds has long been a challenge and is the subject of considerable research in PWC (van der Weide et al. 2008).

Thermal weed control – generally involving heat treatment (fire or flame, infra-red heaters, hot water or steam (Ascard et al. 2007)), but also freezing (liquid carbon dioxide, (Mahoney, Jeffries and Gannon 2014)) – may be considered a PWC practice, but will not be covered in this chapter. Flaming is reviewed by Datta and Knezevic (2013).

It has been nearly a decade since Chicouene (2007) and Cloutier et al. (2007) published their reviews on PWC, and even longer since Mohler's (2001) chapter on the subject. These academic reviews are supported by two excellent and practical publications for farmers by Bowman (2002) and van der Schans (2006) (see also van der Weide et al. 2008) that provide clear and detailed photographs, illustrations and descriptions of the many tools available for PWC. Videos of tools in operation are often an even better way to understand an implement's application, and YouTube is now a first stop for many researching PWC. Given these existing resources, we do not intend to present a comprehensive overview of existing tools and their applications, but, rather, to focus on more recent innovations and advances in mechanistic understanding of PWC.

We start with a brief discussion of tillage, the most aggressive of PWC practices. While tillage remains important and useful for weed management, we consider it foremost a practice for establishing a seedbed. The importance of tillage to subsequent PWC activities, however, should not be underestimated; conditions established at planting can have a large effect on later PWC operations. Next we move on to our focal area – PWC – practices specifically targeting weeds, and notably the key variables of *tools*, *weeds*, *soil conditions* and *weed-crop selectivity*. We discuss several fundamental problems with PWC, some that may be addressed by future innovations in design and application, and others that require PWC to be deployed within the context of a broader systems approach to weed management, notably, strategies to reduce the weed seed bank. We conclude with a call for additional research, suggesting that PWC could ultimately approach herbicides in its efficacy and selectivity and could provide robust and effective weed control for farmers choosing to reduce their reliance on herbicides.

2 Tillage

Tillage can be useful for both annual and perennial weed control. In the case of perennial weeds, deep burial of perennating organs can reduce subsequent shoot density. Properly timed shallow tillage removes and destroys weed shoots, breaking dormancy of vegetative meristems, thereby encouraging plantlet regrowth which draws upon carbohydrate reserves (Håkansson 2003). When repeated at an optimal interval, below-ground reserves are depleted, resulting in plant death. If tillage events are too frequent, heterotrophic shoots are destroyed before using significant carbohydrate reserves; if too infrequent, shoots become sufficiently autotrophic that below-ground carbohydrate reserves are replenished (Håkansson 2003). Melander et al. (2012) demonstrated that post-harvest, fall rotary tillage, followed by spring moldboard ploughing, offered effective control of a mixed perennial weed stand, noting, however, the undesirable nature of such disturbance from a soil quality perspective.

In the case of annual weed management, tillage is commonly used to prepare a stale seedbed, a foundational weed management strategy for many organic vegetable farmers. In practice, a seedbed is prepared, but crop planting is delayed, allowing weeds to germinate and establish; they are subsequently killed just prior to planting (Caldwell and Mohler 2001). Ideally, weeds are killed with minimal soil disturbance to reduce the density of a subsequent cohort of seedlings. Flaming (Boyd et al. 2006) or glyphosate herbicide (Riemens et al. 2007) can be especially useful in this regard. Rasmussen (2003) described a 'punch planting' system that minimized planter-caused soil disturbance following flaming. Compared to normal planting, this minimal disturbance technique reduced weed density by 30%. Mohler (2001) distinguished *stale* and *false* seedbed approaches, terms that are sometimes used interchangeably by farmers. Both practices aim to reduce the density of weed seedlings emerging with a sown crop. The false seedbed approach simply refers to multiple shallow disturbance events, generally about one week apart, which aim to more exhaustively deplete the germinable portion of the seedbank by encouraging germination (Peruzzi et al. 2007). Using a depth-structured model, Lamour et al. (2007) predicted that initial deep burial followed by timely shallow tillage would reduce weed density by 32% compared to a similar sequence concluding with deep burial.

Exploiting the false seedbed approach requires suitable soil moisture and temperature to promote germination. Schutte et al. (2013) examined shallow soil disturbance effects on emergence of *Chenopodium album* over 16 site-years in North America and Europe. Based on thermal time, peak emergence of *C. album* occurred 500 d°C (base 3C) following disturbance, and thus, flaming or subsequent disturbance would be most effective at this interval. Temperature can be increased with black plastic or vinyl tarps, so-called 'occultation' (see Fortier 2014), or using clear plastic to 'solarize' the soil (Cohen and Rubin 2007). Recently there has been a surge of interest in these practices among small-scale organic vegetable producers in northern New England.

Tillage and seedbed preparation can also have a large impact on subsequent PWC events (Mohler 2001). Intuition and experience suggest that most PWC tools perform optimally in soils that are level and relatively free of clods, residue and stones. It is possible, even likely, that PWC tools vary in their efficacy depending on these soil conditions. As discussed in the following sections, tool-soil condition relationships have only rarely been studied.

3 Physical weed control: overview

3.1 Aims and principles of PWC

At first glance, PWC appears as a relatively simple operation: push or drag a tool though the soil to kill weeds. In the field, however, it is evident that the key variables – including tools, weeds, soil and crop (selectivity) – are more complicated than they may appear.

Tools vary widely in their design, adjustment and speed of operation. *Weeds* vary in species, size and density. *Soil* conditions logically affect the action of tools, especially moisture content, 'tilth', organic matter, textural class or residues. While efficacy is usually the focus of PWC research, *selectivity* is equally important, affected not only by crop species and size, but by these same tool, weed and soil variables. Presently, optimization of these many variables requires both knowledge of principles, and considerable experience, leading some to conclude that successful PWC is as much 'art' as it is 'science' (Bowman, 2002).

On the basis of his review of the literature, and his own considerable field experience, Mohler (2001) established seven operational principles to guide PWC:

1   Row-oriented cultivators should work the same number of rows as the planter, or a simple fraction of this number.
2   The action of the cultivator must be appropriate for the growth stages of the weeds and the crop.
3   Creation and maintenance of a size differential between the crop and the weeds facilitates effective mechanical weed control.
4   The effectiveness of cultivation decreases as weed density increases.
5   Effective cultivation requires good tilth, careful seedbed preparation and adequate soil drainage.
6   Cultivation (and tillage) in the dark stimulates germination of fewer weed seeds than cultivation in daylight.
7   Attentive timing relative to changing weather and soil conditions can improve the effectiveness of cultivation.

These operational guidelines highlight fundamental tool-weed-soil-crop components, showing the scope and complexity of PWC.

The control of inter-row weeds in row crops is a routine PWC operation, easily achieved with a wide range of tools including sweeps, knives, spiders, rolling baskets, rotating tines and brushes. Pullen and Cowell (1997) conducted a thorough field evaluation of inter-row cultivators, both new designs (e.g. brush hoe) and long-used tools (e.g. sweep). Of note was the fact that selected treatments performed equally, or better, than their herbicide reference treatment. The top performing tool was a 'sweep hoe', an implement reportedly with excellent depth control (25 mm) and with large L- and/or A-shaped blades that run just below, but largely parallel to, the soil surface. Surrogate weeds, rapeseed in this case, and natural weeds were reduced in density by 90% and 88%, respectively. The sweep hoe is a particularly useful design when it is important to avoid soil movement into the crop row, a problem noted with the ducksfoot in these experiments (Pullen and Cowell 1997).

A key goal in inter-row PWC operations is to cultivate as close to the crop row as possible, that is, to minimize the size of the intra-row zone. At smaller scales, this may be achieved using specially designed cultivating tractors that offer the operator improved visibility and precise steering of tools close to crop rows while optimizing speed (working rate). The now antique Allis Chalmers G remains a desirable cultivation tractor for small-scale vegetable farmers. Many models of Farmall cultivating tractors from the 1950s and 1960s are also seen on smaller-scale farms. Other examples, long out of production, include the Ford 1710, IH Cub, Case IH 265 and Kubota L245H, all off-set cultivating tractors with good operator visibility for centre-mounted tools. Several European manufacturers offer tool carriers with a longer wheelbase and greater centre clearance, offering more space for tools, including the ability to stack several tools in sequence (see examples from HAK, Baertschi-Fobro, Mazzotti and Terratek). Two encouraging developments in the United States include Tuff-bilt and Oggún tractors, both relatively affordable at $12 500. The Oggún is unique in its open-source manufacturing model, constructed using commonly available components and parts for easy user repair. The Tuff-bilt and Oggún are modelled after the Allis Chalmers G, an innovative tractor for its time, but lacking contemporary design elements present on the modern European tool carriers.

Control of intra-row weeds is far more challenging owing to relatively crude weed-crop selectivity mechanisms being exploited (see Section 5). Two areas of innovation have improved intra-row weed control in recent years. First, guidance systems, both camera- and GPS-based, permit closer adjustment of tools to the crop row, and thus, higher working rates (Fennimore et al. 2010). Second is the development of so-called 'intelligent' weeders for crops with relatively wide intra-row spacing. The Sarl Radis from France was one of the first commercially available machines to use cameras, precise ground speed measurement and computer-actuated hoes to disturb intra-row soil between crop plants. Presently there are four such weeders commercially available in the marketplace (Robovator, F. Poulsen Engineering; Robocrop, Garford Farm Machinery; Intra-row Hoe 'IC', Steketee BV; and Remoweed, Ferrari). Their performance in the field, however, has only recently been independently tested. Fennimore et al. (2010) found that the intelligent weeders delivered better weed control and net return than herbicide-based strategies in lettuce, due presumably to herbicide injury to the crop. However, rainy weather reduced the performance of the cultivator, requiring supplemental herbicide application to maintain crop yields and profitability. Melander et al. (2015) recently compared a state-of-the-art intelligent weeder, the Robovator (F. Poulsen Engineering, Denmark) to the best available intra-row tools (harrow, finger and torsion weeders) in transplanted onion and cabbage. The Robovator worked well, but was not superior to the simpler tools, despite a considerable difference in machinery cost. These intelligent intra-row tools rely on a single moving blade or knife for disturbance. It may be necessary to 'stack' additional cultivation tools on these weeders to improve efficacy (see Section 4.1 subsection 'Stacking' Tools and 'Synergy').

3.2 Mechanisms of mortality

PWC generally targets weeds at the seedling stage, causing death by three distinct, but often difficult-to-measure mechanisms: burial, cutting and uprooting with subsequent desiccation (Terpstra and Kouwenhoven 1981). The dominant mechanism of weed mortality has been shown to vary by tool. Early investigations of spring tine harrowing found that burial was the main cause of mortality (Habel 1954; Kees 1962; Koch 1964). A laboratory study by Kurstjens et al. (2000) digitized surrogate weed seed locations prior to harrowing to allow for evaluation of mortality of emerging white thread stage plants. Plants loosened or moved by harrowing were categorized as uprooted. Fifty-one percent of emerging plants and 21% of seedlings were uprooted, demonstrating that mechanism may depend on plant size and that previous studies ignoring emerging plants were underestimating the importance of uprooting. Furthermore, 70% of all uprooted plants were completely buried by soil, highlighting the difficulty of determining cause of death and also the potential for increased burial of loosened weeds.

Fogleberg and Dock Gustavsson (1999) separated burial and uprooting mechanisms by vacuuming loose soil following brush hoeing, and counting still-anchored seedlings. Uprooting was determined to be the major mechanism of control for weeds at the two- to four-leaf stage, but severing was not evaluated, and it is plausible that the brush may have severed many of the weeds categorized as uprooted. Mortality of the carrot (*Daucus carota*) test crop was mostly the result of burial. Similarly, in laboratory experiments using 2.5–9.0 cm garden cress (*Lepidium sativum* L.) as surrogate weeds, burial was the only source of mortality (45%) for seedlings on either side of a ducksfoot hoe, whereas in the path of the hoe, 57% and 33% were killed by burial and uprooting, respectively. Perhaps the effectiveness of the hoe is predicated on its ability to utilize multiple mortality mechanisms. Hoes can sever weeds just below the soil surface, lift and finally bury seedlings with soil. The relative importance of these mechanisms may depend on soil conditions, seedling and tool characteristics, but the combination of the three sources likely serves to reduce variability in efficacy. In an evaluation of several hoes, most weeds were killed by burial, but it was acknowledged that buried weeds likely also incurred some severing and uprooting (Evans et al. 2012).

4 Tools, weeds and soil conditions

4.1 Tools

PWC efficacy and selectivity are perhaps most affected by choice of tool, its fundamental design, operating speed and adjustment. In the following section, we discuss these topics and introduce the concept of tool 'stacking' used by innovative manufacturers and farmers but only recently, to our knowledge, subjected to objective scientific study.

Design

Tool design can be considered from the perspective of the crop and selectivity, with some designs disregarding the crop, and others specifically targeting inter- and intra-row zones. Narrow-row crops, for example, cereals, canola, drilled soybean and grain legumes, rely primarily on crop competition to minimize yield loss and ideally suppress weeds. Rows are generally spaced such that the inter-row area is sufficiently narrow to preclude cultivation that is routinely used in more widely spaced row crops. In these crops PWC tools are limited to tine harrows and rotary hoes, pre- and post-emergence. These are the so-called 'blind' cultivation tools as they disturb the entire soil surface, intra- and inter-row alike, and rely on a large weed/crop size differential to provide selectivity. Ideally, small weeds are uprooted or buried by the soil disturbance action of the tines or rotating hoes while a more deeply rooted, larger crop survives. Under ideal conditions, including very small 'white-thread' weeds, large robust crop, hot sun and dry soil surface, efficacy of these tools may approach that of a reference herbicide programme, for example, 95% or greater (Gallandt, unpublished). Under more typical, less-than-ideal conditions, efficacy is much lower and more variable.

There are, however, farmers who have increased the inter-row distance in cereals to permit a more aggressive PWC practice of hoeing, essentially treating the cereal like a more traditional wide-row crop (Kolb and Gallandt 2012).

Inter-row hoeing, often combined with harrowing, was developed in northern Europe in response to challenging weed problems (e.g. tap-rooted, erect species including volunteer rape, *Brassica napus*, *Papaver rhoeas*, and *Tripleurospermum inodorum*) that require aggressive disturbance (Melander et al. 2003). Compared to a tine harrow, the aggressive sweeps of the inter-row hoe can control larger weeds, providing a wider window of opportunity for use, and with modern camera guidance systems, working rates are reportedly as high as 12 km hr^{-1} (Garford 2016). Field experiments conducted over four years in Maine, USA, demonstrated that, in years of heavy weed pressure, inter-row hoeing plus tine harrowing offered improved weed control and profitability compared to reference systems relying on enhanced crop competition and harrowing alone (Kolb et al. 2010 and 2012).

Speed

Increased forward speed of spring tine harrowing operations has been linked to increased efficacy (Rasmussen and Svenningsen 1995; Rydberg 1994) possibly due to increased weed uprooting (Kurstjens et al. 2000), increased soil movement (reviewed by Kouwenhoven and Terpstra 1979), or increased plant bending that results in more burial (Kurstjens and Perdok 2000). However, the relationship between forward speed and efficacy of spring tine harrowing is sometimes unclear (Cirujeda et al. 2003; Rasmussen 1990).

Increased speed was correlated with greater weed mortality for sweeps because it resulted in increased soil movement and soil dispersion; in contrast, two novel hoes that slice through soil with minimal soil movement, efficacy was not affected by speed (Evans et al. 2012). Torsion weeders were less effective at lower speed (Ascard and Fogelberg 2002). In an evaluation of the effect of working speed on the efficacy of several PWC tools using rapeseed (*B. napus*) as a surrogate weed, efficacy improved with increased speed for a spring tine harrow, sweeps and ducksfoot hoes but decreased for a powered rotary hoe and a powered brush hoe, likely because at increased speed, the rotation of the powered tools is less complete over a given area (Pullen and Cowell 1997).

Camera-guided intra- and inter-row weeding equipment can be operated at speeds of up to 5 and 12 km h^{-1}, respectively (Garford 2016). Increasing forward speed from 2 to 4 km h^{-1} had no effect on efficacy of a camera-guided intra-row weeder (Melander et al. 2015). However, even if there is no increase in efficacy, it is a practical advantage to be able to cultivate more land per hour. Small-scale farmers using human-steered cultivators limited to 6 km h^{-1} are not necessarily confined to low efficacy because ample soil movement can be observed at low speed with many tools, tools can be adjusted to throw more soil (Terpstra and Kouwenhoven 1981), or tools unaffected by forward speed may be used (Evans et al. 2012; Pullen and Cowell 1997).

Depth

Increasing tool working depth generally increases efficacy. In laboratory studies, increasing working depth of a tine harrow from 1 cm to 3 cm resulted in a doubling of uprooted plants (Kurstjens et al. 2000). Likewise, in field experiments, the most aggressive setting of a spring tine harrow resulted in greatest weed control (Raffaelli et al. 2002), likely due to a deeper working depth. Bleeker et al. (2004) showed that increasing working depth may increase the efficacy of torsion weeders, but with greater associated crop damage. Increasing working depth of a ducksfoot hoe from 2.5 to 4.0 cm only increased efficacy by 10% (Terpstra and Kouwenhoven 1981) possibly since increasing the depth of a hoe may not increase the soil disturbance in the top soil layer and may actually undercut weeds below the bulk of their root mass.

Soil disturbance to control weeds also stimulates the emergence of new weeds (Mohler 1993). Such stimulation caused by pre-emergence harrowing may increase weed competition during crop growth (Kees 1962). Ideally, a cultivation tool would control existing weeds while creating soil conditions that minimize the number of safe sites so that subsequent weed emergence would be reduced (Gallandt et al. 1999). Such establishment of new weeds was evaluated two weeks after cultivation using several different hoe designs and no difference was detected; however, it was expected that shallower operating tools would stimulate less germination, or after repeated use, exhaust germination from the surface soil layer (Evans et al. 2012).

'Stacking' tools and 'synergy'

Cultivation tools differ in their effects on soil and, consequently, weeds. Conceptually, it may be useful to consider PWC tools as differing in their primary 'mechanism of action'. Thus, perhaps tools can be combined, or 'stacked', to act in an additive or possibly synergistic manner that results in improved efficacy. This concept is evident in the design of the HAK® S-Series cultivator (HAK Schoffeltechniek, Moerkapelle, Holland), a thoughtfully designed machine including three intra-row tools in sequence: torsion weeders, finger weeders, and a tine rake (like a spring tine harrow, but utilizing a drop weight for downward pressure). In 2015, we conducted a trial examining the potential of stacked cultivation using each intra-row tool singly as well as in every feasible two- and three-tool combination. Field corn (*Zea mays* cv. Waipsie Valley) was used as a test crop, condiment mustard (*Sinapis alba* cv. Idagold) was seeded as a surrogate weed, and cultivation treatments were conducted while the mustard was in the cotyledon stage. Pre- and post-censuses of surrogate weeds were completed in a 6 m long by 10 cm wide intra-row census zone for each plot. Stacked tools resulted in a substantial improvement in efficacy. For example, efficacy was 44%, 29% and 54% for the torsion weeders, finger weeders and tine rake, respectively. Mean

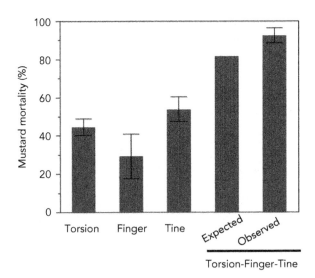

Figure 1 Efficacy of intra-row cultivation tools using condiment mustard (*Sinapis alba* cv. Idagold) as a surrogate weed. Expected efficacy of the Torsion-Finger-Tine combination was calculated using the Colby (1967) method. Standard error bars shown for actual results. Expected efficacy of the Torsion-Finger-Tine combination is a single value (Brown and Gallandt, unpublished).

efficacy for two- and three-tool combinations was 81% and 84%, respectively. Efficacy followed a logistic function of the number of tools used ($R^2 = 0.80$), indicating that the most substantial increases in efficacy may be achieved by stacking merely two or three tools. An additional benefit of tool stacking was that as efficacy increased, variance in efficacy decreased ($R^2 = 0.49$, $P = 0.015$). Using the method of Colby (1967), which is often used to test for synergy of herbicide combinations, expected efficacy of each tool combination was calculated based on an assumed additive relationship (data not shown). None of the tool combinations had an antagonistic effect, most had an additive effect, and several had a synergistic effect. For example, the torsion-finger-tine rake combination had an efficacy of 92%, which was greater than the expected 82% ($P = 0.026$, Fig. 1). To our knowledge, this is the first documented evidence of a synergistic combination of cultivation tools.

4.2 Weeds

Weeds vary in their response to PWC based on species-specific attributes and individual size. Farmers are often advised to cultivate when weeds are small and more susceptible to damage. This rule of thumb is supported by field experiments with spring tine harrows (Cirujeda and Taberner 2004) and laboratory experiments that have demonstrated a steep drop in effectiveness of tine harrows when plant size increased from emerging- to cotyledon stage (Kurstjens et al. 2000). Efficacy decreased with larger weeds for a GPS-guided intra-row hoe (Griepentrog et al. 2007) and an inter-row hoe (Terpstra and Kouwenhoven 1981); however, efficacy of aggressive, inter-row tools is typically less sensitive to weed size (Evans et al. 2012; Pullen and Cowell 1997). Indeed, when we evaluated inter-row weed control with gangs of 10 cm sweeps (4-row Case International Model 183 with Danish S-tines and gage wheels) using four growth stages of condiment mustard (*S. alba* cv. Idagold) as surrogate weeds, efficacy declined gradually from above 90% for cotyledon-stage plants to around 70% for plants with 5–6 leaves (Fig. 2).

Mohler et al. (2016) conducted detailed pre- and post-cultivation censuses to measure cultivation efficacy in corn and soybean. As expected, based on seedling growth rates,

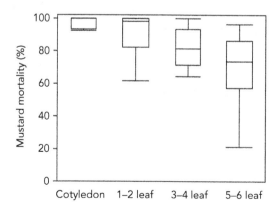

Figure 2 Efficacy of inter-row weed control with gangs of 10-cm sweeps (4-row Case International Model 183 with Danish S-tines and gage wheels) using four growth stages of condiment mustard (*Sinapis alba* cv. Idagold) as surrogate weeds (Brown and Gallandt, unpublished).

cultivation was more effective against annuals and perennials from seeds compared to perennial weeds originating from roots and rhizomes. Furthermore, cultivation efficacy was poorer for larger-seeded annuals, common ragweed in this study (Mohler et al. 2016).

4.3 Soil conditions

Several soil parameters have the potential to impact cultivation efficacy. Soil moisture is likely the most critical soil parameter related to cultivation efficacy. In laboratory tine harrowing experiments, weed uprooting was promoted by higher soil moisture content (Kurstjens et al. 2000), but mortality of uprooted weeds increased from 36% to 91% when soil moisture decreased from 16% to 5% (Kurstjens 2002). Similarly, in laboratory experiments with a hoe-ridger, using garden cress (*Lepidium sativum* L.) as a surrogate weed, efficacy was reduced from 90% to 78% with the addition of wetting after cultivation (Terpstra and Kouwenhoven 1981). In an evaluation of optimal conditions for spring tine harrowing in winter cereals, dry soils and sunshine during and after treatment resulted in the highest efficacy (Cirujeda and Taberner 2004). Perhaps improved tool design can limit the negative effect of soil moisture on cultivation efficacy. For example, soil moisture was found to be correlated with increased weed survival for cultivation with sweeps and a stirrup hoe, but a more aggressive 'block cultivator' maintained high efficacy even in moist conditions (Evans et al. 2012).

Soil texture is likely also an important factor, but the effect of soil texture may depend on the mechanism of weed mortality imposed. For example, spring tine harrowing was more effective on sandy soils than clay soils (van der Weide and Kurstjens 1996) but when burying weeds, as would happen with hilling operations, laboratory experiments have shown that effective control was attained with less than 1.0 cm of sand with a particle size of 0.10 mm, whereas 1.5 cm was needed with a particle size of 0.95 mm (Baerveldt and Ascard 1999).

It is typically advised that level, friable soil is ideal for cultivation. However, in some cases these attributes may be mutually exclusive. For example, cultivation was more effective in a seedbed that was chiselled and disked, than a more level but more compacted seedbed that was chiselled, disked, and culti-mulched (Mohler et al. 2000). Similarly, Evans et al. (2012) found an inverse relationship between soil surface levelness and efficacy, likely due to conditions being irregular and loose for trials immediately following tillage and more level and compact for trails later in the season. Perhaps the level and loose conditions created by a rototiller would provide the best of both worlds. However, this tactic should not be used with soil with high clay content and low soil organic matter that are prone to compaction (Kooistra and Tovey 1994) since the pulverization of soil structure can promote crusting after rainfall events.

5 Weed–crop selectivity

In simplest terms, the 'selectivity' of a weed management practice refers to its ability to kill weeds without damaging the crop. The concept of selectivity is particularly critical when weed and crop seedlings emerge synchronously and in close spatial proximity. Therefore, it is not surprising that discussion of selectivity of PWC tools has occurred primarily in studies involving tools targeting weeds in the intra-row zone, where some level of crop damage often occurs. When used in close proximity to sensitive crops, cultivation tools may kill crops through the same mechanisms by which they kill weeds. But subtler negative impacts may also occur. For example, cultivation tools may prune roots, create entry points in plant stems for disease or adversely affect soil physical properties which ultimately reduce crop yield or quality (Ascard and Mattson 1994). However, it should be noted that, in some cases, cultivation (even under weed free conditions) improves crop growth and yields through reductions in disease, increases in nitrogen availability, or improvements in soil moisture retention (Perry 1983; Werf et al. 1991).

Given the practical importance of minimizing damage to a crop, efficacy per se is an inadequate measure of the practical value of a tool, unless it is evaluated at a fixed level of crop damage (Vanhala et al. 2004). To help address this issue, Rasmussen (1990) defined selectivity as the ratio of percent weed control to the percent crop burial (proxy for crop damage), and observed that greater efficacy was typically associated with lower selectivity. As a result, optimal yields are often associated with less than full weed control when cultivation is used (Rasmussen 1991; Mohler 2001). Therefore, quantification and modelling of the relationship between weed control efficacy, crop damage, and crop yield have been emphasized as important for evaluating optimal tool selection and settings (Kurstjens et al. 2000 and 2004).

5.1 Selective ability vs. selective potential

A useful framework for understanding and improving selectivity of PWC is presented in Kurstjens et al. (2004) in the context of their work relating plant anchorage forces to the selectivity of harrow cultivators (Fig. 3). In their study, an important distinction was made between 'selective potential' and 'selective ability', which together determine the actual selectivity observed in the field. Selective potential is the maximum achievable selectivity for a given crop-weed-soil combination, assuming an idealized weeder that can apply the same optimal precise force to both crop and weed populations. In contrast, selective ability is a measure of an actual tool's ability to attain selective potential. Therefore, selective potential depends on the relative tolerance of crops and weeds to hypothetical forces applied by cultivation tools, whereas selective ability depends on a tool's ability to apply the forces that optimize selectivity.

Using this framework, selective potential of a crop-weed combination depends critically not only on the mean, but also on the variance of plant population tolerances of cultivation forces, including uprooting (Kurstjens et al. 2004; Fogelberg and Dock Gustavsson 1998) and burial (Mohler et al. 2016). When distributions of weed and crop tolerance to cultivation overlap, as shown in Fig. 3, attainment of complete weed control is impossible without some level of crop damage (assuming the same force is applied to both weeds and crops), and selective potential is low. This framework also helps to clarify that the selective ability of a cultivation tool under field conditions is almost always less than selective potential because cultivation tools are imprecise, and the forces they apply are variable. Therefore, selectivity of a given weed-crop-tool combination depends on the probability distributions of tolerance traits of weeds and crops, as well as the probability distribution of the forces applied by the tool.

Based on these distinctions, strategies for improving selectivity can be separated into those which improve selective potential and those that improve selective ability. Specific tactics may include those that improve selective potential by increasing the mean (Fig. 3A) or reducing the variance (Fig. 3B) of crop tolerance to cultivation forces; or those that

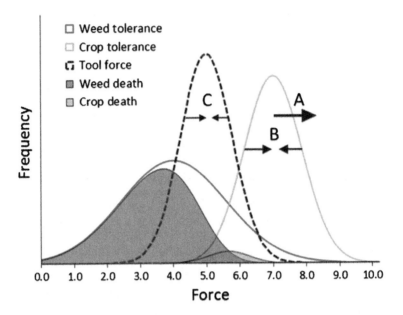

Figure 3 The selectivity of a tool that uproots weeds depends on 1) the distribution of anchorage forces of the crop population (green line) and the weed population (red line) and 2) the frequency distribution of the uprooting force applied by the tool (black dashed line). Together, these distributions determine crop death (shaded green line) and weed death (shaded red area) associated with the tool-weed-crop combination. Selectivity can be defined as the ratio of weed death to crop death. General strategies for improving tool selectivity include (A) increasing the mean tolerance of the crop population, (B) reducing the variance of crop tolerance or (C) reducing the variance of the force applied by the tool. Adapted from Kurstjens et al. (2004).

improve tool selective ability by reducing the variance or shifting the distribution of forces applied by those tools to optimize selective ability (Fig. 3C).

5.2 Improving selective potential

Weed–crop size differential

Improvements in selective potential can most obviously be obtained by increasing the 'size differential' between crops and weeds, and many authors have emphasized the importance of this approach (Kurstjens and Perdok 2000; Mohler 2001). This approach follows logically from the observation that large plants are more tolerant of a given level of disturbance (uprooting or burial) than small plants (Baerveldt and Ascard 1999; Mohler 2016). To attain a size differential between weeds and crops, a wide range of approaches have been suggested including transplanting rather than direct seeding and seed priming to give crops a head start (Bond et al. 1998; Zhao et al. 2007); targeted fertilization or irrigation to promote crop growth relative to weed growth (Blackshaw et al. 2004); and frequent and intense early weed suppression (e.g. stale seed bed) to retain crop-weed size differential at the time of each cultivation (Mohler 2001).

Since variation in seed size is often correlated with variation in seedling size (e.g. Jurado and Westoby 1992; Wulff 1986), and seedling size often correlates with cultivation tolerance, seed-sieving or air-column separation may provide a simple approach to reducing variance in cultivation tolerance, and hence improving selective potential. Similarly, selection of the most vigorous and uniform crop cultivars and seedlots may significantly improve the selectivity of PWC. For example, Rasmussen and Rasmussen (2000) found that artificially reduced seed vigor in barley reduced weed competitiveness; although they did not detect an improvement in selectivity of cultivation with more vigorous seeds, they suggested that use of high quality seeds was an important potential component of successful PWC.

Cultivator-tolerant traits and crops

Although the size differential of crops and weeds is a critical determinant of selective potential, differences in more subtle, non-size factors can also play an important role in promoting crop tolerance to cultivation. For example, root architecture has a major influence on plant tolerance to vertical and rotational forces, with fibrous root systems protecting plants from vertical tensile forces, and tap roots improving tolerance of rotational forces (Ennos and Fitter 1992). Fogelberg and Dock Gustavsson (1998) found that anchorage forces of carrots and various weed species increased with crop size (leaf stage), but also that these forces differed substantially independent of size, presumably due to differences in root architecture. Variation in plant tolerance to burial also sometimes appear to be unrelated to size. For example, Mohler et al. (2016) reported several instances in which Powell amaranth (*Amaranthus powellii*) recovered more effectively from burial than lambsquarters despite having similar or lower leaf number and height. Although the mechanism for such variation is unclear, traits such as desiccation tolerance, or ability to rapidly reallocate carbon reserves to new tissue likely play an important role. Plant tolerance to cultivation tools with slicing mechanisms have not been studied extensively, but traits such as stem strength or increased branching after release of apical dominance, which have benefits for improving tolerance to lodging (Piñera-Chavez et al. 2016) or herbivory (Strauss and Agrawal 1999) may also explain differences in plant tolerance to cultivation tools that slice.

These studies suggest that identification and adoption of crop cultivars with cultivation-tolerant traits may be a useful strategy for improving selective potential of PWC. For example, growers who rely on flextine cultivators for weed control may want to choose crop cultivars known to have greater anchorage force for a given crop size (e.g. wheat varieties bred for lodging resistance). Similarly, growers relying on sweeps to bury weeds in the crop row may improve selective potential by planting cultivars with greater early carbon allocation to height (increased specific stem length), or seedlings with longer or stronger hypocotyls. Unfortunately, limited information is available on which traits may be most important for conferring cultivation tolerance; how commercial varieties vary in these traits; and whether potential tradeoffs associated with these traits reduce their viability as practical options for growers.

A second general approach to improving selective potential which has received relatively little attention involves reducing the variance of cultivation tolerance of crop populations (Fig. 3B). This approach includes the relatively common suggestion that growers implement practices which promote uniform establishment and early growth of crops including use of precision bed preparation and planting equipment that places seeds with uniform depth and spacing (Mohler et al. 2001). Another potentially useful tactic for reducing crop variability in tolerance to cultivation forces is use of cultivars or seed lots with less variability in cultivation-tolerant traits.

5.3 Improving selective ability

Strategies for improving selective ability of PWC tools through improvements in mean efficacy and reduction in variance of cultivation tools have been discussed extensively in other parts of this chapter (see Section 3). Within the framework used by Kurstjens et al. (2004), these approaches include tactics which reduce the variability of forces which are applied, and those which adjust the mean force to optimize selectivity (Fig. 3C). Strategies for reducing the variability of cultivator forces include most obviously improvements in depth control (e.g. gauge wheel adjustments; uniform flat seedbeds) and steering (e.g. computer guidance systems; straight rows).

Improvements in adjustment in the mean force applied by a tool (Fig. 3C) through careful calibration are also critical for improving selective ability. Identifying the optimal force (settings and speed) of a cultivation tool is complicated by the fact that optimal forces change with soil, crop and weed conditions (Fogelberg and Dock Gustavsson 1998) and typically requires a trial-and-error calibration process which may take more time and may cause more crop injury than is practical for a grower during the busy growing season. Calibration is particularly challenging when crop and weed growth stages or soil conditions are variable across a field, as may often be the case in growing regions with non-uniform topography, even when precise planting and bed preparation tools are used. Uncertainty regarding the long-term impact of cultivation injury on crops is also a major factor inhibiting growers from selecting the optimal level of cultivation intensity. A crop which is uprooted is clearly undesirable, but threshold levels of disturbance that do not adversely impact crop yield are often unknown without years of production experience. To some extent, research and extension programmes can reduce these uncertainties by providing growers with suggested initial settings and steps for calibration of cultivation tools and by evaluating the yield consequences of different levels of crop disturbance.

6 Fundamental problems with cultivation

The foremost aim of PWC is to reduce weed density. Research results, logic and our experience suggest that efficacy is (1) inversely related to weed size; (2) greater in dry soil and bright sun; (3) greatest inter-row, but decreasing the closer you get to the crop row; and (4) usually independent of seedling density. Suboptimal conditions, both within a given field and over time, often result in low and highly variable efficacy. In our review of 55 previous investigations of mechanical cultivation tool efficacy (Fig. 4), the overall mean was 66% (95% CI: 61–71%) and the range was 21% to 90%. Clearly, efficacy of mechanical cultivation tools is low and highly variable. However, for most of the tools, the highest efficacy was greater than 80%, perhaps demonstrating their potential when used in ideal conditions with optimal adjustment. Factors that impact the efficacy of tools include design, forward speed, working depth, and mechanism of mortality. Tool efficacy may also be constrained by the necessity of avoiding crop damage. This is a challenge in interpreting PWC studies as conditions may require less aggressive settings to reduce crop injury, thereby reducing efficacy.

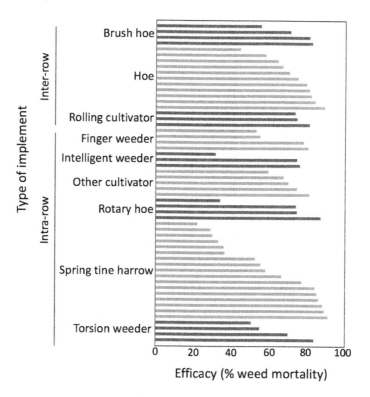

Figure 4 Mean physical weed control efficacy from published studies including 55 tests with various inter- and intra-row cultivation tools.

In 2010, we conducted, field-scale assays of PWC, conducting pre- and post-cultivation censuses to quantify efficacy, and measurements of soil texture, surface roughness, organic matter and moisture. Mortality was measured over inter- and intra-row zones, at 140 locations, randomly selected in a 2-ha, uniformly appearing field of silage corn. The cultivator was a Case International with Danish S-tines and 10 cm sweeps. Condiment mustard – 'Idagold' (*S. alba*) – was sown as a surrogate weed and was uniformly in the cotyledon to early 1-leaf stage at the time of cultivation. The extent of the variability in the dataset was unexpected. Efficacy ranged from 7 to 100%, with mean and median values of 67 and 68%, respectively (Fig. 5). Soil moisture, surface roughness, texture and bulk density failed to explain the variation in efficacy despite relatively large ranges in these explanatory variables. The dataset presents two important questions: What are the sources of variation? Could mean efficacy be increased and variance reduced with more sophisticated implement(s) or better adjustment/use of the test cultivator? The variation could represent the random movement of tools, something that could be improved with more modern precision equipment. Alternatively, variable efficacy may have been due in part to variation in the size of individual plants, even though they appeared to be a uniform cohort, or variation in cultivation-tolerant traits (e.g. anchorage force; hypocotyl length or strength). For example, an interesting ecological

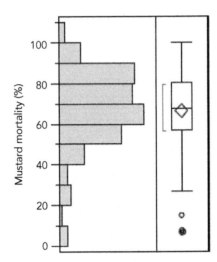

Figure 5 Condiment mustard (*Sinapis alba* cv. Idagold) mortality following cultivation with an inter-row cultivator (4-row Case International Model 183 with Danish S-tines, gangs of 10-cm sweeps, and gage wheels). Mortality was measured by pre- and post-cultivation censuses, inter- and intra-row, at 140 randomly chosen locations within a 2 ha field of corn (*Zea mays*). Census areas were centred on the crop row and included both inter- and intra-row areas. The experiment was conducted in Stillwater, Maine, in 2012 (Gallandt, unpublished).

explanation was offered by Mohler (1996), who noted that, although most small-seeded weeds emerge from the top 2–3 cm of soil, there are some, albeit fewer, that successfully emerge from greater depths, and thus may be better anchored and better able to survive a PWC event.

Mohler (2001) highlighted an important limitation to PWC related to weed density. Put simply, with increasing weed density, the density of so-called 'escapes', that is, seedlings surviving cultivation, increases proportionally. In other words, PWC efficacy is density independent. In most circumstances this is a logical consequence of the PWC tool's mechanism of action. The tool moves through the soil, uprooting, severing or burying small weed seedlings. The area of soil disturbed and the action of the tool are affected more by the tool design and operation and the soil than by the weed seedlings. A surviving weed seedling was simply fortunate enough to establish in a zone that the tool either missed completely or insufficiently disturbed; neighbouring seedlings would not change this. Indeed, plotting the same efficacy data shown previously (Fig. 5) against pre-cultivation density values demonstrates the independence of this relationship (Fig. 6). Density-independent efficacy is most simply managed with repeated PWC events, each killing some proportion of the established seedlings, with declining efficacy as escapes from a particular cohort increase in size. Another response is to improve efficacy, for example, by stacking existing tools (Fig. 1), and, hopefully in the future, advanced, next-generation tools. Lastly, management can focus on reducing seedling densities with system-level strategies to reduce the weed seedbank (Gallandt 2006 and 2014).

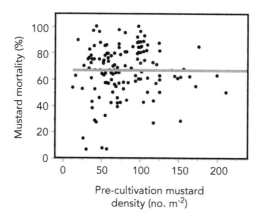

Figure 6 Effect of initial, pre-cultivation density of condiment mustard (*Sinapis alba* cv. Idagold) on mortality following cultivation with an inter-row cultivator (4-row Case International Model 183 with Danish S-tines, gangs of 10-cm sweeps, and gage wheels). Mortality was measured by pre- and post-cultivation censuses at 139 randomly chosen locations within a 2-ha field of corn (*Zea mays*). Census areas were centred on the crop row and included both inter- and intra-row areas. The experiment was conducted in Stillwater, Maine, in 2012 (Gallandt, unpublished).

While we consider density-independent efficacy in PWC to be the rule, we have observed exceptions in the field, suggesting that, in certain tool-weed-soil conditions, efficacy may be inversely related to density. Consider, for example, a very dense stand of 1-leaf crabgrass (*Digitaria sanquinalis*) seedlings. Even at this early growth stage, an individual crabgrass seedling has an impressive root system, making it difficult to dislodge from the soil. At higher densities the roots intertwine and the individuals, although competing for soil resources, benefit one another if subject to soil disturbance. The crabgrass essentially forms a sod that is simply lifted by a sweep, fracturing in segments or clumps, some ending up inverted, burying the seedlings, but others returning as they started, allowing the seedlings to survive. It is likely that other species exhibiting high seedling growth rates and root:shoot partitioning respond similarly. Indeed, as noted above, weed seedling growth analyses may offer insight into weed species and densities where tool–soil combinations are likely to result in density dependent efficacy. It should be noted that proportional survivorship relative to initial seedling densities was also observed by Dieleman et al. (1999).

Cultivation is often criticized as a soil-degrading practice not compatible with farmers' aims to maintain or improve soil quality. Unfortunately, there are few published studies that have focused on this topic. Tillage effects on soil quality are clear, with improving soil quality parameters resulting from decreasing tillage intensity (Karlen et al. 1994). Cultivation in the context of PWC, however, is decidedly different than tillage. PWC operations are generally shallow, 2–4 cm deep, and ideally performed when soil is dry and thus aggregates more stable (Cousen and Farres 1984). Compaction is an inherent negative effect of all field operations, including PWC. Several steps may be taken to minimize compaction, including controlled traffic, avoiding field work when the soil is too wet; compaction may be ameliorated with deep tillage or deep-rooted rotation crops (Bowman 2002).

7 Future research priorities

Innovations in equipment targeting weed seeds recently expanded the domain of PWC. Inspired by intractable herbicide resistance problems in Australia, the Harrington Seed Destructor is a powered rolling cage mill, towed behind a combine (Walsh et al. 2012; Jacobs and Kingwell 2016). Residues and weed seeds cleaned from the harvested grain exit the combine and are processed by the mill, effectively killing weed seeds before they are returned to the field. Walsh et al. (2013) suggest that such 'harvest weed seed control' is a new weed management paradigm. Indeed, the case for seed-focused management of annual weeds has been widely recommended (Jordan 1996; Norris 1999; Forcella 2003; Gallandt 2006).

The need for advanced PWC technologies targeting weed seedlings is greatly outpacing research. Continued growth in the organic food sector has created demand for crops grown without herbicides, but effective and profitable weed control prevents many farmers from entering this sector. Climate change, particularly in areas facing more extreme and greater variability in precipitation events, will reduce the performance of existing PWC options. Farmers will need to increase working rates and, ideally, have new tools designed to provide improved efficacy over suboptimal weed-soil conditions. Recent evidence of synergy with stacked tools (Fig. 1) is particularly promising in this regard, but is only a small step in this direction.

Improved working rates and automation with robotic weeders will increasingly be used to overcome low and variable efficacy by allowing farmers to automatically deploy multiple cultivation events. These tools, although innovative, are perhaps more aptly called self-guided tools, or vision-guided intra-row weeding tools, not truly robots (Merfield 2016). Fully autonomous weeding remains an elusive goal, and Merfield (2016) outlined an ambitious set of nine criteria for true robotic weeding, one of which is most essential, in our opinion: the ability to adjust tool depth, angle and speed for optimal soil disturbance and efficacy. This will require a considerable investment in mechanistic studies of the tool-weed-soil-selectivity variable space. Co-robotics represents a transitory approach to solving this problem. Pérez-Ruíz et al. (2014) developed such a system that used GPS precision transplanting, odometry and, as needed, human adjustments to optimize operation of an intra-row hoe. Nevertheless, robots are being developed widely and have recently entered the marketplace (e.g. Oz Weeding Robot, Naïo Technologies). Young and Pierce (2014) recently published an edited volume that provides a comprehensive overview of this rapidly changing field. They suggest that 'Precision Weed Management' is the future, whereby plants are individually identified and an autonomous platform deploys an appropriate control tool, physical, thermal or chemical.

While technology advances are inspiring, it is unfortunate that Mohler's (2001) research priorities of 15 years ago remain priorities today: (1) develop scientific bases for use of available tools – speed, angles, depths, position relative to the crop row, selectivity and use in various crop species; (2) identify sensitive growth stages of crops and weeds, during cultivation with available tools, to avoid crop injury and maximize efficacy of weed control; and (3) document the role of soil tilth, texture and clod size on tool performance. Fundamental research in these areas would lead to the rational design of new tools and their optimal use. In the tool domain, PWC continues to rely on sweeps, knives or tines based on intuitive designs and empirical field testing. Applying contemporary engineering technologies with expert farmer evaluation should be used to develop next-generation tools, iteratively improving mean efficacy, reducing variability, and optimizing tool choice and operation over the large variable space of soil conditions, weeds and crops.

Another underexplored approach to improving performance of PWC involves identification and development of crop cultivars with greater tolerance to PWC. Just as herbicide efficacy has been improved through identification and development of crops with herbicide-tolerant traits (e.g. glyphosate resistance), cultivation efficacy may be improved by targeting traits that confer tolerance to burial, uprooting or severing. Studies identifying key cultivation-tolerant traits of importance, quantifying genetic variability in these traits among existing crop cultivars, and testing the impact of this variation on tolerance would be helpful in gaining insight into the potential impact of this approach. Preliminary studies demonstrate that commercially available carrot cultivars vary substantially in their tolerance of cultivation tools including flextine and finger weeders, but the traits responsible for this variability are unclear (Tilton and Brainard, unpublished).

A first step to advance our mechanistic understanding of seedling mortality would be to develop a repeatable assay protocol, using uniform, standardized soil-bin systems to characterize (i) existing tools over reasonable adjustment settings and operational speeds, (ii) soil textural classes and moisture profiles, and (iii) weed species and growth stages. Secondly, a focused effort to develop models, perhaps using discrete element analysis, to simulate tool/soil interactions, and to relate this information iteratively to controlled environment assays. Lastly, coordinated field testing of top-performing tools and combinations would validate and guide a next round of modelling and controlled environment testing. While the need for advances in PWC remains great, we consider there to be great potential as well. PWC is complex. We contend, however, that the 'art' of PWC can be deconstructed, offering insights that will guide a next-generation of tools and practices.

8 Where to look for further information

PWC research is occasionally featured in selected academic journals, including *Weed Research*, *Weed Science* and *Weed Technology*, as well as more broadly focused agronomic or horticultural journals including *Agronomy Journal*, *HortScience* and *Field Crops Research*, for example. Many scientists participating in this area of research are active in the Physical and Cultural Weed Control Working Group of the European Weed Research Society (http://www.ewrs.org/pwc/). Proceedings from this group's annual meetings are available online and are an excellent source of contemporary research on PWC.

9 References

ASAE. (2005), 'Terminology and definitions for soil tillage and soil-tool relationships', *American Society of Agricultural Engineers Standards* 2005: 129–32.

Ascard, J. and Fogelberg, F. (2002), 'Mechanical intra-row weed control in organic onion production', 5th EWRS Workshop on Physical Weed Control, Pisa, Italy, 11–13 March.

Ascard, J. and Mattsson, B. (1994), 'Inter-row cultivation in weed-free carrots: the effect on yield of hoeing and brush weeding', *Biological Agriculture & Horticulture*, 10(3): 161–73.

Ascard, J., Hatcher, P., Melander, B. and Upadhyaya, M. (2007), 'Thermal weed control', In *Non-Chemical Weed Management*, Upadhyaya, M. and Blackshaw, R. (eds), CAB International, pp. 155–75.

Baerveldt, S. and Ascard, J. (1999), 'Effect of soil cover on weeds', *Biological Agriculture & Horticulture*, 17(2): 101–11.

Blackshaw, R., Molnar, L. and Janzen, H. (2004), 'Nitrogen fertilizer timing and application method affect weed growth and competition with spring wheat', *Weed Science*, 52: 614–22.

Bleeker, P., van der Weide, R. and Kurstjens, D. A. G. (2002), 'Experiences and experiments with new intra-row weeders'. 5th EWRS Workshop on Physical Weed Control, Pisa, Italy, 11–13 March 2002.

Bond, W., Burston, S., Bevan, J. R. and Lennartsson, M. E. K. (1998), 'The optimum timing of weed removal in drilled salad onions and transplanted bulb onions grown in organic and conventional systems', *Biological Agriculture and Horticulture*, 16:191–201.

Bowman, G. (2002), *Steel in the Field: A Farmer's Guide to Weed Management Tools*, Beltsville, MD 20705: Sustainable Agriculture Network.

Boyd, N. S., Brennan, E. B. and Fennimore, S. A. (2006), 'Stale seedbed techniques for organic vegetable production', *Weed Technology*, 20(4): 1052–7.

Caldwell, B. and Mohler, C. L. (2001), 'Stale seedbed practices for vegetable production', *Crop Protection*, 36(4): 703–5.

Chicouene, D. (2007), 'Mechanical destruction of weeds: A review', *Agronomic Sustainable Development*, 27(2): 19–27.

Cirujeda, A. and Taberner, A. (2004), 'Defining optimal conditions for weed harrowing in winter cereals on Papaver rhoeas L. and other dicotyledoneous weeds'. 6th EWRS Workshop on Physical and Cultural Weed Control. Lilliehammer, Norway, 8–10 March.

Cirujeda, A., Melander, B., Rasmussen, K. and Rasmussen, I. A. (2003), 'Relationship between speed, soil movement into the cereal row and intra-row weed control efficacy by weed harrowing', *Weed Research*, 43: 285–96.

Cloutier, D. C., van der Weide, R. Y., Peruzzi, A. and Leblanc, M. L. (2007), 'Mechanical weed management', In *Non-Chemical Weed Management: Principles, Concepts and Technology*, 111–34.

Cohen, O. and Rubin, B. (2007), 'Soil solarization and weed management', In *Non-Chemical Weed Management*, Upadhyaya, M. K. and Blackshaw, R. E. (eds),CABI Publishing Wallingford, GBR, pp. 177–200.

Colby, S. R. (1967), 'Calculating synergistic and antagonistic responses of herbicide combinations', *Weeds*, 15: 20–2.

Cousen, S. M. and Farres, P. J. (1984), 'The role of moisture content in the stability of soil aggregates from a temperate silty Soil to raindrop impact', *Catena*, 11(4): 313–20.

Datta, A. and Knezevic, S. Z. (2013), 'Flaming as an alternative weed control method for conventional and organic agronomic crop production systems: A review', In *Advances in Agronomy*, 118: 399–428.

Dieleman, J. A., Mortensen, D. A. and Martin, A. R. (1999), 'Influence of velvetleaf (*Abutilon theophrasti*) and common sunflower (*Helianthus annuus*) density variation on weed management outcomes', *Weed Science*, 47(1): 81–9.

Ennos, A. R. and Fitter, A. H. (1992), 'Comparative functional morphology of the anchorage systems of annual dicots', *Functional Ecology*, 71–78.

Evans, G. J., Bellinder, R. R. and Hahn, R. R. (2012), 'An evaluation of two novel cultivation tools', *Weed Technology*, 26(2): 316–25.

Fennimore, S., Tourte, L., Rachuy, J., Smith, R. and George, C. (2010), 'Evaluation and economics of a machine-vision guided cultivation program in broccoli and lettuce', *Weed Technology*, 24(1): 33–8.

Fogelberg, F. and Dock Gustavsson, A. M. (1998), 'Resistance against uprooting in carrots (*Daucus carota*) and annual weeds: a basis for selective mechanical weed control', *Weed Research*, 38: 183–90.

Forcella, F. (2003), 'Debiting the seedbank: Priorities and predictions', *Aspects of Applied Biology*, 69: 151–62.

Fortier, J. (2014), *The Market Gardener*, New Society Publishers, Gabriola Island, BC, Canada. 240pp.

Gallandt, E. R. (2006), 'How can we target the weed seedbank?', *Weed Science*, 54(3): 588–96.

Gallandt, E. R. (2014), 'Weed management in organic farming', In *Recent Advances in Weed Management*, Chauhan, B. and Mahajan, G. (eds),Springer Science+Business Media, New York, pp. 63–85.

Gallandt, E. R., Liebman, M. and Huggins, D. R. (1999), 'Improving soil quality: Implications for weed management', *Journal of Crop Production*, 2(1): 95–121.

Garford (2016), http://www.garford.com/products_robocrop.html (accessed 18 November 2016).

Griepentrog, H. W., Gulholm-Hansen, T. and Nielsen, J. (2007), 'First field results from intra-row rotor weeding'. 7th EWRS Workshop on Physical and Cultural Weed Control. Salem, Germany, 11–14 March.

Habel, W. (1954), Uber die Wirkungsweise der Eggen gegen Samenunkrauter sowie die Empfndlichkeit der Unkrautarten und ihrer Altersstadien gegen den Eggvorgang. Ph.D. Thesis. Landwirtschaftlichen Hochschule Hohenheim, Germany.

Håkansson, S. (2003), 'Soil tillage effects on weeds', In *Weeds and Weed Management on Arable Land: An Ecological Approach*, CAB International, pp. 158–96.

Heap, I. (2014), 'Global perspective of herbicide-resistant weeds', *Pest Management Science*, 70(9): 1306–15.

Jacobs, A. and Kingwell, R. (2016), 'The Harrington seed destructor: Its role and value in farming systems facing the challenge of herbicide-resistant weeds', *Agricultural Systems*, 142: 33–40.

Jordan, N. (1996), 'Weed prevention: Priority research for alternative weed management', *Journal of Production Agriculture*, 9(4): 485–90.

Jurado, E. and Westoby, M. (1992), 'Seedling growth in relation to seed size among species of arid Australia', *Journal of Ecology*, 80(3): 407–16.

Karlen, D. L., Wollenhaupt, N. C., Erbach, D. C. and Berry, E. C. (1994), 'Long-term tillage effects on soil quality', *Soil and Tillage Research*, 32: 313–27.

Kees, H. (1962), Untersuchungen zur Unkrautbekampfung durch Netzegge und Stoppelbearbeitungsmassnahmen unter besonderer Berucksichtigung des leichten Bodens. Ph.D. Thesis. Landwirtschaftlichen Hochschule Hohenheim, Germany.

Koch, W. (1964), 'Unkrautbekampfung durch Eggen, Hacken und Meisseln in Getreide I. Wirkungsweise und Einsatzzeitpunkt von Egge, Hacke und Bodenmeissel'. *Zeitschrift fur Acker- und Pfanzenbau*, 120: 369–82.

Kolb, L. N. and Gallandt, E. R. (2012), 'Weed management in organic cereals: Advances and opportunities', *Organic Agriculture*, 2(1): 23–42.

Kolb, L. N., Gallandt, E. R. and Mallory, E. B. (2012), 'Impact of spring wheat planting density, row spacing, and mechanical weed control on yield, grain protein, and economic return in Maine', *Weed Science*, 60(2): 244–53.

Kolb, L. N., Gallandt, E. R. and Molloy, T. (2010), 'Improving weed management in organic spring barley: Physical weed control vs. interspecific competition', *Weed Research*, 50(6): 597–605.

Kooistra, M. J. and Tovey, N. K. (1994), 'Effects of compaction on soil microstructure', In *Soil Compaction in Crop Production: Developments in Agricultural Engineering*, Soane, B. D. and van Ouwerkerk, C. (eds). Elsevier, Amsterdam, pp. 91–111.

Kouwenhoven, J. K. and Terpstra, R. (1979), 'Sorting action of tines and tine-like tools in the field', *Journal of Agricultural Engineering Research*, 24(1): 95–113.

Kurstjens, D. A. G. (2002), Mechanisms of Selective Mechanical Weed Control by Harrowing. PhD Thesis. Wageningen University.

Kurstjens, D. A. G. and Perdok, U. D. (2000), 'The selective soil covering mechanism of weed harrows on sandy soil', *Soil and Tillage Research*, 55(3–4): 193–206.

Kurstjens, D. A. G., Kropff, M. J. and Perdok, U. D. (2004), 'Method for predicting selective uprooting by mechanical weeders from plant anchorage forces', *Weed Science*, 52(1): 123–32.

Kurstjens, D. A. G., Perdok, U. D. and Goense, D. (2000), 'Selective uprooting by weed harrowing on sandy soils', *Weed Research*, 40: 431–48.

Lamour, A. and Lotz, L. A. P. (2007), 'The importance of tillage depth in relation to seedling emergence in stale seedbeds', *Ecological Modelling*, 201 (3–4): 536–46.

Mahoney, D. J., Jeffries, M. D., and Gannon, T. W., (2014), 'Weed control with liquid carbon dioxide in established turfgrass', *Weed Technology*, 28(3), 560–68.

Melander, B., Cirujeda, A. and Jorgensen, M. H. (2003), 'Effects of inter-row hoeing and fertilizer placement on weed growth and yield of winter wheat', *Weed Research*, 43(6): 428–38.

Melander, B., Holst, N., Rasmussen, I. A. and Hansen, P. K. (2012), 'Direct control of perennial weeds between crops – implications for organic farming', *Crop Protection*, 40: 36–42.

Melander, B., Lattanzi, B. and Pannacci, E. (2015), 'Intelligent versus non-intelligent mechanical intra-row weed control in transplanted onion and cabbage', *Crop Protection*, 72: 1–8.

Merfield, C. N. (2016), 'Robotic Weeding's False Dawn? Ten requirements for fully autonomous mechanical weed management,' *Weed Research*, 56(5): 340–4.

Mohler, C. L. (1993), 'A model of the effects of tillage on emergence of weed seedlings', *Ecological Applications*, 3(1): 53–73.

Mohler, C. L., DiTommaso, A. and Joslin, K. R. M. (2000), 'The effect of soil tilth on weed control by cultivation', Toward Sustainability Foundation Report. http://www.organic.cornell.edu/research/tsfsumms/organicpdfs/3tilthcult1.pdf. (accessed 3 November 2009).

Mohler, C. L. (1996), 'Ecological bases for the cultural control of annual weeds', *Journal of Production Agriculture*, 9(4): 468–74.

Mohler, C. L. (2001), 'Mechanical Management of Weeds', In *Ecological Management of Agricultural Weeds*, Liebman, M., Mohler, C. L. and Staver, C. P. (eds), Cambridge University Press, Cambridge, pp. 139–209.

Mohler, C. L., Iqbal, J., Shen, J. and DiTommaso, A. (2016), 'Effects of water on recovery of weed seedlings following burial', *Weed Science*, 64(2): 285–93.

Mohler, C. L., Marschner, C. A., Caldwell, B. A. and DiTommaso, A. (2016), 'Weed mortality caused by row-crop cultivation in organic corn-soybean-spelt cropping systems', *Weed Technology*, 30(3): 648–54.

Norris, R. F. (1999), 'Ecological implications of using thresholds for weed management', *Journal of Crop Production*, 2(1): 31–58.

Pérez-Ruíz, M., Slaughter, D. C., Fathallah, F. A., Gliever, C. J. and Miller, B. J. (2014), 'Co-robotic intra-row weed control System', *Biosystems Engineering*, 126: 45–55.

Perry, D. A. (1983), 'Effect of soil cultivation and anaerobiosis on cavity spot of carrots', *Annals of Applied Biology*, 103(3): 541–7.

Peruzzi, A., Ginanni, M., Fontanelli, M., Raffaelli, M. and Bàrberi, P. (2007), 'Innovative strategies for on-farm weed management in organic carrot', *Renewable Agriculture and Food Systems*, 22(4): 246–59.

Piñera-Chavez, F. J., Berry, P. M., Foulkes, M. J., Jesson, M. A. and Reynolds, M. P. (2016), 'Avoiding lodging in irrigated spring wheat. I. Stem and root structural requirements', *Field Crops Research*, 196: 325–36.

Pullen, D. W. M. and Cowell, P. A. (1997), 'An evaluation of the performance of mechanical weeding mechanisms for use in high speed inter-row weeding of arable crops', *Journal of Agricultural Engineering Research*, 67(1): 27–34.

Raffaelli, M., Barberi, P., Peruzzi, A. and Ginanni, M. (2002), Options for mechanical weed control in grain maize – effects on weeds. 5th EWRS Workshop on Physical Weed Control, Pisa, Italy, 11–13 March.

Rasmussen, J. (1990), 'Selectivity-an important parameter on establishing the optimum harrowing technique for weed control in growing cereals', Proceedings of an EWRS symposium, Helsinki, Finland, 4–6 June 1990, pp. 197–204, European Weed Research Society.

Rasmussen, J. (2003), 'Punch planting, flame weeding and stale seedbed for weed control in row crops', *Weed Research*, 43(6): 393–403.

Rasmussen, J. and Svenningsen, T. (1995), 'Selective weed harrowing in cereals', *Biological Agriculture and Horticulture*, 12: 29–46.

Rasmussen, K. and Rasmussen, J. (2000), 'Barley seed vigour and mechanical weed control', *Weed Research*, 40(2): 219–30.

Riemens, M. M., van der Weide, R. Y., Bleeker, P. O. and Lotz, L. A. P. (2007), 'Effect of stale seedbed preparations and subsequent weed control in lettuce (cv. Iceboll) on weed densities', *Weed Research*, 47(2): 149–56.

Rydberg, T. (1994), 'Weed harrowing – the influence of driving speed and driving direction on degree of soil covering and the growth of weed and crop plants', *Biological Agriculture and Horticulture*, 10: 197–205.

Schutte, B. J., Tomasek, B. J., Davis, A. S., Andersson, L., Benoit, D. L., Cirujeda, A., Dekker, J., Forcella, F., Gonzalez-Andujar, J. L., Graziani, F., Murdoch, A. J., Neve, P., Rasmussen, I. A., Sera,B., Salonen, J., Tei, F., Tørresen, K. S. and Urbano, J. M. (2013), 'An investigation to enhance understanding of the stimulation of weed seedling emergence by soil disturbance', *Weed Research*, 54(1): 1–12.

Strauss, S. Y. and Agrawal, A. A. (1999), 'The ecology and evolution of plant tolerance to herbivory', *Trends in Ecology & Evolution*, 14(5): 179–85.

Terpstra, R. and Kouwenhoven, J. K. (1981), 'Inter-row and intra-row weed control with a hoe-ridger', *Journal of Agricultural Engineering Research*, 26(2): 127–34.

van der Schans, D., Bleeker, P. O. and Moledijk, L. (2006), *Practical Weed Control in Arable Farming and Outdoor Vegetable Cultivation Without Chemicals*, PPO publication 532, Applied Plant Research Wageningen University, Lelystad, The Netherlands, 77pp.

van der Weide, R. and Kurstjens, D. (1996), 'Plant morphology and selective harrowing', Physical Weed Control, 2nd EWRS Workshop.

van der Weide, R. Y., Bleeker, P. O., Achten, V. T. J. M., Lotz, L. A. P., Fogelberg, F. and Melander, B. (2008), 'Innovation in mechanical weed control in crop rows', *Weed Research*, 48(3): 215–24.

Vanhala, P., Kurstjens, D. A. G., Ascard, J., Bertram, A., Cloutier, D. C., Mead, A., Raffaelli, M. and Rasmussen, J. (2004), 'Guidelines for physical weed control research: Flame weeding, weed harrowing and intra-row cultivation', In *Proceedings 6th EWRS Workshop on Physical and Cultural Weed Control*, pp. 194–225.

Walsh, M. J., Harrington, R. B. and Powles, S. B. (2012), 'Harrington Seed Destructor: A new nonchemical weed control tool for global grain crops', *Crop Science*, 52(3): 1343–7.

Walsh, M. J., Newman, P. and Powles, S. (2013), 'Targeting weed seeds in-crop: A new weed control paradigm for global agriculture', *Weed Technology*, 27(3): 431–36.

Werf, H. M. G., Klooster, J. J., Schans, D. A., Boone, F. R. and Veen, B. W. (1991), 'The effect of inter-row cultivation on yield of weed-free maize', *Journal of Agronomy and Crop Science*, 166(4): 249–58.

Wulff, R. D. (1986), 'Seed size variation in *Desmodium paniculatum*: II. Effects on seedling growth and physiological performance', *The Journal of Ecology*, 74(1): 99–114.

Young, S. L. and Pierce, F. J., eds (2014), *Automation: The Future of Weed Control in Cropping Systems*. Springer Science+Business Media Dordrecht. 265pp.

Zhao, D. L., Bastiaans, L., Atlin, G. N. and Spiertz, J. H. J. (2007), 'Interaction of genotype × management on vegetative growth and weed suppression of aerobic rice', *Field Crops Research*, 100(2): 327–40.

Chapter 14

Flame weeding techniques

Stevan Z. Knezevic, University of Nebraska-Lincoln, USA

1 Introduction

Organic crop production in the United States is increasing due to strong demand for organic food and an attractive income potential for organic farmers (Johnson, 2004). An example is the growing demand for organically grown corn (*Zea mays* L.): there was a 300% increase in the production of organic corn (in hectares) from 1995 to 2005 in the United States, as the total organic corn production grew from 13 212 to 52 881 ha (AgMRC, 2009). Corn was the second largest organic grain/seed crop after organic wheat (*Triticum aestivum* L.), which was grown on nearly 112 550 ha in 2005 (AgMRC, 2009). Organic crops can be used for human consumption or livestock feed.

Weeds are one of the major pests and are responsible for significant reductions in crop yields (Stopes and Millington, 1991). Organic production systems lack effective weed management because herbicides are not allowed (Gianessi and Reigner, 2007; Hiltbrunner et al., 2007; Wszelaki et al., 2007). Therefore, organic producers rely extensively on mechanical cultivation and hand weeding (Hiltbrunner et al., 2007). Mechanical cultivation is one of the oldest and most commonly used weed control practices in row crops (Bond and Grundy, 2001). For example, organic farmers typically conduct three to five cultivations per season for weed control in corn. Cultivation, however, leaves a strip of uncontrolled weeds within the crop row, directly influencing crop yield. In addition, repeated cultivation causes loss of soil organic matter, destroys soil aggregates, increases the chance for soil erosion and promotes emergence of new weed flushes (Hiltbrunner et al., 2007; Wszelaki et al., 2007). The labour required for hand weeding costs from $300 to $800 ha⁻¹, is time consuming and can be difficult to find and organize (Kruidhof et al., 2008). Hence, systems-oriented approaches to weed management that make better use of alternative weed

http://dx.doi.org/10.19103/AS.2017.0025.16

management tactics need to be developed (Kruidhof et al., 2008). Propane flaming is one of the most promising alternatives for weed control in organic cropping systems, which also has a potential for use in non-organic crops (Knezevic, 2009). In order to optimize the use of propane flaming as a weed control tool, the biologically effective dose (ED) of propane for tolerance of major crops must be determined. Depending on the tolerable crop injury level, a propane dose could be selected to either control weeds, or reduce their competitive ability against the crop.

2 Flaming specifications, effectiveness and equipment

2.1 Propane flaming

Propane flame weeding is made possible with the use of propane burners that generate combustion temperatures of up to 1900°C, which raises the temperature of plant tissues rapidly (Ascard, 1998). Heat denatures membrane proteins and disrupts cell function (Parish, 1990; Pelletier et al., 1995; Rifai et al., 1996; Lague et al., 2001) resulting in weed death or drastically reduced competitive ability.

A variety of fuels were used in early 1940s, including diesel, kerosene and oils. However, they were replaced by liquefied petroleum gas (LPG; mixture of propane and butane). During the 1960s, flaming was widely used in the United States for controlling weeds in cotton (*Gossypium hirsutum* L.), corn, sorghum (*Sorghum bicolor* L.), soybean, potato (*Solanum tuberosum* L.) and other crops (Lague et al., 2001).

During the 1970s, flaming was replaced by the use of chemical herbicides due to escalated LPG prices and availability of less expensive residual herbicides. In recent years, increasing concerns about leaching of pesticide into surface- and groundwater and residues in drinking water and food have sparked public awareness and restrictions on herbicide use. Flaming is used in organic systems because it is cheaper than hand weeding. There is potential for use in other crops, especially to manage herbicide-resistant weeds.

2.2 Plant response to heat

In general, heat can damage any plant tissue as propane-fuelled burners can generate combustion temperatures as high as 3400°F (1900°C) (Ascard et al., 2007), which rapidly changes the internal temperature of plant cells. This rapid temperature change expands the cell's contents (e.g. 95% water) causing the cell walls to rupture. This primary cause of cellular death is followed by the evaporation of water that is released when the cell walls burst, which rapidly dries out the affected plant tissue.

Direct heat injury also causes cell proteins to denature (starts at 113°F = 45°C). Temperatures in the range of 203–212°F (95–100°C) have been lethal to weed leaves and stems when applied for at least 0.1 seconds, which further results in cell desiccation and ultimately the loss of cell function (Lague et al., 2001). The loss of water and denaturing of proteins drastically reduce the weed's competitive ability and kill the plant.

The lethal damage of the plant tissues can also be a function of a 'time–temperature' relationship. As temperature increases linearly, tissue damage increases exponentially and the 'time to kill' between sensitive and insensitive plants declines exponentially. The critical temperature reported for effective leaf mortality ranges from 55 to 70°C, with an exposure time of 65–130 milliseconds (Ascard, 1998).

Morphological differences of plant species are important in determining sensitivity to heat.

Leaf type, leaf thickness, leaf cuticle, location of growing point, nature of storage organs, leaf orientation, shape, presence of hair, growth stage and degree of stress (both moisture and nutrient) will affect sensitivity.

Environmental factors can also affect flaming. Any kind of moisture, including heavy dew and rain droplets on the leaves, will reduce efficacy. Moisture on plant tissue can reduce the amount of heat (temperature) that reaches inside the plant tissue, as the portion of the heat is wasted on evaporating the surface water present on the leaf. However, the presence of moisture can be useful for (i) reducing the crop injury and (ii) wetting the crop residue present on the soil, thus reducing the fire hazards. Irrigation an hour before flaming can help reduce fire hazards. Irrigation also promotes early weed emergence, which can be controlled with flaming. Organic vegetable producers often irrigate prepared seedbeds to stimulate weed germination, and then utilize flaming to kill weed seedlings before crop planting, or after planting but before crop emergence.

Time of day has also been observed to affect flaming efficacy (Ulloa et al., 2012). The basis for the differential plant response is not well understood; however, daily variation in leaf relative water content (RWC) is thought to contribute to the response. Leaf RWC is the ratio of the amount of water in the leaf tissue to that when fully turgid. Ulloa et al. (2012) flamed two crops [4-leaf corn (*Zea mays*) and second trifoliate soybean (*Glycine max*)] and two weed species [5-leaf velvetleaf (*Abutilon theophrasti*) and 6-leaf green foxtail (*Setaria viridis*)] at 0, 4, 8 and 12 h after sunrise-HAS. Leaf RWC was also measured before treatment. All plants were more susceptible to flaming during the afternoon when they had lower leaf RWC, 8 and 12 HAS. For example, green foxtail flamed at 0, 8 and 12 HAS had injury of 62, 76 and 82%, respectively. Similar trends occurred for velvetleaf, corn and soybean suggesting that leaf RWC could be one of the factors affecting plant response to flaming. The practical implication is that flaming conducted in the afternoon will be more effective and will reduce propane consumption.

2.3 Assessing flaming effectiveness: a rapid field test

The effectiveness of flaming can be assessed by fingerprint test few minutes after treatment. To conduct the test, place a treated leaf between thumb and index finger. If a darkened impression is visible after firmly pressing on the leaf surface (Fig. 1), it is immediate evidence of loss of internal pressure due to water leakage from ruptured cells.

Propane

LPG is made of at least 90% propane plus other hydrocarbons (e.g. butane).

Given the high percentage of propane, LPG properties can be estimated as pure propane (Williams and Lom, 1974). The one-step chemical reaction for propane (C_3H_8) combustion is given by:

$$C_3H_8 + 5O_2 \rightarrow 3CO_2 + 4H_2O$$

Given the very similar molecular weights of propane and carbon dioxide (44.096 kg kmol^{-1} and 44.011 kg kmol^{-1}, respectively), three kilograms of carbon dioxide are produced for every kilogram of LPG burned. Using the lower heating value for propane, each kilogram of LPG produces 46.4 MJ of heat.

2.4 Flame weeding equipment

Propane-fuelled flamers have several essentials (Fig. 2, Knezevic et al., 2014b).

Supply Tank: A propane supply tank (Part 1, Fig. 2). The tank capacity ranges from 5 kg to greater than 2500 kg. Tanks must have a liquid or vapour source safety valve (Part 2, Fig. 2).

Supply Network: The supply network is a combination of parts that control propane flow and distribution (Parts 3–10, Fig. 2). Manual or electronic valves control flow from the tank to a pressure regulator, which reduces the pressure to 100–150 pounds per square inch (PSI) (7–11 bars) from the tank to an operating pressure of 5–80 PSI (0.3–5.5 bars). Additional valves control flows to the torches. Pressure compatible tubing or hose carries the propane from the supply tank to the torch.

Figure 1 The 'thumbprint' test demonstrates whether plant tissue has died during treatment.

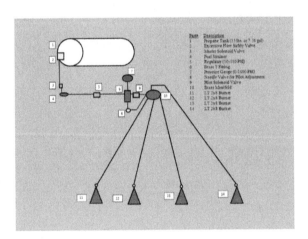

Figure 2 Diagram of basic flame weeding system components (Knezevic et al., 2014b).

Figure 3 An 8-row flamer with 250 gal. tank.

Figure 4 (a) Velvetleaf before flaming. (b) Velvetleaf at 14 days after flaming with 12GPA.

Torch or Burner: The torch (also known as the burner) is the point of propane combustion (Parts 11–14, Fig. 2). The combustion reaction begins at ignition. Torches are generally flat or tubular and may accept liquid or gaseous propane.

Frame: The frame supports the tank, supply network and torches and allows proper orientation of the torches.

Mobilizer: The mode of mobilization of a flame weeding system depends on the type of flamer. A large row crop flaming unit requires a tractor or other vehicle, while a smaller garden handheld flaming unit is either pushed, pulled or carried by the operator.

In addition, propane-fuelled flamers can include *flame shields*, which are hoods over the torch flame or shields on the sides of the torches to direct the heat and increase fuel efficiency. Shields are especially helpful when the wind blows.

Emergency Shutdown Switch: All propane-fuelled flame weeding systems should be equipped with an emergency shutdown switch that would be activated in the event of a malfunctioning burner.

Remote Monitoring and Ignition: It is helpful if the system immediately informs the operator of an extinguished torch. This increases system efficiency and protects operators.

Several flame weeders are commercially available, ranging from small, handheld units to multi-row, tractor-pulled systems (Fig. 3 and 4) (Knezevic et al., 2007). Many farmers have designed flamers to meet their economic and crop requirements.

3 Weed response to heat

Ascard (1995) classified weed species into four groups based on their susceptibility to heat. The first group of species with unprotected growing points and thin leaves [e.g. common lambsquarters (*Chenopodium album* L.)] are controlled with 20–50 kg/ha. The second group of species with protected growing points and leaves that are moderately heat-tolerant [e.g. common knotgrass (*Polygonum aviculare*)] require more than 50 kg/ha. The third group with more protected growing points [e.g. shepherd's purse (*Capsella bursa-pastoris*)] are controlled at early stages (up to 4 leaves), but complete control is difficult at later growth stages. The fourth group of plants with creeping growth habit and well-protected growing points (e.g. perennial broadleaf and grasses) regrow from below-ground meristems. Control of species from the third and the fourth group requires more than one treatment.

The most important factor affecting the success of flame weeding is the life cycle of the weed. In general, perennials are more tolerant to flaming than annuals (Rifai et al., 2002). Flaming has no residual soil activity and the heat cannot penetrate the soil; therefore, flaming does not provide control of root structures of perennials. However, some biennials and perennials, such as *Cirsium arvense* (L.) Scop. *Carduus crispus* L. and *Carduus acanthoides* L., are extremely sensitive to heat, and their above-ground tissue can be desiccated. A propane dose as low as 15 kg ha⁻¹ turned leaves of *C. arvense* completely black 24 hours after flaming; however, the plants regrew from their below-ground meristems within 2–3 weeks. Therefore, flaming must be repeated several times during the season to control perennials and biennials (Knezevic et al., 2009a,b; Knezevic et al., 2014a,b). Rifai et al. (1999) also reported less than 60% reduction in perennial weed numbers after one, two or three treatments of 35, 23 or 17 kg uf propane ha⁻¹ and only 75% after 320 and 216 kg ha⁻¹ over four flaming treatments. Hakansson (2003) and Rask and Andreasen (2007) also reported that two to three treatments should be carried out after an initial regrowth, but before regrowing shoots become too large. Therefore, multiple flaming conducted within a year and repeated over several years can deplete root nutrient reserves and kill 5- to 6-year-old perennials (Knezevic et al., 2014a).

Ascard (1994) reported that plant size was a key factor in controlling annual weeds. It was much more important than plant density (plant number per area), with smaller weeds being more sensitive to flaming than larger ones. Larger plants could have leaves and stems with protective layers (the lignification level) and multiple growing points. Therefore, flaming controls most annual weeds best during early growth stages (e.g. 5–10 cm tall) (Knezevic et al., 2014b).

Figure 5 (a) Green foxtail before flaming. (b) Green foxtail at 14 days after flaming with 12GPA, note green regrowth.

Knezevic et al. (2009c, 2014a,b) determined that 60–80 kg/ha of propane controlled 90% control of major broadleaf species (velvetleaf (Fig. 4), ivyleaf morningglory (*Ipomoea hederacea* Jacq.), redroot pigweed (*Amaranthus retroflexus* L.), common waterhemp (*Amaranthus rudis* Sauer), lambsquarters, field bindweed (*Convolvulus arvensis* L.), kochia (*Kochia scoparia* (L.) Schrad.) and Venice mallow (*Hibiscus trionum* L.)) and 80% control of several grass species (barnyardgrass (*Echinochloa crus-galli* (L.) Beauv.), green foxtail (*Setaria glauca* (L.) Beauv.; Fig. 5) and yellow foxtail (*Setaria pumila* (Poir.) Roemer and J.A. Schultes)).

There was a visual difference in response to heat between broadleaf and grassy species (Ulloa et al., 2010a,b). Leaves of most broadleaf species are completely desiccated within a few days after flaming, and there is no regrowth (Fig. 4a).

In contrary, leaves of grassy species typically turn white shortly after flaming, leaving an appearance of a dead plant; however, within a week or two grasses recover with new leaves (Fig. 5a). Control of grassy species can be improved by second flaming 7–10 days later. If an additional flaming is not feasible, an aggressive cultivation conducted within 24 hours after flaming to push soil onto the top of the flamed grass will be helpful in preventing the regrowth (Stepanovic et al., 2016a,b).

4 Uses of flame weeding

4.1 Flame weeding in agronomic crops

Crop tolerance (e.g. less injury) to heat varies with species and growth stages (Knezevic and Ulloa, 2007; Domingues et al., 2008; Teixeira et al., 2008; Knezevic et al., 2009a,b,c,

2013). For example, corn and sorghum flamed at earlier stages were more tolerant than soybean. Also, tolerance level varied among corn types, their growth stages and propane dose (Knezevic et al., 2009b).

Field corn flamed at the 5-leaf stage (V5) was the most tolerant, while the 2-leaf (V2) was the most susceptible with the highest crop injury and the largest yield loss (Ulloa et al., 2011a). Visual crop injury symptoms included initial whitening and then browning of leaves. Stunting of growth was especially evident when the plants were flamed with higher propane doses (44 and 85 kg ha^{-1}). Most visual injuries were transient because corn recovered within a few weeks. Most growth stages showed higher injury at early evaluation dates compared to later rating dates, suggesting that flamed plants were able to recover. Among the yield components, number of plants m^{-2} and number of kernels cob^{-1} were the most affected parameters followed by 1000-kernel weight. A propane dose of 60 kg ha^{-1} was highly efficient in providing 80–90% control of many grasses and broadleaf weeds (Ulloa et al., 2010a,b). A propane dose of 60 kg ha^{-1} caused yield losses of 3%, 9% and 13% for the V5, V7 and V2 growth stages, respectively. From a practical standpoint, the 3% yield reduction of field maize flamed broadcast at the V5 stage was below the arbitrary assigned acceptable yield reduction of about 5% (e.g. threshold level) and would be acceptable by organic producers.

In addition, Datta et al. (2013) studied the effect of single and repeated flaming on yield components and yield of field corn. They concluded that, flaming conducted once at V2, V4 or V6 growth stages resulted in 38%, 30% and 20% crop injury, respectively, 7 DAT compared to significantly lower injury levels of 8%, 6% and 6% for the corresponding growth stages by 28 DAT. Similar trends occurred when plants were flamed two times (e.g. V2 and V6 as well as V4 and V6) exhibiting injury rating about 6% 28 DAT. However, corn plants did not recover well over time when flamed three times (e.g. at V2, V4 and V6 growth stages), which resulted in over 50% injury 7 DAT and more than 30% injury by 28 DAT. Corn yields in their study ranged from 11.1 to 11.6 t ha^{-1} in the plots flamed once or twice, which were statistically similar to the yield obtained from the non-flamed control (11.7 t ha^{-1}). Flaming conducted three times (at V2, V4 and V6 growth stages) was the only treatment that resulted in significantly lower yield (9.9 t ha^{-1}) compared to the non-flamed control. Maize flamed three times produced about 8.5% lower yield than the non-flamed control. From a practical standpoint, an 8.5% yield reduction may not be acceptable. These results indicate that corn can tolerate a maximum of two flaming operations per season. Corn plants fully recovered when flamed twice under weed-free situations, suggesting that broadcast flaming two times in the season could be an acceptable alternative for weed control in corn.

Soybean flamed at the VC (unfolded cotyledons) stage was the most tolerant, whereas the VU (fully unrolled unifoliate leaves) stage was the most susceptible resulting in the highest crop injury and the largest yield loss (Ulloa et al., 2010f). Soybean flamed at the VU stage was unable to recover over time from the injuries. Plants flamed at other growth stages [e.g. VC, V2 (second trifoliate stage) and V5 (fifth trifoliate stage)] recovered over time. Unlike corn, where the growing point is protected beneath the soil surface for several weeks, the growing point of soybean is between the cotyledons and moves above the soil surface at emergence. The growing point is partially protected by both cotyledons at the VC stage resulting in the least effect of flaming. In contrast, if the growing point was exposed at the VU stage it results in severe damage. Consequently, many plants did not get the chance to regrow. Of all yield components, number of plants m^{-2} was the most affected by broadcast flaming. A significant number of plants were lost when flaming was conducted at the VU and V2 stages. They also reported that a propane dose of 60 kg ha^{-1} caused yield losses of 6%, 67%, 30% and 10% for the VC, VU, V2 and V5 growth stages, respectively. The 6% yield reduction in soybean flamed at the VC stage was above the arbitrary assigned acceptable yield reduction value of 5%; therefore, it might not be acceptable by organic growers. Furthermore, they explained that these yield reductions were the result of the torches intentionally positioned directly over crop rows. However, positioning flames below the soybean canopy would reduce the exposure time and should reduce yield loss.

In addition, Knezevic et al. (2013) studied soybean yield as influenced by single and repeated flaming. They concluded that soybean response varied with growth stage at flaming and number of treatments. For example, they suggested that VC and V5 were the most tolerant growth stages for broadcast flaming with the least crop injury. Flaming conducted once at VC or V5 stages, or twice (at VC and V5 stages) resulted in 23%, 19% and 24% crop injury 7 DAT, respectively, whereas the injury levels were reduced to 8%, 10% and 12% for the corresponding growth stages by 28 DAT. These results suggested that soybean plants flamed at VC or V5 stages were able to recover. In contrast, any treatment combination that included flaming at V2 stage resulted in over 70% injury and no recovery. Flaming conducted once (at V2), twice (at VC and V2) and thrice (at VC, V2 and V5) resulted in 72%, 83% and 92% injury, respectively, at 28 DAT. These results indicated that soybean is extremely sensitive to heat when flamed at V2, regardless of the number of subsequent flamings. This sensitivity is primarily as a result of the heat-induced damage to the growing point of soybean (Ulloa et al., 2010f), as the soybean plants were not tall enough to avoid the heat, despite the presence of the hoods. Furthermore, they reported that all yield components were affected by flaming with the exception of number of seeds pod^{-1}. For example, treatments that included broadcast flaming at VC and/or V5 stages resulted in more plants m^{-2}, fewer branches plant^{-1}, fewer pods plant^{-1} and greater 1000-seed weight compared to any of the treatments with flaming at V2 stage. With the exception of plants m^{-2}, there were no significant differences among yield components when flaming was conducted at VC and/or V5 stages compared to the non-flamed control. For instance, flaming once at V5 stage resulted in similar plant number m^{-2} compared to the non-flamed control. Any flaming treatment that included V2 stage resulted in significant loss of soybean stand, which provided greater space for the survived plants to develop more branches and consequently more pods plant^{-1} compared to the non-flamed control. For example, flaming treatment conducted three times (at VC, V2 and V5) produced 1.59 more branches plant^{-1} and 54.1 more pods plant^{-1} than the non-flamed control. In contrary, the 1000-seed weight for the same treatment decreased by 27.5 g compared to the non-flamed control. From practical standpoint, this increase in branch number plant^{-1} and pods plant^{-1} could not compensate for the loss in plant number area^{-1}; thus the yields were the lowest in all treatments that contained flaming at V2 stage.

Finally, Knezevic et al. (2013) also reported that the highest crop yields were obtained from the non-flamed control (3.45 t ha^{-1}), which were not statistically different from the plots flamed once at VC (3.35 t ha^{-1}) and V5 stages (3.32 t ha^{-1}), and two times at VC and V5 stages (3.24 t ha^{-1}). Significantly, lower yields were in all plots flamed at V2 stage (1.03 t ha^{-1} at V2, 0.46 t ha^{-1} at VC and V2, and 0.38 t ha^{-1} at V2 and V5). The lowest yields were in soybean flamed three times (VC, V2 and V5 stages), which yielded only 0.36 t ha^{-1}. These results suggest that soybean could tolerate a maximum of two flaming applications per season (e.g. at VC and V5 growth stages). This differential response is attributed to the position of the growing point relative to the heat source during flaming. At VC stage, the growing point of soybean is between the cotyledons

(Ritchie et al., 1997), which are full of moisture and swollen, thus providing a physical barrier that protects the growing point from the flame (Ulloa et al., 2010f). In contrast, the growing point was exposed to heat during flaming at V2 stage resulting in many dead or severely stunted plants, with little or no potential to regrow. At V5 stage, the growing point was at the top of the soybean plant; thus it was physically at least 30 cm above the specially designed hoods, and away from the heat.

Hoods kept the heat primarily close to the bottom 20 cm of soybeans; therefore, the heat does not damage the growing point. Despite the fact that the bottom leaves were damaged by the heat, the plants were able to continue growing and produce yields statistically similar to the non-flamed plots. Therefore, from practical standpoint, flaming two times at VC and V5 stages would be an acceptable practice in the toolbox of integrated weed management that can be used by soybean producers. Flaming can also be conducted after V5 stage (e.g. until canopy closure); however, the heat from the flame has potential to damage soybean flowers. Additional studies are needed to determine the extent of such damage.

4.2 Flame weeding at archaeological sites

Knezevic et al. (2014b) suggested that flaming can also be used in non-agricultural settings, including city parks, sidewalks or even at archaeological sites. Archaeological sites are often large open spaces with a variety of monuments and statues. Presence of weed between the monuments is not desirable because weeds can (1) conceal the monuments, (2) interfere with regular maintenance and restoration, (3) obstruct visitor site access, (4) impair the aesthetics of the site and (5) increase the risk of fire during hot and dry summers (Lisci et al., 2003; Mishra et al., 1995; Zahos, 1998).

Currently there are no official guidelines for vegetation management around archaeological sites in the Mediterranean region, including Greece (Kanellou et al., 2015; Zahos, 1998). Greek national law (8197/90920/B/1883/01.08.2013) prohibits the use of pesticides in and around archaeological sites. Therefore, there is a need to develop alternative tools for vegetation control. Eradication is not the intent at archaeological sites. Management of vegetation at a desirable level is the goal. It is a challenging task. The most common weed control methods are string trimming and hand weeding (Catizone, 1998; Kanellou et al., 2015), which are expensive and labour intensive (Arvanitis, 1998; Zahos, 1998). Mechanical cultivation, a common weed management method in organic farming systems (Hiltbrunner et al., 2007; Ulloa et al., 2010a), cannot be used due to an uneven landscape and potential mechanical damage of unexcavated archaeological sites. Therefore, there was interest in propane-fuelled flame weeding (Datta and Knezevic, 2013). Knezevic led a research team to examine the effects of propane dose and multiple flaming operations on weed control at several archaeological sites in Greece from 2011 to 2014. The two propane rates of 99 kg ha^{-1} and 129 kg ha^{-1} were applied two, three or four times at three archaeological sites of Greece (Kolona, Ancient Messene and Early Christian Amfipolis). Results from the study suggested that propane flaming can be used to suppress vegetation (Electra et al., 2017). Because short vegetation presence is very important to provide a ground cover, the results from this study are encouraging as they showed that multiple flaming (3–4 times) with the dose of 129 kg ha^{-1} has the potential to keep vegetation suppressed for the first 2–3 months of the growing season and keep it at the desirable height (e.g. 5–10 cm).

Electra et al. (2017) showed that flaming can control vegetation for the first 100 d of the growing season, which is appropriate in the Mediterranean climate. Flaming maintained the desirable vegetation height (5–10 cm), and provided pleasant visuals of the site, as well as protection of the monuments from fires. The research team also recommended that flaming should be initiated at about 10 cm of vegetation height and repeated at least four times in 2- to 3-week intervals (at 10–15 cm regrowth height) to provide at least 3 months of vegetation suppression (Electra et al., 2017) and maintain desirable vegetation height. Vegetation height was identified by the managers of archaeological sites as one of the most important factors contributing to the overall enjoyment of the site (Kanellou, 2016). Site managers agreed that a 5–10 cm height is the ideal height. Taller vegetation can interfere with monument restoration, and will visually impair the monuments and obstruct free access. Flame weeding could be combined with hand weeding and trimming in an integrated weed management programme, especially in those years when weather conditions may not be favourable for flaming (e.g. rainy or extremely wet or dry periods), or if the cost of propane makes flaming too expensive. Cost (propane + labour) can range from €200 to €350 ha^{-1} ($220–400) compared to string trimming at €450 ha^{-1} ($500) or hand weeding at €1.800 ha^{-1} ($2000). Flame weeding could become an alternative tool for weed management in cropping systems across Greece.

5 Advantages, disadvantages and environmental impacts

5.1 Economics of flame weeding

From an economic standpoint, the costs of a single flaming operation applied broadcast below the crop canopy could be $30–$40/ha, without taking into account the cost of the equipment and labour (current price of propane = $0.5/kg × 60–80 kg). Banded application (over the crop row) can cost $12–$20/ha due to lower propane use rates (30–40 kg/ha).

5.2 Advantages and disadvantages of flaming

Flaming does not disturb soil structure as repeated mechanical weeding does. It has been reported that repeated cultivation promotes loss of organic matter and soil erosion induced by wind and heavy rains (Wszelaki et al., 2007; Stepanovic et al., 2016a,b). Flaming can be carried out on wet or stony soils, does not disrupt the soil surface and does not bring buried weed seeds to the soil surface (Ascard et al., 2007; Wszelaki et al., 2007).

Flame weeding is less expensive than hand weeding and organic herbicides (Nemming, 1994), and there is no chance for weeds to develop resistance to high levels of instant heat. Cost of hand weeding can range from $100 to $300 per acre ($300–$800 per hectare).

The main disadvantages is fire in fields with heavy crop residue. Because heat is non-selective, it has the potential to injure healthy crops.

The disadvantage of flame weeding when compared to conventional herbicides include higher cost of equipment compared to herbicide applicators, lack of selectivity for crop safety, low speed of application due to smaller coverage (e.g. most flamers can treat only 4–12 rows) and lack of residual weed control (Ascard, 1995; Ascard et al., 2007).

From a resource and environmental point of view, the high energy requirement, the ineffective use of fossil fuels and the release of carbon emissions in relation to climate change could be seen as disadvantages; however, propane combustion is relatively clean compared to other fossil fuels, for example, diesel (Ascard et al., 2007).

5.3 Propane and the environment

Life cycle analysis

Life cycle analysis (LCA) has been used to compare the full range of environmental and social effects assignable to products and services, to be able to select the least burdensome ones. Alternative weed management practices available for conventional and organic farmers such as flaming, conventional herbicide (glyphosate) and tillage were compared for CO_2 emission and energy use.

CO_2 emission and energy use

Ulloa et al. (2010a,b) determined that a propane dose of 60 kg ha^{-1} provided up to 80–90% control of many annual broadleaf and grass species in Nebraska. (Such a propane dose could produce 189.0 kg CO_2 ha^{-1} (180.0 kg CO_2 ha^{-1} from propane combustion plus 8.9 kg CO_2 ha^{-1} from the diesel consumption) from an energy use of 2.9 GJ ha^{-1}). However, it should be pointed out that the 60 kg ha^{-1} flaming treatment covers 100% of the treatment area. When flaming is conducted only over a crop row (narrow band), the heat over the crop row can reduce the required propane dose to as little as 20 kg ha^{-1}. This reduces the energy use and CO_2 emission by approximately 67%; the area not treated can be controlled with cultivation. This produces 90.8 kg CO_2 ha^{-1} (60.0 kg CO_2 ha^{-1} from propane combustion and 30.8 kg CO_2 ha^{-1} from diesel consumption), and energy use of 1.35 GJ ha^{-1}.

Manufacturing, transporting and applying the recommended dose of glyphosate could produce 98.1 kg CO_2 ha^{-1} (95.9 kg CO_2 ha^{-1} from manufacturing and 2.3 kg CO_2 ha^{-1} from the diesel consumption) and use 0.51 GJ ha^{-1} of energy. The comparable figures for cultivation were diesel consumption of 8.2 L ha^{-1}, which produced 21.9 kg CO_2 ha^{-1} from 0.30 GJ ha^{-1} of energy.

The energy use and CO_2 emission of flaming are higher than glyphosate and mechanical methods (Table 1). However, a banded flaming treatment can reduce energy use and

Table 1 Comparison of energy use and CO_2 emission among three different weed control methods available for conventional and organic crop production systems

Weed control method	Energy use (GJ ha^{-1})	CO_2 emission (kg CO_2 ha^{-1})
Propane flaming[a]	3.0	178.0
Glyphosate spraying[b]	1.7	100.1
Crop cultivation[c]	0.3	19.4

[a]Based on liquefied petroleum gas (LPG) combustion and diesel fuel consumption. The energy use and CO_2 emission are calculated from a propane dose of 60 kg ha^{-1}.
[b]Based on glyphosate manufacturing process and diesel fuel consumption. The energy use and CO_2 emission are calculated from the recommended field application dose of glyphosate (1059 g ae ha^{-1}).
[c]Based on diesel fuel consumption (7.2 L ha^{-1}).

CO_2 emission. Therefore, banded flaming combined with cultivation treatment would emit slightly less CO_2 than glyphosate treatment.

Additional contribution to CO_2 emission comes from the diesel fuel consumed by the tractor utilized for the flamer. A 5325N Diesel John Deere tractor was used for sample calculations. Assuming the tractor is operating at 50% of maximum power while carrying the flamer, it will consume 6.6 L diesel h^{-1} operating at a constant engine speed of 1700 rpm (Nebraska Tractor Test Laboratory [NTTL], 2006).

At the operating speed of 6.4 km h^{-1}, a four-row flaming unit covers one hectare in 0.51 hours. The US Environmental Protection Agency [EPA] (2010) uses 2.7 kg of CO_2 emission per litre of diesel fuel, and the lower heating value of conventional diesel fuel is 42.8 MJ kg^{-1} (Wright et al., 2009). Each litre of diesel fuel has a mass of 0.845 kg (NTTL, 2006).

In general, there are both benefits and concerns with the use of flame weeding. Environmental benefits are for soil conservation and water quality, while the concerns include higher energy use, and the release of greenhouse gases (Lague et al., 2001; Ascard et al., 2007). Because of the above-mentioned benefits, we believe that flaming can be used in an integrated weed management programme in conventional and organically produced agronomic crops.

6 Future research and practical recommendations

6.1 Future directions

Flaming is a non-selective method of vegetation control that will negatively affect growth of any plants. There are several areas that deserve further research.

Knezevic et al. (2014b) developed recipes of flaming techniques for six agronomic crops. Recipes are also needed for various vegetables, small fruits, orchards, grapes and other agronomic crops.

Collaboration with mechanical engineers and equipment manufacturing companies would help test and build custom-designed flamers. For example, examining different positioning of the burners, designing torch nozzles that could provide different flame patterns and designing crop-specific flaming hoods could all help avoid crop damage and yield loss. Having automated electronic ignition systems and flame detection sensors can stabilize the flame and reignite the torches if the flame is lost due to wind. Custom designing flame weeding equipment for various crops would speed up the process of adoption of flame weeding.

Recent advances in precision agriculture techniques such as GPS tracking, row detection, weed identification and detection systems, and robotic weeding could help lower crop injury risks and propane consumption.

6.2 Practical implications of flaming

Flaming can be used in an integrated pest management for weeds and insect control. A series of manuscripts from flame weeding research (Knezevic et al., 2014b) demonstrated that flame weeding has a potential for use in at least six agronomic crops (sweet corn, field corn, popcorn, soybean, sorghum and sunflower). A 60 kg propane ha^{-1} (12 gal/acre) was the most effective field application dose to control many annual broadleaf weeds and grasses. Grass-type crops (e.g. field corn, popcorn, sweet corn and sorghum) are more tolerant to propane flaming than soybean. Post-emergent flaming is not recommended in winter wheat due to high injury and unacceptable yield reduction. Corn and sorghum can be safely flamed between VE (emergence) and V10 (10-leaf) stage. Soybean is tolerant to flaming only at the VE–VC stage (emergence-unfolded cotyledon) and at the V4–V5 stage (4–5 trifoliate). It is not recommended at the VU, V1 (first trifoliate), V2 and V3 (third trifoliate) stages because of very high crop injury and yield reduction (Knezevic et al., 2014b).

Corn and soybean can tolerate a maximum of two post-emergence flaming operations per season. Flaming conducted three times in field corn at the V2, V4 and V6 growth stages exhibited more than 30% injury with as high as 15% yield reduction compared to the weed-free control plots (Nedeljkovic et al., 2011). Flaming soybean three times at the VC, V2 and V5 growth stages resulted in more than 90% crop injury and as high as 90% yield reduction (Tursun et al., 2011).

Flame weeding is more effective during the afternoon, but crop injury will be higher, regardless of the propane dose. Thus, flaming should be done around noon to obtain maximum weed control with minimum crop damage (Ulloa et al., 2012).

Flaming has a potential to be used effectively in organic crop production systems of three corn types, sorghum, soybean and sunflower, when conducted properly at the most tolerant growth stage. It is important to emphasize that propane flaming should not be the only method for non-chemical weed control; however, it could be part of an IWM programme. Other measures are still needed to control weeds that emerge later during the growing season. More research is needed to develop new equipment and methods of flaming. Information from such research would expand flaming options as part of an IWM programme for both organic and conventional crop production systems.

7 Where to look for further information

Additional information and details can be found within the references listed below.

8 References

AgMRC: Agricultural Marketing Resource Center (2009). Organic Corn Profile. *Web page:* http://www.agmrc.org/commoditiesproducts/grainsoilseeds/corn/organiccornprofile.cfm (accessed 1 February 2010).

Arvanitis, P. (1998). Flora and fauna management in the archaeological sites of second Ephorate of Prehistorical and Classical Antiquities, pp. 45–47. In *Proceedings of Symposium on Spontaneous vegetation in Archaeological sites*. Athens: Acropolis.

Ascard, J. (1994). Dose–response models for flame weeding in relation to plant size and density. *Weed Res.* 34, 377–85.

Ascard, J. (1995). Effects of flame weeding on weed species at different developmental stages. *Weed Res.* 35, 397–411.

Ascard, J. (1998). Comparison of flaming and infrared radiation techniques for thermal weed control. *Weed Res.* 38, 69–76.

Ascard, J., Hatcher, P. E., Melander, B. and Upadhyaya, M. K., (2007). Thermal weed control. In M. K. Upadhyaya and R. E. Blackshaw (Eds), *Non-chemical Weed Management*, Chapter 10, pp. 155–75. CAB International: Wallingford, UK.

Bond, W. and Grundy, A. C. (2001). Non-chemical weed management in organic farming systems. *Weed Res.* 41, 383–405.

Catizone, P. (1998). Pompeii and Selinunte: *Two examples of vegetation management in Mediterranean archaeological sites*, pp. 48–53. In Proceedings of Symposium on Spontaneous vegetation in Archaeological sites. Athens: Acropolis Friends Association.

Datta, A. and Knezevic, S. Z., (2013). Flaming as an alternative weed control method for conventional and organic agronomic crop production systems: a review. *Adv. Agron.* 118: 399–428.

Domingues, A. C., Ulloa, S. M., Datta, A. and Knezevic, S. Z., (2008). Weed response to broadcast flaming. *RURALS*. Web page: http://digitalcommons.unl.edu/rurals/vol3/iss1/2 (accessed 15 March 2012).

Electra, K., Economou, G., Papafotiou, M, Ntoulas, N., Lyra, D., Kartsonas, E. and Knezevic, S. (2017). Flame Weeding at Archaeological Sites of the Mediterranean Region. *Weed Technol..* In Print.

Gianessi, L. P. and Reigner, N. P. (2007). The value of herbicides in U.S. crop production. *Weed Technol.* 21, 559–66.

Hakansson, S. (2003). *Weeds and Weed Management on Arable Land – An Ecological Approach.* CABI publishing: Uppsala, Sweden.

Hiltbrunner, J., Liedgens, M., Bloch, L., Stamp, P. and Streit, B., (2007). Legume cover crops as living mulches for organic wheat: components of biomass and the control of weeds. *Eur. J. Agron.* 26, 21–9.

Johnson, W. C. (2004). Weed control with organic production. Proceedings of the *Southeast Regional Fruit and Vegetable Conference*, Savannah, GA, USA, pp. 13–14.

Kanellou, E, Papafotiou, M., Economou, F., Dogani, I. and Galanos, A. (2015). *Unwanted vegetation management in archaeological sites*, pp. 87–9. In Book of Abstracts of the 4th Panhellenic Conference on Restoration. Association for Research and Promotion of Restoration of Monuments: Thessaloniki, Greece.

Kanellou, E. (2016). Development of native vegetation management methods based on indicative recording of vegetation and associated problems, as well as investigation of vegetation design principles aiming to protect the monument and enhance the historical landscape. *Ph.D. dissertation*. Agricultural University of Athens: Athens.

Knezevic, S. Z. and Ulloa, S. M. (2007). Flaming: potential new tool for weed control in organically grown agronomic crops. *J. Agric. Sci.* 52, 95–104. https://doi.org/10.2298/jas0702095k

Knezevic, S. Z., Dana, L., Scott, J. E. and Ulloa, S. M., (2007). Building a research flamer. In R. G. Hartzler and A. N. Hartzler (Eds), *Proceedings of the North Central Weed Science Society Conference* (), 62:32. St. Louis, MO, USA.

Knezevic, S. Z. (2009). Flaming: A new weed control tool in organic crops. *Crop Watch*, University of Nebraska-Lincoln Extension. Web page: http://cropwatch.unl.edu/archives/2009/crop17/organic_flaming.htm (accessed 12 March 2012).

Knezevic, S. Z., Datta, A. and Ulloa, S. M., (2009a). Growth stage impacts tolerance to broadcast flaming in agronomic crops. In D. C. Cloutier (Ed.), *Proceedings of the 8th European Weed Research Society Workshop on Physical and Cultural Weed Control*, pp. 86–91, Zaragoza, Spain.

Knezevic, S. Z., Costa, C. M., Ulloa, S. M. and Datta, A., (2009b). Response of maize (Zea mays L.) types to broadcast flaming. In D. C. Cloutier (Ed.), *Proceedings of the 8th European Weed Research Society Workshop on Physical and Cultural Weed Control*, pp. 92–7, Zaragoza, Spain.

Knezevic, S. Z., Datta, A. and Ulloa, S. M., (2009c). Tolerance of selected weed species to broadcast flaming at different growth stages. In "Proceedings of the 8th European Weed Research Society Workshop on Physical and Cultural Weed Control" (D. C. Cloutier, Ed.), pp. 98–103, Zaragoza, Spain.

Knezevic S., Stepanovic. S, Datta A, Nedeljkovic D and Tursun N., (2013). Soybean yield and yield components as influenced by the single and repeated flaming. *Crop Protection* 50:1–5.

Knezevic S., Stepanovic S, and Datta A., (2014a). Growth Stage Affects Response of Selected Weed Species to Flaming Propane. *Weed Technology*, 28(1):233–42.

Knezevic, S., A. Datta, C. Bruening and G. Gogos., (2014b). Propane Fueled Flame Weeding in Corn, Soybean and Sunflower. *A 38 page manual, free downloads from* http://www.agpropane.com/ContentPageWithLeftNav.aspx?id=1916.

Kruidhof, H. M., Bastiaans, L., and Kropff, M. J., (2008). Ecological weed management by cover cropping: effects on weed growth in autumn and weed establishment in spring. *Weed Res.* 48, 492–502.

Lague, C., Gill, J., and Peloquin, G., (2001). Thermal control in plant protection. In *"Physical Control Methods in Plant Protection"* (C. Vincent, B. Panneton, and F. Fleurat-Lessard, Eds.), pp. 35–46, Springer-Verlag, Berlin, Germany.

Lisci M, Monte M, Pacini E., (2003). Lichens and higher plants on stone: a review. *International Biodeterioration and Biodegradation* 51, 1–17.

Mishra, A. K., Jain, K. K. and Garg, K. L. (1995). Role of higher plants in the deterioration of historic buildings. *Sci. Total Environ.* 167, 375–92.

Nebraska Tractor Test Laboratory (2006, March). Nebraska Tractor Test 1867: *John Deere 5325 Diesel.* Retrieved23 July 2010, from http://tractortestlab.unl.edu/Deere%20test%20reports.htm.

Nedeljkovic, A., Stepanovic, S., Neilson, B., Datta, A., Bruening, C., Gogos, G. and Knezevic, S. Z. (2011). Corn tolerance to multiple flaming. In R. G. Hartzler and A. N. Hartzler (Eds), *Proceedings of the North Central Weed Science Society Conference*, 66:11, Milwaukee, WI, USA.

Nemming, A. (1994). Costs of flame cultivation. *Acta Hort.* 372, 205–12.

Parish, S. (1990). A review of non-chemical weed control techniques. *Biol. Agric. Hort.* 7, 117–37.

Pelletier, Y., McLeod, C. D. and Bernard, G. (1995). Description of sub-lethal injuries caused to the Colorado potato beetle by propane flamer treatment. *J. Econ. Entomol.* 88, 1203–5.

Rask, A. M. and Andreasen, C. (2007). Influence of mechanical rhizome cutting, rhizome drying and burial at different developmental stages on the regrowth of *Calystegia sepium*. *Weed Res.* 47, 84–93.

Rifai, M. N., Lacko-Bartosova, M. and Puskarova, V. (1996). Weed control for organic vegetable farming. *Rostl. Vyroba* 42, 463–6.

Rifai, N. M., Lacko-Bartosova, M. and Somer, R. (1999). Weed control by flaming and hot steam in apple orchards. *Plant Prot. Sci.* 35, 147–52.

Rifai, M. N., Astatkie, T., Lacko-Bartosova, M. and Gadus, J. (2002). Effect of two different thermal units and three types of mulch on weeds in apple orchards. *J. Environ. Eng. Sci.* 1, 331–8.

Ritchie, S. W., Hanway, J. J. and Benson, G. O. (1996). How a corn plant develops. Spec. Rep. 48. *Iowa State Uni. Coop. Ext. Serv.*, Ames, IA, USA.

Stepanovic, S., Datta A, Neilson B., Bruening C., Shapiro C., Gogos G., Knezevic S., (2016a). The effectiveness of flame weeding and cultivation on weed control, yield and yield components of organic soybean as influenced by manure application. *Renew. Agr. Food Sys.* 31(4):288–99.

Stepanovic, S., Datta, A., Neilson, B., Bruening, C., Shapiro, C., Gogos, G. and Knezevic S. (2016b). Effectiveness of flame weeding and cultivation for weed control in organic maize. *Biol. Agr. Hortic.* 32(1), 47–62.

Stopes, C. and Millington, S. (1991). Weed control in organic farming systems. In *Proceedings of the Brighton Crop Protection Conference-Weeds*, pp. 185–92. Brighton, UK.

Teixeira, H. Z., Ulloa, S. M., Datta, A., and Knezevic, S. Z., (2008). Corn (*Zea mays*) and soybean (*Glycine max*) tolerance to broadcast flaming. *RURALS*. Web page: http://digitalcommons.unl.edu/rurals/vol3/iss1/1 (accessed16 March 2012).

Tursun, N., Datta, A., Neilson, B., Stepanovic, S., Bruening, C., Gogos, G., and Knezevic, S. Z., (2011). Soybean tolerance to multiple flaming. In (R. G. Hartzler and A. N. Hartzler (Eds.), *Proceedings of the North Central Weed Science Society Conference*, 66:16, Milwaukee, WI, USA.

Ulloa, S. M., Datta, A. and Knezevic, S. Z. (2010a). Growth stage influenced differential response of foxtail and pigweed species to broadcast flaming. *Weed Technol.* 24, 319–25.

Ulloa, S. M., Datta, A. and Knezevic, S. Z. (2010b). Tolerance of selected weed species to broadcast flaming at different growth stages. *Crop Prot.* 29, 1381–8.

Ulloa, S. M., Datta, A., Malidza, G., Leskovsek, R. and Knezevic, S. Z. (2010f). Yield and yield components of soybean [*Glycine max* (L.) Merr.], are influenced by the timing of broadcast flaming. *Field Crops Res.* 119, 348–54.

Ulloa, S. M., Datta, A., Bruening, C., Neilson, B., Miller, J., Gogos, G. and Knezevic, S. Z. (2011a). Maize response to broadcast flaming at different growth stages: Effects on growth, yield and yield components. *Eur. J. Agron.* 34, 10–19.

Ulloa, S. M., Datta, A., Bruening, C., Gogos, G., Arkebauer, T. J. and Knezevic, S. Z. (2012). Weed control and crop tolerance to propane flaming as influenced by the time of day. *Crop Prot.* 31, 1–7.

US Environmental Protection Agency (2010). Emission Facts: Average Carbon Dioxide Emissions Resulting from Gasoline and Diesel Fuel. *Web page:* http://www.epa.gov/otaq/climate/420f05001.html (accessed 28 July 2010).

Williams, A. F., and Lom, W. L., (1974). *Liquefied Petroleum Gases: A Guide to Properties, Applications, and Usage of Propane and Butane.* Halsted Press a division of John Wiley & Sons, Inc: New York.

Wright, L., Boundy, B., Badger, P. C., Perlack, B. and Davis, S. (2009, December). *Biomass Energy Data Book: Edition 2.* Retrieved 9 August 2010, from http://cta.ornl.gov/bedb/index.shtml.

Wszelaki, A. L., Doohan, D. J. and Alexandrou, A. (2007). Weed control and crop quality in cabbage [*Brassica oleracea* (capitata group)] and tomato (*Lycopersicon lycopersicum*) using a propane flamer. *Crop Prot.* 26, 134–44.

Zahos K. (1998). The spontaneous flora in archaeological sites. The Greek experience, pp. 9–17. In *Proceedings of: Symposium on Spontaneous vegetation in Archaeological sites.* Acropolis Friends Association: Athens.

Chapter 15

Soil solarization: a sustainable method for weed management

Baruch Rubin, The Hebrew University of Jerusalem, Israel; and Abraham Gamliel, The Volcani Center, Israel

1 Introduction

The need for non-chemical, effective and sustainable weed management methods has become critical in light of increased public concern for the environment and the increasingly rigorous regulation of pesticide use (Upadhyaya and Blackshaw, 2007). The recent ban of numerous agrochemicals in Europe, where almost 75% of the previously registered pesticides have been removed from the market (Kudsk et al., 2013), necessitates the wider adoption of an integrated approach to weed management. The global phase out of methyl bromide and the lack of an appropriate replacement fumigant have directed much attention to alternative practices for chemical pest control, particularly in commercial vegetable crops (Besri et al., 2012).

Soil solarization or solar heating is a non-chemical method that has become an important weed management practice, not only in organic farming, but also in conventional horticultural (vegetable and ornamental) crops. All aspects of solarization were recently reviewed in the book *Soil Solarization: Theory and Practice* (Gamliel and Katan, 2012). The material presented in the weed management sections of that book was reviewed by Cohen and Rubin (2007) and later in Rubin et al. (2007) and Rubin (2012). Future technologies might provide improved methods for soil heating with less environmental effect. Plant protection specialists worldwide are making every effort to minimize pesticide use by developing environmentally acceptable methods of pest control to be used either

http://dx.doi.org/10.19103/AS.2017.0025.17

alone or in combination with other methods, including the application of conventional pesticides at reduced rates. Soil solarization constitutes a substantial contribution to these efforts.

Soil solarization uses solar radiation to heat moist soil that is covered with a transparent plastic film, usually polyethylene (PE), for several weeks, thereby increasing soil temperatures. Those high temperatures help to control germinating weeds (Cohen and Rubin, 2007; Rubin, 2012) and other pests. Maintaining soil moisture during the solarization process (generally achieved by drip irrigation) significantly increases temperature conductivity of mulched and sealed soil, and pest control efficacy.

The temperature in the upper layer of soil under the PE may reach 40–55°C depending on soil type and moisture level but temperature declines with depth.

The main factor involved in the weed control process is the physical mechanism of thermal killing, which apparently triggers additional chemical and biological processes such as breaking the dormancy of some weed seeds that contribute to the weed control process (Rubin and Benjamin, 1983, 1984; Cohen and Rubin, 2007; Katan and Gamliel, 2012b).

Because no chemicals are involved, soil solarization can be used for weed management in organic farming systems. It can also be incorporated into integrated pest management programmes in conventional farming systems, alone or in combination with other weed control techniques such as cultivation or herbicides.

2 Solarization: mode of action, effect on weeds, benefits and limitations

2.1 Mode of action

Soil solarization increases the temperature of moist soil by covering it with a transparent, thin (0.025–0.05 mm) layer of PE for several weeks during the summer (Mahrer, 1991; Mahrer and Shilo 2012). It converts solar energy to thermal energy in a day/night cycles (Fig. 1). Soil water influences the temperature conductivity of the soil, with greater amounts of soil water leading to higher peak temperatures, which increases the efficacy of the treatment even in deeper soil layers. The combination of high soil temperatures and the duration of solarization (Duration × Temperature) determine the level of control of weeds and other pests, including insects, pathogens and nematodes (Stapleton et al., 2002; Cohen and Rubin, 2007; Dahlquist et al., 2007; Hanson et al., 2014). The effect of mulching duration will be discussed later.

While farmers cannot regulate the status of the weed seedbank and the environmental conditions, the properties of the plastic film and soil moisture are manageable and can be modified. The type of PE used strongly affects the heat conductivity of the wet soil. Several types of low-density PE (LDPE) are available, differing in thickness, durability, degradability, price and efficacy (Gamliel, 2012b). Using two-ply PE (Patel et al., 2009), glazed PE (Hanson et al., 2014), UVA PE, anti-drip PE and aged PE (Avissar et al., 1986; Gamliel et al., 2009) for mulching generally results in increased radiation influx, higher soil temperatures and better control of various soil pests and weeds (Rubin, 2012). Candido et al. (2011, 2012) tested the efficacy of different films including a biodegradable cornstarch-based film for solarization in greenhouse and field-grown lettuce. No significant differences were found

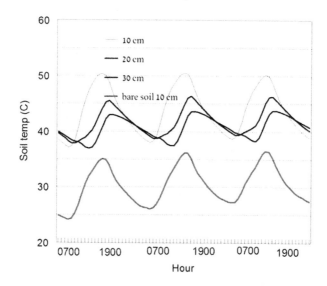

Figure 1 Daily cycles of temperature in solarized soil at depths of 10, 20 and 30 cm and in bare soil at a depth of 10 cm.

among these materials and all of them provided good control of annual weeds, but failed to control most perennial weeds.

The commonly used single-sheet LDPE (0.025 mm thick, often sheeting that has been used previously), combined with drip irrigation that maintains a continuous supply of moisture and heat conductivity, is the most cost-effective choice (Gamliel, 2012b; Rubin, 2012). Future improvements in this technology might provide better soil heating with less environmental effects, which would enable us to minimize pesticide use. Soil solarization makes a substantial contribution to these efforts.

Solarization should be conducted for several weeks during the warm and high-radiation period, which generally coincides with the warmer season of the year. Soil temperatures may peak in the afternoon hours (~16:00) with the lowest temperatures reached early in the morning (~06:00; Fig. 1).

The effect of transparent PE on the soil-heating process resembles to some extent the greenhouse effect. A portion of the solar radiation is transmitted through the transparent PE, absorbed by the soil and transformed to conserved heat. The PE largely prevents the escape of long-wave radiation and the evaporation of water from the soil to the atmosphere, consequently exerting a greenhouse effect. In addition, the water vapour that accumulates on the underside surface of the PE sheet further enhances the greenhouse effect, resulting in higher soil temperatures (Stevens et al., 1999). Black PE film, however, absorbs most of the intercepted solar radiation, while the energy is transmitted from the film by convection, which is a short-distance process, resulting in inferior soil heating and poorer weed control (Horowitz et al., 1983; Rubin and Benjamin, 1984; Standifer et al., 1984; Singh, 2006).

As mentioned above, most weed seeds are more sensitive to wet heating than dry heating, presumably, because moist seeds are likely to be carrying out more intense metabolic activity (Delouche and Baskin, 1973; Egley, 1990). This assumption is also valid for other soil pests (Shlevin et al., 2003). Therefore, all soil pre-treatments that improve the water content of the soil, such as soil cultivation or drip irrigation during mulching, may improve the efficacy of solarization treatments. Moreover, appropriate soil preparation, which leads to a levelled and smooth soil surface, facilitates the placement of the PE and helps prevent tearing (Elmore, 1991; Zaragoza, 2006).

To achieve effective pest control, the soil generally needs to be covered for 4–5 weeks. The required period varies with radiation level, the weed population in the seedbank, soil characteristics, climatic conditions and properties of the PE (Katan, 1981; Rubin and Benjamin, 1983, 1984; Stevens et al., 1990; Katan and DeVay, 1991; Cohen and Rubin, 2007; Rubin, 2012). In a study in northeast Brazil, Marenco and Lustosa (2000) reported that, although soil mulching for nine weeks resulted in only 60% control of the weed species, the marketable carrot yield was greater in solarized than in unsolarized soil.

The efficacy of solarization is highly dependent on the status of the seedbank, as dormant weed seeds are not controlled by this process (Johnson et al., 2012). The temperature cycles during solarization may enhance the germination of some dormant seeds, exposing them to the lethal process. Hence, soil pre-treatment such as irrigation combined with appropriate PE technology may be useful for overcoming unfavourable environmental conditions like low radiation or low temperatures prevailing in some regions or in certain seasons, increasing weed sensitivity and shortening the period that the soil must be kept covered (see recent review by Gamliel, 2012a).

2.2 Spectrum of weed control

A wide range of soil pests are effectively controlled by soil solarization, including soil-borne bacterial and fungal pathogens and nematodes (Gamliel and Katan, 2012), as well as numerous annual and very few perennial weeds (Cohen and Rubin, 2007). Weeds with hard seeds (i.e. weeds that have an impermeable testa or seed coat) such as small-seeded legumes, *Chenopodium* spp. and *Cuscuta* spp. are more tolerant to soil solarization (Cohen and Rubin, 2007). The sensitivity of weeds that propagate by rhizomes, bulbs and tubers varies as the high temperatures in upper soil layers may affect these propagules, but not those in deeper layers, which may sprout later (Rubin and Benjamin, 1983).

As reviewed by Cohen and Rubin (2007), most studies have confirmed that solarization is not sufficiently effective against purple nutsedge (*Cyperus rotundus* L.), but Stevens et al. (1990) demonstrated that mulching for 98 days resulted in excellent nutsedge control. Kumar et al. (2012) showed that solarization followed by the application of glyphosate during the summer fallow of a soybean–wheat system significantly reduced nutsedge density and led to higher soybean and wheat yields. In addition, Wallace et al. (2013) demonstrated that the sharp diurnal fluctuations in temperature caused by solarization increase the sprouting of the tubers of purple nutsedge, which might make them more vulnerable to other control measures.

Similar temporal control may be possible for seeds of the root parasites (e.g. broomrapes, the Orobancaceae family; Rubin, 2012). Mauro et al. (2015) improved the control of broomrape by combining the application of a humus product with solarization. Rathore et al. (2014) regarded solarization as an economical, simple and unhazardous method for broomrape control that does not involve toxic materials and is suitable for organic farming. Those authors claimed that solarization killed ~95% of buried viable *Orobanche* spp. seeds; no broomrape shoots emerged and no haustoria or underground tubercles of the parasite weed were found in their solarized soil. In contrast, Rubiales and Fernandez-Aparicio (2012) and Fernandez-Aparicio et al. (2016), who reviewed the methods currently available for cost-effective control of *Orobanche* spp. and *Phelipanche* spp., concluded that solarization is not economically feasible for low-value and low-input legume crops.

2.3 Benefits and limitations of solarization

Soil solarization has several benefits and limitations (Katan and DeVay 1991; Gamliel and Katan 2012). First, solarization is simple and can be introduced and adopted quickly. Second, as a non-chemical method, it is safer for the user, the crop grown on solarized soil and other non-target terrestrial organisms. Third, soil solarization rarely involves disturbances or undesirable side effects to the biological balance. Fourth, solarization can feasibly be used in combination with other weed control methods in various crops. Finally, added economic value over the long term has frequently been observed in solarized soils.

Soil solarization also has limitations and impediments. First, the high cost of the plastic mulching is justified only where and when cash crops are grown, that 'can' cover the cost with higher yield. Second, soil is mulched during solarization for as long as 3–9 weeks. However, shortening the solarization period by using combined treatments can overcome this limitation (as discussed later). Third, because it is climate-dependent, its use is restricted to certain regions and months of the year. However, improving heating efficiency or combining solarization with other methods may help. Fourth, certain heat-tolerant pests and weeds, as well as most perennial weeds are not sufficiently controlled by solarization alone. Therefore, the use of solarization in combination with chemical, biological or physical weed management techniques can significantly improve solarization performance (Rubin and Benjamin, 1984; Stevens et al., 1990; Golzardi et al., 2014; Jabran and Chauhan, 2015; Dai et al., 2016). A major component of solarization is the disposal of used PE after completion of solarization. The film might remain and used throughout the cropping season in the open field or inside greenhouse. Usually, the used plastic sheets are removed from the plots and taken for recycling, burial or incineration, all of which add to the cost of solarization. Another 'expensive' way is to use degradable or sprayable films, which eliminate the need of removal (Gamliel, 2012b).

Pfeifer-Meister et al. (2012) successfully used solarization and herbicides to restore wetlands by decreasing the cover of exotic plants and reducing their seedbank in the soil. Re-infestation, particularly with perennial weeds and some annuals, may occur and this necessitates monitoring treated fields to detect, identify and develop management methods for new invaders as early as possible. Additional uses and benefits of soil solarization are continually being discovered, for example, its use in the control of invasive plants (Cohen et al., 2008), its use for weed management in aerobic rice (Jabran and Chauhan, 2015) and new film materials for outdoor and greenhouse use (Candido et al., 2011; Hanson et al., 2014).

3 Plastic mulching technologies

3.1 Overview

Soil solarization is defined as covering soil with transparent plastic sheets, mainly LDPE, with only limited effect of film thickness on soil heating. Therefore, a thin plastic sheet

Figure 2 Broadcast application of soil solarization in a broomrape-infested field. Transparent polyethylene film is laid with a tarp layer and attached to the previous laid strip by hot-air welding.

(20 μm) can be used provided that it possesses the necessary physical properties to enable the solarization process. PE technology offers a wide range of types of sheeting that differ in thickness, durability, UV protection and anti-drip properties (i.e. preventing water condensation on the underside PE surface), which improve the efficacy of solarization. In general, all types of transparent PE sheets are appropriate for weed management.

Avissar et al. (1986) reported that aged (previously used) PE causes soil temperatures to rise higher than new PE. They concluded that significant changes occur in the photometric properties of the PE during ageing, resulting in increased radiation influx at the soil surface. Later, it was found that the dust accumulates on aged film provides anti-drip properties that improve the transmittance of solar irradiation (Arbel et al., 2003).

Continuous mulching is needed for effective soil solarization over a wide area, as it reduces the 'border effect' (Grinstein et al., 1995) and improves control. Moreover, broadcast soil solarization reduces the risk of re-infestation by pests. The optimal sheeting width for a solarized plot would be the whole field rather than just over the crop row. Broadcast field solarization improves crop yield and provides long-term effects, spreading the cost of disinfestation over several crops. The economic advantages resulting from complete covering of the plot should be evaluated against the expenses involved.

Broadcast soil solarization of small plots can be carried out manually. A technology to fasten PE sheets together for large-scale application (Fig. 2) has been developed in Israel (Grinstein and Hetzroni, 1991). Specifically, sheets are fused together with hot air emitted from a combustion chamber and directed onto the sheet margins. The welding system is assembled on a tractor-mounted tarping machine and its capacity is similar to that of the gluing method for other fumigation purposes (Fig. 2).

3.2 Improving plastic formulation and performance

Double-layer films: two transparent films

The use of a single layer of plastic film is the most common method of soil solarization. Transparent films allow the transition of the solar irradiation, heating the soil that is adjacent to the film. Since these films are thin and lack an insulating layer, some of the heat energy is lost to the atmosphere, thereby reducing the efficiency of the soil heating.

One way to increase soil heating is to use a two-layer PE mulch that heats the soil to higher levels than a single layer. This was demonstrated by Raymundo and Alcazar (1986), who achieved a temperature increase of 12.5°C at a depth of 10 cm (60°C vs 47.5°C) using a double-layer film, as opposed to a single-layer film. Ben-Yephet et al. (1987) achieved a similar increase in soil temperature by using double-layer films. They observed a 98% reduction in the viability of *Fusarium oxysporum* f. sp. *vasinfectum* after 30 days under the double mulch, as compared with a 58% reduction under a single-layer mulch, at a depth of 30 cm.

Double-layer films form a static, insulating air space between the plastic layers. This prevents the loss of heat to the atmosphere especially during night. The use of a double-layer film also offers opportunities for the use of solarization in areas and climates that are not favourable for single-layer solarization techniques. In central Italy, an area that is climatically marginal for solarization, a double-film solarization process effectively reduced the viability of *Pythium* spp., *Fusarium* spp. and *Rhizoctonia solani* Kuhn in a forest nursery (Annesi and Motta, 1994). This approach has also been used successfully to treat with nursery potting mix in Australia (Duff and Connelly, 1993). Using a double-layer mulch should improve weed control as well.

Soil solarization in a closed greenhouse can be regarded as a modified version of the double-layer film proposed for improved disease control (Garibaldi and Gullino, 1991). Hanson et al. (2014) further improved the method by mulching the soil surface within a walk-in plastic tunnel with a thin PE (25.4 µm), while the tunnel itself was covered with a thick (152.4 µm) IR/AC PE. This experiment was conducted in southern Arizona in June, the ideal time for soil solarization in that region with highest solar radiation. Soil temperature was elevated in the mulched plots to an average daily maximum of 63.4°C at a depth of 5 cm and 52.8°C at a depth of 15 cm, with those temperatures maintained for 14 h. Under these conditions, even a short, one-week long solarization treatment is sufficient for effective control of weeds and other pests.

Double-layer films: transparent over black double film

A different approach to the double-layer technique involves the use of a black sprayable polymer as the lower layer, to increase solar heating. Such technology is applied commercially in the solar panels of solar water heaters. Arbel et al. (2003) achieved an increase in soil temperature by mulching a transparent PE sheet over a layer of sprayable black coating. The use of a double mulch that consists of a transparent film over a black polymer coat very effectively increases soil temperatures. However, the cost of this double mulching and the technology that needs to be developed for simultaneous tarping limit the implementation of such an approach. However, this approach can be successfully applied in strips and in small plots.

Improved plastic films

The use of the new films, which intensify solar heating, offer improved control of weeds and pathogens in a variety of crop-management systems. The new films prevent heat loss through the film with infrared (IR) blocking material. Chase et al. (1999) observed improved heating when IR films were used under rainy and cloudy conditions in Florida. Hanson and her co-workers (2014) showed that covering a high tunnel with thick IR/AC PE significantly improved the results of solarization. A PE film formulated with the addition of anti-drip components has also been successfully tested for solarization (Arbel et al., 2003). This formulation prevents the condensation of water droplets on the film surface, leading to a 30% increase in irradiation transmittance, as compared with older films. Soil temperatures under the anti-drip film were 2–7°C higher than those under regular film.

A sprayable polymer for coating with a black polymer formulation can also be used for solarization purposes (Gamliel and Katan 2009). The soil-heating process is faster with sprayable mulch than with plastic film, but the soil also cools down to lower levels at night. Overall, soil temperatures under sprayable mulch are somewhat lower than those obtained under regular plastic film. Nevertheless, soil solarization using sprayable mulches was effective in controlling *Verticillium* wilt and potato scab in potato and a wide spectrum of weeds. Other studies have aimed to improve and further develop this approach. The advantages of this approach include rapid application with no need to remove the film at the end of the process. However, it is still an 'expensive' way to use this technology (Gamliel, 2012b).

4 Effects of solarization on soil nutrients and pesticides

4.1 Effect on soil nutrients

Solarization has been shown to increase soil nutrient availability, thereby reducing the need for fertilizer (Stapleton and DeVay, 1986). An improved growth response beyond weed and pathogen control has been observed in plants grown in solarized soils, and in other disinfested soils (Stapleton and DeVay, 1986; Chen et al., 1991; Gamliel and Katan, 1991; Gruenzweig et al., 1993). This improved plant development, beyond visible pest control, can be attributed to presently unknown chemical, biological and physical changes that occur during and following the solarization.

One of the biggest advantages of solarization is an increase in the population of beneficial organisms. Stevens et al. (1990) reported that the rhizosphere in solarized soil contained more heat-tolerant fungi and bacteria than the rhizospheres of comparable unsolarized soil. Populations of beneficial, growth-promoting bacterial and fungal competitors can quickly recolonize solarized soil (Chen et al., 1991; Gamliel and Katan, 1991; Stevens et al., 2003), while plant pathogenic fungi that have been weakened by high soil temperatures may become more sensitive to the effects of soil competitors (Freeman and Katan, 1988). Along with the increased availability of essential nutrients and improved soil tilth, the biological recolonization of solarized soil may also stimulate plant growth. Abdel-Rahim et al. (1988) described the release of a wide range of nutrients upon solarization, including Ca, Mg and N (as also found in other studies), as well as a 30–50% reduction in salinity as a result of solarization. The latter effect alone could account for growth responses above and beyond those associated with effective pathogen control. However, one should remember that all of these advantages might be available to uncontrolled weeds as well.

4.2 Solarization and accelerated degradation of pesticides

The degradation of pesticides in soil and their consequent persistence is affected by microbial activity. Hence, any agricultural practice that affects microbial activity will also affect the fate of pesticides in the soil. This is especially true for soil fumigation, which usually suppresses microbial activity, with potential negative or positive effects. This is another aspect of the interaction of solarization with pesticides that is often overlooked. For example, solarization can modify the dissipation of pesticide residues in soil (Aharonson et al., 1982). Solarization (as well as soil fumigants) may slow herbicide degradation, thereby increasing phytotoxicity. This has been demonstrated with the combination of solarization with the herbicide chlorthal-dimethyl (dacthal) in collard greens (Stevens et al., 1990). Similarly, the half-life of the herbicide terbutryn was extended from 15 days in untreated soil to 77 days in solarized soil (Avidov et al., 1985). Thus, herbicide doses should be adjusted when those chemicals are used in combination with solarization or fumigation. On the other hand, extending the persistence of a soil fungicide by solarization, as shown for benomyl (Yarden et al., 1985), is desirable. Similarly, solarization is a potential tool for curbing the undesirable phenomenon of accelerated degradation, as demonstrated for carbamothioate herbicides (Tal et al., 1990) and metam sodium (Di Primo et al., 2003). In a recent study, terbuthylazine was degraded more quickly during solarization combined with the application of sugarbeet vinasse, as opposed to composted sheep manure (Navarro et al., 2009; Fenoll et al., 2014). Soil mulching with LDPE film has been shown to enhance the dissipation of sulphentrazone as compared to bare soil, whereas the opposite was found for halosulfuron and metolachlor (Grey et al., 2007). Field studies have shown that solarization and straw mulching each enhance the dissipation of lactofen in soil, reducing the half-life of that herbicide from 30 days to 15 days, while a combination of both treatments has been shown to reduce the half-life of lactofen to only ten days (Mukherjee and Das, 2007).

5 Solarization and integrated pest management

5.1 Combining solarization with pesticides

Combining solarization with other control methods can have positive effects (Katan, 1996). Using solarization in combination with pesticide applied at a reduced rate can improve pest control and protect the environment. Enhancing the effects of solarization by combining it with other methods will increase its reliability, widen the spectrum of weeds controlled and enable its use under marginal climatic conditions and may also shorten the length of the period that the soil needs to be covered (Fig. 3).

Many studies have combined pesticides with solarization, frequently with good results. Such a combination may result in either a synergistic or an additive effect. The weakening of a weed by one agent may potentially lead to synergy, or at least improved control when both are applied simultaneously. Combining pesticides with solarization is especially promising because the plastic tarp prevents the escape of volatiles, enhances the chemical's activity at high temperatures and increases the potential vulnerability of soil-borne pests that may be weakened after exposure to sub-lethal doses of the control agent or other stress factors (Rubin and Benjamin, 1983; Freeman and Katan, 1988). Thus, the population of weakened propagules declines faster in solarized than in unsolarized soil, due to the activity of soil organisms. This can be regarded as induced biocontrol.

Figure 3 Soil solarization combined with metam sodium fumigation in a broomrape-infested field. Solarization applied as broadcast mulching using transparent PE film. Metam sodium was applied two weeks after mulching through a drip-irrigation system.

5.2 Combining solarization with organic amendments

The combination of solarization and organic amendments, also termed as 'biosolarization', is another attractive option. For example, Gamliel and Stapleton (1993) reported improved pest control when solarization was combined with the application of chicken manure. Similarly, in a study in California, the use of tomato pomace, a by-product of the tomato-processing industry, for 'biosolarization', reduced the primary energy demand and the global warming potential of a crop, as calculated by Oldfield et al. (2017).

The use of an organic amendment in combination with solarization has been investigated previously in many studies (Gamliel and Katan, 2012; Rubin, 2012). The combined use of solarization and organic amendments leads to the generation of biotoxic volatile compounds that accumulate under the plastic mulch, enhancing the vulnerability of soil organisms to soil heating (Gamliel and Stapleton, 2012). Using residues of species of Poaceae (Stapleton et al., 2010) or Brassicaceae (Stapleton and Bañuelos, 2009) for biofumigation together with solarization has been shown to improve weed control due to the release of allelopathic decomposition products such as isothiocyanates (Matthiessen and Shackleton, 2005; Price et al., 2005; Matthiessen and Kirkegaard, 2006). Saha et al. (2007) showed that the use of a cowpea cover crop followed by raised bed solarization was more effective than methyl bromide for nematode control and weed suppression and increased pepper yields. Solarization for 6–7 weeks with or without chicken manure or olive-processing waste has been shown to successfully control common purslane (*Portulaca oleracea* L.), *Amaranthus* spp. and *Setaria* spp. in greenhouse production (Boz, 2009). Furthermore, a combination of a short solarization period (3 weeks) with a half-rate of dazomate has also been shown to provide excellent weed control (Gamliel 2012a).

Mallek et al. (2007) reported that residues of onion and garlic reduced the germination of barnyardgrass (*Echinochloa crus-galli* (L.) Beauv.), common purslane, London rocket

(*Sisymbrium irio* L.) and black nightshade (*Solanum nigrum* L.) seeds, particularly when the soil temperature was elevated to 39°C (as opposed to 23°C). Thimmegowda et al. (2007) showed that farmyard manure applied at a rate of 10 t/ha after irrigation to field capacity before solarization for 60 days, proved better in terms of weed control and sunflower yield than a shorter solarization (45 days), or irrigation to 50% field capacity.

6 Recent applications of organic amendments and solarization in weed control

Combination of solarization and organic amendments has expanded in recent years to include different amendments. Achmon et al. (2016) studied the use of tomato processing waste as a soil amendment together with solarization under controlled conditions and achieved the killing of weed seeds. In addition, biosolarization provided effective control of corky root and reduced weed pressure in tomato plots (Diaz Hernandez et al., 2016).

Soil solarization in combination with the application of aqueous suspensions of *Trichoderma* sp. or earthworm compost increased soybean yield and significantly reduced weed infestation ~90%. The weed suppression lasted through harvest (McGovern et al., 2013). Similarly, the incorporation of sunn hemp (*Crotalaria juncea* L.) before solarization significantly reduced weed emergence in an onion in Hawaii (Quintanilla-Tornela et al., 2016). The latter studies also demonstrated additional benefits, including the suppression of nematodes and above-ground arthropods.

A commercially formulated humus product used in combination with solarization enhanced the efficacy of solarization against the broomrape seedbank in tomato greenhouses and increased tomato yields (Mauro et al., 2015). Another study found that the use of vermicompost and neem cake together with solarization suppressed weed infestation in a chilli pepper nursery (Maheswari, 2016).

As mentioned above, solarization treatments affect the persistence of pesticides in soil. The long-term effectiveness of the integration of solarization with chemical, biological and cultural means was thoroughly studied by Stevens et al. (2003). All of these data indicate that combining solarization with other methods can improve pest control, plant growth and environment quality.

7 Concluding remarks

Soil solarization is an additional tool in the plant protection toolbox, particularly, but not only for organic farming. It is a feasible technique for sustainable weed management that is both effective and economical under the right circumstances, but which must be adapted to each situation (Katan, 1987). Soil solarization aims to improve plant and soil health, goals which extend far beyond mere pest control. Any new control option, even if it is of limited use, should be welcomed and thoroughly investigated, adapted and adopted. It is most encouraging that, despite its limitations, solarization has been successfully applied to thousands of acres worldwide. It should be emphasized that not only does solarization introduce a variety of possibilities for the most sophisticated agricultural systems, it also provides possibilities for the simplest of systems such as those found in the Third World.

8 Where to look for further information

There are two books that contain much information on various aspects of soil solarization including weed management. The first one is edited by Katan and DeVay (1991) and was written during the early development of soil solarization, and a much more recent one, edited by Gamliel and Katan (2012) that compiles many more aspects of soil solarization and assembles detailed basic and applies information.

9 References

Abdel-Rahim, M. F., Satour, M. M., Mickail, K. Y., El Eraki, S. A., Grinstein, A., Chen, Y. and Katan, J. (1988). Effectiveness of soil solarization in furrow-irrigated Egyptian soils. *Plant Dis.* 72: 143–6.

Achmon, Y., Fernandez-Bayo, J. D., Hernandez, K., McCurry, D. G., Harrold, D. R., Su, J., Dahlquist-Willard, R. M., Stapleton, J. J., VanderGheynst, J. S. and Simmons, C. W. (2017). Weed seed inactivation in soil microcosms via biosolarization with mature compost and tomato processing waste amendments. *Pest Manag. Sci.* 73:862–79. DOI:10.1002/ps.4354.

Aharonson, N., Rubin, B., Katan, J. and Benjamin, A. (1982). Effect of methyl bromide or solar heating treatments on the persistence of pesticide in the soil, pp. 189–94. In *Pesticide Chemistry: Human Welfare and the Environment* (J. Miyamoto and P. C. Kearney (Eds)). Pergamon Press: Oxford, UK.

Annesi, T. and Motta, E. (1994). Soil solarization in an Italian forest nursery. *Eur. J. For. Pathol.* 24: 203–9.

Arbel, A., Siti, M., Barak, M., Katan, J. and Gamliel, A. (2003). Innovative plastic films enhance solarization efficacy and pest control, pp. 91.1–91.3. In *the Proceedings of 10th Annual International Research Conference on Methyl Bromide Alternatives and Emission Reduction*, San Diego, CA.

Avidov, E., Aharonson, N., Katan, J., Rubin, B. and Yarden, O. (1985). Persistence of terbutryn and atrazine in soil as affected by soil disinfestation and fungicides. *Weed Sci.* 33: 457–61.

Avissar, R., Naot, O., Mahrer, Y. and Katan, J. (1986). Field aging of transparent polyethylene mulches. II. Influence on the effectiveness of soil heating. *Soil Sci. Soc. Amer. J.* 50: 205–9.

Ben-Yephet, Y., Stapleton, J. J., Wakeman, R. J. and DeVay, J. E. (1987). Comparative effects of soil solarization with single and double layers of polyethylene on survival of *Fusarium oxysporum* f. sp. *vasinfectum. Phytoparasitica* 15: 181–5.

Besri, M., Pizano, M. and Porter, I. J. (2012). Soil solarization and the methyl bromide phase-out, pp. 77–87. In *Soil Solarization: Theory and Practice* (A. Gamliel and J. Katan (Eds)). APS Press: St. Paul, MN, USA.

Boz, O. (2009). Effect of olive-processing waste, chicken manure, and dazomet on weeds with or without soil solarization. *Afr. J. Biotechnol.* 8: 4946–52.

Candido, V., D'Addabbo, T., Miccolis, V. and Castronuovo, D. (2011). Weed control and yield response of soil solarization with different plastic films in lettuce. *Sci. Hort.* 130: 491–7.

Candido, V., D'Addabbo, T., Miccolis, V. and Castronuovo, D. (2012). Effect of different solarizing materials on weed suppression and lettuce response. *Phytoparasitica* 40: 185–94.

Chase, C. A., Sinclair, T. R., Chellemi, D. O., Gilreath, J. P., Locassio, S. T. and Olson, S. M. (1999). Heat-retentive films for increasing soil temperature in a humid, cloudy climate. *HortScience* 34: 1085–9.

Chen, Y., Gamliel, A., Stapleton, J. J. and Aviad, T. (1991). Chemical, physical, and microbial changes related to plant growth in disinfested soils, pp. 103–29. In *Soil Solarization* (J. Katan and J. E. DeVay (Eds)). CRC Press: Boca Raton, FL.

Cohen, O., Riov, J., Katan, J., Gamliel, A., and Bar, P. (2008). Reducing persistent seed banks by soil solarization – The case of *Acacia saligna. Weed Sci.* 56:860–5.

Cohen, O. and Rubin, B. (2007). Soil solarization and weed control, pp. 177–200. In *Non-Chemical Weed Management: Principles, Concepts and Technology* (M. K. Upadhyaya and R. E. Blackshaw (Eds)). CAB International: Wallingford, UK.

Dahlquist, R. M., Prather, T. S. and Stapleton, J. J. (2007). Time and temperature requirements for weed seed thermal death. *Weed Sci.* 55: 619–25.

Dai, Y., Senge, M., Yoshiyama, K., Zhang, P. and Zhang, F. (2016). Influencing factors, effects and development prospect of soil solarization. *Rev. Agric. Sci.* 4: 21–35.

Delouche, J. C. and Baskin, C. C. (1973). Accelerated aging temperatures for predicting the relative storability of seed lots. *Seed Sci. Technol.* 1: 427–52.

Diaz Hernandez, S., Gallo Lobel, A., Dominguez-Correa, P. and Rodriguez, A. (2016). Effect of repeated cycles of soil solarization and biosolarization on corky root, weeds and fruit yield in screen house tomatoes under subtropical climate conditions in Canary Islands. *Crop Prot.* 94: 20–7

Di Primo, P., Gamliel, A., Austerweil, M., Steiner, B., Beniches, M., Peretz-Alon, I. and Katan, J. (2003). Accelerated degradation of metam sodium and dazomet in soil: Characterization and consequences of pathogen control. *Crop Prot.* 22: 635–46.

Duff, J. D. and Connelly, M. I. (1993). Effect of solarization using single and double layers of clear plastic mulch in *Pythium*, *Phythophthora*, and*Sclerotium* species in a potting mix. *Australas. Plant Pathol.* 22: 28–35.

Egley, G. H. (1990). High temperature effects on germination and survival of weed seeds in soil. *Weed Sci.* 38: 429–35.

Elmore, C. L. (1991). Effect of soil solarization on weeds. In *the Proceedings of the First International Conference on Soil Solarization*, Amman, Jordan, 19–25 February 1990. FAO Plant Production and Protection, Paper 109. Rome. http://www.fao.org/docrep/T0455E/T0455E0c.htm#5.

Fenoll, J., Hellin, P., Flores, P., Lacasa, A. and Navarro, S. (2014). Solarization and biosolarization using organic wastes for the bioremediation of soil polluted with terbuthylazine and linuron residues. *J. Environ. Manag.* 143: 106–12.

Fernandez-Aparicio, M., Reboud, X. and Gibot-Leclerc, S. (2016). Broomrape weeds, underground mechanisms of parasitism and associated strategies for their control: A review. *Front. Plant Sci.* 7: 1–23. https://doi.org/10.3389/fpls.2016.00135

Freeman, S. and Katan, J. (1988). Weakening effect on propagules of Fusarium by sublethal heating. *Phytopathology* 78: 1656–61.

Gamliel, A. (2012a). Combining soil solarization with pesticides, pp. 99–108. In *Soil Solarization: Theory and Practice* (A. Gamliel and J. Katan (Eds)). APS Press: St. Paul, MN, USA.

Gamliel, A. (2012b). Plastic films for soil disinfestation: Chemistry and technology, pp. 165–74. In *Soil Solarization: Theory and Practice* (A. Gamliel and J. Katan (Eds)). APS Press: St. Paul, MN, USA.

Gamliel, A., Siti M., Arbel, A. and Katan, J. (2009). Soil solarization as a component of the integrated management of fusarium crown and root rot in tomato. *Acta Hort.* 808: 321–6.

Gamliel, A. and Katan, J. (1991). Involvement of fluorescent pseudomonads and other microorganisms increased growth response of plants in solarized soils. *Phytopathology* 81: 494–502.

Gamliel, A. and Katan, J. (2009). Control of plant disease through solarization, pp. 196–220. In *Disease Control in Crops, Biological and Environmentally Friendly Approach* (D. Walters (Ed.)). Wiley-Blackwell: Oxford.

Gamliel, A. and Katan, J. (2012). *Soil Solarization: Theory and Practice*. APS Press: St. Paul, MN, USA.

Gamliel, A. and Stapleton, J. J. (1993). Effect of chicken compost or ammonium phosphate and lettuce growth. *Plant Dis.* 77: 886–91.

Gamliel, A. and Stapleton, J. J. (2012). Combining soil solarization with organic amendments, pp. 109–20. In *Soil Solarization: Theory and Practice* (A. Gamliel and J. Katan (Eds)). APS Press: St. Paul, MN, USA.

Garibaldi A. and Gullino, M. L. (1991). Soil solarization in southern European countries, with emphasis on soilborne disease control of protected crops. In *Soil Solarization* (J. Katan and J. DeVay (Eds.)). CRC Press: Boca Raton, FL, USA.

Golzardi, F., Vaziritabar, Y., Vaziritabar, Y., Sarvaramini, S. and Zohreh Ebadi, S. Z. (2014). Solarization period and thickness of polyethylene sheet effects on weed density and biomass. *Indian J. Fund. Appl. Life Sci.* 4(S3): 587–93.

Grey, T. L., Vencill, W. K., Mantripagada, N. and Culpepper, S. (2007). Residual herbicide dissipation from soil covered with low-density polyethylene mulch or left bare. *Weed Sci.* 55: 638–43.

Grinstein, A. and Hetzroni, A. (1991). The technology of soil solarization, pp. 160–71. In *Soil Solarization* (J. Katan and J. E. DeVay (Eds)). CRC Press: Boca Raton, FL, USA.

Grinstein, A., Kritzman, G., Hetzroni, A., Gamliel, A., Mor, M. and Katan, J. (1995). The border effect of soil solarization. *Crop Prot.* 14: 315–20.

Gruenzweig, J. M., Rabinowitch, H. D. and Katan, J. (1993). Physiological and developmental aspects of increased plant growth in solarized soils. *Ann. Appl. Biol.* 122: 579–91.

Hanson, K., Mahato, T. and Schuch, U. K. (2014). Soil solarization in high tunnels in the semiarid southwestern United States. *HortScience* 49: 1165–70.

Horowitz, M., Regev, Y. and Herzlinger, G. (1983). Solarization for weed control. *Weed Sci.* 31: 170–9.

Jabran, K. and Chauhan, B. S. (2015). Weed management in aerobic rice. *Crop Prot.* 78: 151–63.

Johnson III, W. C., Langston Jr., D. B., MacLean, D. D., Sanders, F. H., Torrance, R. L. and Davis, J. W. (2012). Integrated systems of weed management in organic transplanted Vidalia sweet onion production. *Hort Technol.* 22: 64–9.

Katan, J. (1981). Solar heating (solarization) of soil for control of soilborne pests. *Annu. Rev. Phytopathol.* 19: 211–36.

Katan, J. 1987. Soil solarization, pp. 77–105. In *Innovative Approaches to Plant Disease Control* (I. Chet (Ed.)). John Wiley & Sons: New York.

Katan, J. 1996. Soil solarization: Integrated control aspects, pp. 250–78. In *Principles and Practices of Managing Soilborne Pathogens* (R. Hall (ed.)). APS Press: St Paul.

Katan, J. and DeVay, J. E. (1991). *Soil Solarization*. CRC Press: Boca Raton, FL, USA.

Katan, J. and Gamliel, A. (2012a). Soil solarization for the management of soilborne pests: The challenges, historical perspective, and principles, pp. 45–52. In *Soil Solarization: Theory and Practice* (A. Gamliel and J. Katan (Eds)). APS Press: St. Paul, MN, USA.

Katan, J. and Gamliel, A. (2012b). Mechanisms of pathogen and disease control and plant-growth improvement involved in soil solarization, pp. 135–45. In *Soil Solarization: Theory and Practice* (A. Gamliel and J. Katan (Eds)). APS Press: St. Paul, MN, USA.

Kudsk, P., Bàrberi, P., Bastiaans, L., Brants I., Bohren, C., Christensen, S., Economou, F., Gerowitt, B., Hatcher, P. E., Melander, B., Neve, P., Pannacci, E., Rubin, B., Streibig, J. C., Torresen, K. and Vurro, M. (2013). Weed management in Europe at a crossroads – Challenges and opportunities, p. 10. In the Proceedings of the 16th EWRS Symposium 2013, Samsun Turkey.

Kumar, M., Das, T. K. and Yaduraju, N. T. (2012). An integrated approach for management of *Cyperus rotundus* (purple nutsedge) in soybean-wheat cropping system. *Crop Prot.* 33: 74–81.

Maheswari, T. U. (2016). Weed management in a chili nursery by soil solarization – A novel approach. *Internatl. J. Innov. Agric. Sci.* 1: 37–40.

Mahrer, Y. (1991). Physical principles of solar heating of soils by plastic mulching in field and in glasshouses are simulation models, pp. 75–86. In *Soil Solarization* (J. Katan and J. E. DeVay (Eds)). CRC Press: Boca Raton, FL.

Mahrer, Y. and Shilo, T. (2012). Physical principles of solar heating of soil, pp. 147–52. In *Soil Solarization: Theory and Practice* (A. Gamliel and J. Katan (Eds)). APS Press: St. Paul, MN, USA.

Mallek, S. B., Prather, T. S. and Stapleton, J. J. (2007). Interaction effects of *Allium* spp. residues, concentrations and soil temperature on seed germination of four weedy plant species. *Appl. Soil Ecol.* 37: 233–9.

Marenco, R. A. and Lustosa, D. C. (2000). Soil solarization for weed control in carrot. *Pesq. Agropec. Bras.* 35: 2025–32.

Matthiessen, J. and Kirkegaard, J. (2006). Biofumigation and enhanced biodegradation: Opportunity and challenge in soilborne pest and disease management. *Crit. Rev. Plant Sci.* 25: 235–65.

Matthiessen, J. N. and Shackleton, M. A. (2005). Biofumigation: Environmental impacts on the biological activity of diverse pure and plant-derived isothio-cyanates. *Pest Manag. Sci.* 61: 1043–51.

Mauro, R. P., Lo Monaco, A., Lombardo, S., Restuccia, A. and Mauromicale, J. (2015). Eradication of Orobanche/Phelipanche spp. seedbank by soil solarization and organic supplementation. *Sci. Hort.* 193: 62–8.

McGovern, R. J., Chaleeprom, W., Chaleeprom, W., McGovern, P. and To-Anun, C. (2013). Evaluation of soil solarization and amendments as production practices for lettuce and vegetable soybean in northern Thailand. *J. Agric. Technol.* 9: 1863–72.

Mukherjee, I. and Das, M. G.K. (2007). Soil amendment: A technique for soil remediation of lactofen. *Bull. Environ. Contam. Toxicol.* 79: 49–52.

Mukherjee, I. and Das, M. G.K. (2007). Soil amendment: A technique for soil remediation of lactofen. *Bull. Environ. Contam. Toxicol.* 79: 49–52.

Navarro, S., Bermejo, S., Vela, N. and Hernandez, J. (2009). Rate of loss of simazine, terbuthylazine, isoproturon, and methabenzthiazuron during soil solarization. *J. Agric. Food Chem.* 57: 6375–82.

Oldfield, T. L., Achmon, Y., Perano, K. M., Dahlquist-Willard, R. M., VanderGheynst, J. S., Stapleton, J. J., Simmons, C. W. and Holden, N. M. (2017). A life-cycle assessment of biosolarization as a valorization pathway for tomato pomace utilization in California. *J. Cleaner Produc.* 141: 146–56.

Patel, P. P., Patel, M. M., Patel, D. M., Patel, M. M., Patel, G. N. and Bhatt, R. K. (2009). Soil solarization – An eco-friendly alternative for weed control in groundnut–potato sequence. *Res. Crops* 10: 566–72.

Pfeifer-Meister, L., Roy, B. A., Johnson, B. R., Krueger, J. and Bridgham, S. D. (2012). Dominance of native grasses leads to community convergence in wetland restoration *Plant Ecol.* 213: 637–47.

Price, A. J., Charron, C. S., Saxton, A. M. and Sams, C. E. 2005. Allyisothicyanate and carbon dioxide produced during degradation of *Brassica juncea* tissue in different soil conditions. *HortScience* 40: 1734–9.

Quintanilla-Tornela, M. A., Wang, K. H., Tavares, J. and Hooks, C. R. R. (2016). Effects of mulching on above and below ground pests and beneficials in a green onion agroecosystem. *Agric. Ecosys. Environ.* 224: 75–85.

Rathore, S. S., Shekhawat, K., Premi, O. P., Kandpal, B. K. and Chauhan, J. S. (2014). Biology and management of the fast-emerging threat of broomrape in rapeseed–mustard. *Weed Biol. Manag.* 14: 145–58.

Raymundo, S. A. and Alcazar, J. (1986). Increasing efficiency of soil solarization in controlling root knot nematode by using two layers of plastic mulch (abstract). *Nematology* 18: 626.

Rubiales, D. and Fernandez-Aparicio, M. (2012). Innovations in parasitic weeds management in legume crops. A review. *Agron. Sust. Dev.* 32: 433–49.

Rubin, B. (2012). Soil solarization as a tool for weed management, pp. 71–6. In *Soil Solarization: Theory and Practice* (A. Gamliel and J. Katan (Eds)). APS Press: St. Paul, MN, USA.

Rubin, B. and Benjamin, A. (1983). Solar heating of the soil: Effect on weed control and on soil-incorporated herbicides. *Weed Sci.* 31: 819–25.

Rubin, B. and Benjamin, A. (1984). Solar heating of the soil: Involvement of environmental factors in the weed control process. *Weed Sci.* 32: 138–42.

Rubin, B., Cohen, O. and Gamliel, A. (2007). Soil solarization: An environmentally-friendly alternative. http://www.fao.org/3/a-i0178e/i0178e02.pdf.

Saha, S. K., Wang, K. H., McSorley, R., McGovern, R. J. and Kokalis-Burelle, N. (2007). Effect of solarization and cowpea cover crop on plant-parasitic nematodes, pepper yields, and weeds *Nematropica* 37: 51–63.

Shlevin, E., Saguy, S., Mahrer, M. and Katan, J. (2003). Modeling the survival of two soilborne pathogens under dry structural solarization. *Phytopathology* 93: 1247–57.

Singh, R. (2006). Use of soil solarization in weed management in soybean under Indian conditions. *Tropic. Sci.* 46: 70–3.

Standifer, L. C., Wilson, P. W. and Sorbet R. P. (1984). Effect of solarization on soil weed seed populations. *Weed Sci.* 32: 569–73.

Stapleton J. J. (2000). Soil solarization in various agricultural production systems. *Crop Prot.* 19: 837–41.

Stapleton, J. J. and Bañuelos, G. S. (2009). Biomass crops can be used for biological disinfestation and remediation of soils and water. *Calif. Agric.* 63: 41–6.

Stapleton, J. J., and DeVay, J. E. (1986). Soil solarization: A nonchemical approach for the management of plant pathogens. Crop Prot. 5:190–8.

Stapleton, J. J., Parther, T. S., Mallek, S. B., Ruiz, T. S. and Elmore, C. L. (2002). High temperature solarization for production of weed-free container soils and potting mixes. *Hort Technol.* 12: 697–700.

Stapleton, J. J., Summers, C. G., Mitchell, J. P. and Prather, T. S. (2010). Deleterious activity of cultivated grasses (Poaceae) and residues on soilborne fungal, nematode and weed pests. *Phytoparasitica* 38: 161–9.

Stevens, C., Khan, V. A., Okoronkuro, T., Tang, A. Y., Wilson, M. A., Lu, J. and Brown, J. E. (1990). Soil solarization and dachtal: Influence on weeds, growth, and root microflora of collards. *HortScience* 25: 1260–2.

Stevens, C., Khan, V. A., Rodriguez-Kabana, R., Proper, L. D., Backman, P. A., Colins, D. J., Brown, J. E., Wilson, M. A. and Iqebe, E. C. K. (2003). Integration of soil solarization with chemical, biological and cultural control for the management of soilborne diseases of vegetables. *Plant Soil* 253: 493–506.

Stevens, C., Khan, V. A., Wilson, M. A., Brown, J. E. and Collins, D. J. (1999). Use of thermo film-IR single-layer and double-layer soil solarization to improve solar heating in a cloudy climate. *Plasticulture* 118: 20–34.

Tal, A., Rubin, B., Katan, J. and Aharonson, N. (1990). Involvement of microorganisms in accelerated degradation of EPTC in soil. *J. Agric. Food Chem.* 38:1100–5.

Thimmegowda, M. N., Nanjappa, H. V. and Ramachandrappa, B. K. (2007). Effect of soil solarization and farmyard manure application on weed control and productivity of sunflower (*Helianthus annuus*) bell pepper (*Capsicum annuum*) sequence. *Indian J. Agron.* 52: 204–7.

Upadhyaya, M. K. and Blackshaw, R. E. (2007). *Non-Chemical Weed Management: Principles, Concept and Technology*. CAB International: Wallingford, UK.

Wallace, R. D., Grey, T. L., Webster, T. M. and Vencill, W. K. (2013). Increased purple nutsedge (Cyperus rotundus) tuber sprouting with diurnally fluctuating temperatures. *Weed Sci.* 61: 126–30.

Yarden, O., Katan, J., Aharonson, N. and Ben-Yephet, Y. (1985). Delayed and enhanced degradation of benomyl and carbendazim in disinfested and fungicide-treated soils. *Phytopathology* 75: 763–7.

Zaragoza, C. (2006). Weed management in vegetables. In *Weed Management for Developing Countries* (R. Labrada (Ed.)). FAO Plant Production and Protection: Rome. http://www.fao.org/docrep/006/y5031e/y5031e0b.htm.

Chapter 16

Weed management in organic crop cultivation

Greta Gramig, North Dakota State University, USA

1 Introduction

Weeds have always plagued farmers. In an ecological sense, most weeds are plant species that have evolved to occupy resource niches left open by disturbance and lack of species diversity. Because most agriculture production systems are characterized by species-poor crop monocultures and some degree of soil disturbance, these systems present an open door to weedy species. Since the advent of herbicide technology in the 1940s, weed management has become increasingly reliant on synthetic chemicals. However, in spite of the undeniable success of herbicides, weeds remain problematic for all crop production systems. Continued evolution of herbicide-resistant weeds, and concerns about human health and environmental effects of synthetic chemical herbicides, have led to a resurgence of interest in non-chemical weed management tactics. The enormous recent growth of the organic food industry has also spurred renewed focus on the need to reduce effects of weedy plants without using synthetic herbicides. Organic crop production systems present unique and difficult weed management challenges, because organic farmers cannot use synthetic herbicides. Because of these challenges, organic farmers typically acknowledge that weed management is a top concern. Furthermore, weed problems may deter some farmers from transitioning to organic, or may drive producers who have transitioned back to conventional approaches. Because consumer demand for organically produced food is strong and continues to increase, weed management challenges in organic systems must be solved to ensure that farmers can continue to meet this demand. Additionally, escalating problems with herbicide-resistant weeds will increase the need for non-chemical weed management approaches, even in conventional production systems currently dominated by chemical weed management.

http://dx.doi.org/10.19103/AS.2017.0025.18

2 Tools and tactics used in organic systems

Organic weed management includes several tactics. The main categories and examples of specific weed control tactics are briefly summarized in Table 1. Discussion in this chapter will focus on the primary tools upon which most organic growers rely to provide weed suppression without the use of synthetic herbicides.

2.1 Preventative weed management

Preventative weed management is extremely important when one manages weeds without synthetic herbicides. Some weed problems, especially those associated with perennial weeds, are difficult to address without synthetic herbicides, so modest efforts to prevent new infestations can be a good investment. Foremost, an awareness of what weed species are most problematic on a particular farm is necessary. A farmer must be able to correctly identify each weed species and understand their biology to effectively manage them. Frequent scouting will enable a farmer to identify problems as they appear and find appropriate solutions.

A farmer should also scout and take early action to prevent small weed infestations from growing larger. Especially in the case of highly problematic perennial weeds, such as Canada thistle and field bindweed, catching an infestation early may allow the farmer to quickly eradicate the weed, thus preventing future headaches. This approach is more feasible for the manager of a small farm. After visiting a field infested with highly troublesome weeds, one should carefully clean equipment before entering adjacent fields that do not yet contain these weeds. Monitoring and managing field edges will also help prevent the migration of weeds into cultivated fields. Conventional farmers may use broadleaf herbicides to treat field edges, but options for organic farmers are more limited. The farmer may choose to mow edges before weeds produce seed. Alternatively,

Table 1 Examples of preventative, chemical, physical (mechanical), cultural and other weed management tactics that can be used in organic crop production systems

Preventative	Chemical	Physical	Cultural	Other
Scouting	Acetic acid	Hand pulling	Planting density	Weed seed predation
Education/weed ID	D-limonene	Hand hoeing	Row spacing	Flaming/burning
Clean seed/feed	Cinnamon oil	Tillage	Crop rotations	Solarization
Patch treatment	Other essential oils	Mowing/cutting	Stale seedbed	Livestock grazing
Prevent reproduction	Citric acid	Chaining/pulling	Cover crops	Flooding
Clean equipment	Corn gluten meal	Mulching	Intercropping	Freezing
Quarantine livestock	Salts	Girdling (trees)	Competitive varieties	Desiccation
Maintain field edges	Fatty acids/soaps	Abrasive blasting	Fertility manipulation	Seed decomposition

organic-approved chemicals might provide some spot control for weeds growing within field borders. Preventing weeds from producing seed is always critical – many farmers know that 'one year's seeding equals many years weeding'. For this reason, preventing weeds from producing seed is absolutely imperative. New infestations may be borne in purchased or saved seed, so planting weed seed-free crop seed is important. Composts or livestock manure can also be a source of new, viable weed seeds. One must ensure that any compost or manure spread on field has been properly treated to destroy weed seeds. Weed seeds are killed by exposure to high temperatures generated by the composting process. The exact temperature and duration of exposure required to kill seeds varies by species, but one study found that all weed seeds were killed by three minutes of exposure to 13°C (Grundy et al., 1998). Similarly, when importing livestock, animals should be quarantined in a contained area for a period of time to limit the spread of new weed species. The amount of time needed for weed seed passage varies by seed type and amount ingested, and typically ranges from 30 to 70 hours (Hogan and Phillips, 2011).

2.2 Chemical weed management

Few pesticides are approved by the U.S. Organic Standards for use in certified organic production systems (Tworkoski, 2002). Approved products are generally formulated from natural, often plant-derived, compounds. Most natural herbicides available for weed control need to be applied in relatively large quantities (L or kg ha^{-1}) and are usually applied post-emergence and are non-selective, which limits their usefulness and practicality. The following discussion gives examples of some of the most commonly used and approved chemicals available for organic weed management. One caveat – the reader should bear in mind that requirements of organic certifiers vary widely among various states and countries, so the examples discussed here may or may not satisfy requirements for organic certification among all certifiers (Dayan et al., 2009). Moreover, U.S. Organic Standards allow chemical weed management only if cultural and physical methods prove insufficient (U.S. Code of Federal Regulations, 2000).

One example of a 'natural' herbicide is corn gluten meal (CGM), which was developed following observations by researchers that corn crop residue appeared to inhibit grass seedling growth. Subsequent research demonstrated that the protein fraction of corn grain, or CGM, which is extracted during a wet milling process, effectively suppressed the growth of seedlings of several weed species. Root growth of susceptible species is inhibited, but some species are more susceptible (Bingaman and Christians, 1992). Thus, CGM acts like a pre-emergence herbicide. Due to the expense of this product, most testing has been conducted in high-value horticulture crops and turfgrass. Results have been mixed. A study in onions found that a relatively high rate (4000 kg ha^{-1}) provided 72% total weed suppression 46 days after planting, but by harvest weed cover in CGM-treated plots did not differ from the weedy control plots, indicating that CGM provided only temporary suppression of weed emergence (Webber et al., 2008). Another study showed that CGM, when applied at 900 kg ha^{-1} on a 30-cm band centred on the crop row, did not adequately control in-row weeds in peanuts, even when applied sequentially at pre-emergence, two weeks and four weeks post-emergence (Johnson et al., 2013).

Acetic acid has a long history as an herbicide. Typically, acetic acid is prepared in aqueous solutions ranging from 10 to 20% concentration. Several commercially available acetic acid preparations are marketed as natural herbicides approved for use in certified organic systems. The mechanism of acetic acid is essentially to burn the aerial portions of unwanted plants. It is non-selective and has limited use for annual crop production. It affects only the aerial portions of plants, leaving the roots intact, and therefore it is not effective for control of perennial weeds. However, repeated spot spraying may slow the spread of small patches of perennial weeds such as Canada thistle (*Cirsium arvense*) (Main, 2003). Finally, high cost limits its use (Dayan et al., 2009).

Chemical preparations derived from plants, such as essential oils, have served as pesticides for millennia. Essential oils are typically extracted or distilled from plant parts and contain complex mixtures of aromatic or volatile compounds (Bakkali et al., 2008). Historically, essential oils were primarily used as insecticides or fungicides. Recently, however, various essential oils have been tested for herbicidal activity. Tworkoski (2002) screened essential oils from twenty-five different plant species for herbicidal effects using an assay wherein oils were applied to dandelion leaf surfaces. Results showed that red thyme (*Thymus vulgaris*), summer savory (*Satureja hortensis*), cinnamon (*Cinnamomum zeylanicum*) and clove (*Syzgium aromaticum*) essential oils were most phytotoxic, causing electrolyte leakage resulting in cell death. Boyd and Brennan (2006) also tested herbicidal properties of clove essential oil on common purslane (*Portulaca oleracea*) and burning nettle (*Urtica urens*). They found that extremely high concentrations (>12 L ha^{-1}) of clove oil were required for adequate control. The main mechanism of action of most essential oils is cytotoxicity, or cell wall destruction. Therefore, essential oils act primarily as contact herbicides, which burn aerial portions on contact. However, large quantities are typically required to damage weeds enough to completely destroy them, and in the case of perennial weeds, the plants regrow readily from root reserves. Consequently, essential oils have found limited use as agricultural herbicides. In fact, compounds approved for use as herbicides in certified organic systems seldom are used because they are expensive, non-selective, require repeat applications and work best on newly emerged broadleaf weed species (McErlich and Boydston, 2013).

2.3 Physical weed management

Unlike chemical tactics, physical (or mechanical) weed management is widely used in organic crop production systems. Such tactics include tillage, hand pulling, hand hoeing, mowing/cutting, chaining/pulling, mulching, girdling for tree weeds and blasting with abrasive grits (Table 1). All of these methods use physical force to remove or destroy weeds, or to prevent them from emerging. Of these tactics, many organic annual crop production systems rely primarily on tillage to achieve crop protection from weed competition. In the United States, the National Organic Standards stipulate that a farmer must develop an 'Organic System Plan' that documents practices used on the farm, including tillage procedures (NOP section §205.201(a)(1)).

Tillage modifies soil structure mechanically. It is used before planting to prepare a weed-free seedbed, within the growing season between rows to control weeds emerging with or after the crop, and after harvest to remove later-emerging weeds. Primary tillage is used to loosen soil and incorporate amendments and produces a relatively rough soil surface. Secondary tillage is used to create a smoother soil surface before planting or to control weeds within crops. Primary tillage is more aggressive and results in considerable soil disturbance and mixing at greater depths (>12 cm) than secondary tillage. Secondary tillage produces less soil mixing and typically involves more shallow depths of the soil profile. A wide variety of tillage implements are available to farmers including disk cultivators, rotary hoes, mouldboard ploughs, chisel ploughs, finger weeders, brush weeders, harrows, sweep cultivators, strip tillers, ridge tillers, para-ploughs, subsoil tillers and cultipackers.

Using tillage to manage weeds is as much an art form as a science. Considerable skill is needed to combine the correct implement with proper adjustment, timing, operating speed and soil conditions to achieve the desired outcome (Gruver and Wander, 2015). Also, weed management may not be the only consideration when deciding how and when to till the soil. Tillage influences other factors such as the amount of residue left on the soil surface, soil organic matter, soil nitrogen (Malhi et al., 2006) and soil water content. Therefore, many complex variables must be considered when making a decision to till soil. Organic farmers employ tillage more than conventional farmers, but many organic farmers are striving to reduce tillage to conserve beneficial soil attributes.

Weeding by hand pulling or using hand-operated implements such as hoes is as old as agriculture. Tactics of this sort are uncommonly employed in conventional crop production, but are common in small-scale organic production and in high-value crop production (Mohler and Baker, 2016). Hand removal of smaller annual weeds can be an effective, albeit laborious, weed management approach. Small infestations of perennial weeds that reproduce vegetatively can sometimes be eliminated successfully by careful digging to remove all pieces of plant roots from which the plants might reproduce. As with mechanized tillage, one must be aware of soil conditions. Wet soil might allow weeds to be more easily uprooted via hand pulling, but hand weeding in waterlogged fields may cause soil compaction and clodding. Hand weeding is costly, and is used primarily in small-scale horticultural systems and in high-value crops where contamination with weeds must be avoided (e.g., herb and salad green crops) (McErlich and Boydston, 2013).

Mulching is widely used in organic systems, particularly in fruit and vegetable production. Mulching creates a physical barrier on top of the soil surface that prevents weeds from emerging and growing. Some mulch materials enhance water or heat retention. Many different materials can be used in organic production, but each material must be approved for use during the organic certification process. Even though plastic mulches are made of synthetic material, they are common in organic production. Plastic mulches suppress weeds effectively, can be installed mechanically and are economical. However, they must be removed at the end of each season, creating a disposal problem. Biodegradable plastic mulches are also available, but are currently prohibited for use under the U.S. Organic Standards (U.S. Code and Regulations, 2000). Organic mulches such as hay, straw, paper or wood chips can also suppress weeds fairly effectively. Organic mulches biodegrade and add valuable organic matter and nutrients to soil, but applying them is laborious and they are more expensive than plastic mulch. Another important drawback of organic mulches is that they can harbour weed seeds, and thus create new weed problems. The effectiveness of most organic mulches is dependent on the amount of material applied. Small-seeded broadleaf weeds will be suppressed for several weeks by mulch applications ranging from 6.7 to 11.2 Mg ha^{-1}, whereas 15.7–24.4 Mg ha^{-1} might be required to suppress grasses or large-seeded broadleaf species. Perennial weeds are seldom deterred by organic mulches, and even plastic can sometimes be penetrated by particularly aggressive perennials (Schonbeck, 2015).

As the demand for organically grown food increases, so does the demand for innovative and cost-effective mechanical tools for weed management. The continued evolution of herbicide-resistant weeds has encouraged the development of automated mechanical weed-removal machines for use in conventional production systems. Intensive research has recently been directed towards developing extremely precise robotic weeders that physically remove weeds without damaging crops. A particular challenge is removing weeds that grow very close to crop plants. This task requires sensors that are able to differentiate small weeds from similarly sized crop plants growing directly adjacent to the weeds. Some progress has been made developing sensors capable of this feat, but further development is required before such devices will be able to sufficiently differentiate weeds from crops with similar size and appearance (Merfield, 2016). However, regardless of the remaining challenges, numerous commercial robotic weeders have been developed and marketed, particularly by European companies. In the United States, interest in and development of such tools has lagged, but momentum may be increasing. Devices that spray air-propelled abrasive grit are also currently in the research and development phase and might soon provide an additional option for in-row early season weed management. These devices perform like sandblasters to spray small annual weeds with an abrasive grit. Research demonstrated that 2–3 abrasion events with corn cob grit coupled with between-row cultivation provided season-long weed control of up to 90% in corn (*Zea mays*) (Forcella, 2012). Though the device works well to kill smaller broadleaf weeds, it is not as effective for grass weeds, for which the growing point is located beneath the soil surface. More thorough and detailed discussion of recent advances in mechanical weed control is further discussed in this volume.

2.4 Cultural weed management

Cultural weed management approaches focus on designing and manipulating cropping systems to achieve field conditions that create advantages for crop plants while hindering weed emergence, growth or reproduction (Mohler, 1996). Examples of cultural weed management include planting competitive crop varieties, increasing planting density, intercropping, growing cover crops, practising specific crop rotations, using stale seedbed techniques and manipulating the timing and placement of fertilizer to benefit the crop more than the weeds (Table 1). Use of cultural approaches is common among organic producers (Tautges et al., 2016), whereas cultural approaches are less commonly relied upon for weed management in conventional production systems. Although cultural practices are diverse and numerous, most of these practices are designed to take advantage of fundamental ecological differences between crop and weed plants. Most crop plants are large-seeded annuals, with greater initial absolute growth rates than smaller-seeded annual weed species. Farmers have found numerous methods for exploiting the initial greater size and absolute growth rate of crop species to favour the crop (Mohler, 1996).

Achieving canopy closure as soon as possible vastly deters the growth of weeds that emerge with or after the crop. Spatial arrangement and density of the crop is one factor that governs how quickly the canopy will close. Therefore, many organic growers will increase planting density within the row and/or reduce spacing between rows to create a dense crop canopy that reaches closure faster. The approach has limitations because at some point the crop plants will compete with each other for resources and yield will decrease. Consequently, much research has been devoted to testing planting configurations for various crops. Many studies have shown that crop yield increases substantially with increased crop density (Mohler, 1996). However, the effect of planting density on crop yield is more variable when weeds are absent. Consequently, because organic systems usually contain substantial weed populations, organic systems may benefit more than conventional systems from increased crop density. For instance, one study compared 12 versus 24-cm row spacing for organic winter wheat (*Triticum aestivum*) production and found that the

12-cm spacing resulted in a more even crop spatial arrangement that led to increased leaf are a index and light interception when compared with the 24-cm row spacing (Drews et al., 2009). Conversely, Hiltbrunner et al. (2005) found that organic winter wheat yield did not differ between rows spaced 18.75 versus 37.5 cm apart, but the wider row spacing was associated with greater wheat protein content. These results illustrate that effects of planting density on crop yield are variable and that planting density may affect other crop responses besides yield (Bàrberi, 2002). For this reason, many organic farmers experiment to find the planting density that produces optimal results for their cropping system and unique environment.

Competitive crop cultivars can produce a more competitive canopy. Usually, differences in competitive ability among crop cultivars are related to differences in seedling emergence timing and early growth rates. A variety or cultivar that emerges quickly and has a fast early growth rate will lead to faster canopy closure and increased competitiveness. However, perceived differences in competitive ability may be caused by allelopathy (Bàrberi, 2002). Unfortunately, most crop varieties and cultivars have been developed to perform optimally under conventional, not organic, management (Lammerts van Bueren et al., 2011; Murphy et al., 2004). Traits such as dwarf stature that confer increased small grain yield under conventional management may result in reduced root development, which leads to reduced nutrient and water acquisition, reduced nutrient use efficiency, greater disease susceptibility and reduced competitive ability in organic systems (Lammerts van Bueren et al., 2011). Numerous studies have reported differences among crop cultivars or varieties with respect to competitive ability (e.g., spring wheat (Huel and Hucl, 1996), sunflower (*Helianthus annuus*) (Latify et al., 2016), rice (*Oryza sativa*) (Ni et al., 2000) and sweet corn (Williams et al., 2007)), though few of these studies have been conducted under organic management. Also, given the substantial amount of time needed to improve crop varieties (10 years or more), development of varieties with increased weed competitive ability lags behind the need.

Other approaches that may enhance competitive ability include precision placement of fertilizer, variation of planting time and using a stale seedbed. Many annual weed species evolved to capitalize on excess nitrogen or other nutrients. Therefore, spatially targeting fertilizer application within the crop row can provide an advantage to crops over weeds (Di Tomaso, 1995). Most conventional farmers prefer to plant crops as early as possible; however, by delaying planting, an organic farmer can wait for more weeds to emerge and destroy them with tillage before planting, thereby reducing overall weed pressure. Even though this approach leaves a shorter growing season for the crop, the reduction in weed pressure is well worth the delay. A stale seedbed relies on early light tillage to stimulate emergence of small-seeded annual weeds, which are subsequently destroyed before planting. All three of these approaches are widely used in organic production and examples of each are included in Section 3. An additional tactic that is common in organic high-value crops is using transplants to establish the crop instead of planting seeds. This approach gives the crop a head start similar to delayed planting.

Crop rotation is one of the most important cultural practices in the weed management toolbox of organic producers. It consists of planting a sequence of crops in different years in a field. The practice has many benefits, including nutrient and pest management. With respect to weed management, planting crops with different life cycles, planting and harvesting dates creates temporal diversity which provides a variety of opportunities to reduce populations of various weed species, which have different life cycles (Liebman and Dyck, 2003; Teasdale et al., 2004). For instance, winter annual weeds such as downy brome (*Bromus tectorum*) can be notoriously difficult to control in winter wheat, because the crop and weed plants emerge at approximately the same time. If a farmer-planted winter wheat in the same field every year, downy brome populations would increase over time. By alternating winter wheat with a summer annual crop that is planted later in the spring, a farmer can use the time period during which the crop is absent to control downy brome. Research in the northern Great Plains has shown that a 4-year rotation – 2 years of cool season followed by 2 years of warm season crops, produced the lowest overall weed density. Moreover, crop species diversity within a life cycle category was important. For instance, planting dry pea (*Pisum sativum*) one year instead of wheat in the 'cool' phase of the rotation permitted more effective weed control than planting winter wheat both years (Anderson, 2010). Although crop rotation is a key strategy for reducing weeds in organic production, the benefits of the rotation must be balanced with marketing opportunities associated with various crops. Fortunately, some leading chefs have recently recognized this disconnection between optimal agricultural production and demand for various foodstuffs. By championing more diverse diets, these chefs hope to create more demand for diverse crops (Barber, 2014).

Related to crop rotations are the practices of growing cover crops and intercropping. Cover crops are crops that are sown within the rotation sequence not to be harvested for yield, but to provide various benefits such as adding organic matter and nutrients to the soil, preventing nitrogen leaching, reducing soil erosion and suppressing weeds. A wide variety of cover crop species can be chosen for specific functions (Snapp et al., 2005). Different species have varying abilities to contribute to weed management. The primary mechanisms of weed suppression are competition and physical suppression of emergence. Cover crops that are planted out of phase with the cash crop or within the cash crop compete with emerged weeds for resources and limit weed emergence via shade signals that impose seed dormancy. Another tactic is to plant a cover crop, then terminate it, leaving the residue on the soil. This approach has been successful in no-till organic production. The degree of weed suppression is related to the amount of surface residue produced. For example, Smith et al. (2011) showed that greater than 8 Mg dry matter ha^{-1} of roller-killed rye (*Secale cereale*) was required to adequately suppress weeds in soybean. Other mechanisms associated with cover crops may also act to reduce weed pressure. Some cover crops (e.g., rye) can be allelopathic (Barnes and Putnam, 1987). When cover crop residue is incorporated, the addition of organic matter may increase pathogenic microorganisms that attack weed seeds, increasing mortality (Davis et al., 2006).

Perennial pasture crop (ley or sod crops) are commonly employed in organic but only rarely in conventional crop production. Pasture crops are typically a mixture of perennial grasses and legumes that are planted and allowed to grow for a number of years before termination (Liebman and Dyck, 1993). Planting a pasture crop provides benefits to the cropping system above what an annual cover crop can provide. If one incorporates perennial legumes, a pasture crop can provide a substantial amount of nitrogen for annual crops that follow. This approach is also critically important for managing otherwise intractable weed problems that can plague organic production systems. Many of the tactics that are widely used to control weeds in organic systems, such as tillage, crop rotation and annual cover crops, do not sufficiently control perennial weeds (Mohler, 1996). The dense cover created by the pasture cover in tandem with periodic mowing and ploughing to terminate the pasture can substantially reduce perennial weeds (Liebman and Dyck, 1993). For this reason, many organic producers incorporate perennial pasture crops into their rotations as a critical aspect of weed management. One limitation to using this approach is the lack of livestock on many modern farms. Without livestock to consume hay produced by pastures, farmers may choose to cut and bale the hay for market. Section 3 provides numerous examples of the use of pasture to manage weeds and fertility on organic farms.

Some organic producers also use intercropping to provide weed management and other agro-ecosystem services (e.g., nitrogen provision). Whereas crop rotations achieve temporal crop diversification, intercropping provides spatial diversity by mixing different crop species in close proximity (Willey, 1990). Intercropping strategies are highly diverse and can include mixing annuals or mixing annuals with perennials (Liebman and Dyck, 1993). Intercrops may also include mixtures of cash crops or mixtures of cash and cover crops. Intercropping can provide many beneficial effects, including disruption of pest cycles (Trenbath, 1993), more efficient use of soil resources and increased yields (Willey, 1990). More efficient use of resources is possible because different crop species may acquire soil water and nutrients at different times or from different soil depths (Bulson et al., 1997). Resource complementarity is likely the major reason that intercrops have often been found to suppress weeds more than monocultures of each constituent crop. With resource complementarity, a farmer may sow greater overall plant densities, leading to increased weed suppression. Bulson et al. (1997) tested the effects of growing mixtures and monocultures of spring wheat and field beans (*Phaseolus vulgaris*) at varying densities under organic management on resource complementarity and weed suppression. They found that increasing densities of wheat and bean monocrops were associated with greater resource complementarity and greater weed suppression. Another study assessed the effects of monocultures of field bean, cowpea (*Vigna unguiculata*) and maize to field bean-maize and cowpea-maize intercrops on canopy light interception and weed suppression in an organic production system. Results indicated that both intercrop mixtures produced increased light interception and weed suppression compared to monocultures. These effects were attributed to the increased plant density (Bilalis et al., 2010). However, not all studies have shown that intercropping enhanced weed suppression compared to monocultures. The mechanisms underlying these interactions are not completely understood (Liebman and Dyck, 1993).

2.5 Other weed management tactics

Flame weeders comprise arrays of liquid fuel (usually propane) torches that are passed over a field to destroy small weed seedlings with high heat (Bond and Grundy, 2001). The effectiveness of flame cultivation is dependent on temperature, exposure time and energy output (Ascard, 1997). Flaming is not used much in conventional production, but is typically less expensive than hand weeding, so it is fairly common in organic production. Flame weeding can be done when soil is too wet for tillage, which is a distinct advantage (Bond and Grundy, 2001). A survey conducted among organic growers in upstate New York found that several growers used flame weeding successfully. Most stated that it did not work equally well for all crops and worked better during pre- than post-emergence for most crops. Vegetable crops such as onions and carrots were typically amenable to flame weeding, but some producers also used the technique for corn pre-emergence (Baker and Mohler, 2014). Research also showed that dicot weeds responded more favourably than monocots (Sivesind et al., 2009).

Solarization is another method that employs heat to reduce weed infestations. It involves covering a moist soil surface with plastic sheets for several weeks, during which time solar heat is concentrated in the soil to kill weed seeds and inhibit germination and emergence. Research has shown that clear plastic performs better than black plastic and that annual weeds can be reduced, whereas perennial weeds are unaffected (Horowitz et al., 1983). Solarization is best suited to high-value crops grown in hot sunny climates. A problem is disposal of non-biodegradable plastic.

Grazing or other animal behaviours can be used to reduce weed infestations in organic production. Sheep, goats and cattle graze on vegetation, including weeds (Popay and Field, 1996). However, even non-grazing animals such as pigs can assist because they forage through the soil, removing weed roots and rhizomes. Animals have been used to manage weeds in pastures. Animals can reduce weeds in annual crops. They can graze before planting or after harvest to reduce weed cover. All weeds are not equally palatable and different animals prefer different plants. Therefore, grazing can result in uneven removal of weeds. Only farms that have livestock can use grazing. Animals can move weed seeds from one site to another via manure, thus creating new weed problems.

Enhancing weed seed predation may produce substantial reductions in weed populations and subsequent herbicide use (Westerman et al., 2006). Therefore, manipulating weed seed predation by changing agronomic practices could be an economically and environmentally sound way to manage weeds (Westerman et al., 2008). Substantial quantities of newly dispersed weed seeds are destroyed by seed predators such as birds, rodents and insects. Cromar et al. (1999) estimated that 82% of barnyardgrass (*Echinochloa crus-galli*) seed is consumed in the fall by predators. Eliminating or delaying tillage operations in the fall exposes newly dispersed seeds on the soil surface to longer predation periods. Agronomic practices that influence the amount of residue or cover on the soil surface can also influence levels of weed seed predation by providing habitat and protection for vertebrate and invertebrate seed predators (Cromar et al., 1999; Gallandt et al., 2005). Also, increased vegetative cover associated with various cover cropping systems led to enhanced post-dispersal weed seed predation, particularly predation by ground beetles (*Harpalus rufipes*) (Gallandt et al., 2005).

3 Farmer case studies

Effectively managing weeds in organic production systems consists of much more than implementing a laundry list of various tactics. Through listening to the stories of many organic farmers, one theme that emerges is that organic weed management is not a one-size-fits-all endeavour. Organic farmers and conventional farmers who seek to limit pesticide inputs typically engage in much experimentation to arrive at individual farm management strategies that work for their particular production system, region, climate and market. Understanding how to manage weeds without synthetic herbicides requires more than a large toolbox holding 'many small hammers' (Liebman and Gallandt, 1997). Effective organic weed management depends on selecting the optimal tool for a particular task, and using the tool correctly at the right time. Organic weed management is not a disconnected set of tools and tactics. All process and tools must be interwoven to create a functioning holistic system that takes advantage of natural processes and synergies.

Several farmers were asked to provide case studies about their farm and weed management approaches for this chapter. They are leaders with long experience pushing the envelope of organic weed management while simultaneously protecting or improving their natural resource base. They manage farms that differ widely in size, location and crops (and in some cases, livestock).

3.1 Carmen Fernholz, A-Frame Farm, Madison, MN

Carmen Fernholz owns and operates a 140-ha certified organic cash grain farm. He has been farming using organic methods since the 1970s and has experimented with many approaches to weed management. He has grown a wide variety of crops, but generally employees a three-year rotation consisting of wheat under-seeded with red clover (*Trifolium pratense*) or alfalfa (*Medicago sativa*), followed by corn and then soybean. His approach is based on the principles of integrated weed management, including the use 'many little hammers' instead of a single tactic.

He stresses selecting the best tillage implement for each problem. He relies on low-disturbance implements: spring-tooth harrows, disc cultivators or rotary hoes. He does not use a mouldboard plough because it disturbs and exposes the soil. He employs deeper tillage with a chisel plough and sweeps only when terminating perennial legumes such as clover or alfalfa. Carmen stresses timeliness to manage weeds. He advocates split tillage, a stale seedbed approach, which entails initial tillage with a disc or field cultivator once the soil warms and dries in mid- to late April in Minnesota. About two weeks later, after weeds emerge, a second tillage is followed by planting. Waiting until the soil temperature warms to a consistent 10°C, gives the crop a chance to get ahead of the weeds.

Another important consideration for establishing a competitive crop canopy is the correct seeding rate. Carmen emphasizes that, because seeds are of different sizes, one should calculate seeding rates based on seed number, not seed weight. For instance, small grains such as wheat or oat can vary from 24 000 to 40 000 seeds kg^{-1}. The correct seeding rate ensures that the crop canopy will close as quickly as possible, which is key to outcompeting later-emerging weeds. He also notes that other tactics such as flamer and hand weeding can be important tools to manage weeds organically. Finally, he recognizes that biological processes, such as weed seed predation, can be encouraged by strategic timing of various operations such as tillage. Delaying or eliminating fall tillage will allow substantial numbers of weed seeds to be removed by insects, rodents and birds.

Although Carmen is a successful organic farmer, he continues to refine his approaches and acknowledges that many challenges remain. For instance, winter wheat was found to encourage Canada thistle, so this crop is not included in his rotation. He claims that expanded use of cover crops would be facilitated by inclusion of more livestock. He also hopes to experiment more with using a modified seed drill to inter-seed cover crops into standing corn at midseason. By continuing to develop and refine his farming approaches, Mr Fernholz will continue to successfully manage weeds and produce high-quality organic grain crops for an expanding market.

3.2 Lynn Brakke, Lynn Brakke Organic Farm, Moorhead, MN

Lynn Brakke converted his conventional farm to a certified organic farm in 1993. He farms 930 ha near Moorhead, MN, focusing on organic grass-fed beef, annual grain and alfalfa. Lynn strongly advocates rotation as the most important tool for weed management in organic production systems. His crop rotation includes three years alfalfa, followed by corn alternated with soybean (*Glycine max*) for four years. Lynn credits alfalfa for lack of perennial weeds. During some years he might include a small grain. In the past he has included a 'green year', which consists of two or three plantings of various cover crops, which are ploughed down. With lower commodity prices in 2017, he is planning on reintroducing green years on his farm by growing a variety of cover crops. An example of a green year approach would be a May planting of tame mustard (*Brassica juncea*), turnip (*Brassica rapa* var. *rapa*) and buckwheat, which would be tilled under then planted to rye, daikon radish (*Raphanus sativus* var. *Longipinnatus*) and berseem clover (*Trifolium alexandrinum*) in July. Lynn notes that buckwheat (*Fagopyrum esculentum*) is a great cover crop for increasing soil phosphorus availability. Mustard, a nutrient scavenger, roots at depths below that at which most crops root. Glucosinolates produced by mustard act as biofumigants in the soil to reduce nematodes and other pathogens. Radish and turnip break up heavy soils and add organic matter. Berseem clover adds nitrogen, while rye helps hold nitrogen and has allelopathic activity.

Lynn also notes that he has observed a reduction in weed problems since he started paying attention to 'balancing' soil nutrients. He uses the Albrecht System of Soil Fertility (Albrecht and Smith, 1942), also known as the 'base cation saturation ratio' (BCSR) approach, to achieve an optimal balance of soil nutrients. This approach emphasizes the importance of micronutrients as well as macronutrients, so Lynn tests for N, P, K, Cu, Zn, Mn, S, Bo, Co and Mg and adds these nutrients as necessary. He has noticed a reduction in weed problems since adopting this approach. He also asserts that attention to soil nutrient balancing has produced alfalfa hay with superior nutrition and palatability for dairy cattle. Although BCSR soil balancing has gained favour among many organic producers, a review of the science underlying the concept suggested that it is not adequately supported by previous research (Kopittke and Menzies, 2007). In spite of this, many growers remain convinced that BCSR is beneficial. Consequently, a research project at the Ohio State University has been initiated to investigate the effects of BCSR on crop growth and weed community dynamics (Matthew Kleinhenz, pers. comm., 2017).

For grain production, Lynn uses 56-cm rows and crop densities greater than recommended (e.g., 494 000 seeds ha^{-1} for soybean). Tillage is an integral part of weed management. His farm is in the Red River Valley of MN, an area dominated by heavy clay soils. He emphasizes using the tool best adapted to soil conditions when tilling to control weeds. Tilling when soil is too wet can destroy soil structure and cause organic matter losses. He recommends purchasing equipment as large as possible so that ground can be covered quickly when the soil and weather conditions are right. He favours adjustable equipment that can be fine-tuned to respond to the desired task and field conditions. He even uses a digital camera mounted on his tillage equipment to obtain a close view of the tillage action and make incremental adjustment to perfect the operation. A tine weeder is preferred over a rotary hoe for early season weed control. Tillage implements need to be carefully adjusted to work at the correct depth. Soybeans are planted 3.2 cm deep and tine weeder operates at 2.5 cm. Precise adjustment varies with firmness of the seedbed. Once soybean seeds have imbibed, moving them might result in death, hence the need to precisely control the depth of operation. Tillage (harrow, or preferably, rotary hoe) occurs 3 d after planting, and every 5–7 d thereafter, until the beans bloom. Corn tillage ceases at approximately 30 cm tall. Flaming is used for pre-emergence weed control in corn before spiking and in soybean up to cotyledon stage, if soil conditions are too wet for tillage. Finally, hand weeding is used in soybean. Zero tolerance for weed seed production reduces future weed problems.

3.3 Bob Quinn, Quinn Farm and Ranch, Big Sandy, MT

Bob Quinn's experience managing weeds on a regenerative organic farm goes back over 30 years since his first experiment comparing organic and chemical agriculture in 1986.

Bob's 1620-ha dryland farm is located on the short grass prairie of north central Montana at 945 m elevation with a sandy loam soil. This region receives 30–36 cm precipitation yr^{-1}, primarily during May and June. Frost-free days range from about the middle of May to the middle of September. Prior to organic transition, winter wheat, spring wheat, barley and oats were alternated with summer fallow – a traditional farming approach in this region prior to the advent of chemical no-till.

Two overriding principles governing Bob's farm management are 1) soil building and 2) diversity (both natural diversity and crop diversity). These principles aid weed management. The goal is not to eradicate or control all weeds, but rather to manage the effect of weeds while tolerating some as part of the natural diversity of the system. Currently, he has a nine-year rotation – five years of cash crops interspersed with four years of soil building crops. The cash crops are winter wheat (including both modern varieties and ancient/heritage varieties such as spelt (*Triticum spelta*), spring wheat (Kamut brand khorasan wheat, *Triticum turanicum*), hull-less barley (*Hordeum vulgare*), hi-oleic safflower (*Carthamus tinctorius*) and alfalfa hay. Sometimes garbanzo beans or peas may be substituted for some of the barley. The soil-building plants include alfalfa, yellow blossom sweet clover (*Melilotus officinalis*), field peas and buckwheat. The legumes contribute nitrogen while the buckwheat helps make soil phosphorus more available and is an extremely good competitor against late summer annual weeds. These are terminated by undercutting or disking.

This long, diverse rotation builds the soil; breaks disease, insect and weed cycles; and disrupts regeneration niches. Rotating soil-building crops is important because each of these crops roots at a different depth and therefore competes with different weed species. During the late 1990s, when alfalfa and clover failed because of drought, peas became the sole source of green manure. However, soon the peas developed root rot because there was not enough time to break the pest cycle. Now peas are planted at least four years apart on the same field.

The most important weed management tactic for cash crop phases is swift and complete canopy closure, achieved by reducing row spacing to 18 cm rows. This can also be accomplished with seed spreaders on the drill boot or double seeding at an angle by those who do not have narrow spaced drills. The importance of a clean seedbed also cannot be underestimated. Typically, as soon as conditions permit in early spring, fields are tilled to eliminate winter annuals. Then, after early summer annuals have emerged, another tillage pass is completed shortly before seeding the crop. This gives the crop a slight advantage over the weeds, but soon rain brings more weeds. If rains come within 5–7 d after planting, a pass with a spike harrow is conducted when weeds reach the two-leaf stage to reduce weed pressure and give the cash crop a chance to move ahead. For fall-seeded crops, tillage is performed shortly before planting to eliminate winter annual weeds.

Bob's best solution for managing tough weeds such as Canada thistle or wild oat is growing a solid stand of alfalfa. If a weed infestation is especially severe, the alfalfa may be left and cut for hay for an additional year or two until weeds have nearly disappeared. Management of perennial field bindweed (*Convolvulus arvensis*) remains particularly challenging, however, and further research is needed to determine the best approach. Ironically, some annual weed species that are easy to control with herbicides, such as field pennycress, represent the greatest weed management challenges in an organic system. Conversely, kochia (*Bassia scoparia*), which due to resistance is increasingly difficult to control with herbicides, has largely disappeared on the Quinn farm. Bob's theory is that kochia may thrive on large amounts of highly soluble nitrogen, which allows this species to outcompete the crop. But when the soluble nitrogen is greatly reduced with green manure crops, kochia loses its competitive advantage and declines.

Bob strongly advises that organic farmers should do everything in their power to reduce the number of weeds that produce seed. Also, he suggests walking through fields to note the effects of activities and rotations on the dynamics of the weed and crop populations. A farm is alive and changing from year to year, so the farmer must always assess the best way to respond to these changes.

3.4 Mark Doudlah, Doudlah Farms LLC, Stoughton, WI

Mark Doudlah operates a 647-ha farm in Rock County, WI, which produces a diverse array of crops including corn, soybean, cover crop (rye, hairy vetch (*Vicia villosa*)), forage pea seed, dry edible beans and vegetables for canning. The farm also produces free-range chickens and eggs and grass-fed hogs. Mark started the gradual transition to organic practices in 2008 after becoming concerned about the effects of synthetic pesticides on the health of his family. Like many organic farmers, Mark is committed to building soil and human health through his farming practices. Following his interest in improving soil health, Mark was inspired by research conducted at the Rodale Institute and the University of Wisconsin-Madison to investigate developing cover crop-based no-till and reduced tillage approaches for his farm. Gradually, Doudlah Farms has transitioned more of their total acreage to certified organic production, with approximately 567 ha in organic or transition during 2017.

Mark employs cereal rye and other cover crops to replace herbicides. After experimenting over a number of years, he developed two different rotations – one suited for the portion of his acreage that can be tilled and another for highly erodible soils that are too hilly, clayey or gravelly. The tillable lands rotation consists of 1) spring disc tillage to incorporate the rye + vetch cover crop, corn main crop followed with cereal rye cover crop; 2) roller-crimped rye, dry edible beans or soybean main crop, disc after harvest followed by oat + mustard + buckwheat cover crop; 3) spring disc tillage, followed by a canning pea or forage pea main crop, followed by dry edible beans then a winter wheat + daikon radish cover crop, possibly aerial-seeding medium red clover; and 4) harvest winter wheat, followed by vetch + rye cover crop. The non-tillable rotation consists of 1) roller-crimp cereal rye, soybean main crop, with cereal rye + hairy vetch cover crop; 2) cereal rye and hairy vetch continue as main crop, which is harvested for seed, followed by vertical tillage to incorporate fallen seed as volunteer crop; 3) spring disc tillage of vetch (N source) followed by corn main crop, then cereal rye cover crop.

Mark plants 'Aroostook' cereal rye because it can be planted after a late harvest in regions with short growing seasons. This cultivar also produces consistent, large quantities of biomass, which is critical for weed suppression. 'Purple Bounty' hairy vetch is favoured because it is winter hardy, early maturing and was developed for high nitrogen fixation. Mark advocates the benefits of cover crops, which suppress weeds and add nutrients and organic matter to the soil. He does not refrain from using purchased inputs when necessary, especially for corn production. Encouraging quick emergence and fast growth of corn is critical for closing the canopy quickly to outcompete weeds. Mark accomplishes this task by 'supercharging' his corn planting with seed treatments, biological organisms and nutrients to protect the seed from soil pathogens and boost early growth. He plants corn in 38-cm rows instead of the standard 72-cm rows to further speed canopy closure.

Mark has experimented with using a no-till crimped system using a rye + vetch cover crop for corn production. Thus far, he has noticed that time is inadequate for thorough decomposition of vetch residues. Therefore, inadequate N is available for the corn crop. Next he plans to try planting oat + sorghum (*Sorghum bicolor*) + daikon radish (*Raphanus sativus*) in 38-cm rows, with 'Lynx' winter peas in double rows five inches apart between the 38-cm corn rows. The corn can be strip-tilled into the winter-killed cover crop residue of the oat + sorghum + daikon, with the decomposing daikon roots providing early nitrogen. At V2–V3 stage corn, the winter peas can be crimped to provide a later source of nitrogen.

At Doudlah Farms, effective weed management relies mostly on the weed-suppressive effects of the rye or rye + vetch cover crops. Mark emphasizes that proper cover crop establishment is key to good weed suppression. Timing of planting is critical and he targets early to mid-October for his region (southern WI). He plants 187 kg rye seed ha^{-1}, aiming for even coverage without misses or gaps. He thinks that drilling the cover crop seed is most effective, because this approach ensures soil-seed contact. Aerial seeding is an option, but usually does not result in optimal soil-seed contact.

After several years of experimentation, Mark has had success producing consistent, profitable yield – soybeans 3133 kg ha^{-1} and corn 12 444 kg ha^{-1}. Many challenges remain. Current needs are crop and cover crop varieties that function optimally in cover crop-based reduced tillage system including winter pea varieties with improved emergence. He received a grant that will allow him to explore breeding winter pea and hairy vetch varieties with improved traits.

4 Future trends and conclusion

Managing weeds in organic production is critical to the economic success of organic farmers. Organic farmers may also eventually lead the way for truly integrated weed management approaches to be implemented on a wider scale in conventional production systems. Problems with weeds represent a major reason why organic operations fail, or never get started. A wide range of tools and tactics can be used to contend with weeds in organic systems, but seldom does one tactic suffice. Organic weed management is complex, and relies on integration of several tools and tactics. Because every system is unique, each farmer must combine the best advice from scientists and other farmers with trial and error to develop an individual management plan. The case studies illustrate different management plans and unifying themes and principles. For most organic farmers, cultural practices such as rotation are extremely important. A recent survey showed that, among organic farmers who relied on 4–7 weed management practices (identified as the 'medium management group'), 95% reported using crop rotation (Tautges et al., 2016). So, crop rotation was almost always important. Mechanical weed management tools are prevalent among organic farmers as well, but the case studies presented in Section 3 demonstrate that effective tillage requires substantial knowledge and experience. Farmers are cognizant of the damaging effects of tillage and seek to minimize these. Many farmers mentioned the need to own a variety of tillage implements and to spend time fine-tuning the machines for optimal effect. Another commonality was the use of either annual cover crops or perennial pasture crops, or both, for weed suppression and nutrient management. Most of the farmers included a perennial pasture phase, and agreed that it is critical for managing intractable weeds. Mark Doudlah does not include a long perennial pasture phase, but does employ roller-crimped rye and hairy vetch to achieve weed suppression. All the farmers interviewed expressed their commitment to using practices that led to preservation or enhancement of the soil resource, one of the bedrock goals of organic farming. Another constant among the farmers was a long history of experimentation to arrive at best practices for their climate, soils and crops. Because of the complexity of organic crop production systems, absolute prescriptive recommendations are likely to be less important than experimentation. However, at least one farmer mentioned that he saw a need for older, more experienced farmers to mentor and guide younger farmers. This knowledge from experienced farmers, combined with information gained via scientific research, will help future organic farmers manage weeds while conserving and building soil resources.

5 Where to look for further information

For more information about managing weeds in organic crop production systems, please consult the references listed at the end of this chapter. The Rodale Institute website (http://www.rodale.com) is an additional source of information about organic weed management. North Carolina State University maintains an excellent selection of extension publications about managing weeds in organic systems (https://content.ces.ncsu.edu/weed-management-on-organic-farms). Finally, a text titled *Ecological Management of Agricultural Weeds* (Liebman et al. 2001) is a comprehensive treatise on using ecological principles to manage agricultural weeds. Methods and approaches described therein are relevant to managing weeds in every type of production system, but are particularly relevant for organic production systems.

6 References

Albrecht, W. A. and Smith, G. E. (1942) Biological assays of soil fertility. *Soil. Sci. Am. J.* 6:252–58. doi:10.2136/sssaj1942.036159950006000C0046x.

Anderson, R. L. (2010) A rotation design to reduce weed density in organic farming. *Renew. Agr. Food. Syst.* 25:189–95. doi:10.1017/S1742170510000256.

Ascard, J. (1997) Flame weeding: the effects of fuel pressure and tandem burners. *Weed Res.* 37:77–86.

Baker, B. P. and Mohler, C. L. (2014) Weed management by upstate New York organic farmers: Strategies, techniques and research priorities. *Renew. Agr. Food. Syst.* 30:418–27. doi:10.1017/S1742170514000192.

Bakkali, F., Averbeck, S., Averbeck, D. and Idaomar, M. (2008) Biological effects of essential oils – A review. *Food Chem. Toxicol.* 46:446–75.

Barber, D. (2014) What farm-to-table got wrong. *New York Times Sunday Review*, May 14. Available online at: https://www.nytimes.com/2014/05/18/opinion/sunday/what-farm-to-table-got-wrong.html. Accessed 24 February 2017.

Bàrberi, P. (2002) Weed management in organic agriculture: are we addressing the right issues? *Weed Res.* 42:177–93.

Barnes, J. P. and Putnam, A. R. (1986) Evidence for allelopathy by residues and aqueous extracts of rye (*Secale cereale*). *Weed Sci.* 34:384–90.

Bilalis, D., Papastylianou, P., Konstantas, A., Patsiali, S., Karkanis, A. and Efthimiadou, A. (2010) Weed-suppressive effects of maize–legume intercropping in organic farming. *Int. J. Pest Manag.* 56:173–81.

Bingaman, B. R. and Christians, N. E. (1996) Greenhouse screening of corn gluten meal as a natural control product for broadleaf and grass weeds. *HortScience* 30:1256–9.

Bond, W. and Grundy, A. C. (2001) Non-chemical weed management in organic farming Systems. *Weed Res.* 41:383–405.

Boyd, N. S. and Brennan, E. B. (2006) Burning nettle, common purslane, and rye Response to a clove oil herbicide. *Weed Technol.* 20:646–50.

Bulson, H. A. J., Snaydon, R. W. and Stopes, C. E. (1997) Effects of plant density on intercropped wheat and field beans in an organic farming system. *J. Agr. Sci.* 128:59–71.

Cromar, H. E., Murphy, S. D. and Swanton, C. J. (1999) Influence of tillage and crop residue on postdispersal predation of weed seeds. *Weed Sci.* 47:184–94.

Davis, A. S., Anderson, K. L., Hallett, S. G. and Renner, K. A. (2006) Weed seed mortality in soils with contrasting agricultural management histories. *Weed Sci.* 54:291–7.

Dayan, F. E., Cantrell, C. L. and Duke S. O. (2009) Natural products in crop protection. *Bio. Med. Chem.* 17:4022–34.

Drews, S., Neuhoff, D. and Köpke, U. (2009) Weed suppression ability of three winter wheat varieties at different row spacing under organic farming conditions. *Weed Res.* 49, 526–33. doi:10.1111/j.1365-3180.2009.00720.x.

Forcella, F. (2012) Air-propelled abrasive grit for postemergence in-row weed control in field corn. *Weed Technol.*, 26:161–4.

Gallandt, E. R., Molloy, T., Lynch, R. P. and Drummond, F. A. (2005) Effect of cover-cropping systems on invertebrate seed predation. *Weed Sci.* 53:69–76.

Grundy, A. C., Green, J. M. and Lennartsson, M. (1998) The effect of temperature on the viability of weed seeds in compost. *Compost Sci. Util.* 6:26–33.

Gruver, J. and Wander, M. (2015) Use of tillage in organic farming systems: The basics. Available online at: http://articles.extension.org/pages/18634/use-of-tillage-in-organic-farming-systems:-the-basics. Accessed 25 February 2017.

Hiltbrunner, J., Liedgensa, M., Sampa, P. and Streitb, B. (2005) Effects of row spacing and liquid manure on directly drilled winter wheat in organic farming. *Eur. J. Agron.* 22:441–7.

Hogan, J. P. and Phillips, C. J. (2011) Transmission of weed seed by livestock: a review. *Anim. Prod. Sci.* 51:391–8.

Horowitz, M., Regev, Y. and Herzlinger, G. (1983) Solarization for weed control. *Weed Sci.*, 31:170–9.

Huel, D. G. and Hucl, P. (1996), Genotypic variation for competitive ability in spring wheat. *Plant Breed.*, 115:325–9. doi:10.1111/j.1439-0523.1996.tb00927.x.

Johnson, W. C., Boudreau, M. A. and Davis, J. W. (2013) Combinations of corn gluten meal, clove oil, and sweep cultivation are ineffective for weed control in organic peanut production. *Weed Technol.* 27:417–21.

Kopittke, P. M. and Menzies, N. W. (2007) A review of the use of the basic cation saturation ratio and the 'ideal' soil. *Soil. Sci. Am. J.* 71:259–65.

Lammerts van Bueren, E. T., Jones, S. S., Tamm, L., Murphy, K. M., Myers, J. R., Leifert, C. and Messmer, M. M. (2011) The need to breed crop varieties suitable for organic farming, using wheat, tomato and broccoli as examples: A review. *Wagen. J. Life Sci.* 58:193–205. doi.org/10.1016/j.njas.2010.04.001.

Latify, S., Yousefi, A. R. and Jamshidi, K. (2016) Integration of competitive cultivars and living mulch in sunflower (*Helianthus annuus* L.): a tool for organic weed control. *Organic Agriculture.* doi:10.1007/s13165-016-0166-2.

Liebman, M. and Dyck, E. (1993) Crop rotation and intercropping strategies for weed management. *Ecol. Appl.* 3:92–122.

Liebman, M. and Gallandt, E. R. (1997) Many little hammers: ecological management of crop-weed interactions, pp. 291–343. In L. E. Jackson (Ed.), *Ecology in Agriculture*. San Diego, CA: Academic.

Liebman, M., Mohler, C. L., and Staver, C. P. (2001) *Ecological Management of Agricultural Weeds*. Cambridge University Press.

Mahli, S. S., Lemke, R., Wang, Z. H. and Chhabra, B. S. (2006) Tillage, nitrogen and crop residue effects on crop yield, nutrient uptake, soil quality, and greenhouse gas emissions. *Soil Tillage Res.* 90:171–83.

Main, M. (2003) Evaluation of acetic acid as a thistle top-killer on pastures. Available online at: http://oacc.info/DOCs/ResearchPapers/res_thistle_acetic_mm.pdf. Accessed 23 February 2017.

McErlich, A. F. and Boydston, R. A. (2014) Current state of weed management in organic and conventional cropping systems. In S. L. Young and F. J. Pierce (Eds), *Automation: The Future of Weed Control in Cropping Systems*. Dordrecht: Springer Science+Business Media.

Merfield, C. M. (2016) Robotic weeding's false dawn? Ten requirements for fully autonomous mechanical weed management. *Weed Res.* 56, 340–4.

Mohler, C. L. (1996) Ecological bases for the cultural control of annual weeds. *J. Prod. Agric.* 9: 468–74.

Mohler C. and Baker B. 2016. Weeds your way: How upstate New York organic farmers manage weeds. Available online at: http://articles.extension.org/pages/72963/weeds-your-way:-how-upstate-new-york-organic-farmers-manage-weeds. Accessed 25 February 2017.

Ni, H., Moody, K., Robles, R. P., Paller Jr., E. C. and Lales, J. S. (2000) *Oryza sativa* plant traits conferring competitive ability against weeds. *Weed Sci.* 48:200–4. doi.org/10.1614/00431745(2000)048[0 200:OSPTCC]2.0.CO;2.

Popay, I. and Field, R. (1996) Grazing animals as weed control agents. *Weed Technol.* 10:217–31.

Schonbeck, M. (2015) Mulching for weed management in organic vegetable production. Available online at: http://articles.extension.org/pages/62033/mulching-for-weed-management-in-organic-vegetable-production. Accessed 25 February 2017.

Sivesind, E. C., LeBlanc, M., Cloutier, D. C., Seguin, P. and Stewart, K. A. (2009) Weed response to flame weeding at different developmental stages. *Weed Technol.*, 23:438–43.

Snapp, S. S., Swinton, S. M., Labarta, R., Mutch, D., Black, J. R., Leep, R., Nyiraneza, R. and O'Neil K. (2005) Evaluating cover crops for benefits, costs and performance within cropping system niches. *Agron. J.* 97:322–32.

Tautges, N. E., Goldberger, J. R. and Burke, I. C. (2016) A survey of weed management in organic small grains and forage systems in the northwest United States. *Weed Sci.* 64:513–22. dx.doi.org/10.1614/WS-D-15-00186.1.

Teasdale, J. R., Mangum, R. W., Radhakrishnan, J. and Cavifelli, M. A. (2004) Weed seedbank dynamics in three organic farming crop rotations. *Agron. J.* 96:1429–35.

Trenbath, B. R. (1993) Intercropping for the management of diseases. *Field Crops Res.* 34:381–405.

Tworkoski, T. (2002) Herbicide effects of essential oils. *Weed Sci.* 50:425–31.

U.S. Code of Federal Regulations (2000) National Organic Program Rule. U.S. Government Printing Office. Available at Web site: http://www.ecfr.gov/cgi-bin/textidx?c=ecfr&tpl=/ecfrbrowse/Title07/7cfr205_main_02.tpl.

Webber, C. L., Shrefler, J. W. and Taylor, M. J. Corn gluten meal as an alternative weed control option for spring-transplanted onions. *Int. J. Veg. Sci.* 13:17–33.

Westerman, P. R., Liebman, M., Heggenstaller, A. H. and Forcella, F. (2006) Integrating measurements of seed availability and removal to estimate weed seed losses due to predation. *Weed Sci.* 54:566–74.

Westerman, P. R., Borza, J. K., Andjelkovic, J., Liebman, M. and Danielson, B. (2008) Density-dependent predation of weed seeds in maize fields. *J. Appl. Ecol.* 45:1612–20.

Williams, M. M., Boydston, R. A. and Davis, A. S. (2007) Wild proso millet (*Panicum miliaceum*) suppressive ability among sweet corn hybrids. *Weed Sci.* 55:245–51.

Willey, R. W. (1990) Resource use in intercropping systems. *Agric. Water Manag.* 17:215–31.

Part 5

Biological methods for weed control

The use of allelopathy and competitive crop cultivars for weed suppression in cereal crops

James M. Mwendwa, *Charles Sturt University, Australia; Jeffrey D. Weidenhamer, Ashland University, USA; and Leslie A. Weston, Charles Sturt University, Australia*

1 Introduction: key issues and challenges

Weeds are a persistent problem in agriculture, increasing production costs but reducing crop yields (Wu 2016). Worldwide, yield losses of approximately 34% are caused by crop weeds in broadacre crops and are higher than the losses caused by other crop pests (Jabran et al. 2015). Herbicides are the most widely used method to manage weeds in commercial crops. To date, weeds have globally evolved resistance to 23 of the 26 known herbicide sites of action and to 161 different herbicides (Heap 2017). Herbicide resistance in weeds restricts control options, thereby escalating economic loss and threatening agricultural sustainability (Wu 2016). This threat comes at a time when increasing global populations require greater agricultural productivity, and environmental concerns have resulted in significant restrictions on use of some herbicides.

http://dx.doi.org/10.19103/AS.2017.0025.19

In Australia, herbicide resistance in both grasses and broadleaf weeds is on the rise, with resistance to multiple herbicides being reported for an increasing number of weeds (Owen et al. 2013). However, diversity in weed management tools could potentially provide sustainable weed control and slow the development of herbicide resistance in weeds. To combat the challenges of environmental pollution and herbicide resistance, crop types which release natural herbicides or produce significant biomass to reduce weed growth (i.e. allelopathic or competitive crops) are suggested (Jabran et al. 2015). Both competition and allelopathy as mechanisms of plant interference have been well documented under controlled conditions (Weston 2005). The combined effects of allelopathy and crop competition determine the total weed-suppressive potential of a given cultivar, and research groups worldwide have been working to improve both traits simultaneously to achieve maximum gains in weed suppression (Bertholdsson et al. 2012; Worthington and Reberg-Horton 2013), particularly in cereal crops.

1.1 Plant interference with weed growth

Plant interference can be defined as any physical or chemical mechanism that results in the reduction of plant growth over time due to the presence of another plant. Competition is usually described as the process whereby plants interfere with the growth of neighbouring plants by utilization or competition for growth-limiting resources, including light, nutrients or moisture (Weston and Duke 2003; Weston 2005). Highly competitive wheat cultivars (Triticum aestivum L.) typically have the ability to access better light, nutrients and water resources in a limited space, thus suppressing the growth and reproduction of neighbouring weed species (Bertholdsson 2011; Mwendwa et al. 2016a; Worthington et al. 2015). However, it has also been suggested that chemical interference (allelopathy) may prove to be as important as competition for resources in modulating plant community function and dynamics (Fernandez et al. 2016).

One strategy for integrated weed management is to use the inherent ability of many cereal crops to suppress weeds through a combination of high early vigour (competition) and allelopathic activity to further reduce weed interference (Bertholdsson 2005; Mwendwa et al. 2016a,b). Some crops including rice (Oryza sativa L.), sunflower (Helianthus annuus L.), sorghum (Sorghum bicolor L.), wheat, rye (Secale cereale L.), maize (Zea mays subsp. Mays L.), barley (Hordeum vulgare L.), alfalfa (Medicago sativa L.) and Brassica spp. can exhibit strong allelopathic potential (Jabran and Farooq 2013). Allelopathic interference mechanisms are often difficult, if not impossible, to distinguish from interference due to competition in a cropped field (Weidenhamer 1996; Weston 2005). Determination of the mechanism(s) associated with weed suppression is essential to determine if the use of crop cultivars for allelopathic and competitive weed suppression in cereal crops is going to provide sustainable solutions for weed management. Allelopathy and competition most likely act separately and interactively, and this may prove important for highly competitive crop cultivars.

Although allelopathic effects are mainly considered to be negative interactions, positive (stimulatory) interactions have also been reported depending on the mixture of allelochemicals in an extract, the target plant and the concentration tested (Eichenberg et al. 2014; Muzell Trezzi et al. 2016). Regardless of whether the interactions are positive or negative, maximizing the allelopathic potential of crops to reduce pest pressure and assist judicious nutrient management by crop sequences via rotation cover crops or intercrops requires a better understanding of such interactions (Jabran and Farooq 2013).

1.2 Chapter overview

This chapter focuses on competitive cereal crops and cultural strategies for weed management, including the use of weed-suppressive cultivars, post-harvest crop residues and cover crops for management of the weed seedbank and eventual weed suppression. This chapter also addresses important factors influencing the effect of allelopathy on weeds, including soil and environmental conditions which limit or intensify the efficacy of allelochemicals. The response of some weeds to secondary metabolites released by living cereal crops and/or crop residues (selectivity) is also reviewed, since this ability may limit the use of some crop residues for weed control. Finally, recommendations for future research are required to address the knowledge gap regarding the fate of these compounds in the environment and their role in important physiological processes in both plants and microbes in the soil rhizosphere.

2 Competitive crops and cultural strategies in weed management

In Australia, Europe and North America, non-chemical methods for weed control such as harvest weed seed destruction and crop competition are being increasingly adopted across grain-growing areas and are proving successful in controlling weeds that escape pre-sowing herbicide control (Harker et al. 2011; Norsworthy et al. 2012). A diversified herbicide-use pattern and integration of appropriate non-chemical methods are envisaged to minimize the pace of herbicide resistance evolution (Manalil 2014). Cultural weed management is also employed and refers in part to agronomic practices that use competitiveness of the crop to maximize crop growth while diminishing the growth and subsequent competitiveness of associated weeds or mechanical, biological or other practices to remove or restrict establishment of weeds in cropping areas (Vencill et al. 2012).

One alternative weed management strategy involves production of cereal crop cultivars with increased competitive ability with weeds. Crop competitive ability can be specified in terms of either crop tolerance against weeds or growth inhibition of weeds themselves by resource limitation and/or allelopathy (Bertholdsson 2010). Cultural strategies including crop rotation or successional planting, improved crop competition through cultivar selection and planting date, and optimized seeding rates are often considered but not always used by commercial producers (Beckie and Gill 2006). Crop sequences taking advantage of different planting times and production practices allow a variety of cultural techniques to be used to optimize crop competition with weeds at the expense of weed growth and reproduction (Vencill et al. 2012).

The ability to suppress weeds clearly appears to be cultivar dependent (Bertin et al. 2003; Wu et al. 2007). Crop cultivars with higher early vigour are generally capable of extracting more soil moisture which in turn enables them to maintain lower canopy temperatures on warm days (Zerner et al. 2008), an essential trait in dryland broadacre farming. Certain wheat cultivars will produce acceptable yields in the absence of herbicides under ideal conditions and suppress weeds (Andrew et al. 2015; Mwendwa et al. 2016a; Worthington et al. 2015).

Cultivar-dependent weed suppression has been observed in a variety of cereal crops. For instance, Seavers and Wright (1999) showed that the suppressive ability of oat (*Avena sativa* L.) cultivars at early stages of growth was greater than could be accounted for by canopy structure alone. The oat cultivars had low early ground cover and were the slowest of the three cereal species (oats, barley, wheat) to develop a closed canopy, but their suppressive ability was higher throughout the growing season. In a recent study carried out at two ecologically different locations, differences in weed suppression by selected Australian commercial winter wheat cultivars were largely determined by crop architecture and phenology early in the growing season (Mwendwa et al. 2016a; Fig. 1).

Andrew et al. (2015) state that in comparison with aboveground canopy measurements, belowground traits have received relatively little attention in cereal crop–weed interactions. This is partly due to the difficulties associated with measuring root traits, particularly when incorporating them into a screening protocol for new cultivars. It has been proposed that belowground traits determine the degree to which crop and weeds share resource pools (Smith et al. 2010), meaning more tolerant cultivars may be those with belowground traits which avoid resource pool overlap (Andrew et al. 2015). However, at this stage there is relatively little information available on the root traits among cereal cultivars in relation to weed suppression, resulting in a knowledge gap in technical information to select cultivars for competitive ability based on root traits.

Competitive traits in cereal crops also vary in their effects between years (Vandeleur and Gill 2004; Coleman et al. 2001) and locations due to crop or weed species response to weather (Andrew et al. 2015). These variations may cause challenges in selecting the best cultivars. However, Lemerle et al. (1996) found that the best cultivars were generally consistent across years and sites, despite different weather patterns, and suggested that the similarity in soil type may have had a bearing on this. However, the lack of correlation between the yield of genotypes and weed-suppressive ability indicates that selection for weed suppression may not be equally efficient in all environments and that the performance of genotypes identified as highly weed suppressive may be affected by cultural practices such as planting date as well as environmental conditions (Worthington et al. 2015).

Development of wheat cultivars with increased inherent competitiveness against herbicide-resistant weeds is a potential supplement to in-crop herbicide use and in some cases an alternative management strategy, particularly for organic producers (Weston 1996). Competitive genotypes can better access light, nutrients and water resources in limited space, thus suppressing the growth and reproduction of nearby weed species

Figure 1 Differences in canopy closure between two wheat cultivars (cv. Condo (left) and cv. Espada (right)) at 113 days after emergence (DAE) at Wagga Wagga in 2015. Note the presence of greater canopy closure in the crop on the left (cv. Condo) which later resulted in significantly reduced weed biomass in this crop compared to the crop on the right (cv. Espada).

(Worthington et al. 2015; Fig. 1). Our most recent three-year studies evaluated six replicates of 12 genetically diverse wheat cultivars in standard plots (12 m × 2 m) grown in two ecologically different study sites. It was conducted using seed of each cultivar produced in the same location to eliminate variation associated with seed production differences. Results also showed that wheat cultivar competitive traits were influenced by both genotype and environmental factors, as evidenced by clear differences in cultivar performance, yield and weed suppression among both locations. However, only a few selected cultivars, such as Condo (Fig. 1) performed well at both sites in all years (Mwendwa et al. 2016a).

In Greece, the use of competitive cultivars alone resulted in a 50% reduction in the total amount of herbicides used for weed control in commercial wheat (Travlos 2012; Andrew et al. 2015). Thus, developing cereal cultivars with superior competitive ability against weeds can complement cultural methods for weed control while maintaining acceptable yields and suppressing weeds (Andrew et al. 2015; Mwendwa et al. 2016a; Worthington and Reberg-Horton 2013).

Breeding for more competitive crops

In the past century, it was thought that the competitive ability of wheat had been reduced by selection based on yield potential. Research has shown that ancient cereal cultivars or landraces are often more competitive with weeds than the higher yielding, semi-dwarf modern cultivars (Bertholdsson et al. 2012; Vandeleur and Gill 2004). Enhanced suppression was thought to be associated with crop height, root architecture or allelopathy. Today, selective breeding programmes for improvement of the competitive ability of modern cereal crops without compromising yielding ability exist. They focus on incorporation of morphological traits that enhance early crop vigour (size of leaf 1 and 2) and light interception without affecting harvest index (Andrew et al. 2015; Vandeleur and Gill 2004).

To realize the potential of competitive crop cultivars as a tool in integrated weed management, a quick and simple-to-use protocol for assessment of the competitive potential of new cultivars is required. It is likely that assessment will not be based on a single trait, but the combined effect of multiple traits for interference (Bertholdsson 2011; Andrew et al. 2015). Because weed-suppressive ability is strongly and positively correlated with wheat competitive traits, including vigour and erect growth habit during tillering (Zadoks GS 29), high leaf area index (LAI) at stem extension (GS 31), plant height at tillering and stem extension (GS 29, 31), grain yield in weedy conditions and grain yield tolerance (Bertholdsson 2010; Worthington et al. 2015), these traits will be particularly important to incorporate into an assessment protocol. Our recent field studies of wheat competition attempted to incorporate an evaluation of many of the traits described above (Mwendwa et al. 2016a).

3 The effect of allelopathy on weed suppression

Allelochemicals, typically considered to be plant secondary metabolites, are produced by plants for a variety of purposes including chemical defence and communication. Along with microbial decomposition products, they are the active agents of allelopathy (Cheng and Cheng 2015). Cereal crops can frequently suppress weeds through the release of allelochemicals from intact roots of living plants and/or by decomposition of phytotoxic plant residues (Bertin et al. 2003; Belz 2007; Ferreira and Reinhardt 2010). The incidence of growth inhibition of certain weeds and the induction of phytotoxic symptoms by plants and their residues have been well documented for many crops, including all major grain crops such as rice, rye, barley, sorghum (*Sorghum bicolor* (L.) Moench] and wheat (Belz 2007; Ferreira and Reinhardt 2010; Weston 1996), and is density, location and weed dependent (Weston 1996). Varietal autotoxicity occurs when plants of a given cultivar release chemical substances that inhibit or delay germination and growth of the same cultivar (Wu et al. 2007).

Allelochemical production and mode of action

Allelochemicals tend to exhibit mechanism(s) of action in plants that differ from the sites targeted by herbicides (Muzell Trezzi et al. 2016). Duke and Dayan (2011) note that some of the most potent phytotoxins are synthesized by microbes. A few phytotoxins or allelochemicals share molecular target sites with synthetic herbicides but molecular target sites for allelochemicals are frequently unidentified or unique. This implies that the development of allelopathic crop cultivars may not be severely affected by existing herbicide resistance in weed populations. In addition, allelochemicals are present in all plant species and tissues and are typically released into the rhizosphere by a variety of mechanisms, including decomposition of residues, volatilization and root exudation (Weston 2005). Expression of allelopathic traits may allow plants to strengthen their defence system against biotic and abiotic stressors, and aid in regulating nutrient transformation (Jabran and Farooq 2013).

Plant secondary metabolites are essential for the interaction of plants with their biotic environment, help attract pollinators or seed dispersers, act in defence against natural enemies and inhibit potential competitors (Muzell Trezzi et al. 2016). The ability of a plant to produce and release allelopathic metabolites into the environment and/or to tolerate the presence of allelochemicals released by neighbouring plants including weeds can be crucial to the ability of a species to survive and reproduce (Bertholdsson et al. 2012, Worthington et al. 2015). Allelopathy is generally regulated by a dynamic mixture of allelochemicals and their metabolites, which is affected by genotype and developmental stage of the producing plant, the environment, cultivation and the rate of chemical or microbial degradation in the rhizosphere (Belz 2007; Mwendwa et al. 2016a).

Bertholdsson (2005, 2011) has noted that the weed-suppressive ability of today's commercial cereal crop wheat is typically lower than that of other commercially produced cereals including rye, oats and barley. It has been suggested that selection for enhanced suppression may improve these traits substantially through focused breeding. Two factors are likely to be important for expression of weed-suppressive ability: 1) early season biomass production by the crop and 2) potential allelopathic activity of the crop (Bertholdsson 2011).

For example, cereal rye (*Secale cereale* L.) is thought to be allelopathic due to the presence of phytotoxic benzoxazinoids (BXs) whose biosynthesis is developmentally regulated, with the greatest accumulation in young shoot tissue. Concentration is influenced by cultivar and environment (Schulz et al. 2013). In wheat seedlings, BX biosynthesis begins shortly after germination and reaches a maximum of 7–10 days (Argandoñaa et al. 1981; Bertholdsson et al. 2012), but in rye, synthesis was highest 60 days following germination (Burgos and Talbert 2000; Bertholdsson 2012). Therefore, periods of maximal allelochemical production vary with crop and field location and should be targeted and monitored to achieve optimal weed suppression over time.

BXs are characteristic secondary compounds produced not only by rye but also by several other species of Poaceae, including maize, triticale (*Triticosecale* Wittm. ex A. Camus.), and wheat, and some dicots belonging to the *Acanthaceae*, *Scrophulariaceae*, and *Lamiaceae* (Schulz et al. 2013). The BXs present in wheat, barley and rye, and their suppressive effects on weeds, pest and diseases are of great interest in sustainable agriculture (Bertholdsson et al. 2012; Tanwir et al. 2013). Rye produces numerous BXs but [2,4-dihydroxy-1,4(2*H*)-benzoxazin-3-one (DIBOA) and 2(3*H*)-benzoxazolinone (BOA)] are generally associated with its allelopathic potential. Other important allelochemicals have also been reported (Jabran et al. 2015). These products are also associated with enhanced human health, especially potential suppression of prostate cancer (Adhikari et al. 2015; Steffensen et al. 2016).

Application of BOA to developing weed or crop seedlings has been shown to significantly affect the transcriptome, proteome and metabolome of germinating seedlings, resulting in inhibition of both germination and growth, sometimes even death of sensitive species and up-regulation of plant defence and stress genes (Reigosa et al. 1999; Reigosa and Pazos-Malvido et al. 2007). The inhibition of germination and reduction of seedling growth were observed in many plant species exposed to BOA and DIBOA. Often, radicles and root tips were more affected than shoots. Both DIBOA and BOA inhibited emergence of barnyardgrass (*Echinochloa crusgalli* L. Beauv.), cress (*Lepidium sativum* L.) and lettuce (*Lactuca sativa* L.) when applied to Petri dish bioassays and in field soil, indicating their potential to cause allelopathic inhibition of plant growth (Barnes and Putnam 1986, 1987; Schulz et al. 2013).

In barley, allelochemical toxicity increased after release of BXs by roots, between day 0 and day 6 (in a lab seed-after-seed protocol). The allelopathic potential of barley root exudates was also dependent on the receiver weed species (Bouhaouel et al. 2015). Plant part and rhizosphere location (distance from root) also affected BX concentration in wheat and rye (Mwendwa et al. 2016b) with release of these allelochemicals dependent upon genotype and environmental conditions. For example, in rye, genotypes with varying contents of BOA and DIBOA were identified (from 0.52 to 1.15 mg/g dry tissue). Typically, those with the highest content of BXs were the most phytotoxic in laboratory assays (Schulz et al. 2013).

While BXs are important allelochemicals present in both wheat and rye, and their suppressive effects on weeds, pest and diseases are of great interest in sustainable agriculture (Carlsen et al. 2009, Bertholdsson et al. 2012), they do not occur in sorghum or rice, which produce other highly active allelochemicals (Schulz et al. 2013). The chemistry implicated in allelopathic interactions of wheat and rye is thus believed to be based on the activity of benzoxazolinones, a class of BXs, in contrast to the diverse chemistry produced by other cereals including rice (Duke et al. 2005; Wang and Kong 2013) and sorghum (Ferreira and Reinhardt 2010; Weston et al. 2013).

Studies performed with suppressive rice have found that the synthesis of two compounds phytotoxic to barnyardgrass [a flavone (*5, 7, 4-trihydroxy-3, 5-dimethoxyflavone*) and a cyclohexanone (*3-isopropyl-5-acetoxycyclohex-2-en-1-one*)] is induced in rice plants by the presence of weeds (Kong et al. 2004; Duke et al. 2005). The mechanism of this induction is unknown. In addition, momilactone A and B have been identified as potent allelochemicals in rice seedlings (Kato-Noguchi et al. 2008; Kato-Noguchi and Ino 2005; Kato-Noguchi and Peters 2013) and its production has been shown to be enhanced by the presence of barnyardgrass seedlings in cultivated rice. Induction of sorgoleone production in sorghum has also been reported following exposure of sorghum seedlings to velvetleaf (*Abutilon theophrasti* Medik.) extracts and exposure to weeds (Dayan 2006; Weston et al. 2013). Understanding how to induce allelochemical expression in field crops is critical to maximizing allelopathic activity. Therefore, additional research is needed to determine environmental effects on allelochemical production.

There is evidence that the allelopathic potential of barley has been impaired by modern breeding which has emphasized selection for other traits, including yield. Using a bioassay with perennial ryegrass (*Lolium perenne L.*) as the model weed (receiver), Bertholdsson (2005) showed that the allelopathic activity measured as root growth inhibition of perennial ryegrass decreased 14–31% in the Nordic barley germplasm collection since the start of selection and breeding over 100 years before. Differences in the sensitivity of cultivars and ecotypes may be due to different weed species-dependent strategies that have evolved to cope with allelochemicals (Schulz et al. 2013). However, the allelopathic activity of barley is still considered to be generally high, in contrast to other cereals such as wheat. Despite the negative effects of selective breeding for other traits, it may contribute to the effectiveness of barley as a cover crop (Bertholdsson 2010).

Sorghum's allelopathic properties were first suggested by reports of reduced growth of crops grown in rotation with sorghum (Weston et al. 2013). Its use as a green manure or a cover crop for suppressing the growth of weeds is likely also related to its allelopathic properties. Certain species of *Sorghum*, such as Sudangrass (*Sorghum × drummondi* Nees ex Steud.), are very allelopathic, making them practical to grow as weed-free monocultures (Ferreira and Reinhardt 2010). Sorghum-Sudangrass hybrids (Sudex) are also often used as green manures in nursery and horticultural cropping systems. In recent years, sorghum phytotoxicity and allelopathic interference have been well described in greenhouse and laboratory settings. Many observations of weed suppression in diverse locations and with various sorghum plant parts and residues have been reported (Weston et al. 2013).

A diverse group of sorghum allelochemicals, including numerous phenolics, a cyanogenic glycoside (*dhurrin*) and a hydrophobic p-benzoquinone (*sorgoleone*) were isolated and identified in recent years from sorghum shoots, roots, root exudates and soil, as our capacity to analyse and identify complex secondary products in trace quantities in the plant and in the soil rhizosphere has improved (Weston et al. 2013; 2015; Weidenhamer 2005). Sorgoleone (2-hydroxy-5-methoxy-3- [(8′Z, 11′Z)-8′, 11′, 14′pentadecatriene]-p-benzoquinone) was identified as the most phytotoxic compound produced by sorghum. It occurs as a mixture of several hydroquinones as root exudates and inhibits growth and establishment of many weeds (Ferreira and Reinhardt 2010; Weston et al. 2013). The allelochemicals present in sorghum tissues also vary with plant part, age and cultivar evaluated (Weston et al. 2013).

4 The effect of soil and environment on plant metabolites (allelochemicals)

4.1 Overview

In recent years, hundreds, if not thousands, of wheat cultivars have been screened for their weed-suppressive potential on several weed species in laboratory studies (Wu et al. 2000; Duke et al. 2005). Related wheat species such as *Triticum durum*, *T. spelta* and *T. speltoides*, as well as triticale and rye, have been considered as possible sources of allelopathic germplasm, but despite the fact that some have exhibited strong allelopathic potential, these sources have been relatively under investigated (Belz and Hurle 2005;

Duke et al. 2005; Schulz et al. 2013). In general, phytotoxicity has primarily been attributed to root exudates or secretions or decomposing residues (Duke et al. 2005).

Root exudates represent one of the largest direct inputs of plant chemicals into the rhizosphere, and therefore also likely represent the largest source of allelochemical inputs into soil (Bertin et al. 2003). Wheat root exudation, as measured by excretion of phenolics in agar media, was shown to vary with wheat accession (Wu et al. 1999, 2000). Analysis by multiple regression showed that accessions with strong allelopathic potential were associated with production of significantly higher levels of allelochemicals in shoots and roots (Wu et al. 1999, 2000, 2001).

However, actual weed suppression in the field is dependent on the behaviour of these compounds in the soil. To understand an allelopathic interaction process within a given soil, the process must be viewed and understood within the specific environmental constraints of that soil (Blum 2006). Upon release from a source organism, the soil is the main vehicle that mediates contact between allelochemicals and their target plants (Muzell Trezzi et al. 2016). Therefore, assessment of allelopathy must account for the effect of soil on the behaviour and activity of allelochemicals (Teasdale et al. 2012; Blum 1995, 2006). In addition, allelopathy should be assessed in a range of soil types as several climatic and edaphic factors affect soil microflora (Inderjit 2005).

The role of soil microorganisms in chemically mediated interactions between plants is poorly understood, and appropriate methodologies are needed to assess the role of soil microbial ecology in allelopathy. Soil properties including organic matter, reactive mineral surfaces, ion exchange capacity, inorganic ions and abiotic and biotic factors of the soil environment significantly influence allelochemical activity (Inderjit 2001; Blum 2006). Microorganisms can chemically alter the allelochemicals released into an ecosystem, highlighting their key role in chemical plant–plant interactions and suggesting that allelopathy is likely to shape the vegetation composition and participate in the control of ecological biodiversity (Fernandez et al. 2013). For example, allelochemicals released into the environment inhibited germination and growth of neighbouring plants by altering their metabolism or affecting their soil community mutualists (Fernandez et al. 2016). Wang et al. (2013) further suggested that the shift in the microbial community composition induced by barnyardgrass might generate a positive feedback in rice growth and reproduction in a given paddy system.

However, the mechanism of the release of the associated allelochemicals from residues and by root exudation and their interaction with the soil microbial community are not well understood. Therefore, additional research on their release mechanisms is suggested due to the complexity of these interactions. Currently, we are evaluating the effect of soil microbial communities on the activity of cereal and other plant residues in suppression of weeds in projects funded by the Australian Grains Research and Development Corporation. Allelochemicals are altered or degraded by soil microbes over time in the field; however, soil microbes can also degrade residues to release additional metabolites (Weston and Duke 2003). Therefore, studies on microbial influence on allelochemical activity will result in fundamental knowledge about degradation of plant residues and associated metabolites in various soil types and the role of soil microbial communities in phytotoxicity.

Under the appropriate environmental conditions, phytotoxins may be released into the environment in sufficient quantities to affect the growth of neighbouring plants (Weidenhamer 1996; Weston 1996). For example, Bertholdsson (2010) found that highly allelopathic spring wheat lines derived from a cross between allelopathic and non-allelopathic parents suppressed weed biomass 24% more than the non-allelopathic parent in a dry year and 12% more in a wet year. However, the effects of temperature and drought on the toxicity of allelochemicals, their rates of microbial breakdown and other factors could have a major effect on the efficacy of these compounds in controlling weed growth in extreme environments such as those encountered in Australia.

According to Einhellig (1996), allelopathy is strongly coupled with exposure to other crop stressors, including insects and disease, temperature extremes, nutrient and moisture variables, radiation and herbicides. These specific stressors can enhance both allelochemical production and toxicity, thus increasing the potential for allelopathic interference. Therefore, environmental effects on release of allelochemicals or residue degradation and release over time are critical to study under controlled conditions.

4.2 Analysis of allelochemicals in plant and soil

Precise metabolic profiling of allelochemicals in the plant and, at the same time, in the soil rhizosphere could provide strong insight into the dynamics of release of bioactive metabolites following incorporation of plant material or living root exudates into soil (Krogh et al. 2006; Weston et al. 2015). Chen et al.'s (2010) study quantified DIMBOA and MBOA in the wheat rhizosphere and analysed the soil microbial community structure. MBOA rather than DIMBOA was found in the wheat rhizosphere, and its concentration varied with cultivars, plant densities and growth conditions. Recent studies have incorporated the use of metabolomics to study release rates of various metabolites from plant to soil or aboveground environment. These studies have shown that certain plant species not only compete for resources; they produce allelochemicals which further interfere with plant growth. Work in the Mediterranean forests has shown the effect of allelochemicals and competition on seedling growth of local trees (Fernandez et al. 2016). We suggest similar studies be done in agroecosystems for cereal production.

Studies reported by Rice et al. (2012) and Teasdale et al. (2012) demonstrated relatively low concentrations of the most toxic BX compounds, 2-aminophenoxazin-3-one (APO), DIBOA and DIMBOA from incorporated rye residues when cover crops were soil incorporated, whereas the less toxic compounds, BOA and MBOA, and the non-toxic compounds, HBOA and HMBOA, were predominant BX species in amended soils. Growth assays with lettuce and smooth pigweed (*Amaranthus hybridus* L.) species showed inhibition whether rye residue was left on the surface or incorporated in soil during the first two weeks after rye applications; however, there were not sufficient concentrations of any one BX in the soil to explain these effects (Rice et al. 2012). This suggests that the activity of BXs may depend on interactions of mixtures and the soil environment as influenced by soil microbiota. Hence inconsistent results for bioassays using only single applications of pure compounds to test for allelopathy activity are often observed.

In addition, Teasdale et al. (2012) removed soil from beneath a field site maintained with coverage of rye residue and assayed this soil in pots but observed little phytotoxicity against the weeds encountered in the field. This could be due to degradation of allelochemicals over time under field conditions. When BOA and MBOA were exogenously added to soils to maintain extractable levels of up to 10 μg g^{-1} soil (100–500 times higher than measured BX in field soils), no significant inhibition of pigweed was observed (Teasdale et al. 2012).

However, Rice et al. (2012) found that movement of these compounds into the soil column was minimal, with more than 70% of BOA and 97% of MBOA remaining in the top 1 cm of soil profiles, and complete dissipation was noted in less than 24 h. Because communication between plants and other organisms below ground drives community dynamics (Inderjit 2005), these results suggest that the movement and the activity of these compounds in the soil may be facilitated by specific dynamics mediated by microbes and the soil medium through chemical signalling, as the presence of these metabolites at significant concentrations in the upper soil profile had no effect on weeds. In addition, this also suggests that rainfall or conditions affecting soil bioavailability of these metabolites may be required for allelopathic activity.

During degradation in the soil, different metabolites accumulate in a manner that is both concentration and soil type dependent. The differences may be related to the density and diversity of microorganisms associated with the plant and those in soil (Glenn et al. 2001, Schulz et al. 2013) or the relative mobility of allelochemicals in soil which is associated with their chemical structure, including their polarity and lipophilicity. Wheat seedlings have been shown to be capable of detecting the presence of competing weeds and responding by increasing MBOA production in the rhizosphere (Chen et al. 2010). Metabolites released from the crop plants depend on the crop species, and may in turn also influence the microflora associated with the root system (Schulz et al. 2013). However, for allelopathic interference to occur, significant and dynamic concentrations of particular allelochemicals are required for uptake by plants or microbial communities so solubility in the soil/water in field soils is critical.

For instance, Chen et al. (2010) found there was positive linear relationship between the MBOA level in the wheat rhizosphere and soil fungi/bacteria. When DIMBOA was applied to soil, MBOA concentrations were increased, resulting in enhanced levels of soil fungi, suggesting that DIMBOA and MBOA could affect soil microbial community structure to their advantage through the change in fungal population structure and growth. Microorganisms are not only responsible for the degradation of allelochemicals in soil, but the common mycorrhizal network of soil can also be involved in transporting compounds away from the zones of highest microbial activity and expanding the zones of bioactivity for allelochemicals in soil to areas some distance away from the plant (Barto et al. 2011, 2012). Sorting out the dynamics of complex mixtures of allelochemicals in soil remains a major challenge.

Various analytical approaches including HPLC, GC, MS and MS/MS can be used in concert to analyse the presence of plant metabolites in tissue and in soil. Both environmental and microbial factors can affect the degradation of plant metabolites in the soil and thereby effect their efficacy. In non-sterilized soil, for instance, DIBOA showed a half-life of 43 h. However, APO, the final microbial degradation product of DIBOA, has a low mineralization rate and, therefore, a half-life greater than 90 days (Macías et al. 2005). In addition, it is up to one thousand times more active as a growth inhibitor than DIBOA or BOA due to its half-life in soil. Therefore, its build up over time in the soil may be correlated with allelopathic activity in contrast to the presence of temporal BXs typically released by plants and rapidly degraded. In contrast, some flavonoid glycoside molecules exuded by rice plants can suffer high rates of mineralization by soil microorganisms, resulting in accumulation of aglycosylated compounds. These flavonoid glycosides and aglycosides have a half-life of 2 h and 30 h, respectively, suggesting the potential for greater allelopathic activity associated with the presence of the second group of compounds (Muzell Trezzi et al. 2016), which are also often more biologically active (Weston and Mathesius 2013).

In summary, to better understand allelopathy, three main areas require focused research in the future: 1) the role of soil microorganisms in chemically-mediated interactions between plants; 2) the development of new analytical tools to enable detection and quantification of low, dynamic concentrations of metabolites and their resulting catabolites; and 3) experimental designs to elucidate the underlying dynamics of complex mixtures of allelochemicals in soil.

5 Use of crop residue mulches and cover crops in weed suppression

5.1 Crop residues as mulches in weed suppression

Crop residues, when present in uniform and dense stands under conservation farming can suppress weed seedling emergence, delay emergence and allow the crop to gain an initial advantage in terms of early vigour (Chauhan et al. 2012). For instance, the seedling emergence rate of littleseed canarygrass (*Phalaris minor* Retz.) was reduced for wheat planted no-till compared with conventional ploughing and sowing (Franke et al. 2007; Bajwa et al. 2015).

Crop residues interfere with weed development and growth in several ways that include physical and chemical effects including the alteration of soil physical, chemical and biological characteristics based on two possible sources of allelochemicals: secondary metabolites can be released directly from crop litter or they can be produced by microorganisms that use plant residues as a substrate (Ferreira and Reinhardt 2010). McCalla and Norstadt's (1974) review on phenolic acids showed that levels required for strong phytotoxicity to successive crops often greatly exceed those typically observed in the soil following residue degradation.

This suggests that a combination of factors may be associated with weed suppression, including physical presence of a mulch or stubble and the presence of allelochemicals, which are most often rapidly degraded once a threshold concentration for inhibition is reached. The inclusion of specific crops or cultivars with allelopathic properties in the cropping rotation may result in more effective weed management. Weston and Duke (2003) reported that cereal rye residues apparently reduce weed seed germination and seedling growth by shading, lowering soil temperature, moderating diurnal temperature fluctuations and acting as a physical barrier to prevent light from reaching the soil surface. In addition, rye and its residues also release secondary metabolites that accumulate near the soil surface to further inhibit weed seed germination and growth (Weston and Duke 2003).

There are multiple approaches to the management of weeds in crop residues or stubble depending upon environmental conditions and management. Teasdale et al. (2012) reported that surface rye residue was highly inhibitory to small-seeded broadleaf seedlings throughout an experimental period of 4 weeks. Residues left on the soil surface have generally led to decreased soil temperature fluctuations and reduced light penetration, both of which have been shown to reduce weed germination (Liebman and Mohler 2001; Schulz et al. 2013). Gavazzi et al. (2010) found that grass weeds were reduced 61%, whereas broadleaf weeds were reduced 96% when rye mulch was used in a no-tillage system. The level of weed suppression was dependent on weed species and the thickness of the mulch layer, with an exponential relationship between mulch biomass and weed emergence (Schulz et al. 2013).

Trends in wheat yield responses to conservation cropping in Australia were analysed using data from 33 medium (3–5 year) and long-term (>5 year) agronomic experiments. The overall effect of tillage (direct-drilled vs. cultivated) was small in all regions (−0.18 to +0.06 t/ha), while stubble retention (stubble retained vs. stubble burnt) reduced yield in all regions (−0.31 to −0.02 t/ha). Reduced early seedling growth of direct-drilled crops was a major factor underlying the yield response at most sites, and yield reduction was rarely associated with the lack of available water or nitrogen (Kirkegaard 1995; Kirkegaard et al. 2014). However, the reduction of early seeding growth and in yield was attributed to reduced soil temperature associated with the presence of residues or the effects of either autotoxicity and/or phytotoxicity associated with residues (Kirkegaard et al. 2014). Further studies on these field-based interactions are required to determine best stubble management strategies and crop choice for the subsequent season and are now the subject of additional research in Australia.

Crop residues retained on the soil surface have been reported to decompose more slowly than residues incorporated in soil, which may result in a slower release rate but longer supply of allelochemicals (Kruidhof et al. 2009). For example, when the residue material is retained on the soil surface, effective weed control has been observed for 4–8 weeks after mulching (Gavazzi et al. 2010; Weston et al. 2014). Previous studies also reported that the concentration of BX compounds released from rye and wheat residues in soil peaked 1–4 days after stubble incorporation in soil and declined to negligible amounts within 10 days (Krogh et al. 2006).

However, in the absence of incorporation, it is difficult to predict how long allelochemicals may be released from residues on the soil surface. When stubble remained on the soil surface, stubble build-up leading to poor seed/soil contact resulted in increased numbers of grass weeds but reduced broadleafs (Scott et al. 2010). This is thought to be due to both the physical presence of residues and release of allelochemicals over a 60-day period following harvest/kill of the cover crop. In Australian broadacre cropping regions, crops are planted 5–6 months after harvest into the remaining crop stubble (Weston et al. 2014). This practice could potentially reduce the physical effects on crop seedlings as adequate decomposition of the stubble may occur before sowing the following season, but chemical interference in Australia is often mediated by the presence of rainfall or soil moisture.

Teasdale and others have shown that incorporated residues of rye inhibited lettuce and pigweed growth for approximately two weeks after incorporation (Teasdale et al. 2012). Emerging weed seedlings are sometimes effectively controlled by allelopathic mulches through leaching or timed release of allelochemicals; however, a well-established weed flora is difficult to eradicate this way (Farooq et al. 2013). In either case, both allelopathy and competitive weed-suppressive ability are complex, quantitatively inherited traits that are heavily influenced by environmental factors (Worthington and Reberg-Horton 2013). In some cases, with residue incorporation, it is difficult to determine if suppression is due to physical and/or chemical suppression of weeds.

To date studies have not been able to elucidate the independent contributions of these traits to weed suppression. The soil has proven to be an important barrier in allelopathic interactions because allelochemicals must survive transit through soil in sufficient concentrations to affect target plants (Barto et al. 2012) and allelochemicals may serve as attractants to common soil microbes (Shi et al. 2011). Further research on these interactions is required to gain a better understanding of the mechanism of the release and activity of allelochemicals in soil.

In addition, the choice of species or cultivars to utilize for desired weed suppression, or the weed biotype encountered in field settings, may affect the ability of the plant or its residues to interfere with plant growth (Weston and Duke, 2003). For example, Barnes and Putnam (1987) as well as Gavazzi et al. (2010) found that broadleaf weeds were approximately 30% more sensitive to DIBOA and BOA compared with grass weeds. Further studies have indicated that larger-seeded weed species are less sensitive to allelochemicals (Weidenhamer et al. 1987; Tabaglio et al. 2008) and that seed size and mass affect selective suppression of weeds with crop residues (Liebman and Davis 2000). The quantity of crop residues also varies among crops. For example, oilseeds and pulses typically produce less biomass than cereals. In rain-fed areas, the crop biomass will also depend on the amount and pattern of rainfall. Therefore, depending on the region, crop and rainfall, the effects of crop residue on the weed population will vary (Chauhan et al. 2012). Campiglia et al. (2015) reported that various mulch strips caused differences in weed species composition dominated by perennial ruderal weeds in mulched areas, while in tilled soil weed flora was dominated by annual weeds.

5.2 Use of cover crops to suppress weeds

Cover crops provide another strategy to minimize weed populations while maintaining seasonal vegetative ground cover to prevent soil erosion. A cover crop is usually a 'noncash' crop that can be grown before, or in the case of a living mulch or smother crop, with a cash crop so that vegetative cover remains on the field for as long as possible during the year (Melander et al. 2005). Cover crops provide several advantages. They typically assist producers meet conservation-tillage requirements for year-round vegetation cover; aid in soil erosion prevention; improve soil structure and, often, organic matter content; protect plants in sandy areas from sand-blow injury; fix nitrogen if the cover crop is a legume; and possibly suppress weed emergence and growth (Melander et al. 2005; Norsworthy et al. 2011; Vencill et al. 2012). Cover crops are frequently used in temperate or subtropical crop production areas where moisture is not limiting production, such as the United States, Canada, Brazil and Queensland or northern Australia.

Suppression of weeds by cover crops depends partly on biomass production of the crop (Vencill et al. 2012). In a field study conducted in the southern United States, rye, crimson clover (*Trifolium incarnatum* L.), hairy vetch (*Vicia villosa Roth.*), barley and a mixture of the four species suppressed the emergence of eastern black nightshade (*Solanum ptycanthum* Dun.). Crimson clover inhibited the emergence of eastern black nightshade beyond what could be attributed to physical suppression alone. The emergence of yellow foxtail [*Setaria glauca* (L.) Beauv.] was inhibited by rye and barley but not by the other cover crops or the cover crop mixture (Creamer et al. 1996). The use of cover crops to suppress weeds is also influenced by type of weed, environmental factors and farming system.

A field study by Gavazzi et al. (2010) examined the allelopathic effects of rye cover crop on grass and broadleaf weeds in maize grown with two tillage systems (no-tillage, conventional tillage) at three nitrogen rates (0, 250, 300 kg N ha^{-1}). Mulching significantly reduced the density of grass and broadleaf weeds by 61% and 96%, respectively. Linear regressions between the concentrations of DIBOA and DIBOA-glycoside in the rye mulch and weed inhibition (%) were statistically significant, with R^2 values of 0.59 and 0.65 for grass and broadleaf weeds, respectively. Another study reported that white mustard (*Sinapis alba* L.) reduced both seedling establishment, by 51–73%, and biomass, by 59–86%, of small-seeded annual broadleaf weeds in a greenhouse (Dion et al. 2014). In addition, under field conditions white mustard was also the most effective cover crop, reducing weed survival 21–57%.

Other factors influencing the effect of cover crops in weed suppression include season, location, type of cover crop, soil cover and density of the resulting mulch. For instance, Dorn et al. (2015) sowed cover crops directly after harvesting cereals and before next year's main crop (grain maize or sunflower). The presence of cover crops caused a 96–100% reduction of weed dry matter at the four sites managed under integrated production, while effects were lower at the four sites managed under organic production, ranging from 19 to 87%. Cover crops that covered soil quickly and produced greater dry matter provided the greatest weed suppression. However, their weed-suppressing effect was difficult to predict, and was typically dependent on the year of the investigation, experimental site, cover crop species, the speed of soil cover in autumn and the density of the resulting mulch layer in spring.

Altieri et al. (2011) and Schulz et al. (2013) assessed the effects of various combinations of rye, vetch (*V. villosa* Roth.), fodder radish (*Raphanus sativus* subsp. *oleiferus*), black oats (*Avena strigosa*) and ryegrass (*Lolium multiflorum*) in reducing winter and summer weed populations in bean crops. Results indicated that the best cover crop mixtures included a significant proportion of rye, vetch and fodder radish (Schulz et al. 2013). The main advantages of these mixtures were the generation of higher crop biomass, improved spectrum of target weeds, broader adaptability to pedo-climatic conditions, complementary effects on soil quality (N fixation for legumes, nematocidal effect and improved soil tilth and soil structure) (Altieri et al. 2011, Schulz et al. 2013). Bradow and Connick (1990) showed that cover crops will release volatile germination inhibitors, so if the crop is disked or killed by rolling, a pulse of volatiles will be released, sufficient to provide effective weed control in the subsequent vegetable crop (Altieri et al. 2011).

6 Case studies: production of benzoxazinoids in cereal crops

In recent experimentation using metabolic profiling in wheat tissues to examine metabolite levels of diverse BXs representing several chemical groups, lactams and hydroxamic acids predominated and were detected in various plant parts (Fig. 2). Fourteen BXs and their derivatives have been reported in tissues, rhizoplane and rhizosphere bulk soil including their respective BX glycosides (Adhikari et al. 2015; Tanwir et al. 2013; Mwendwa et al. 2016b). When produced in great quantities, BXs are typically stored as glycosides which are enzymatically converted to aglycone forms under stress conditions (Tanwir et al. 2013). Their biosynthesis involves nine enzymes thought to form a linear pathway leading to the storage of DIBOA and DIMBOA as the glucoside conjugates (Dutartre et al. 2012). The aglycones and their derivatives are thought to be responsible for the phytotoxic effects of rye residues, but they may also act in concert with other compounds, such as ferulic acid and related phenolics, luteoline glucuronides, β-phenyllactic acid and β-hydroxybutyric acid (Schulz et al. 2013).

The concentrations of BOA and DIBOA varied depending on plant organ, age, genotype and the fertilization regime, as well as on temperature, water supply, photoperiod, UV irradiation and light intensity (Niemeyer 2009; Schulz et al. 2013). Selected Australian wheat genotypes were also at the same time evaluated for their ability to suppress annual

Benzoxazolinones		Lactams			Hydroxamic acids		
R₁		R₁	R₂		R₁	R₂	
H	BOA	H	H	HBOA	H	H	DIBOA
OCH₃	MBOA	H	Glc	HBOA-Glc	H	Glc	DIBOA-Glc
		OCH₃	H	HMBOA	OCH₃	H	DIMBOA
		OCH₃	Glc	HMBOA-Glc	OCH₃	Glc	DIMBOA-Glc
		H	Glc-Hex[a]	HBOA-Glc-Hex	H	Glc-Hex[a]	DIBOA-Glc-Hex

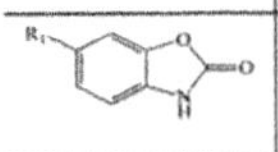

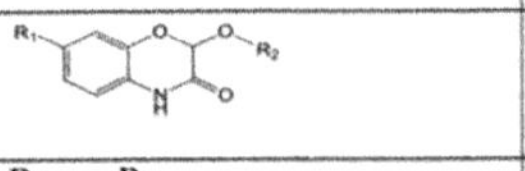

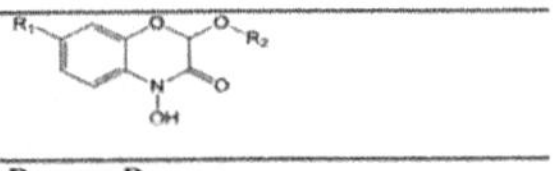

Figure 2 Chemical structures of the BXs most commonly found in cereal grains and bakery products. BOA, benzoxazolin-2-one; MBOA, 6-methoxy-benzoxazolin-2-one; HBOA, 2-hydroxy-1,4-benzoxazin-3-one; HMBOA, 2-hydroxy-7-methoxy-1,4-benzoxazin-3-one; HBOA-Glc, 2-β-D-glucopyranosyloxy-1,4-benzoxazin-3-one; HMBOA-Glc, 2-β-D-glucopyranosyloxy-7-methoxy-1,4-benzoxazin-3-one; HBOA-Glc-Hex, double-hexose derivative of HBOA; DIBOA, 2,4-dihydroxy-1,4-benzoxazin-3-one; DIMBOA, 2,4-dihydroxy-7-methoxy-1,4-benzoxazin-3-one; DIBOA-Glc, 2-β-D-glucopyranosyloxy-4-hydroxy-1,4-benzoxazin-3-one; DIMBOA-Glc, 2-β-D-glucopyranosyloxy-4-hydroxy-7-methoxy-1,4-benzoxazin-3-one; DIBOA-Glc-Hex, double-hexose derivative of DIBOA. [a]Structure not fully elucidated (Adhikari et al. 2015; Tanwir et al. 2013).

weeds in the field. Metabolites were extracted from the plant tissues and soil rhizosphere and profiled in the Liquid chromatography mass spectrometry quadrupole-time of flight (LCMS QToF). There was a clear difference in the distribution and abundance of metabolites in wheat tissues and on the root surface or rhizoplane depending on genotype, growth stage and time of harvest (Mwendwa et al. 2016b; Fig. 3). With increasing age of wheat seedlings in the field, metabolite levels increased in roots but generally remained stable in shoots, up to 75 days following seeding.

For example, wheat cv. Condo exhibited higher relative abundance of DIMBOA-Glc and HMBOA-Glc in its roots in July compared to June, when it was 10 weeks old. In addition, the wheat cultivars and cereal rye (cv. Grazer) showed significantly higher levels of MBOA and BOA in the soil rhizoplane in July compared to June, showing that as the plant matures it is releasing higher concentrations of allelochemicals into the soil and rooting zone. The distribution of BX secondary metabolites in wheat cultivar tissues suggested differential production of certain key bioactive metabolites among cultivars. Interestingly, Condo, the most weed-suppressive cultivar in aboveground field assessments conducted in several years of field experimentation (Fig. 1) also exhibited the greatest abundance of the four major BXs in its root tissues and on the rhizoplane or root surface (Fig. 3). Further metabolic analysis of wheat tissue, rhizoplane and rhizosphere bulk soils is currently underway to evaluate the potential role of these metabolites, as well as microbially altered metabolites such as APO and AMPO, in weed interference by various wheat cultivars.

Generally, BOA and MBOA are more stable in field soil than DIBOA and DIMBOA, and rapid soil degradation is associated with microbial activity (Schulz et al. 2013). For example, when MBOA was placed in a previously sterilized soil, its concentration did not change over a period of 4 days (Macías et al. 2004). MBOA, an intermediate in the degradation pathway from DIMBOA to 2-amino-7-methoxy-3H-phenoxazin-3-one (AMPO), was resistant to biodegradation in the soil (Chen et al. 2010). However, AMPO was the final degradation

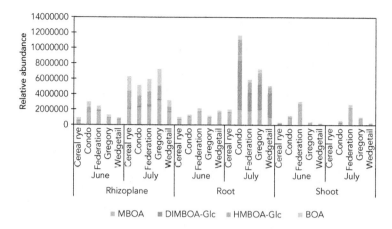

Figure 3 The relative abundance (each bar represents an ion having a specific mass-to-charge ratio (m/z) and the length of the bar indicates the relative abundance of the ion) of four main benzoxazinoid metabolites (BXs) detected in the wheat cultivar tissue and rhizoplane in June, 25 days after crop emergence (DAE) and July (57 DAE) 2016 (Mwendwa et al. 2016b).

product observed for DIMBOA in non-sterile soils (Macías et al. 2004). BX concentration in wheat and rye and release to the soil rhizosphere are therefore significantly affected by plant tissue and rhizosphere location (distance from root) (Mwendwa et al., 2016b), with the release and toxicity of these allelochemicals being influenced by genotype and environmental conditions (Weidenhamer 1996; Weston 1996).

Studies on the structural requirements for phytotoxicity on the benzoxazinone skeleton revealed that the oxygen atom at N-4 is crucial for the phytotoxic effect, since all 4-hydroxy benzoxazinones (DIBOA, DIMBOA, D-DIBOA and D-DIMBOA) are more active than the corresponding lactams (HBOA, HMBOA, D-HBOA, and D-HMBOA) (Macías et al. 2006; Schulz et al. 2013). In a comprehensive screening of the activity of many BXs and related compounds, Macías et al. (2005, 2006) found that APO was the most active compound requiring 0.05–0.1 mM to inhibit root length elongation of several species by 50%. In contrast, DIBOA and DIMBOA were less inhibitory requiring 0.5 mM for similar activity while BOA and MBOA were least phytotoxic, requiring 1.0 mM. HBOA and HMBOA had minimal inhibitory activity. Teasdale et al. (2012) reported that the most toxic BX compounds – APO, DIBOA and DIMBOA – were present at relatively low levels, compared to the less toxic compounds, BOA and MBOA, and the non-toxic compounds, HBOA and HMBOA, were the predominant BX species in amended soils. This suggests that concentrations of these metabolites are dynamic in the soil rhizosphere and are dependent on soil microbial community and local environment.

Together these findings suggest that allelopathic activity is dependent upon the presence of a mixture of bioactive compounds at concentrations that result in actual growth inhibition of specific species and that these metabolites are persistent for sufficient periods of time to result in toxicity. However, the specific activity, rate of degradation and persistence of BXs and associated microbially produced metabolites needs to be further explored individually and in mixtures for them to be used effectively for weed and pest suppression in sustainable agriculture. Clearly their activity in soil varies from that observed in laboratory assays.

7 Case studies: competitive cereal cultivars as a tool in integrated weed management

Several plant traits associated with early wheat vigour (early canopy cover, greater leaf width and tiller number) have been reported to be positively correlated with crop competitive ability (Mwendwa et al. 2016a; Vandeleur and Gill 2004; Zerner et al. 2008). Worthington et al. (2015) demonstrated that weed-suppressive ability was correlated with competitive traits, including vigour and erect growth habit during tillering (Zadoks GS 29), high LAI at stem extension (GS 31), plant height at tillering and stem extension (GS 29, 31), grain yield in weedy conditions, and grain yield tolerance.

Recent studies have shown that early leaf area formation in cereal crops is an important indicator of their weed-suppressive ability (Coleman et al. 2001). Mwendwa et al. (2016a) reported that wheat (cv. Federation and Condo) and cereal rye (cv. Grazer) cultivar crops with the highest LAI were also the most weed suppressive. Cultivar light interception was also highly positively correlated to LAI at 57 days after crop emergence ($r^2 = 0.97$, $P < 0.001$; Fig. 4).

Bertholdsson (2005, 2011) used early crop mass as an indicator of vigour in wheat and barley and found it to be one of two traits (along with allelopathy) that significantly contributed to suppression of *Lolium perenne* L. and volunteer *B. napus* (oilseed rape) across all years of study. Bread wheat and durum wheat (*Triticum durum* Desf.) cultivars that were more competitive against *L. rigidum* also had high vigour, acquiring higher biomass at the seedling stage (Lemerle et al. 1996). Mwendwa et al. (2016a) demonstrated that crop phenology early in the season may be particularly important to weed suppression throughout the season and crop yields (Fig. 5).

As previously stated, some cereal cultivars have been reported to produce high yields in the absence of herbicides under ideal conditions while also suppressing weed populations (Andrew et al. 2015; Mwendwa et al. 2016a; Worthington et al. 2015). Table 1 shows significant differences in weed suppression and yield of six wheat cultivars and a cereal rye

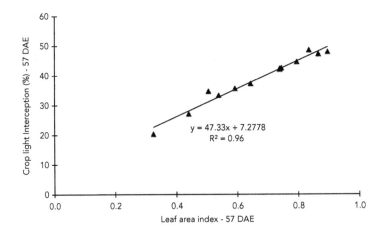

Figure 4 The relationship between wheat cultivar ground surface PAR light interception and leaf area index at 57 DAE in Wagga Wagga in 2015 ($r^2 = 0.97$; $P < 0.001$) from Mwendwa et al. 2016a.

Figure 5 Differences in early crop vigour, canopy closure and weed infestation between control treatment; rye (cv. Grazer; left), wheat (cv. Gregory; middle and cv. Federation, right) 80 days after crop emergence at Wagga Wagga (NSW, Australia) field trials in 2016. There are visible weeds in Gregory but not in Federation.

Table 1 Differences in weed biomass (g/m²) 110 and 130 days after crop emergence (DAE) respectively and crop yield (t/ha) at field trials Wagga Wagga (NSW, Australia) in 2014 and 2015. In some cases, low weed biomass was associated with high yielding ability, as in cv. Condo

Year	2014		2015	
Crop	Biomass (130 DAE) g/m²	Yield t/ha	Biomass (110 DAE) g/m²	Yield t/ha
Rye	1.1	2.2	0.0	2.5
Janz	2.5	2.9	0.3	5.2
Gregory	5.7	3.3	3.2	5.1
Mace	13.5	3.4	1.2	5.6
Livingstone	7.3	3.6	3.1	5.4
Espada	3.3	3.9	0.6	5.4
Condo	5.1	4.0	0.1	5.7
P Valuea	NS	***	*	***
LSD 5%	10.9	0.4	2.5	0.3

Wagga Wagga biomass and yield – 2014 and 2015

ª P-Value: NS= not significant, * P<0.05, ** P<0.01, *** P<0.001

over two growing seasons. Although, rye was the most suppressive in both years, some wheat cultivars such as cv. Condo, Espada and Janz Cl were equally weed suppressive and high yielding.

This suggests that breeding cultivars with early growth vigour and competitive ability may potentially affect weed suppression and reduce weed propagule numbers in the seedbank at harvest. For example, in 2014 Condo produced 21% greater yield than

Gregory and reduced weed biomass by 10% compared to Gregory. In addition, in 2015 Condo produced 10% greater yield than Gregory and reduced weed biomass by 97%.

Plant height has also been identified as one of the traits most commonly associated with competitiveness (Vandeleur and Gill 2004). While lower yielding in weed-free situations, taller cultivars were typically better tolerators of weed pressure and suppressors of weed growth (Lemerle et al.1996). Although, the advantages of plant height in terms of shading weeds are clear, it alone cannot explain variation in competitive ability (Andrew et al. 2015).

In organic fields, increased plant height and early maturity were associated with reduced weed biomass, while strong early season vigour was related to increased yield, increased spikes m^{-2} and reduced weed biomass (Mason et al. 2007). However, timing of emergence also influenced light interception, as the same weed species may be relatively tall or short depending on the emergence time relative to the crop. It is important to consider these weed and cultivar-specific traits when selecting a cultivar for weed suppression (Mwendwa et al. 2016a).

8 Summary and future trends

8.1 Summary: how research can contribute to enhanced and sustainable crop production

In Europe, Australasia and North America, non-chemical methods for weed control such as harvest weed seed destruction and crop interference or competition are being increasingly adopted across the grain-growing areas and are proving successful in controlling weeds that escape pre-sowing herbicide control (Norsworthy et al. 2012). The use of cereal crops with both strong competitive ability and allelopathic effects to control weeds will be a key step towards diversification of weed control and management tools that could both provide sustainable weed control and reduce chances of herbicide resistance development in weeds (Jabran et al. 2015). However, one of the great challenges to the effective use of allelopathy as a tool for weed management is the lack of knowledge about the active allelopathic constituents, their toxicity and their availability in the soil. This lack of knowledge has led to scepticism about the utility of this approach.

As knowledge of the active chemistry of allelopathic crops grows, their allelopathic potential can be further exploited through selection of cultivars with improved weed-suppression capability. Similarly, the allelopathic potential of crops can be further strengthened through conventional breeding and use of modern tools of biotechnology and genomics. Improving the allelopathic potential of crops against weeds, insect pests and disease pathogens through conventional breeding, molecular genetics and biotechnology offers promise for effective pest suppression (Jabran and Farooq 2013). In addition, metabolic profiling of secondary metabolites in soil and plant tissues will provide important physiological information regarding crop competitive traits and biosynthesis and activity of related allelochemicals that may be important in long-term weed suppression in crops (Mwendwa et al. 2016b; Weston et al. 2015).

Breeding of BX-resistant cereal crops with high BX content, as well as a better understanding of the soil persistence of these compounds and their microbial metabolites will be important areas of future research. Knowledge about the selectivity of the BXs in managing diverse populations of weeds, and influence of cultural practices on weed management through crop suppression may provide greater impetus to include allelopathic cereal cultivars in broadacre and/or organic cropping systems for weed control (Schulz et al. 2013). However, climatic and edaphic factors clearly influence soil microflora and allelochemical activity, and therefore allelopathy should be assessed in a range of soil types. To fully understand interference by allelopathic crops one must understand soil microbial ecology and evaluate the roles of soil microorganisms in chemically mediated interactions between plants (Inderjit 2005).

A better understanding of allelopathic effects in field situations (soil, climate and agronomic conditions), and of the dependency on cultural practices (cultivar, mixtures, fertilization level, conventional tillage or no till systems, timing, and system of termination), may also provide the opportunity to profitably include cereal crops as cover crops in both broadacre and organic cropping systems and to use them as a complementary strategy in weed management (Schulz et al. 2013). The recommendation of Dorn et al. (2015) to support weed management in conservation-tillage systems by use of locally adapted cover crops with rapid establishment, good soil coverage and high dry matter production is logical. However, additional post-emergent weed management measures coupled to diverse crop rotations are likely to be needed for reliable weed control on farms.

From an agronomic perspective, selectivity between crop and weeds is another important consideration that should be addressed. Studies regarding the effect of cover crop residues on weed flora generally concern the amount of aboveground biomass and reduction in the number of the weeds, while limited information regarding the variation in tolerance of weed species exists in the literature (Campiglia et al. 2015). In addition, although high residue biomass may facilitate weed suppression, it often interferes with planting and establishment of the crop, or with crop growth by raising the soil C/N ratio, leading to lack of N availability in crops (Schulz et al. 2013). Therefore, the management of cereal cover crops and/or the use of allelochemicals in weed management must be optimized to provide maximal weed suppression, and a limited or no effect on crops.

8.2 Future research trends

The current knowledge of cereal crop allelopathy shows some striking deficits that should be considered in future research. One of the primary deficits is the lack of knowledge about the dynamics of these metabolites in soil, and whether soil concentrations are sufficient to inhibit germination and subsequent growth of target weeds. To better understand and improve the role of allelopathy against weeds, we recommend that future research focus on three main areas: 1) the role of soil microorganisms in chemically mediated interactions between plants; 2) the application of new analytical tools such as high-resolution mass spectrometry coupled to HPLC to enable detection and quantification of low dynamic concentrations of metabolites and the resulting degradation compounds; and 3) experimental designs to elucidate the underlying dynamics of complex mixtures of allelochemicals in soil (Weston et al. 2015).

Research addressing these issues will help resolve the role of BXs in the allelopathic activity of cereal grains such as wheat and rye. Most studies indicate that the allelopathic activities of these crops are associated with the production of higher BX concentration, yet the amounts found in soil seem inadequate in most cases to account for the observed toxicity. The effect of BXs on soil microbial ecology is a particularly important question, as is the question of whether soil microbes in the presence of soil available BXs may generate other as yet unidentified phytotoxins.

Therefore, systematic screening for crop detoxification strategies and their differences in crop cultivars will help to unravel various detoxification pathways in plants and microbes. Cultivar-dependent variation in BX composition in cereal rye root exudates has rarely been investigated, and the pathways of exudation and potential for long-distance transport of BXs are unknown (Schulz et al. 2013). Further studies on the chemical signalling of plants in response to other plant and microbial populations and directed allelochemical delivery to target plants are required to fully elucidate the role of allelochemicals in the rhizosphere.

9 Where to look for further information

Further information on allelopathy in cereal crops including physiological processes, ecological implications, non-chemical weed management and biological weed and pest control; practices and environmental impact can be found in the following books:

Inderjit, K. M. M. Dakshini, Frank A. Einhellig (eds). *Allelopathy: Organisms, Processes, and Applications*. Volume 582, Publication Date (Print): 09 December 1994 Copyright © 1995 American Chemical Society.
Reigosa, M. J., Pedrol, N. and González, L. (eds) (2006). *Allelopathy: A Physiological Process with Ecological Implications*. Springer Science & Business Media.
Upadhyaya, M. K. and Blackshaw, R. E. (eds). (2007). *Non-chemical Weed Management: Principles, Concepts and Technology*. CAB International.
Rice, E. L. (2012). *Allelopathy*. Academic Press.
Rice, E. L. (2013). *Allelopathy*. Academic Press.
Rizvi, S. J. (ed.). (2012). *Allelopathy: Basic and Applied Aspects*. Springer Science & Business Media.
Travlos, I. S., Bilalis, D. and Chachalis, D. (eds). (2016). *Weed and Pest Control: Molecular Biology, Practices and Environmental Impact*. Nova Science Publishers, Inc.

10 References

Adhikari, K. B., Tanwir, F., Gregersen, P. L., Steffensen, S. K., Jensen, B. M., Poulsen, L. K. and Fomsgaard, I. S. (2015). Benzoxazinoids: Cereal phytochemicals with putative therapeutic and health-protecting properties. *Molecular Nutrition and Food Research*, 59 (7), 1324–38.
Altieri, M. A., Lana, M. A., Bittencourt, H. V., Kieling, A. S., Comin, J. J. and Lovato, P. E. (2011). Enhancing crop productivity via weed suppression in organic no-till cropping systems in Santa Catarina, Brazil. *Journal of Sustainable Agriculture*, 35(8), 855–69.
Andrew, I. K. S., Storkey, J. and Sparkes, D. L. (2015). A review of the potential for competitive cereal cultivars as a tool in integrated weed management. *Weed Research*, 55(3), 239–48.
Argandoña, V. H., Niemeyer, H. M. and Corcuera, L. J. (1981). Effect of content and distribution of hydroxamic acids in wheat on infestation by the aphid Schizaphis graminum. *Phytochemistry*, 20(4), 673–6.
Bajwa, A. A., Mahajan, G. and Chauhan, B. S. (2015). Nonconventional weed management strategies for modern agriculture. *Weed Science*, 63(4), 723–47.
Barnes, J. P. and Putnam, A. R. (1986). Evidence for allelopathy by residues and aqueous extracts of rye (Secale cereale). *Weed Science*, 384–90.
Barnes, J. P. and Putnam, A. R. (1987). Role of benzoxazinones in allelopathy by rye (Secale cereale L.). *Journal of Chemical Ecology*, 13(4), 889–906.

Barto, E. K., Hilker, M., Müller, F., Mohney, B. K., Weidenhamer, J. D. and Rillig, M. C. (2011). The fungal fast lane: Common mycorrhizal networks extend bioactive zones of allelochemicals in soils. *PLoS One*, 6(11), e27195.

Barto, E. K., Weidenhamer, J. D., Cipollini, D. and Rillig, M. C. (2012). Fungal superhighways: Do common mycorrhizal networks enhance below ground communication? *Trends in Plant Science*, 17(11), 633–637.

Beckie, H. and Gill, G. (2006). Strategies for managing herbicide-resistant weeds. In *Handbook of Sustainable Weed Management* (Eds: Singh, H. P., Batish, D. R. and Kohli, R. K.), pp. 581–626. Food Products Press: Binghamton, NY.

Belz, R. G. (2007). Allelopathy in crop/weed interactions – an update. *Pest Management Science*, 63(4), 308–326.

Belz, R. G. and Hurle, K. (2005). Differential exudation of two benzoxazinoids one of the determining factors for seedling allelopathy of *Triticeae* species. *Journal of Agricultural and Food Chemistry*, 53(2), 250–261.

Bertholdsson, N. O. (2005). Early vigour and allelopathy – two useful traits for enhanced barley and wheat competitiveness against weeds. *Weed Research*, 45(2), 94–102.

Bertholdsson, N. O. (2011). Use of multivariate statistics to separate allelopathic and competitive factors influencing weed suppression ability in winter wheat. *Weed Research*, 51(3), 273–283.

Bertholdsson, N. O., Andersson, S. C. and Merker, A. (2012). Allelopathic potential of *Triticum spp.*, *Secale spp.* and *Triticosecale spp.* and use of chromosome substitutions and translocations to improve weed suppression ability in winter wheat. *Plant Breeding*, 131(1), 75–80.

Bertholdsson, N. O. (2010). Breeding spring wheat for improved allelopathic potential. *Weed Research*, 50, 49–57.

Bertin, C., Yang, X. and Weston, L. A. (2003). The role of root exudates and allelochemicals in the rhizosphere. *Plant and soil*, 256(1), 67–83.

Blum, U. (1995). The value of model plant – microbe – soil systems for understanding processes associated with allelopathic interaction. In *Allelopathy*, pp. 127–31. *ACS Symposium Series*, Vol. 582, American Chemical Society.

Blum, U. (2006). Allelopathy: A soil system perspective. In *Allelopathy*, pp. 299–340. Dordrecht, Springer Netherlands.

Bouhaouel, I., Gfeller, A., Fauconnier, M. L., Rezgui, S., Amara, H. S. and Du Jardin, P. (2015). Allelopathic and autotoxicity effects of barley (*Hordeum vulgare* L. ssp. vulgare) root exudates. *BioControl*, 60(3), 425–36.

Bradow, J. M. and Connick, W. J. (1990). Volatile seed germination inhibitors from plant residues. *Journal of Chemical Ecology*, 16(3), pp. 645–66.

Burgos, N. R. and Talbert, R. E. (2000). Differential activity of allelochemicals from *Secale cereal* L. in seedling bioassays. *Weed Science*, 48(3), 302–10.

Campiglia, E., Radicetti, E. and Mancinelli, R. (2015). Cover crops and mulches influence weed management and weed flora composition in strip-tilled tomato (*Solanum lycopersicum*). *Weed Research*, 55(4), 416–25.

Chauhan, B. S., Singh, R. G. and Mahajan, G. (2012). Ecology and management of weeds under conservation agriculture: A review. *Crop Protection*, 38, 57–65.

Chen, K. J., Zheng, Y. Q., Kong, C. H., Zhang, S. Z., Li, J. and Liu, X. G. (2010). 2, 4-Dihydroxy-7-methoxy-1, 4-benzoxazin-3-one (DIMBOA) and 6-methoxy-benzoxazolin-2-one (MBOA) levels in the wheat rhizosphere and their effect on the soil microbial community structure. *Journal of Agricultural and Food Chemistry*, 58(24), 12710–16.

Cheng, F. and Cheng, Z. (2015). Research Progress on the use of plant allelopathy in agriculture and the physiological and ecological mechanisms of allelopathy. *Frontiers in Plant Science*, 6, 1020.

Coleman, R. K., Gill, G. S. and Rebetzke, G. J. (2001). Identification of quantitative trait loci for traits conferring weed competitiveness in wheat (*Triticum aestivum* L.). *Crop and Pasture Science*, 52(12), 1235–46.

Creamer, N. G., Bennett, M. A., Stinner, B. R., Cardina, J. and Regnier, E. E. (1996). Mechanisms of weed suppression in cover crop-based production systems. *HortScience*, 31(3), 410–13.

Dayan, F. E. (2006). Factors modulating the levels of the allelochemical sorgoleone in *Sorghum bicolor*. *Planta*, 224(2), 339–46.

Didon, U. M., Kolseth, A. K., Widmark, D. and Persson, P. (2014). Cover crop residues-effects on germination and early growth of annual weeds. *Weed Science*, 62(2), 294–302.

Dorn, B., Jossi, W. and Heijden, M. G. A. (2015). Weed suppression by cover crops: Comparative on-farm experiments under integrated and organic conservation tillage. *Weed Research*, 55(6), 586–97.

Duke, S. O. and Dayan, F. E. (2011). Modes of action of microbially-produced phytotoxins. *Toxins*, 3(8), 1038–64.

Duke, S. O., Baerson, S. R., Pan, Z., Kagan, I. A., Sanchez-Moreiras, A., Reigosa, M. J. and Schulz, M. (2005, August). Genomic approaches to understanding allelochemical modes of action and defenses against allelochemicals. In *Proceedings of the Fourth World Congress on Allelopathy, International Allelopathy Society, Wagga Wagga, Australia* (pp. 107–13).

Dutartre, L., Hilliou, F. and Feyereisen, R. (2012). Phylogenomics of the benzoxazinoid biosynthetic pathway of poaceae: Gene duplications and origin of the Bx cluster. *BMC Evolutionary Biology*, 12(1), 1.

Eichenberg, D., Ristok, C., Kröber, W. and Bruelheide, H. (2014). Plant polyphenols – implications of different sampling, storage and sample processing in biodiversity-ecosystem functioning experiments. *Chemistry and Ecology*, 30(7), 676–92.

Einhellig, F. A. (1996). Interactions involving allelopathy in cropping systems. *Agronomy Journal*, 88(6), 886–93.

Farooq, M., Bajwa, A. A., Cheema, S. A. and Cheema, Z. A. (2013). Application of allelopathy in crop production. *International Journal of Agricultural Biology*, 15, 1367–78.

Fernandez, C., Monnier, Y., Santonja, M., Gallet, C., Weston, L. A., Prévosto, B., Saunier, A., Baldy, V. and Bousquet-Mélou, A. (2016). The impact of competition and allelopathy on the trade-off between plant defense and growth in two contrasting tree species. *Frontiers in plant science*, 7, 1–14.

Fernandez, C., Santonja, M., Gros, R., Monnier, Y., Chomel, M., Baldy, V. and Bousquet-Mélou, A. (2013). Allelochemicals of Pinus halepensis as drivers of biodiversity in Mediterranean open mosaic habitats during the colonization stage of secondary succession. *Journal of Chemical Ecology*, 39(2), 298–311.

Ferreira, M. I. and Reinhardt, C. F. (2010). Field assessment of crop residues for allelopathic effects on both crops and weeds. *Agronomy Journal*, 102(6), 1593–600.

Franke, A. C., Singh, S., McRoberts, N., Nehra, A. S., Godara, S., Malik, R. K. and Marshall, G. (2007). Phalaris minor seedbank studies: Longevity, seedling emergence and seed production as affected by tillage regime. *Weed Research*, 47(1), 73–83.

Gavazzi, C., Schulz, M., Marocco, A. and Tabaglio, V. (2010). Sustainable weed control by allelochemicals from rye cover crops: From the greenhouse to field evidence. *Allelopathy Journal*, 25(1).

Glenn, A. E., Hinton, D. M., Yates, I. E. and Bacon, C. W. (2001). Detoxification of corn antimicrobial compounds as the basis for isolating Fusarium verticillioides and some other Fusarium species from corn. *Applied and Environmental Microbiology*, 67(7), 2973–81.

Harker, K. N., O'Donovan, J. T., Blackshaw, R. E., Johnson, E. N., Holm, F. A. and Clayton, G. W. (2011). Environmental effects on the relative competitive ability of canola and small-grain cereals in a direct-seeded system. *Weed Science*, 59(3), 404–15.

Heap, I. (2017). www.weedscience.com.

Inderjit, N. F. (2001). Soil: environmental effects on allelochemical activity. *Agronomy Journal*, 93(1), 79–84.

Inderjit. (2005). Soil microorganisms: an important determinant of allelopathic activity. *Plant and Soil*, 227–36.

Jabran, K. and Farooq, M. (2013). Implications of potential allelopathic crops in agricultural systems. In *Allelopathy*, pp. 349–85. Springer, Berlin Heidelberg.

Jabran, K., Mahajan, G., Sardana, V. and Chauhan, B. S. (2015). Allelopathy for weed control in agricultural systems. *Crop Protection*, 72, 57–65.

Kato-Noguchi, H. and Ino, T. (2005). Possible involvement of momilactone B in rice allelopathy. *Journal of Plant Physiology*, 162(6), 718–21.

Kato-Noguchi, H. and Peters, R. J. (2013). The role of momilactones in rice allelopathy. *Journal of Chemical Ecology*, 39(2), 175–85.

Kato-Noguchi, H., Ota, K. and Ino, T. (2008). Release of momilactone A and B from rice plants into the rhizosphere and its bioactivities. *Allelopathy Journal*, 22(2), 321–8.

Kirkegaard, J. A. (1995). A review of trends in wheat yield responses to conservation cropping in Australia. *Animal Production Science*, 35(7), 835–48.

Kirkegaard, J. A., Conyers, M. K., Hunt, J. R., Kirkby, C. A., Watt, M. and Rebetzke, G. J. (2014). Sense and nonsense in conservation agriculture: Principles, pragmatism and productivity in Australian mixed farming systems. *Agriculture, Ecosystems & Environment*, 187, 133–45.

Kong, C. H., Wang, P., Gu, Y., Xu, X. H. and Wang, M. L. (2008). Fate and impact on microorganisms of rice allelochemicals in paddy soil. *Journal of Agricultural and Food Chemistry*, 56(13), 5043–9.

Krogh, S. S., Mensz, S. J., Nielsen, S. T., Mortensen, A. G., Christophersen, C. and Fomsgaard, I. S. (2006). Fate of benzoxazinone allelochemicals in soil after incorporation of wheat and rye sprouts. *Journal of Agricultural and Food Chemistry*, 54(4), 1064–74.

Kruidhof, H. M., Bastiaans, L. and Kropff, M. J. (2009). Cover crop residue management for optimizing weed control. *Plant and soil*, 318(1–2), 169–84.

Lemerle, D., Verbeek, B., Cousens, R. D. and Coombes, N. E. (1996). The potential for selecting wheat varieties strongly competitive against weeds. *Weed Research*, 36(6), 505–13.

Liebman, M. and Mohler, C. L. (2001). Weeds and the soil environment. In *Ecological Management of Agricultural Weeds* (Eds: Liebman, M., Mohler, C. L. and Staver, C. P.). Cambridge University Press: Cambridge, UK, 210–268.

Liebman, M. and Davis, A. S. (2000). Integration of soil, crop and weed management in low-external-input farming systems. *Weed Research-Oxford*, 40(1), 27–48.

Macías, F. A., Marín, D., Oliveros-Bastidas, A., Castellano, D., Simonet, A. M. and Molinillo, J. M. (2005). Structure-activity relationships (SAR) studies of benzoxazinones, their degradation products and analogues. Phytotoxicity on standard target species (STS). *Journal of Agricultural and Food Chemistry*, 53(3), 538–48.

Macías, F. A., Marín, D., Oliveros-Bastidas, A., Castellano, D., Simonet, A. M. and Molinillo, J. M. (2006). Structure-activity relationship (SAR) studies of benzoxazinones, their degradation products, and analogues. Phytotoxicity on problematic weeds *Avena fatua* L. and *Lolium rigidum* Gaud. *Journal of Agricultural and Food Chemistry*, 54(4), 1040–8.

Macías, F. A., Oliveros-Bastidas, A., Marín, D., Castellano, D., Simonet, A. M. and Molinillo, J. M. (2004). Degradation studies on benzoxazinoids. Soil degradation dynamics of 2, 4-dihydroxy-7-methoxy-(2 H)-1, 4-benzoxazin-3 (4 H)-one (DIMBOA) and its degradation products, phytotoxic allelochemicals from Gramineae. *Journal of Agricultural and Food Chemistry*, 52(21), 6402–13.

Manalil, S. (2014). Evolution of Herbicide Resistance in under Low Herbicide Rates: An Australian Experience. *Crop Science*, 54(2), 461–74.

Mason, H. E., Navabi, A., Frick, B. L., O'Donovan, J. T. and Spaner, D. M. (2007). The weed-competitive ability of Canada western red spring wheat cultivars grown under organic management. *Crop Science*, 47(3), 1167–76.

McCalla, T. M. and Norstard, F. A. (1974). Toxicity problems in mulch tillage. *Agriculture and Environment*, 1(2), 153–74.

Melander, B., Rasmussen, I. A. and Bàrberi, P. (2005). Integrating physical and cultural methods of weed control – examples from European research. *Weed Science*, 53(3), 369–81.

Muzell Trezzi, M., Vidal, R. A., Balbinot Junior, A. A., von Hertwig Bittencourt, H. and da Silva Souza Filho, A. P. (2016). Allelopathy: Driving mechanisms governing its activity in agriculture. *Journal of Plant Interactions*, 11(1), 53–60.

Mwendwa, J. M., Brown, W. B., Haque, S., Heath, G., Wu, H., Quinn, J. C. .. and Weston, L. A. (2016a). Field evaluation of Australian wheat genotypes for competitive traits and weed suppression. In *20th Australasian Weeds Conference, Perth, Western Australia, 11–15 September 2016* (pp. 48–53). Weeds Society of Western Australia.

Mwendwa, J. M., Weston, P. A., Fomsgaard, I., Laursen, B. B., Brown, W. B., Wu, H. and Weston, L. A. (2016b). Metabolic profiling for benzoxazinoids in weed-suppressive and early vigour wheat genotypes. In 20th Australasian Weeds Conference, Perth, Western Australia, 11–15 September 2016 (pp. 353–7). Weeds Society of Western Australia.

Niemeyer, H. M. (2009). Hydroxamic acids derived from 2-hydroxy-2 H-1, 4-benzoxazin-3 (4 H)-one: key defense chemicals of cereals. Journal of Agricultural and Food Chemistry, 57(5), 1677–96.

Norsworthy, J. K., McClelland, M., Griffith, G., Bangarwa, S. K. and Still, J. (2011). Evaluation of cereal and Brassicaceae cover crops in conservation-tillage, enhanced, glyphosate-resistant cotton. Weed Technology, 25(1), 6–13.

Norsworthy, J. K., Ward, S. M., Shaw, D. R., Llewellyn, R. S., Nichols, R. L., Webster, T. M., Bradley, K. W., Frisvold, G., Powles, S. B., Burgos, N. R., Witt, W. W. and Barrett, M. (2012). Reducing the risks of herbicide resistance: Best management practices and recommendations. Weed Science Journal, 60, 31–62.

Owen M., Preston C. and Walker S. (2013). Resistance rising across Australia. GRDC groundcover Supplement on Herbicide resistance, May – June 2013, pp 5–6.

Reigosa, M. J. and Pazos-Malvido, E. (2007). Phytotoxic effects of 21 plant secondary metabolites on Arabidopsis thaliana germination and root growth. Journal of Chemical Ecology, 33(7), 1456–66.

Reigosa, M. J., Sánchez-Moreiras, A. and González, L. (1999). Ecophysiological approach in allelopathy. Critical Reviews in Plant Sciences, 18(5), 577–608.

Rice, C. P., Cai, G. and Teasdale, J. R. (2012). Concentrations and allelopathic effects of benzoxazinoid compounds in soil treated with rye (Secale cereal L.) cover crop. Journal of Agricultural and Food Chemistry, 60(18), 4471–9.

Schulz, M., Marocco, A., Tabaglio, V., Macias, F. A. and Molinillo, J. M. (2013). Benzoxazinoids in rye allelopathy-from discovery to application in sustainable weed control and organic farming. Journal of Chemical Ecology, 39(2), 154–74.

Scott, B. J., Eberbach, P. L., Evans, J. and Wade, L. J. (2010). EH Graham Centre Monograph No. 1: Stubble Retention in Cropping Systems in Southern Australia: Benefits and Challenges. (Eds: Clayton, E. H. and Burns, H. M.). Industry & Investment NSW, Orange. Available at: http://www. csu.edu.au/research/grahamcentre/publications

Seavers, G. P. and Wright, K. J. (1999). Crop canopy development and structure influence weed suppression. Weed Research, 39(4), 319–28.

Shi, S., Richardson, A. E., O'Callaghan, M., DeAngelis, K. M., Jones, E. E., Stewart, A. and Condron, L. M. (2011). Effects of selected root exudate components on soil bacterial communities. FEMS Microbiology Ecology, 77(3), 600–10.

Smith, R. G., Mortensen, D. A. and Ryan, M. R. (2010). A new hypothesis for the functional role of diversity in mediating resource pools and weed–crop competition in agroecosystems. Weed Research, 50(1), 37–48.

Tabaglio, V., Gavazzi, C., Schulz, M. and Marocco, A. (2008). Alternative weed control using the allelopathic effect of natural benzoxazinoids from rye mulch. Agronomy for Sustainable Development, 28(3), 397–401.

Tabaglio, V., Marocco, A. and Schulz, M. (2013). Allelopathic cover crop of rye for integrated weed control in sustainable agroecosystems. Italian Journal of Agronomy, 8(1), 5.

Tanwir, F., Fredholm, M., Gregersen, P. L. and Fomsgaard, I. S. (2013). Comparison of the levels of bioactive benzoxazinoids in different wheat and rye fractions and the transformation of these compounds in homemade foods. Food Chemistry 141(1), 444–50.

Teasdale, J. R., Rice, C. P., Cai, G. and Mangum, R. W. (2012). Expression of allelopathy in the soil environment: Soil concentration and activity of benzoxazinoid compounds released by rye cover crop residue. Plant Ecology, 213(12), 1893–905.

Travlos, I. S. (2012). Reduced herbicide rates for an effective weed control in competitive wheat cultivars. International Journal of Plant Production, 6(1), 1–14.

Vandeleur, R. K. and Gill, G. S. (2004). The impact of plant breeding on the grain yield and competitive ability of wheat in Australia. Crop and Pasture Science, 55(8), 855–61.

Vencill, W. K., Nichols, R. L., Webster, T. M., Soteres, J. K., Mallory-Smith, C., Burgos, N. R. and McClelland, M. R. (2012). Herbicide resistance: Toward an understanding of resistance development and the impact of herbicide-resistant crops. *Weed Science*, 60(sp1), 2–30.

Wang, P., Zhang, X. and Kong, C. (2013). The response of allelopathic rice growth and microbial feedback to barnyardgrass infestation in a paddy field experiment. *European Journal of Soil Biology*, 56, 26–32.

Weidenhamer, J. D. (1996). Distinguishing Resource Competition and Chemical Interference: Overcoming the Methodological Impasse. *Agronomy Journal*, 88, 866–75.

Weidenhamer, J. D. (2005). Biomimetic measurement of allelochemical dynamics in the rhizosphere. *Journal of chemical ecology*, 31(2), 221–36.

Weidenhamer, J. D., Morton, T. C. and Romeo, J. T. (1987). Solution volume and seed number: Often overlooked factors in allelopathic bioassays. *Journal of Chemical Ecology*, 13(6), 1481–91.

Weston, L. A. (1996). Utilization of allelopathy for weed management in agroecosystems. *Agronomy Journal*, 88(6), 860–6.

Weston, L. A. (2005). History and current trends in the use of allelopathy for weed management. *HortTechnology*, 15(3), 529–34.

Weston, L. A. and Duke, S. O. (2003). Weed and crop allelopathy. *Critical Reviews in Plant Sciences*, 22(3–4), 367–89.

Weston, L. A. and Mathesius, U. (2013). Flavonoids: Their structure, biosynthesis and role in the rhizosphere, including allelopathy. *Journal of Chemical Ecology*, 39(2), 283–97.

Weston, L. A., Alsaadawi, I. S. and Baerson, S. R. (2013). Sorghum allelopathy – from ecosystem to molecule. *Journal of Chemical Ecology*, 39(2), 142–53.

Weston, L. A., Stanton, R., Wu, H., Mwendwa, J., Weston, P. A., Weidenhamer, J. and Brown, W. B. (2014). Comparison of grain crops and their associated residues for weed suppression in the southern Australian mixed farming zone. In *19th Australasian Weeds Conference, September*, pp. 296–9.

Weston, L. A., Skoneczny, D., Weston, P. A. and Weidenhamer, J. D. (2015). Metabolic profiling: An overview – new approaches for the detection and functional analysis of biologically active secondary plant products. *Journal of Allelochemical Interactions*, 1, 15–27.

Worthington, M. and Reberg-Horton, C. (2013). Breeding cereal crops for enhanced weed suppression: Optimizing allelopathy and competitive ability. *Journal of Chemical Ecology*, 39(2), 213–31.

Worthington, M., Reberg-Horton, S. C., Brown-Guedira, G., Jordan, D., Weisz, R. and Murphy, J. P. (2015). Relative contributions of allelopathy and competitive traits to the weed suppressive ability of winter wheat lines against Italian Ryegrass. *Crop Science*, 55(1), 57–64.

Wu, H (2016). Integrating belowground non-chemical approaches for future weed management. In *Weed and Pest Control: Molecular Biology, Practices and Environmental Impact* (Eds: Travlos, I. S., Bilalis, D. and Chachalis, D.). Hauppauge, Nova Science Publishers, Inc.

Wu, H., Haig, T., Pratley, J., Lemerle, D. and An, M. (2000). Distribution and exudation of allelochemicals in wheat (*Triticum aestivum* L.). *Journal of Chemical Ecology* 26(9), 2141–54.

Wu, H., Haig, T., Pratley, J., Lemerle, D. and An, M. (1999). Simultaneous determination of phenolic acids and 2, 4-dihydroxy-7-methoxy-1, 4-benzoxazin-3-one in wheat (*Triticum aestivum* L.) by gas chromatography–tandem mass spectrometry. *Journal of Chromatography A*, 864(2), 315–21.

Wu, H., Haig, T., Pratley, J., Lemerle, D. and An, M. (2001). Allelochemicals in wheat (*Triticum aestivum* L.): Variation of phenolic acids in shoot tissues. *Journal of Chemical Ecology*, 27(1), 125–35.

Wu, H., Pratley, J., Lemerle, D., An, M. and Li Liu, D. (2007). Autotoxicity of wheat (*Triticum aestivum* L.) as determined by laboratory bioassays. *Plant and Soil*, 296(1–2), 85–93.

Zerner, M. C., Gill, G. S. and Vandeleur, R. K. (2008). Effect of height on the competitive ability of wheat with oats. *Agronomy Journal*, 100(6), 1729–34.

Chapter 18

Bioherbicides: an overview

Erin N. Rosskopf, USDA-ARS, United States Horticultural Laboratory, USA; Raghavan Charudattan, BioProdex, Inc., USA; and William Bruckart, USDA-ARS, Foreign Disease-Weed Science Research Unit, USA

1 Introduction

Among agricultural pests, weeds consistently threaten crop yields and land use patterns. Without proper weed management, crop and animal production and aquatic environment and forestry management will be severely affected. The relative importance of weeds as pests is evident from the large proportion of herbicides applied, relative to other pesticides. Between 2008 and 2012, herbicide sales comprised 45.4% of the world pesticide market versus 27.6% and 25.6% for insecticides and fungicides, respectively, for the same period (Atwood and Paisley-Jones 2017). In the United States, over the same time frame, herbicides accounted for 59.2% of pesticide sales, compared with 24% and 15% for insecticides and fungicides, respectively (Atwood and Paisley-Jones 2017).

Although there are many approaches to the management of weeds, the most commonly used tactics are cultural (e.g. cultivation, hand weeding) and chemical (i.e. herbicides). Considerable effort has been devoted to the development and improvement of these strategies, driven by grower and land manager needs for nearly complete control of multiple plant species within a diversity of crops and landscapes. Until recently, development and utilization have been driven by economic factors, without much consideration for public perspective. More recently, the trend has been towards fewer pesticide applications,

http://dx.doi.org/10.19103/AS.2017.0025.20

including growing interest in organic agriculture and, in particular, a growing concern about the general, widespread use of herbicides. Public awareness of herbicides and concern about their use has increased as a result of the use of glyphosate and glyphosate-resistant crops. Concerns are multifaceted, including the use of a single active ingredient resulting in resistant weed populations (Heap 2017), environmental contamination by herbicides (Stone et al. 2014), worker exposure, and fear of potential contamination of food (Carvalho 2017). These concerns and an interest in more local and sustainable food production systems result in a continued increase in consumer purchases of organic products. Based on the annual Organic Industry Survey for 2016, sales in the organic industry in the United States totalled $47 billion, an increase of more than $3 billion from the year before (Organic Trade Association 2016). Although this represents a relatively small percentage, 5.3%, of the total food sales in the United States, there is a consistent market for organic products. According to the Organic Farming Research Foundation, weed control is the number one issue related to yield limitation. While organic growers would prefer to 'use practices rather than inputs' (http://ofrf.org/organic-priorities, accessed 7 August 2017), weed control is often inadequate when cultural and mechanical control measures are employed alone. Although many organic growers would prefer to use combinations of cover crops, tillage, mulches and barrier films, the number of effective tools for weed control is limited when herbicides are not an option. In addition to organic production systems, either weed control in natural areas, pastureland and specialty crops may have few registered herbicides or those that are available are not economically acceptable. For some of these scenarios the use of bioherbicides, which include natural compounds derived from plants or microorganisms and living biological control agents, would be appropriate.

2　Natural products for targeting weed populations

Generally, herbicides can be selected for use based on their level of specificity to target weed populations, giving conventional growers a substantial amount of flexibility in terms of the crops for which they can be labelled and applied. There are few, if any, natural products that provide a similar level of selectivity. Several commercial formulations of corn gluten meal are available for weed control in both conventional and organic settings. The concept of using this natural product as a herbicide was patented in 1994 (Christians 1991). The original claim utilized applications at approximately 200–1700 lb/A. Currently, there are liquid corn gluten products that can be used in lawns and turf which require only about 10 Gal/acre. This type of product is suitable for home owners because the active ingredient is not regulated as a herbicide under the Federal Insecticide, Fungicide, and Rodenticide Act (FIFRA). Their cost is significantly higher than that of herbicides and they work primarily pre-emergence to prevent root development of young seedlings. Both corn gluten meal and mustard seed meal have been evaluated for control of multiple weed species and results have been positive at very high application rates. While they both will control newly germinating weeds, there is little residual effect although the nitrogen content of each provides fertilization for later emerging weeds (Yu and Morishita 2014).

　　Several other materials have also been used as 'burn down' alternatives to broad-spectrum herbicides such as glyphosate. Tworkoski (2002) conducted experiments on 25 different essential oils and found that red thyme (*Thymus vulgaris* L.), summer savory (*Satureja hortensis* L.), cinnamon (*Cinnamomum zeylanicum* Garcin ex Blume), and clove

(*Syzygium aromaticum* [L.] Merr. &L.M. Perry) oils had high herbicidal activity. More recently, research has been conducted on the herbicidal effects of many other essential oils including those derived from rosemary (*Rosmarinus officinalis* L.), oregano (*Origanum vulgare* L.), and several mint species. Foliar application of essential oils results in rapid (hours to days) weed mortality against a broad spectrum of weed species. Several other plant-derived compounds that are also used in some commercial products are mixtures of citric acid and limonene. Acetic acid is another potential contact herbicide. Household vinegar is approximately 5% acetic acid, which is inadequate for weed control (Young 2004), but products specifically formulated for horticultural use may contain up to 8% without being regulated under FIFRA. Higher concentrations, some as high as 30%, are available for weed control, and these must be registered and labelled as herbicides. Although synthetic acetic acid and vinegar are allowed for cleaning and sanitizing purposes under the National Organic Standards, only natural products may be used in organic production. Many of the natural herbicides currently on the market are a mix of one or more of these active ingredients (Dayan et al. 2009). A recent registration was issued for the product MBI-011, composed of 99% sarmentine, originally isolated from the *Piper longum* L. fruit (Huang et al. 2010). This also has broad-spectrum activity and results in rapid weed mortality. Examples of commercially available bioherbicides and their active ingredients, some of which are included in the Organic Materials Review Institute (OMRI) products list, appear in Table 1. Since these are non-selective materials, they must be applied with care to avoid direct contact with crop plants. Most of the commercially available natural products are significantly more costly than conventional herbicides and are primarily offered on sale for lawns and home gardens.

In addition to the plant-derived natural products, microbial metabolic products may be sources of natural herbicides or the starting points for industrial synthesis of similar compounds. One example is bialaphos (Seto et al. 1983; see Ogawa et al. 1973, in Seto et al. 1983). This compound is produced by *Streptomyces viridochromogenes* (Krainsky) Waksman & Henrici and *Streptomyces hygroscopicus* (Jensen) Waksman & Henrici and acts as a proherbicide, being broken down to phosphinothricin by weeds. This is now the commonly used broad-spectrum herbicide, glufosinate, which is a racemic mixture of inactive D-phosphinothricin and L-phosphinothricin. Applications of this herbicide increased significantly with the introduction of the *bar* glufosinate resistance gene into crops, making it possible to apply the herbicide directly to crops without damaging them (Duke 2005). Further increase in glufosinate use, beginning around 2005, resulted from the widespread occurrence of glyphosate-resistant weeds (Sosnoskie and Culpepper 2014). There are now a few cases of glufosinate-ammonium-resistant grasses (Heap 2017). Two other microbially derived products, one of which has EPA registration, are in development by Marrone Bio Innovations. The first, MBI-010, has not yet been approved by the EPA, but is a novel glutamine synthetase inhibitor derived from the inactivated cells and extracts of *Burkholderia rinojensis* Cordova-Kreylos. Opportune® is a pre- and post-emergent bioherbicide composed of spent fermentation medium containing killed cells of *Streptomyces acidiscabies* Lambert and Loria and its phytotoxin thaxtomin A. Thaxtomin A is the predominant and most active member of the thaxtomin group of phytotoxins. It is a cellulose synthesis inhibitor that causes cell swelling and reduces seedling growth. Although Opportune® is currently not available for sale due to an effort by the company to reduce costs associated with its production, it is registered with the EPA and labelled pre-emergence for multiple specialty crops and post-emergent for cereal grains and sod (Opportune® Label, Marrone Bio Innovations 2015).

Table 1[a] Examples[b] of commercial products containing natural active ingredients used for weed management in organic agriculture[c]

Products	Components
WeedBan™ Corn Weed Blocker™	Corn gluten meal
Bioscape Bioweed®	Corn gluten meal (98%), soybean oil (2%)
Scythe®	Pelargonic acid (nonanoic acid) (57%), related short chain fatty acids (3%), 30% paraffinic petroleum oil (30%)
Burnout®	Clove oil (12–18%), sodium lauryl sulfate (8–10%), acetic acid, lecithin, citric acid (30%), mineral oil (80%)
Phydura™	Citric acid (20%), clove oil (20%), malic acid (10%)
AllDown®	Citric acid (23%), acetic acid (14%)
Weed Zap™	Clove oil or cinnamon oil (30%), vinegar (70%)
Suppress®	Caprylic acid(47%), capric acid (32%)
Moss & Algae Killer™ Naturell WK Herbicide™ DeMoss™ Mosskiller™	Potassium salts of fatty acids (40%)
Organic Weed & Grass Killer™	Citrus oil (70%)
Nature's Avenger™	D-Limonene (70%), castor oil (1 to 4%), emulsifiers (18 to 23%)
GreenMatch EX™	Lemongrass oil (50%) and mixture of water, corn oil, glycerol esters, potassium oleate and lecithin
Matran II®	Clove oil (46%), wintergreen oil, butyl lactate, lecithin
Eco-Exempt™ Eco-Smart™	2-phenethyl proprionate (21.4%), clove oil (21.4%)
Axxe®	Ammonium nonanoate (40%)

[a] Modified from Dayan et al. (2009).
[b] List not intended to be exhaustive.
[c] Inclusion on the list does not guarantee OMRI certification, see https://www.omri.org/omri-lists (31 July 2017).

3 Microbial bioherbicides and classical biological control: an overview

Since the 1970s (Charudattan and Dinoor 2000), substantial effort has been made to develop plant pathogens for biological control of weeds. The focus of biological control has been on the use of plant diseases and insects that damage and/or kill specific weeds. This research grew out of concerns about the widespread use of toxic pesticides following World War II, which was brought to public awareness by Rachel Carson (1962). Research on these agents is justified, because several weeds are either not amenable to or effectively controlled by conventional approaches, that is, where conventional control strategies are inadequate or unacceptable. In many cases, these weed problems develop into intractable pest scenarios. Examples include invasive weeds of rangelands and pastures because of both the low monetary return per unit area that makes conventional control relatively expensive and the extensive size or scale of weed infestations, making conventional practices physically challenging and cost-prohibitive. Physical limitations may be a result of terrain or geography that includes mountains and gorges or other inaccessible areas. Ecological considerations may also limit treatment areas, especially if there are sensitive habitats that contain bodies of water or location of weed infestations near endangered, threatened, native, or commercially important plants.

While some of the bioherbicides are a good fit for these niches, there remains considerable need, and therefore opportunity, to use living organisms which can damage or kill a pest as a part of weed management scheme. There are two basic approaches to the deployment of plant pathogens, categorized as 'classical biological control' and 'bioherbicides' or specifically 'mycoherbicides' when fungi are the active ingredient. These, and a general treatment of biological control process and limitations, are described by Cook et al. (1996) and Charudattan and Dinoor (2000). It is important to note that success in the use of plant pathogens for biological control not just depends upon discovery and development of pathogens that are pathogenic only to target weeds, that is, host-specific (Berner and Bruckart 2005), but also the critical support from stakeholders (Charudattan 2005) and the public. Biological control should have appeal to a broader audience, considering ecological and pesticidal issues of the present times; decisions made, 'at the exclusion of ecological and societal benefits, is a serious limitation that could stifle biological control' (Charudattan and Dinoor 2000).

The classical approach to biological control involves the search for and use of coevolved biological agents, commonly herbivorous arthropods or plant pathogens, that are found in the native range of the weed and are then introduced into the area in which a weed has become invasive. Thus, classical biological control methods range from importation and the use of non-native agents from a different geographic region to the control (in the sense of suppression or management, not eradication) of a non-native invasive weed in its new region.

Application of classical biological control agents has resulted in management of several important exotic weeds. Released agents have played an important role, often with spectacular results, in bringing certain exotic invasive weeds to manageable levels and thereby providing vast savings in control costs while also mitigating environmental damage from the invasive species. Typically, the classical approach relies on deploying a suite or host of agents from those found in nature; it is relatively rare, although possible, that a single agent will reduce a weed population to a satisfactory level. Often, it is the interaction among the agents and their combined effects that provides the necessary plant stress to reduce weed density over time. In this regard, besides interactions among biological control agents, the role of plant-associated, naturally present microorganisms, including secondary parasites and saprophytes, which invade and cause decay of a weed that is under attack from biological control agents is a highly important factor influencing weed control (see Charudattan et al. 1978; Caesar 2011; Ray and Hill 2012).

Between the late nineteenth century and 2014, 551 agents targeting 224 weeds (counting some weed groups such as *Opuntia* Mill. spp. as a single target) have been released worldwide (see Introduction in Winston et al. 2014). During this period, countries that have been the principal recipients of these releases and have led the way in the use of classical biological control have been the United States, Australia, Canada, South Africa, and New Zealand.

Though it will take a decade or more in some cases to research and deploy agents, and several more years to realize the results, the return on investment from successful classical biological control programmes can be quite high, with the derived benefits typically lasting over a long time, providing increasing returns on investment with the passing of time (McFadyen 2000). Successful classical biological control programmes have yielded returns on investment ranging from small to as much as several hundred per cent (Culliney 2005; Jarvis et al. 2006; McFadyen 2008; Morin et al. 2009) for instructive discussions on cost-benefit determinations and success evaluations. Finally, as classical biological control programmes are supported by funds from public entities or stakeholder groups, generally there is no pay-to-use levy on end users.

4 Examples of classical biological control

Internationally, one of the most successful examples of classical biological control is of Port Jackson willow (*Acacia saligna* [Labill.]H.L. Wendul), an Australian tree that is invasive in South Africa, using the gall-forming rust, *Uromycladium tepperianum* (Sacc.) McAlpine introduced from Australia (Morris 1997; Woods and Morris 2007). The pathogen was released in 1987 and within eight years, had reduced tree density by 90–95% (Fig. 1). In 2001, *Melanterius compactus* (Coleoptera: Curculionidae) from Australia was released there and the combined effect of the two agents has been positive.

The classical control of *Chondrilla juncea* L. (rush skeletonweed), a native of Eurasia and North Africa that is invasive in Argentina, Australia, South Africa, and the Western United States and Canada, is another excellent example. Several strains of a rust pathogen, *Puccinia chondrillina* Bubák and Syd., from Italy and Turkey were introduced into Australia. The first strain was very host-specific to the narrow-leaf form of rush skeletonweed and resulted in excellent control of this form, leaving other forms of the weed to multiply in abundance. Two additional strains were introduced for control of the intermediate-leaf form with the first failing to establish and the second becoming widespread. This pathogen was later introduced into the United States, Canada and Argentina for control of rush skeletonweed. The pathogen was evaluated at USDA, ARS, at Fort Detrick, MD by R.G. Emge and colleagues (Emge et al. 1981; Fig. 2) and released in California. It was found to have been very effective among the agents released in California (Supkoff et al. 1988) but not in Washington State (Adams and Line 1984a,b). One form of rush skeletonweed was resistant to the isolates of *Puccinia chondrilla* released in the United States, and to date, no promising rust strain has been discovered. Several insects, including a blister-forming midge (*Cystiphora schmidti*, Dipetera: Cecidomyiidae) and a root-attacking moth, *Bradyrrhoa gilveolella* (Lepidoptera, Pyralidae) subsequently improved weed control in parts of the United States where the rust alone was not adequate (Littlefield et al. 2013).

After extensive risk assessments at Fort Detrick, primarily by host range determinations (Politis et al. 1984; Bruckart et al. 1996; Bruckart 2005), another rust pathogen, *Puccinia carduorum* Jacky was released for the control of musk thistle (*Carduus nutans* ssp. *Leiophyllus* [Petrovič] Stoj. & Stef.). It was released for a limited field study in Virginia (Baudoin et al. 1993), where it became established and spread west (Baudoin and

Figure 1 Classical biological control of *Acacia saligna* using *Uromycladium tepperianum*. The rust forms galls (examples top left), that caused plant mortality (right, left, and bottom). This work was conducted by Dr. Mike Morris, formerly of the Plant Protection Research Institute, Stellenbosch, South Africa (Morris 1997; Wood and Morris 2007). Photo credit: R. Charudattan.

Figure 2 Susceptible rush skeletonweed (*Chondrilla juncea*) infected with *Puccinia chondrilina* (left) and untreated control (right). Work of Dr. Robert Emge, Plant Disease Research Laboratory, USDA, ARS, Fort Detrick, MD. Photo credit: W. Bruckart.

Bruckart 1996; Littlefield et al. 1998) all the way to California (Woods et al. 2002). It can now be easily found on musk thistle throughout the United States, and measurements from the field evaluation in Virginia suggest that it can be damaging under favourable environmental conditions and clearly augments damage caused by insects also released for control of musk thistle.

Another rust pathogen assessed at Ft. Detrick was *Puccinia jaceae* Otth for yellow starthistle (*Centaurea solstitialis* L.) (Bruckart 1989, 2006; Shishkoff and Bruckart 1993). It was released into California (Woods et al. 2009, 2010). All accessions of yellow starthistle were susceptible in greenhouse tests, and a protocol for inoculum increase was developed for subsequent releases into the field (Woods et al. 2009). Establishment and limited spread of this rust pathogen were documented, but it was not effective and did not persist in California, despite releases made in many locations (Woods et al. 2010). Releases were also made in Oregon, and recently, diseased yellow starthistle was found at one of the sites where the rust is well established (Bruckart et al. 2016).

Melaleuca quinquenervia (Cav.) S.T. Blake, an exotic tree from Australia, invaded and severely harmed the unique ecosystem of the Florida Everglades. Thanks to a classical biological control programme spearheaded by the USDA-ARS Invasive Plant Research Laboratory, Fort Lauderdale, FL, this seemingly intractable invasive species is being brought under excellent control. Rayamajhi et al. (2007, 2009, 2010) have provided strong confirmation of the success of a guild of natural enemies consisting of two introduced Australian insects (Coleoptera: Curculionidae), *Boreioglycaspis melaleucae* (Hemiptera: Psyllidae) and a naturalized Brazilian rust pathogen, *Puccinia psidii* sensu lato which was not deliberately introduced. As confirmed by the authors' studies, these insects and the pathogen have been shown in controlled, post-release field studies to suppress the invasive species' growth components, change stand dynamics, prevent regrowth of cut-stump and cause population decline while preserving native plant diversity. Clearly, this biological control project is another highly successful example.

Unlike the above systems in which a pathogen may be working in concert with an intentionally released biological insect control agent, control of *Hamakua pamakani* (mistflower, creeping croftonweed, *Ageratina riparia* [Regel] R.M. King and H. Rob.) in Hawaii was accomplished with only a pathogen, the leaf smut fungus *Entyloma ageratinae* R.W. Barreto and H.C. Evans (previously misidentified as *Cercosporella* sp. and *Entyloma compositarum* f.sp. *ageratinae*) (Barreto and Evans 1988). The weed, a native of Mexico, is invasive in several countries including New Zealand and South Africa, as well as the American state of Hawaii. The pathogen was first found in Jamaica from where it was introduced into Hawaii and later into South Africa and New Zealand. In all the three introduced areas, the pathogen has dramatically reduced the weed population (Fig. 3; Trujillo 2005; Barton et al. 2007; Heystek 2011). In Hawaii, the weed population was

Figure 3 Biological control of Hamakua 'Pa-makani', *Ageratina riparia*, with *Entyloma ageratinae*, the white smut fungus introduced from Jamaica in 1974. (a) Non-septate, hyaline, slender, arcuate conidia (30–40 ×3–3.5 µm) of *Entyloma ageratinae*. Scale bar = 12 µm. (b) Abaxial surface of a diseased 'Pa-makani' leaf showing the characteristic white sporodochia, containing masses of spores of *Entyloma ageratinae*. (c) Infestation of *A. riparia* at 900 m elevation at Palani ranch, North Kona, Hawaii, before inoculations in December 1975. (d) Striking biological control of the 'Pa-makani' weed at Palani ranch, a site at 900 m elevation, eight years after inoculation with the biological control fungus. Reprinted from *Biological Control*, Volume 33, Issue 1, Eduardo E. Trujillo, History and success of plant pathogens for biological control of introduced weeds in Hawaii, pp. 113–122, Fig. 2, 2005, with permission from Elsevier.

reduced to 5% of its original coverage within nine months at its first release site. In a period of less than six years, more than 200 000 ha of infested rangelands in Hawaii, Maui and Oahu had been rehabilitated as a result of this pathogen, thereby saving ranchers more than $30 million (Trujillo 2005).

For comprehensive lists of agents, their target weeds, the success or failure of the agents up to 2014 and for some recent reviews, the readers are encouraged to refer to the works of Winston et al. (2014), Morin (2015) and Watson (2017).

5 Limitations and the effects of climate change

5.1 Limitations of classical biological control

Classical biological control has its limitations. First, while it is a proven method to manage invasive weeds in undisturbed sites such as natural areas, forests, rangelands, certain water bodies, wastelands, etc., it is not suitable for controlling weeds in agricultural lands that are subject to frequent disturbance from management practices and cropping cycles.

Second, the process from search and discovery of agents, through testing, release and establishment, to confirmation of success or failure is slow, typically spanning a decade or more. The desired results, namely a level of weed management to the satisfaction of stakeholders, are gradual and almost never immediate. Hence, classical biological control is often considered unsatisfactory or unsuitable where weed infestations should be cleared immediately, such as aquatic weed infestations in waters that are heavily used, weeds that hinder grazing in rangelands and similar cases.

Third, it can be difficult to agree on 'success' of biological control. Whether it was biological control or another method or factor that was responsible for controlling the weed, infestation can become an issue (Delfosse 2004). Among the reasons for this is the difficulty of isolating and confirming the effect of a single method or factor in a complex interaction that causes decline in the weed population. Moreover, different stakeholders may hold different views on 'control'; some seeking eradication from a site may not agree that mere containment or reduction in population density provided by biological control amounts to 'success'. This is because invasive weeds are often managed simultaneously at different sites by various stakeholders (private land owners, public land managers, governmental agencies, etc.), each having different management goals [eradication albeit temporarily, limiting population size and spread (i.e. weed suppression), etc.]. They may use disparate methods of control (chemical herbicides, mechanical controls, biological control). Consequently, there may be questionable claims of success due to the use of one or the other method of control as well as challenges to valid claims of success. So, the issue of which agent, factor (e.g. environmental) or method (biological, chemical, mechanical, or a combination of methods) was responsible for successful control of a weed infestation becomes blurred and contentious.

Fourth, in some cases, the initial acceptance of and enthusiasm for biological control may change due to concerns over unintended consequences of the released agents and a rejection of further implementation of biological control (Louda 2005). For example, *Rhinocyllus conicus* (Coleoptera: Curculionidae) released in North America in late 1960s to control musk thistle, *Carduus nutans* L., was found, a few decades after release, to feed on a few other exotic *Carduus* spp. and a few native *Cirsium* spp. In fact, it was known prior to release that the weevil would feed on other thistles in the Tribe Cynareae as less-preferred hosts, but these species were considered generally undesirable weeds and therefore not a deterrent to the release of the weevil (see Van Driesche et al. 2008). The unexpected non-target attacks by the weevil are now considered unacceptable and all efforts to multiply and redistribute the weevil within and across state lines have been prohibited.

Another recent example is the unforeseen indirect effect of *Diorhabda elongata* (Coleoptera: Chrysomelidae), the saltcedar leaf beetle. It was released in 2001 to control *Tamarix chinensis* Lour. *T. ramosissima* Ledeb., and their hybrids (tamarisk, saltcedar) that are invasive species in arid and semiarid areas in several western American states. The beetle is considered a highly successful agent to suppress saltcedar in North America. Even so, research on an endangered bird, the southwestern willow flycatcher (*Empidonax traillii extimus* Audubon), revealed that the bird was using *Tamarix* spp. as a nesting plant in some western riparian areas. The presence and suspected removal of the bird's nesting plants in some breeding areas led to the halting of further redistribution of the beetle until the issue of the perceived adverse effects of the beetle on the bird was resolved. This was an interesting case, as the saltcedar is not the natural nesting plant for the flycatcher and while nesting was reduced in the first year after the biological control agent was released, the bird population has since rebound significantly.

Compared to insects used in classical biological control, plant pathogens have had an excellent record of safety to non-target plants. Since 1971, 26 species of fungi originating from 15 different countries have been used as classical biological control agents against 26 weed species in seven countries (Barton 2004). Thus far, there have been no reports of introduced fungi unexpectedly attacking non-target plants after their release. Hence, risk assessments based on rigorous host range testing, combined with a good understanding of the taxonomy, biology and ecology of the agent, the target weed and non-target species, can ensure that the introduction of exotic pathogens is a safe and environmentally benign method of weed control (Barton 2004, 2012).

To reap continued benefits from classical biological control to manage invasive weeds in years to come, certain conditions must be met. First, there should be a coordinated global framework to facilitate the collecting, testing and sharing of biological control agents. Researchers should strictly follow sound, scientifically based pre-release risk assessment, safety and efficacy determination methods and cost-benefit analysis prior to seeking the release of agents into new regions. Regulations governing introduction of the exotic agents into new regions should be clearly defined and uniformly applied to facilitate this ecologically sustainable form of weed control. A proactive stance on the part of governmental agencies and the public is needed as classical biological control is for the common good of the society.

5.2 Climate change and classical biological control

Measured parameters associated with a rapidly changing climate are likely to have a significant effect on the discovery and development of candidate biological control agents, particularly related to the classical approach. Increases in atmospheric CO_2, extreme weather events, temperature increases, and changes in precipitation patterns are all likely to change invasive species distribution and growth. While the many potential interactions are highly complex, with entire books dedicated to describing their trends and possibilities (see Ziska and Dukes 2014), there is no doubt that these changes will have both positive and negative effects on biological control efforts. Extreme weather events may spread already established invasive plant species that are currently controlled using insects or pathogens into areas in which the colder temperatures would have minimized their ability to spread, but warmer conditions at critical times could allow for significant movement of invasive plant species into extended areas in which there are no natural enemies or released agents have not yet spread (Hulme 2017). Since the application of biological control agents, both insects and pathogens are timed with specific plant phenological stages most susceptible to attack or infection, significant changes in plant development, such as earlier flowering, could change timing for release (Zelikova et al. 2013). Based on several climate change models, CLIMEX for example, researchers have identified systems in which increases in biological control efficacy may occur, due to prolonged periods of reproductive capability resulting in increased inoculum as well as environmental conditions that are conducive to disease or to population increases of insect agents. Even within a single order of fungi, the effects of specific changes in carbon dioxide and ozone levels can either promote disease development or inhibit pathogen virulence due to a change in the host physiology (Helfer 2014). Since rust fungi (Uredinales or Pucciniales) are often excellent candidates for classical biological control and, as biotrophs, are highly dependent upon host status and distribution, it will be critical to survey the known host–pathogen relationships in both weed–rust and crop–rust systems and document new ones. Recent observations in Florida, for example, show an increase in weed–rust associations in grass weeds (*Puccinia* spp.) and members of the Malpighiales (*Uromyces* spp.) (Rosskopf, unpublished).

Although few studies have been conducted specifically on weed–pathogen interactions, one system in which CO_2 levels were elevated in a controlled environment resulted in significantly increased growth of the target weed, *Parthenium hysterophorus* L., but was coupled with a greater reduction in plant biomass accumulation resulting from infection by the rust pathogen *Puccinia abrupta* var. *partheniicola* (H.S. Jacks) Parmalee than that under ambient CO_2 (Shabbir et al. 2014). In another study, increased winter precipitation resulted in faster growth rates of *Bromus tectorum* L., while also increasing the incidence of infection by *Ustilago bullata* Berk., causing significant reductions in seed production (Prevéy and Seastedt 2015). In both of these studies, only a single variable was manipulated; therefore the effects of other changes in environmental parameters could result in a completely different outcome. Combined climate change effects may result in unexpected outcomes of biological control agent release, such as the potential of pathogens and insects to spread to unintended areas, as well as unanticipated failures due to less than optimum conditions for survival. While these changes may have a more immediate and direct effect on classical bioherbicides, similar changes could be expected in the application of bioherbicides using the inundative approach.

6 Bioherbicides: inundative applications

The bioherbicide approach utilizing living organisms typically uses large numbers of propagules of native pathogens applied to an introduced or native weed on a repetitive basis. Unlike pathogens released in the classical approach, pathogens in inundative applications are not expected to reproduce and maintain a growing population, but are applied seasonally, or multiple times in a season, similar to a chemical herbicide. The concept is that these pathogens occur on a weed target, but are unable to cause epidemics due to an inadequate quantity of inoculum or inability to spread effectively. Often, a potential biological control agent is identified as a result of a chance observation by a researcher versus identification of a target weed followed by a search for a pathogen. More recently, possibly due to more limited availability of funds and thus a need to increase the potential effectiveness of research programme outcomes, more attention is being paid to the selection of target weeds for bioherbicide development. Considerations in selecting a target weed include the distribution of the weed, the value of the potential market, lack of available tools for control, herbicide resistance or the loss of a chemical control product (Charudattan 2005).

6.1 Fungi

Fungi have been the preferred agents to develop as bioherbicides for several reasons: Fungal pathogens are the most common natural enemies of weeds, including some that can inflict destructive and lethal diseases on their host plants. Of the commercially available bioherbicides, only two are not fungi. Many comprehensive lists are available of the fungi that have been studied as potential bioherbicides (Rosskopf et al. 1999; Charudattan 2001; Green 2003; Stubbs and Kennedy 2012; Bailey 2014; Boyetchko, in this book, see Chapter 21). The development of a bioherbicide involves the discovery of a pathogen for the target weed, confirming pathogenicity following Koch's postulates, and the screening of multiple isolates to identify strains that are highly virulent. As demonstrated in the classical biological control case of using multiple strains of *Puccinia chondrillina*, differences between strains can be critical and this applies to those that will be used for inundative control as well. In the past, a great deal of research was conducted on the pathosystem prior to thinking about potential commercialization. This often resulted in pathogens that were effective agents, performing well in greenhouse and field trials, but failing to perform beyond this stage. As limitations associated with specific pathogens were identified, more research was conducted on culturing strategies that influence efficacy of mass-produced inoculum, as well as formulation and application technology (see Jackson 1997; Auld et al. 2003; Weaver et al. 2007).

Numerous bioherbicides have been registered since the first two, *Phytophthora palmivora* Butler (Bulter), in a product named Devine® for control of strangler vine (*Morrenia odorata* [Hook and Arn.] Lindl.) in citrus in Florida, and *Colletotrichum gloeosporioides* f.sp. *aeschynomene* Daniel,Templeton, R.J. Sm. and Fox (Collego®) for control of northern joint-vetch (*Aeschynomene virginica* (L.) Britton, Sterns and Poggenb.) in rice and soybean. However, the majority have not been sustainable as products as a result of limited market potential or issues with efficacy or competition from herbicides. While these types of products were expected to be host-specific, thus safe for non-target plants, this specificity severely limited their applications in the field, as farmers rarely have only a single weed to control. In addition, many assumed that bioherbicides would control weeds quickly and completely as a chemical herbicide would. While this would be the ultimate goal, fewer resources support their marketing, production efficiency and formulation, and application science that go into the development of a new chemical. Bioherbicides that continue to be registered and whose commercial availability has been maintained have some characteristics in common. They may be broad-spectrum, as is the case for Control® Paste, composed of *Chondrostereum purpureum* (Pers.) Pousar formulated for application to woody trees and shrubs on road right-of-ways and in natural areas, and Sarritor®, *Sclerotinia minor* Jagger, for control of dandelion (*Taraxacum officinale* F.H. Wigg) and other broad-leaved weeds in turf. Others are more specific, but intended for application to golf courses (Camperico®), lawns (Organo-Sol®), or pastures (SolviNix®), providing a significantly larger market opportunity than for a single weed, in a single crop.

Bailey and coworkers with Agriculture & Agri-Food Canada (AAFC), Saskatoon, Canada, have developed a fungal pathogen, *Phoma macrostoma* Mont. and conditionally registered it in Canada and the United States for controlling broad-leaved weeds in turf grass (Bailey et al. 2011a,b), including Canada thistle (*Cirsium arvense* L.) and dandelion while leaving crop plants [wheat (*Triticum aestivum* L.) and pumpkin (*Cucurbita* L. spp.)] unharmed. The fungus, isolated from foliar lesions, did not cause extensive damage or kill Canada thistle when sprayed on the foliage but when agar plates colonized by the fungus were inverted on soil containing roots of Canada thistle, the emerging plants were white (bleached), the root growth was inhibited and eventually the plants died (Bailey and Falk 2011). The bioherbicidal activity of *Phoma macrostoma* is largely due to the production of macrocidins, which cause photobleaching in susceptible plants. Recently, Hubbard et al. (2015) have proved that macrocidins inhibit the carotenoid biosynthetic enzyme phytoene desaturase and one or more other steps in the carotenoid biogenesis. This well-studied system also identified potential interactions with turf grass maintenance practices (Bailey et al. 2013). Full registration of *Phoma macrostoma* is expected in a few years.

6.2 Bacteria

While highly virulent and lethal plant pathogenic bacteria exist, only one pathogen, *Xanthomonas campestris* pv. *poae* Egli and Schmidt (= pv. *Poa annua*) was developed in Japan as a bioherbicide named Camperico® to control *Poa annua* L. (annual bluegrass) in golf turf (Imaizumi et al. 1997; Johnson 1994).

Based on observations of the ability of some soilborne bacteria to inhibit crop plant growth, investigation of groups of soilborne bacteria for weed control is an area that continues to result in the identification of specific isolates that selectively inhibit the growth of specific weeds (Cherrington and Elliott 1987). Referred to as deleterious rhizobacteria (DRB), these are non-parasitic bacteria that colonize root surfaces and result in weed suppression rather than eradication. The development of these organisms for biological control is similar to the development of other bioherbicides in that there is a significant screening effort needed for the identification of a host-specific strain that inhibits weed growth without affecting non-target plants. They must also be mass-cultured to enhance populations when applied to soil or seeds. The benefit of DRB may be difficult to quantify, as activity may be primarily through suppression of germinating weeds or in minor reductions in weed growth that allow an emerging crop to outcompete weeds, resulting in increased crop yields. Through extensive screening programmes, multiple strains of *Pseudomonas fluorescens* Migula have been identified as candidates for biological control. Several have been in development by USDA and ARS scientists working in rangelands, turf, row crops and cereal crops. Strain D7 has been extensively researched as a potential control agent for downy brome (*Bromus tectorum* L.) in winter wheat (Kremer and Kennedy 1996). Many of the most promising DRB strains produce toxins that show promise as sources of natural-compound-based herbicides (Adentunji et al. 2017). Work on DRB has encompassed many weed targets, both dicots and monocots (Li and Kremer 2006), and although many efficacious strains continue to be discovered (Kennedy 2016), none has yet to be commercialized as an inundative bioherbicide.

Boyetchko and colleagues at AAFC, Saskatoon, Canada, have discovered a *Pseudomonas fluorescens* strain, BRG100, that is effective in suppressing green foxtail (*Setaria viridis* [L.] P. Beauv.) when applied pre-emergence but not post-emergence (see Chapter 21 by Boyetchko in this book). BRG100 has broad-spectrum activity and can inhibit other annual grass weeds as well. It also has antifungal activity and therefore may be used to control fungal plant diseases in addition to grass weeds. It produces two biologically active compounds, pseudophomins A and B, of which pseudophomin B has higher antifungal activity against plant pathogens than pseudophomin A. The latter shows a stronger inhibition of green foxtail root germination than pseudophomin B. Other unidentified compounds that may work synergistically are also produced. These findings and recent developments in formulation and application technology suggest a possible near-term registration of this bacterium as a broad-spectrum bioherbicide.

6.3 Viruses

Unlike fungi and bacteria, plant viruses have received scant attention as possible biological weed control agents. This can be explained by the prevailing notions that 1) viruses generally have a broad host range and genetic variability that make them unsafe to use as biological weed control agents; 2) they are difficult to contain as most are spread by insect or other mobile vectors that are hard to confine; 3) unlike fungicides and bactericides, there are no effective virucides to control viruses; 4) generally viruses do not kill plants and therefore are unlikely to be effective for weed control; 5) viruses, like bacteria, require wounds or natural openings (e.g. stomata) to enter plants, which makes it difficult to inoculate weeds with viruses in the field; and 6) as plant viruses cannot be produced *in vitro*, it would be too expensive to mass-produce them for commercial use. Thus, despite the well-established fact that viruses can reduce plant growth and reproduction and thereby lessen the competitive ability of weed populations, the common opinion is that viruses cannot be used for biological control of weeds (e.g., Kazinczi et al. 2006).

Despite this prevalent belief, Charudattan et al. (1980) and Hiebert and Charudattan (1984) characterized a newly discovered species of the *Potyviridae*, *Araujia mosaic potyvirus* (common name: Araujia mosaic virus; ArjMV), with the intent to use it as a biological control agent for strangler vine (Morrenia odorata [Hook. & Arn.] Lindl.) in Florida citrus. However, the petition to release ArjMV, originally collected in Argentina, was not allowed by state regulatory authorities because of its exotic origin; it is vectored by aphids and therefore might spread beyond citrus groves, and as potyviruses are a major cause of losses in Florida vegetable production, introducing a new species to the mix would be unwise.

ArjMV was given another look, this time to control moth plant (*Araujia hortorum*) in New Zealand. In a host range study to supplement previously published results (Charudattan et al. 1980), Elliott et al. (2009) established that the virus caused mosaic, chlorotic spots and leaf distortion symptoms on all three *Gomphocarpus* species, including *G. fruticosus* (L.) W.T. Aiton, screened in addition to a few species in other genera in Asclepiadaceae. As *G. fruticosus* was an important host for the monarch butterfly (*Danaus plexippus* L.), an iconic, highly valued species in New Zealand, the virus was withdrawn from further consideration (Elliott et al. 2009).

In 1979, it came to attention in the West that *Solanum carolinense* L. (horsenettle), a North American species that had become invasive in tea plantations of the former Soviet Republic of Georgia, was being controlled with a tobacco mosaic virus (TMV) strain, referred to as the 'Alke strain', isolated from *Physalis alkekengi* L. (Izhevsky 1979). The virus was said to cause systemic mosaic, necrosis, epinasty and leaf abscission, and the infected horsenettle plants suffered severe drought stress (Izhevsky). The virus was said to have been produced in infected plants, the leaves dried and used for inoculation in the field, but it was not pursued as a biological control agent.

In 1986, Randels (1986) proposed TMV strain U2 (TMV U2 = *Tobacco mild green mosaic tobamovirus*, strain U2 [TMGMV U2]) as a biological control agent for controlling the invasive weed *Echium plantagineum* L. (viper's bugloss, Paterson's curse, salvation Jane) in Australia. Natural TMGMV U2 infections in the field and laboratory inoculations caused foliar yellowing and mosaic, increased leaf senescence and reduced leaf and seed production of the weed (Randles 1986). However, the agent was not studied further as no further work appears to have been done to use TMGMV U2 as a control for *Echium plantagineum*.

In Florida and a few southeastern American states, *Solanum viarum* (tropical soda apple [TSA]) is an exotic plant pest of pastures and wooded areas in cattle ranches as well as in some conservation lands. A native to Brazil, Paraguay, Argentina and Uruguay, TSA has been recorded from 19 other countries around the world of which 11 have reported it as being invasive (CABI 2017). In the United States, it is a noxious weed, and is a Class 2 Regionally Prohibited Weed across New South Wales, Australia.

While searching for a plant pathogen that could be used as a biological control agent for TSA, Pettersen et al. (2000) discovered that TMGMV U2 caused a delayed, hypersensitive, systemic necrosis and killed the plant. Similar lethal interaction was not seen between TSA and TMV or TSA and *Tomato mosaic tobamovirus*. TSA is among a few Solanaceae species that respond with lethal host response (Charudattan and Hiebert 2007). Typically, plant response to virus infection is expressed as immunity (no visible plant response); resistance (necrotic local lesions in infected leaves only) or susceptibility (systemic mosaic, foliar mottling, plant stunting and other debilitating but non-lethal symptoms). Relatively rarely, as in the TSA–TMGMV U2 interaction, the resistance response is expressed as lethal, hypersensitive, systemic necrosis.

TMGMV occurs worldwide in tropical and subtropical regions where *Nicotiana glauca* Graham (tree tobacco), a natural host to this virus, is distributed (ICTVdB 2006). Normally, TMGMV causes a mild, green, systemic mosaic symptom in *N. glauca* and other susceptible hosts in Solanaceae. It can cause severe damage to cultivars of tobacco and pepper, although serious crop losses are uncommon. Further work established that 1) the virus kills TSA plants quickly and consistently; 2) as a mechanically transmitted virus, it has no known, confirmed vector capable of spreading it, and therefore, it could be used in targeted applications without risk of secondary spread; 3) despite its worldwide prevalence, it is genetically stable, as evidenced by the low frequency of emergence of new strains in nature (Fraile et al. 1997); 4) since infected TSA plants are completely killed, no infected yet living plants are left in the field to serve as a virus reservoir; 5) in nature, it has a restricted host range unlike the wide host range reported in the literature from artificial manual inoculations done in the laboratory or greenhouse; 6) unlike fungal foliar bioherbicides that require optimum moisture and humidity for infection and subsequent disease development, the virus infectivity and disease development are independent of humidity and, therefore, field application of this virus is generally unencumbered by the weather; and 7) based on published literature on other tobamoviruses and TMGMV, it was clear that TMGMV could be mass-produced.

The above facts prompted BioProdex, Inc. to develop safety and efficacy data and a scalable industrial mass-production process of the virus. The manufactured virus end-product, when stored properly, is stable for many years, which makes the industrial production cost-effective and expedient. In 2014, the EPA granted a FIFRA Section 3 registration for TMGMV U2 as a herbicide active ingredient and a product, *SolviNix*® LC containing this virus as the active ingredient, as a bioherbicide, both a first in the history of herbicides. The registration allows *SolviNix*® LC to be used in the United States in TSA Best

Management Practices, along with chemical herbicides (e.g. aminopyralid [Milestone®]) and a classical biological control agent, *Gratiana boliviana* (Coleoptera: Chrysomelidae).

SolviNix® LC is a post-emergent herbicide labelled for use in pastures and wooded areas. It is applied with a backpack sprayer or an herbicide wiper, both ideal for treating scattered TSA infestations, particularly in sites inaccessible to large spray equipment or mowers. Just one application of the virus to a few physiologically active leaves on a plant is sufficient to infect and kill the entire plant, including the root system. *SolviNix®* LC has performed consistently in numerous field applications, yielding >85% weed kill in about 3–6 weeks after application. So far, no natural resistance to TMGMV U2 has been found among TSA in the United States or among several TSA accessions from New South Wales, Australia. The selectivity of *SolviNix®* LC to TSA and its exemption from a tolerance requirement are reasons it is eminently suitable for use in biodiversity-rich conservation lands, pastures intercropped with clovers, perennial peanut and other forage plants, and where the appearance of chemical herbicide residues in manure and milk is a concern. By applying the bioherbicide before fruit set, further seed buildup in soil can be slowed or even stopped. Thus, *SolviNix®* LC can play a significant role in the ecology and management of TSA.

It is possible that additional virus-based bioherbicides will be brought to use in the future. With the advent of gene manipulation methods involving RNA interference and clustered regularly interspaced short palindromic repeats, naturally occurring plant viruses can be 'trained' or 'edited' to inflict systemic necrosis in undesirable weeds.

7 Integrating bioherbicides into weed management programmes

With many years of investigation devoted to identifying and developing biological control candidates, the number of these incorporated into production systems is extremely limited. In the case of bioherbicides, a lack of consistency in providing a desired level of control, the likely higher cost of bioherbicides compared to chemical herbicides, difficulty in use and a greater educational effort needed on the part of the users are important factors influencing their incorporation. Limitations of classical biological control have been listed earlier. What is needed to make biological control a very appealing, sustainable approach for management of intractable weed problems? Several proposals may be considered. Perhaps it is unrealistic to expect a biological system to act as a chemical when it is used as a single tactic. While the use of single strains of DRB continues to be a promising approach to weed suppression, the application of these bacteria combined with other weed suppression tools may be an avenue that results in greater success. Kremer (2000) suggested combining multiple strains of DRB applied with cover crop seeds in order to enhance the weed-suppressive activity of both, increasing the population and distribution of the DRB. In organic production systems, cover crops are commonly used for weed suppression and soil stabilization, thus is an excellent opportunity to enhance the applicability of biological control agents. Li and Kremer (2000) found that cropping systems that result in increased organic matter also result in an increase in microbial growth. An increase in the isolation of DRB was associated with soils with increased microbial metabolic products. While there are currently no commercially available microbial bioherbicide products labelled for use in organic production systems, there are many opportunities to combine existing organic weed control methods with biological methods for weed control (Yandoc et al. 2014).

Building on this concept, research on the development of soilborne-disease and weed-suppressive soils has resulted in emphasis on the microbial changes associated with suppression (Mazzola 2004). A system that has evolved from the use of organic amendments for this purpose is biological soil disinfestation, also known as reductive soil disinfestation or anaerobic soil disinfestation (ASD). This method was originally used in Japan and the Netherlands for the control of soilborne diseases (Blok et al. 2000; Shinmura 2004), but expanded to address weeds (Muramoto et al. 2008; Rosskopf et al. 2014). The process involves the application of a labile carbon source such as wheat bran, cover crop residue or molasses, covering of soil with oxygen-impermeable agricultural film and the application of enough water to fill soil pore space without flooding (5–10 cm) (Butler et al. 2012). Significant shifts in the microbial community result from the creation of anaerobic conditions (Mowlick et al. 2013; Hewavitharana and Mazzola 2016), increasing the number of facultative anaerobes and the production of organic acids (Rosskopf et al. 2015). Particular emphasis on weed control has been placed on nutsedge, *Cyperus esculentus* L. and *C. rotundus* L., as these species proliferate in raised-bed, plastic mulch cropping systems where ASD is applied. Application of different carbon amendments and quantities reduced yellow nutsedge (*C. esculentus*) germination (McCarty 2012). Although differing carbon amendments did not result in complete control of nutsedge tubers, application of wheat bran as the carbon source resulted in 58% reduction in tuber germination (Shrestha, unpublished). Non-germinating tubers were highly decomposed and the microbial communities associated with the decomposing tubers are currently being characterized. While there are certainly multiple potential causes of the decrease in nutsedge survival, it is a microbially driven process. Although the weed suppression that results from ASD application is general in nature (Di Gioia et al. 2016), there are several opportunities to incorporate other bioherbicides within the ASD system, either as soil-applied DRB or for example, the incorporation of nutsedge-specific rusts during the crop production season or as a fallow treatment. Although *Puccinia canaliculata*, a rust specific to yellow nutsedge, is no longer commercially available due to production and efficacy issues, possibly related to the composition of the nutsedge populations to which it was applied, other isolates of nutsedge-specific pathogens (Kadir et al. 1999b, Rosskopf, unpublished) could be repeatedly applied to fallow fields, where nutsedge is dominant, prior to cultivation or within the cropping cycle when critical weed control is necessary (Kadir et al. 1999a).

An effective integrated programme for the control of *Pueraria montana* var. *lobata* (Willd.) Maes. And S. Almeida (kudzu) illustrates two important points. When removing particularly well-established invasive weeds, either through classical control or the application of a bioherbicide, eradication of the target weed may allow the introduction of an equally problematic replacement. In experiments comparing a standard herbicide programme for the control of kudzu, Weaver et al (2016) developed a herbicide-free approach that utilized a single kudzu mowing event, followed by a broadcast application of the fungal bioherbicide *Myrothecium verrucaria* (Alb. And Schwein.) Ditmar, two subsequent spot applications of the pathogen, and planting of switchgrass (*Panicum virgatum* L.). This true systems approach resulted in 95% kudzu control after two years and was statistically equal to the control achieved using an intensive herbicide programme, thus illustrating the second point that although integrated systems that include a bioherbicide may meet with resistance because the approach is very different from a spray programme, weed control that is equivalent to chemical-dependent systems is possible.

7.1 Bioherbicides in cropping systems

In developing a bioherbicide for any cropping scenario, it is critical for the bioherbicide researcher to have a high level of familiarity with the production practices of the crop. Introducing a biological control agent during the off-season to reduce the weed population prior to crop establishment is an area in which very little work has been done. In addition, knowing the specific practices and products used for crop production will allow the researcher to identify opportunities for combining bioherbicides with other weed control practices, including herbicides, to increase either the spectrum of weeds controlled or the efficacy of the agent, as well as avoid applications in which other crop protection chemicals could have negative effects on bioherbicide performance. Testing the possible agrochemicals that could be used in the target cropping system, Wyss et al. (2004) found that the pathogen *Phomopsis amaranthicola* Ross., Charu., Shab., & Ben., in the development for control of *Amaranthus* spp. in vegetable cropping systems, was tolerant to the fungicide benomyl, and multiple herbicides, but was completely inhibited by chlorothalanil, a commonly used fungicide. Similarly, the nutsedge pathogen, *Dactylaria higginsii* (Luttr.) M.B. Ellis, was found to be sensitive to a majority of tested agrochemicals, but was compatible with the herbicide imazapyr (Yandoc et al. 2006).

While the previous examples illustrate testing to prevent negative effects of agrochemicals on bioherbicide performance, a different approach is to identify herbicides that can be used with bioherbicides to increase efficacy of either of them alone. Application of the rust pathogen *Puccinia canaliculata* (Schw.) Lagerh. alone significantly reduced yellow nutsedge aboveground biomass, but also resulted in an increase in tuber production when compared to non-infected controls. When the pathogen was combined with imaziquin or bentazon, tuber number was significantly reduced (Callaway et al. 1987). The use of pebulate followed by rust inoculation in a tomato production system did not result in adequate control and tomato yield reflected the lack of efficacy (Beste et al. 1992).

The pathogen *Myrothecium verrucaria*, mentioned above, originally isolated from sicklepod, *Senna obtusifolia* (L.) H.S. Irwin and Barneby delivered as conidia in the surfactant Silwet L-77 controlled kudzu extremely well (Boyette et al. 2002) but when applied to redvine [*Brunnichia ovata* (Walter) Shinners] and trumpetcreeper (*Campsis radicans* Seem.), the efficacy was inadequate for control of these weeds. Various application timings and combinations of the pathogen combined with glyphosate, which did not control these weeds alone, were tested in both a controlled environment and under field conditions. The pathogen combined with glyphosate in a tank mix, or applied two days after glyphosate application (Boyette et al. 2006), provided greater than 80% control of both weeds. In the field, the combination application had no effect on the herbicide-resistant soybean crop in which the weeds occurred (Boyette et al. 2008). Similar results were obtained when two bioherbicide candidates for shattercane [*Sorghum bicolor* ssp. *Arundinaceum* (Desv.) de Wet and Harlan], *Colletotrichum graminicola* Ces. Wils. and *Gloeocercospora sorghi* D. Bain and Edg. were tested as pathogen mixes and with glyphosate applied before or after the fungi. The application of glyphosate prior to the application of the fungi resulted in a greater reduction of plant biomass (Mitchell et al. 2008).

7.2 Bioherbicides for herbicide-resistant weeds

Palmer amaranth (*Amaranthus palmeri* S.Wats.) is well distributed throughout the Southeastern United States. In glyphosate-resistant crops, this weed was originally well controlled with glyphosate. There are now herbicide-resistant populations in at least nine states. This weed produces seeds profusely and can easily cross-pollinate with other *Amaranthus* spp., resulting in continued spread. The mycoherbicide *Myrothecium verrucaria*, described earlier was tested for control of glyphosate-resistant and susceptible Palmer amaranth. When the pathogen was applied as a mycelial preparation using Silwet L-77, young plants were killed within 24–30 h after treatment. Mortality in older plants took up to seven days, but there were no differences in the pathogen's ability to kill glyphosate-resistant or susceptible plants (Hoagland et al. 2013, 2016).

Conyza canadensis (L.) Cronquist (horseweed, mare's tail) is another weed that has developed significant glyphosate-resistant populations, now documented in 28 states (Heap 2017). Boyette and Hoagland (2015) examined the potential bioherbicide candidate *Xanthomonas campestris* (Pammel) Dowson, originally isolated from common cocklebur (*Xanthium strumarium* L.), for the control of glyphosate-resistant and susceptible mare's tail. The pathogen caused equivalent damage to resistant and susceptible weeds, but an extended dew period and temperatures from 25 to 30°C resulted in optimal effect. Efficacy was best when applied at 10^6 cells/ml at the rosette stage. These examples demonstrate the potential utility of bioherbicides for the control of herbicide-resistant weeds.

8 Institutional changes for biological control adoption

Biological control, despite its appeal and promise, has been underutilized as a weed control method; Hallett (2005), Ash (2010), Charudattan (2017) and others have pointed out the various reasons in their reviews. Among the reasons, the slowness of the classical biological control process, mentioned earlier, is primary. Equally important is the changing sensibilities of an ecologically minded public towards protecting biodiversity, which in part is the basis for the recent conflicting views on the perceived environmental risks of using classical biological control to manage invasive weeds versus the scientific data proving that several agents and projects have been in fact environmentally beneficial (Van Driesche et al. 2010). For example, Simberloff (2012) has argued that some well-known unintended consequences of 'several biological control projects have led to concern that possible environmental benefits do not warrant inherent risks' of using biological control. Earlier, Van Driesche and 47 coauthors, all experts in classical biological control of insect and invasive plant pests, had presented evidence that biodiversity protection and, to a lesser extent, ecosystem services were provided by biological control projects. The mere suggestion of potential risks of classical biological control, while they may be founded on some specific, justified examples, can trigger public and a broad-brush stroke against ALL biological control projects. This sets the bar quite high for the researchers and project sponsors to reach while gathering safety and risk assessment data and petitioning for an agent for release before regulatory agencies. It can be equally challenging when trying to secure buy-in from the public.

Bioherbicides as commercial products have had limited success (Table 2), in some cases because of a reduction in the efficacy of candidates when transitioning from the greenhouse tests to the field trials and not meeting the expectation that they would perform as quickly and thoroughly as a chemical herbicide. A significant number of effective agents have been dropped from further development due to the expense of commercial development, partially owing to the need to meet regulatory requirements for registration that are designed for chemical pesticides. The development of bioherbicide-specific regulation

Table 2 Registered microbial bioherbicides, active ingredients, weed targets, crops or sites where used, product names and formulations, and current status of registrations and products[a,b]

Microbial agent registered as active ingredient; agent type and disease or damage caused	Registered product name, formulation[c]	Weed target(s) (common name[s]), crops/use sites, if known	Registrant; country(~ies) where registered; year registered	Current status of registration and product
Alternaria destruens E.G. Simmons; strain 059; stem spots and lesions causing fungal pathogen	Smolder, G Smolder, WP	*Cuscuta approximata, C. gronovii, C. indecora* and *C. pentagona* (smallseed, swamp, largeseed, and field dodders), in cranberry, carrot, pepper, tomato, and others	Loveland Products Inc., USA; USA; 2005	Registration lapsed; product unavailable
Chondrostereum purpureum (Pers.) Pouzar, strain HQ1; *Chondrostereum purpureum* (Pers.) Pouzar, strain PFC 2139; cambium and wound infecting fungal pathogen	Myco-Tech, P Chontrol, P	Weedy trees and shrubs in rights-of-way and forests Weedy trees and shrubs in forests, rights-of-way, and riparian lands	Myco-Forestris Corp., Canada; Canada, 2002; USA; 2005 MycoLogic Inc., Canada; Canada, 2007; USA; 2005	Registered, product available Registered, product available
Colletotrichum gloeosporioides (Penz.) Sacc. f. sp. *hakeae* Lubbe, et al.; stem-canker, shoot dieback and gummosis causing fungal pathogen	Hakatak, G	*Hakea sericea* Schrad. & J.C. Wendl. (silky hakea) in conservation areas	Ag. Res. Council-Plant Protect. Res. Inst, South Africa; South Africa; 1990	Produced on request and offered free of charge by ARC-PPRI, South Africa
Colletotrichum gloeosporioides (Penz.) Penz. & Sacc. f. sp. *aeschynomene* Daniel, Tempelton, Sm. & Fox; strain ATCC 20358; leaf spots, stem lesions and cankers, stem dieback, and other symptoms causing fungal pathogen	Collego, WP Lockdown, LC	*Aeschynomene virginica* (L.) B.S.P. (northern joint-vetch, in rice and soybean)	The UpJohn Co., Encore Technol., USA; USA; 1982, 1992 Ag. Res. Initiatives, Inc., USA; USA; 2006	Registration lapsed; reissued for Lockdown Registered; product available
Colletotrichum gloeosporioides (Penz.) Penz. & Sacc. f. sp. *cuscutae* T.Y. Zhang; stem lesions and cankers, stem dieback, and other symptoms causing fungal pathogen	Luboa No. 1, WP Luboa No. S22, WP	*Cuscuta* spp. (dodder species), in soybean *Cuscuta* spp. (dodder species), in soybean	Acad. of Agric. Sci., Inst. Soil and Fertil., China; China; 1966 China;,1987	Registration lapsed; reissued for Luboa No. S22 Registered, product available

(Continued)

Table 2 (Continued)

Microbial agent registered as active ingredient; agent type and disease or damage caused	Registered product name, formulation[c]	Weed target(s) (common name[s]), crops/use sites, if known	Registrant; country(~ies) where registered; year registered	Current status of registration and product
Colletotrichum gloeosporioides (Penz.) Penz. & Sacc. f. sp. *malvae*; leaf spots, stem lesions, cankers and stem dieback causing fungal pathogen	BioMal, WP	*Malva pusilla* Sm. (round-leaved mallow), in field crops	Philom Bios, Canada; Canada; 1992	Registration lapsed; product unavailable
Cylindrobasidium laeve (Pers.) Chamuris; saprophytic colonizer of newly felled wood and wood-rotting fungus	Stumpout, P	*Acacia mearnsii* De Willd and *A. pycnantha* Benth. (black and golden wattles), in natural areas, along water courses, and catchment areas	Ag. Res. Council-Plant Protect. Res. Inst., South Africa; South Africa; 1997	Produced on request and offered free of charge by ARC-PPRI, South Africa
Lactobacillus Beijerinck spp.; citric acid and lactic acid producing milk bacteria that cause cell damage and tissue necrosis	Organo-Sol, LC**	*Lotus corniculatus* L. (birdsfoot trefoil), *Medicago lupulina* L. (black medic), *Oxalis* spp. (woodsorrel), and *Trifolium repens* L. and *T. pratense* L. (white and red clovers), in lawns	Lacto Pro-Tech Inc., Canada; Canada; 2010	Registered, product available
Phoma macrostoma Mont., strain 94–44B; foliar pathogen, causes chlorosis and plant death when soil-applied	Name and formulation to be determined	Several broad-leaved weeds, in turf; registration in crops pending	The Scotts Co., USA; Canada; 2011 USA; 2012[d] Premier Tech Canada, 2017	Conditionally registered; unconditional registration and product pending; project on hold
Phytophthora palmivora (E.J. Butler) E.J. Butler, strain MWV; root-rot causing fungus	DeVine, LC	*Morrenia odorata* (Hook. & Arn.) Lindl. (stranglervine or milkweed vine), in citrus in Florida	Abbott Labs., Valent BioSciencs Corp., USA; USA; 1981	Registration lapsed; product unavailable
Puccinia canaliculata (Schwein.) Legerh.; leaf infecting rust pathogen	Dr. BioSedge, DP	*Cyperus esculentus* L. (yellow nutsedge), in various crops	Tifton Innovation Corp., USA; USA; 1987	Registration lapsed; product unavailable

Microbial agent registered as active ingredient; agent type and disease or damage caused	Registered product name, formulation[c]	Weed target(s) (common name[s], crops/use sites, if known)	Registrant; country(~ies) where registered; year registered	Current status of registration and product
Puccinia thlaspeos Ficinus & C. Schub., strain woad; leaf and stem infecting rust pathogen	Woad Warrior, Finely ground, rust-infected leaf and stem pieces	Isatis tinctoria L. (Dyer's woad), in rangelands, forests and pastures in western USA, particularly in Idaho, Utah and Wyoming	Greenville Farms, USA; USA; 2002	Product availability uncertain
Sclerotinia minor Jagger; strain IMI 344141; fungal pathogen, causes white mould, blight, and soft rot on several dicots	Sarritor, G	Taraxacum officinale F. H. Wigg. agg. (dandelion), in turf	Sarritor, Inc., Canada; Canada; 2007	Product pulled from the market; status unknown
Streptomyces acidiscabies Lambert and Loria, strain RL-110; bacterium (actinomycete), causes potato scab disease, produces thaxtomins (phytotoxins)	Opportune, LC**	Intended for certain annual grasses, broadleaf and sedge weeds by pre- or post-emergent application, in crop and non-crop sites	Marrone Bio Innovations, USA; USA; 2012	Registered, product pending [e]
Tobacco mild green mosaic tobamovirus strain U2; Solanaceae-adapted plant virus that causes a lethal systemic necrosis in a few Solanaceae species	SolviNix, LC	Solanum viarum Dunal (tropical soda apple), a Noxious Weed in pastures and conservation areas in southern USA, particularly Florida	BioProdex, Inc., USA; USA; 2014	Registered, product available [e]
Xanthomonas campestris pv. poae Egli & Schmidt 1982, strain JT-P482; wilt causing bacterial pathogen	Camperico, LC	Poa annua L. (annual bluegrass) in golf courses	Japan Tobacco Inc., Japan; Japan; 1997	Registration status and product availability uncertain

[a] For further details and references, see Winston et al. 2014.

[b] Modified from Charudattan, 2015.

[c] Formulations: DP – dry powder; G – granular; LC – liquid concentrate; LC** – cells and spent fermentation medium containing phytotoxin(s); P- paste; WP – wettable powder.

[d] Original registrant, new commercial partner, https://www.premiertech.com/global/en/press/premier-techpartners-upwithagricultureandagri-foodcanada/ Accessed8 August 2017.

[e] See Marrone Bio Innovations 2015; BioProdex, Inc. 2014.

could reduce the costs associated with commercialization. Regulations are designed for 'no risk' and follow a chemical, as opposed to an ecological, paradigm, a protocol inadequate for biologicals (Cook et al. 1996).

Hunt et al. (2008) have pointed out that among countries that historically have led in the use of biological control, only Australia has legislation specific to biological control, namely 'The Biological Control Act'. New Zealand holds public hearings to allow public participation and interaction in biological control decisions. In the United States, decisions about permitting use of a biological control agent or a biopesticide are made by a single governmental entity (either USDA-APHIS-PPQ or EPA) on the basis of scientific review by an appointed panel or agency scientists, respectively. There is opportunity for public comment, but not for dialogue that involves regulators, the applicant (scientists) and the public. Furthermore, the US agency with statutory authority to *regulate* plant pests (in the case of biological control, exotic insects and pathogens) does not have a mandate to *facilitate* biological control. Involving the public in both target selection and the specific benefits of individual agent releases or bioherbicides more directly than during an online comment period could potentially inform a broader audience of the contributions of bioherbicides. Increased support for the development of bioherbicides in Canada, as well as other types of biological control products, has been the result of significant changes in the Canadian government's legislation regarding pesticide use. The Pest Control Products Act was updated to reflect the public opinion regarding pesticides, and regulations were put in place to eliminate pesticide application in urban settings, coupled with streamlining of some aspects related to registration. The establishment of the Pest Management Centre facilitates the development of biological pest control products, financially supporting development through adoption. Although this system does not eliminate the profit motive, it does provide incentive for new, non-chemical product availability (see Bailey et al. 2010 for description of drivers). While in the United States, discovery, research and deployment of classical biological control agents are generally supported financially by government entities, there is no similar support for bioherbicide distribution beyond the research stage. However, the Inter-Regional Project 4 (IR-4), a federal government-funded entity, does fund efficacy testing of biopesticide agents and products under development and facilitates data submission to the EPA and during data review for registration. Although there is little economic incentive for the commercial development of products to serve small markets addressing pests with highly specialized organisms, there are examples of growers organizing to manufacture and distribute a product beneficial to their industry. For example, in Costa Rica, strains of *Trichoderma* for disease suppression are propagated on individual farms to which they are applied. Similarly, as a result of research by Cotty et al. (2007), aflatoxin 'minus' strains of *Aspergillus flavus* are made available to cotton and corn growers from a fermentation facility run by the grower organization (Wu et al., 2008). Globally, what is needed is a 'holistic weed management strategy' constituting nations' attitude towards weed control that is embraced politically, culturally, socially, morally, economically, ecologically, scientifically, in a manner completely supportive, encouraging and adequately funded, from discovery to distribution.

9 Conclusion

The direction of weed control in the future will be determined by grower and land manager needs, as well as by growing concerns over the health and safety of humans and the environment. Integrated weed management systems that combine herbicides with tillage, crop rotation, variety selection and mulching have increased in application, yet the emphasis on herbicides remains. Technological advances in precision agriculture that are currently employed in herbicide and fertilizer applications, in computerized mapping systems for problem areas and in determining the ability to differentiate crops from weeds are virtually untapped advancements that could enhance utilization of host-specific pathogens for weed control. In an ideal system, biological control agents could be individually selected for a wide range of weed targets and delivered only where they are needed and applied via spray equipment specifically designed for this purpose. While some works have been done regarding handling and delivery systems specific to pathogens, there is ample opportunity to improve in this area.

As new biotechnology continues to emerge, an increased understanding of plant-associated microbiomes should provide new avenues for the discovery of novel herbicide active ingredients as well as biological control agents. The ability to use metabolomics to investigate allelopathy and weed suppression has the potential to open a new realm of natural product and intact organism research. Unfortunately, the greatest limitation to the use of bioherbicides is not the discovery of potential agents, but the lack of a concerted, stakeholder-driven incentive for their development. While there are large, multinational corporations involved in herbicide development and commercialization, there is much less profit motive associated with bioherbicide advancement. This is a tough issue for researchers to address, but one that is critical for the advancement of more diverse integrated weed management systems that combine multiple management tools and take advantage of an increased understanding of weed and microbial ecology.

10 Where to look for further information

While there are a few company's pursuing the development of bioherbicides, numerous government entities and research institutions are engaged in highly productive bioherbicide research. In North America, excellent programs are located at Agriculture Canada and at the United States Department of Agriculture, Agricultural Research Service. Internationally, similar government entities are engaged in bioherbicide research.

USDA, ARS, Biological Control of Pests Research, Stoneville, MS
https://www.ars.usda.gov/research/programs-projects/project/?accnNo=430266

USDA, ARS, Natural Products Utilization Research, Oxford, MS
https://www.ars.usda.gov/research/project/?accnNo=429936

USDA, ARS, Foreign Disease-Weed Science Research, Frederick, MD
https://www.ars.usda.gov/research/project/?accnNo=429926

USDA, ARS, Overseas Biological Control Labs
https://www.ars.usda.gov/office-of-international-research-programs/overseas-biological-control-labs/

USDA, ARS, Pest Management Research, Sidney, MT
https://www.ars.usda.gov/research/project/?accnNo=432544

USDA, ARS Team Leafy Spurge
https://www.team.ars.usda.gov/v2/publications/manuals.html

Agriculture and Agri-Food Canada, Saskatoon Research and Development Center, Saskatoon, Saskatchewan
http://www.agr.gc.ca/eng/science-and-innovation/research-centres/saskatchewan/saskatoon-research-and-development-centre/scientific-staff-and-expertise/boyetchko-susan-phd/?id=1181931622836

Lethbridge Research and Development Centre, Alberta
http://www.agr.gc.ca/eng/science-and-innovation/research-centres/alberta/lethbridge-research-and-development-centre/?id=1180547946064

Australian Government, Commonwealth Scientific and Industrial Research Organisation, Canberra, Australia
https://www.csiro.au/en/Research/Farming-food/Invasive-pests/Biological-control-of-weeds/Overview

The Weed Science Society of America provides an extensive list of internet resource links http://wssa.net/wssa/weed/biological-control/ related to weed biological control. Additional information on international efforts and collaborations can be found at the International Bioherbicide Group, http://ibg.ba.cnr.it/.

In addition to these sites, several general reviews are available for an introduction to the discipline. These include:

Boyetchko, S. M. and Rosskopf, E. N. 2006. Strategies for developing bioherbicides for sustainable weed management. In: Singh, H. P., Batish, D. R., Kohli, R. K. (Eds) *Handbook of Sustainable Weed Management*, The Haworth Press, Binghampton, NY, pp. 393–430.

Boyetchko, S. M., Rosskopf, E. N., Caesar, A. J. and Charudattan, R. 2002. Biological weed control with pathogens: Search for candidates to applications. *Applied Mycology and Biotechnology*, 2:239–74.

Den Breeÿen, A. and Charudattan, R. 2009. Biological Control of Invasive Weeds in Forests and Natural Areas by Using Microbial Agents. In: Inderjit (Ed) Management of Invasive Weeds. *Invading Nature-Springer Series in Invasion Ecology*, vol. 5. Springer, Dordtecht

The use of trade, firm, or corporation names in this publication is for the information and convenience of the reader. Such use does not constitute an official endorsement or approval by the United States Department of Agriculture or the Agricultural Research Service of any product or service to the exclusion of others that may be suitable.

11 References

Adams, E. B. and Line, R. 1984a. Biology of *Puccinia chondrillina* in Washington. *Phytopathology* 74: 742–5.
Adams, E. B. and Line, R. 1984b. Epidemiology and host morphology in the parasitism of rush skeletonweed by *Puccinia chondrillina*. *Phytopathology* 74: 745–8.

Adentunji, C. O., Oloke, J. K., Prasad, G., Bello, O. M., Osemwegie, O. O., Pradeep, M. and Jolly, R. S. 2017. Isolation, identification, characterization, and screening of rhizosphere bacteria for herbicidal activity. *Organic Agriculture* 7: 1–11.

Ash, G. J. 2010. The science, art and business of successful bioherbicides. *Biological Control* 52: 230–40.

Atwood, D. and Paisley-Jones, C. 2017. *Pesticides Industry Sales and Usage: 2008–2012 Market Estimates.* Biological and Economic Analysis Division, Office of Pesticide Programs, Office of Chemical Safety and Pollution Prevention, U. S. Environmental Protection Agency, Washington, DC, 32p.

Auld, B. A., Hetherington, S. D. and Smith, H. E. 2003. Advances in bioherbicide formulation. *Weed Biology and Management* 3(2): 61–7.

Bailey, K. L. 2014. The bioherbicide approach to weed control using plant pathogens, pp. 245–66. In: D. P. Abrol (Ed.). *Integrated Pest Management. Current Concepts and Ecological Perspectives.* Acad. Press-Elsevier, San Diego, CA.

Bailey, K. L. and Falk, S. 2011. Turning research on microbial bioherbicides into commercial products – a *Phoma* story. *Pest Technology* 5: 73–9.

Bailey, K. L., Pitt, W. M., Falk, S. and Derby, J. 2011a. The effects of *Phoma macrostoma* on nontarget plants and target weed species. *Biological Control* 58: 379–86.

Bailey, K. L., Pitt, W. M., Leggett, F., Sheedy, C. and Derby, J. 2011b. Determining the infection process of *Phoma macrostoma* that leads to bioherbicidal activity on broadleaved weeds. *Biological Control* 59: 268–76.

Bailey, K. L., Boyetchko, S. M. and Längle, T. 2010. Social and economic drivers shaping the future of biological control: A Canadian perspective on the factors affecting the development and use of microbial biopesticides. *Biological Control* 52: 221–9.

Bailey, K. L., Falk, S., Derby, J., Melzer, M. and Boland, G. J. 2013. The effect of fertilizers on the efficacy of the bioherbicide, *Phoma macrostoma*, to control dandelions in turfgrass. *Biological Control* 65: 147–51.

Barreto, R. B. and Evans, H. C. 1988. Taxonomy of a fungus introduced into Hawaii for biological control of *Ageratina riparia* (Eupatoriae;Compositae) with observations on related weed pathogens. *Transactions of the British Mycological Society* 91: 81–97.

Barton, J. 2012. Predictability of pathogen host range in classical biological control of weeds: An update. *BioControl* 57: 289–305.

Barton (née Fröhlich), J. 2004. How good are we at predicting the field host-range of fungal pathogens used for classical biological control of weeds? *Biological Control* 31: 99–122.

Barton, J., Fowler, S. V., Gianotti, A. F., Winks, C. J., de Beurs, M., Arnold, G. C. and Forrester, G. 2007. Successful biological control of mist flower (*Ageratina riparia*) in New Zealand: Agent establishment, impact and benefits to the native flora. *Biological Control* 40: 370–85.

Baudoin, A. B. A. M., Abad, R. G., Kok, L. T. and Bruckart, W. L. 1993. Field evaluation of *Puccinia carduorum* for biological control of musk thistle. *Biological Control* 3: 53–60.

Baudoin, A. B. A. M. and Bruckart, W. L. 1996. Population dynamics and spread of *Puccinia carduorum* in the eastern United States. *Plant Disease* 80: 1193–6.

Berner, D. K. and Bruckart, W. L. 2005. A decision tree for evaluation of exotic plant pathogens for classical biological control of introduced invasive weeds. *Biological Control* 34: 222–32.

Beste, C. E., Frank, J. R., Bruckart, W. L., Johnson, D. R. and Potts, W. E. 1992. Yellow Nutsedge (*Cyperus esculentus*) control in tomato with *Puccinia canaliculata* and Pebulate. *Weed Technology* 6: 980–4.

BioProdex, Inc. 2014. *SolviNix® LC Receives EPA Registration.* Online: www.bioprodex.com. Accessed 31 July 2017.

Blok, W. J., Lamers, J. G., Termorshuizen, A. J. and Bollen, G. J. (2000). Control of soilborne plant pathogens by incorporating fresh organic amendments followed by tarping. *Phytopathology* 90: 253–9.

Boyette, C. D., Walker, H. L. and Abbas, H. K., 2002. Biological control of kudzu (Pueraria lobata) with an isolate of Myrothecium verrucaria. *Biocontrol Science and Technology* 12: 75–82.

Boyette, C. D., Reddy, K. N. and Hoagland, R. E. 2006. Glyphosate and bioherbicide interaction for controlling kudzu (*Pueraria lobata*), redvine (*Brunnichia ovata*), and trumpetcreeper (*Campsis radicans*). *Biocontrol Science and Technology* 16: 1067–77.

Boyette, C. D. and Hoagland, R. E. 2015. Bioherbicidal potential of *Xanthomonas campestris* for controlling *Conyza canadensis*. *Biocontrol Science and Technology* 25: 229–37.

Boyette, C. D., Hoagland, R. E. and Weaver, M. A. 2008. Interaction of a bioherbicide and glyphosate for controlling hemp sesbania in glyphosate-resistant soybean. *Weed Biology and Management* 8: 18–24.

Bruckart, W. L. 1989. Host range determination of *Puccinia jaceae* from yellow starthistle. *Plant Disease* 73: 155–60.

Bruckart, W. L. III. 2005. Supplemental risk evaluations and status of *Puccinia carduorum* for biological control of musk thistle. *Biological Control* 32: 348–55.

Bruckart, W. L. III. 2006. Supplemental risk evaluations of *Puccinia jaceae* var. *solstitialis* for biological control of yellow starthistle. *Biological Control* 37: 359–66.

Bruckart, W. L., Michael, J. L., Coombs, E. M. and Pirosko, C. B. 2016. Rust pathogen *Puccinia jaceae* is established on yellow starthistle (*Centaurea solstitialis*) in Oregon. *Plant Disease* 100: 1009.

Bruckart, W. L., Politis, D. J., Defago, G., Rosenthal, S. S. and Supkoff, D. M. 1996. Susceptibility of *Carduus*, *Cirsium*, and *Cynara* species artificially inoculated with *Puccinia carduorum* from musk thistle. *Biological Control* 6: 215–21.

Butler, D. M., Kokalis-Burelle, N., Muramoto, J., Shennan, C., McCollum, T. G. and Rosskopf, E. N., 2012. Impact of anaerobic soil disinfestation combined with soil solarization on plant–parasitic nematodes and introduced inoculum of soilborne plant pathogens in raised-bed vegetable production. *Crop Protection* 39: 33–40.

CABI (Centre for Agriculture and Bioscience International). 2017. *CABI Invasive Species Compendium. Solanum viarum (tropical soda apple)*. Available online: http://www.cabi.org/isc/datasheet/50562. Accessed12 May 2017.

Caesar, A. J. 2011. The importance of intertrophic interactions in biological weed control. *Pest Technology* 5 (Special Issue 1): 28–33.

Callaway, M. B., Phatak, S. C. and Wells, H. D. 1987. Interactions of *Puccinia canaliculata* (Schw.) Lagerh. with herbicides on tuber production growth of *Cyperus esculentus* L. *Trop. Pest Manag.* 33: 22–6.

Carson, R. 1962. *Silent Spring*. Houghton Mifflin, Boston, MA, 378p.

Carvalho, F. P. 2017. Pesticides, environment, and food safety. *Food and Energy Security* 6: 48–60.

Charudattan, R. 2001. Biological control of weeds by means of plant pathogens: Significance for integrated weed management in modern agro-ecology. *BioControl* 46: 229–60.

Charudattan, R. 2005. Ecological, practical, and political inputs into selection of weed targets: What makes a good biological control target? *Biological Control* 35: 183–96.

Charudattan, R. 2015. Weed control with microbial bioherbicides. In: A. N. Rao and N. T. Yaduraju (Eds) *Weed Science for Sustainable Agriculture, Environment and Biodiversity. Vol. I. Proceedings of the Plenary and Lead Papers of the 25th Asian-Pacific Weed Science Society Conference, Hyderabad, India.*

Charudattan, R. 2017. Biologically based methods of weed management: The next 50 years. In: N. Chandrasena and A. N. Rao (Eds.) *Proceedings of the 26th Asian Pacific Weed Science Society Conference, 19–22 September, Kyoto, Japan*, www.apwss.org.

Charudattan, R. and Dinoor, A. 2000. Biological control of weeds using plant pathogens: Accomplishments and limitations. *Crop Protection* 19: 691–5.

Charudattan, R. and Hiebert, E. 2007. A plant virus as a bioherbicide for tropical soda apple, *Solanum viarum*. *Outlooks on Pest Management* 18: 167–71.

Charudattan, R., Perkins, B. D. and Littell, R. C. 1978. Effects of fungi and bacteria on the decline of arthropod-damaged waterhyacinth (*Eichhornia crassipes*) in Florida. *Weed Science* 26: 101–7.

Charudattan, R., Zettler, F. W., Cordo, H. A. and Christie, R. G. 1980. Partial characterization of a potyvirus infecting the milkweed vine, *Morrenia odorata*. *Phytopathology* 70: 909–13.

Cherrington. C. A. and Elliott, L. F. 1987. Incidence of inhibitory pseudomonads in the Pacific Northwest. *Plant Soil* 101: 159–65.

Christians, N. E. (1991) Preemergence weed control using corn gluten meal. US Patent No. 5,030,268

Cook, R. J., Bruckart, W. L., Coulson, J. R., Goettel, M. S., Humber, R. A., Lumsden, R. D., Maddox, J. V., McManus, M. L., Moore, L., Meyer, S. F., Quimby Jr., P. C., Stack, J. P. and Vaughan, J. L. 1996. Safety of microorganisms intended for pest and plant disease control: A framework for scientific evaluation. *Biological Control* 7: 333–51.

Cotty, P. J., Antilla, L. and Wakelyn, P. J. 2007. Competitive exclusion of Aflatoxin producers: Farmer-driven research and development. In: C. Vincent, M. S. Goettel and Lazarovits, G. (Eds) *Biological Control: A Global Perspective.* CAB International, Oxfordshire, UK. pp. 241–253.

Culliney, T. W. 2005. Benefits of classical biological control for managing invasive plants. *Critical Reviews in Plant Science* 24: 131–50.

Dayan, F. E., Cantrell, C. L. and Duke, S. O. 2009. Natural products in crop protection. *Bioorganic and Medicinal Chemistry* 17: 4022–34.

Delfosse, E. S. 2004. Introduction. In: E. M. Coombs, J. K. Clark, G. L. Pier and A. F. Confrancesco, Jr. (Eds) *Biological Control of Invasive Plant in the Unites States,* pp. 1–11. Oregon State University Press, Corvallis, OR, 467p.

Di Gioia, F., Ozores-Hampton, M., Hong, J., Kokalis-Burelle, N., Albano, J., Zhao, X., Black, Z., Gao, Z., Wilson, C., Thomas, J. and Moore, K., 2016. The effects of anaerobic soil disinfestation on weed and nematode control, fruit yield, and quality of Florida fresh-market tomato. *HortScience* 51(6): 703–11.

Duke, S. O. 2005. Taking stock of herbicide-resistant crops ten years after introduction. *Pest Management Science* 61: 211–18.

Elliott, M. S., Massey, B., Cui, X., Hiebert, Charudattan, R. E., Waipara, N. and Hayes, L. 2009. Supplemental host range of Araujia mosaic virus, a potential biological control agent of moth plant in New Zealand. *Australasian Plant Pathology* 38: 603–7.

Emge, R. G., Melching, J. S. and Kingsolver, C. H. 1981. Epidemiology of *Puccinia chondrillina,* a rust pathogen for the biological control of rush skeleton weed in the United States. *Phytopathology* 71:839–43.

Fraile, A., Escriu, F., Aranda, M. A., Malpica, J. M., Gibbs, A. J. and Garcia-Arenal, F. 1997. A century of tobamovirus evolution in an Australian population of *Nicotiana glauca. Journal of Virology* 71: 8316–20.

Green, S. 2003. A review of the potential for the use of bioherbicides to control forest weeds in the UK. *Foresty* 76: 285–98.

Hallett, S. G. 2005. Where are the bioherbicides? *Weed Science* 53: 404–15.

Heap, I. 2017. *The International Survey of Herbicide Resistant Weeds.* Online. http://www.weedscience.org/, Accessed:27 July 2017.

Helfer, S. 2014. Rust fungi and global change. *New Phytologist* 201: 770–80.

Hewavitharana, S. S. and Mazzola, M., 2016. Carbon source-dependent effects of anaerobic soil disinfestation on soil microbiome and suppression of *Rhizoctonia solani* AG-5 and *Pratylenchus penetrans. Phytopathology* 106: 1015–28.

Heystek, F., Wood, A. R., Neser, S. and Kistensamy, Y. 2011. Biological control of two *Ageratina* species (Asteraceae: Eupatorieae) in South Africa. *African Entomology* 19(2): 208–16.

Hiebert, E. and Charudattan, R. 1984. Characterization of Araujia mosaic virus by in vitro translation analyses. *Phytopathology* 74: 642–6.

Hoagland, R. E., Boyette, C. D., Stetina, K. C. and Jordan, R. H. 2016. Bioherbicidal efficacy of a *Myrothecium verrucaria*-sector on several plant species. *American Journal of Plant Sciences* 7: 2376–89.

Hoagland, R. E., Teaster, N. D. and Boyette, C. D. 2013. Bioherbicidal effects of *Myrothecium verrucaria* on glyphosate-resistant and-susceptible Palmer amaranth biotypes. *Allelopathy Journal* 31: 367–76.

Huang H., Morgan C. M., Asolkar R. N., Koivunen M. E. and Marrone P. G. 2010. Phytotoxicity of sarmentine isolated from long pepper (*Piper longum*) fruit. *Journal of Agricultural Food Chemistry* 58: 9994–10000.

Hubbard, M., R. K. Hynes and K. L. Bailey. 2015. Impact of macrocidins, produced by *Phoma macrostoma,* on carotenoid profiles in plants. *Biological Control* 89: 11–22.

Hulme, P. E. 2017. Climate change and biological invasions: Evidence, expectations, and response options. *Biological Reviews* 92(3): 1297–313.

Hunt, E. J., Kuhlmann, U., Sheppard, A., Qin, T.-K., Barratt, B. I. P., Harrison, L., Mason, P. G., Parker, D., Flanders, R. V. and Goolsby, J. 2008. Review of invertebrate biological control agent regulation in Australia, New Zealand, Canada and the USA: Recommendations for a harmonized European system. *Journal of Applied Entomology* 132: 89–123.

ICTVdB Management. 2006. Tobacco mild green mosaic virus. In: C. Büchen-Osmond (Ed) *ICTVdB – The Universal Virus Database, Version 4*. Columbia Univ., New York. Available online: http://ictvdb.bio-mirror.cn/ICTVdB/00.071.0.01.011.htm. Accessed 27 July 2017.

Imaizumi, S., Nishino, T., Miyabe, K., Fujimori, T. and Yamada, M. 1997. Biological control of annual bluegrass (*Poa annua* L.) with a Japanese isolate of *Xanthomonas campestris* pv. *poae* (JT-P482). *Biological Control* 8: 7–14.

Izhevsky, S. S. 1979. The application of pathogenic microorganisms for control of weeds in the U.S.S.R., p. 35. In: J. R. Coulson (Ed.) *Proceedings of the Joint American-Soviet Conference on Use of Beneficial Organisms in the Control of Crop Pests*, 13–14 August 1979, Washington, DC. Entomological Society of America, College Park, MD, USA. ISBN: 0-938522-08-6.

Jackson, M. A., 1997. Optimizing nutritional conditions for the liquid culture production of effective fungal biological control agents. *Journal of Industrial Microbiology and Biotechnology* 19: 180–7.

Jarvis, P. J., Fowler, S. V., Paynter, Q. and Syrett, P. 2006. Predicting the economic benefits and costs of introducing new biological control agents for Scotch broom *Cytisus scoparius* into New Zealand. *Biological Control* 39: 135–46.

Johnson, B. J. 1994. Biological control of annual bluegrass with *Xanthomonas campestris* pv. *poaannua* in bermudagrass. *Horticultural Science* 29: 659–62.

Kadir, J., Charudattan, R., Stall, W. M. and Bewick, T. A. 1999a. Efficacy of *Dactylaria higginsii* on interference of *Cyperus rotundus* with *L. esculentum*. *Weed Science* 47: 682–6.

Kadir, J., Charudattan, R., Stall, W. M. and Brecke, B. J. 1999b. Field efficacy of *Dactylaria higginsii* as a bioherbicide for the control of purple nutsedge (*Cyperus rotundus*). *Weed Technology* 14: 1–6.

Kazinczi, G., Horváth, J. and Takács, A. P. 2006. On the biological decline of weeds due to virus infections. *Acta Phytopathologica et Entomologica Hungarica* 41: 213–21.

Kennedy, A. C. 2016. *Pseudomonas fluorescens* strains selectively suppress annual bluegrass (*Poa annua* L.). *Biological Control* 103: 210–17.

Kremer, R. J. 2000. Growth suppression of annual weeds by deleterious Rhizobacteria integrated with cover crops. In: *Proceeding of the X International Symposium on Biological Control of Weeds*, 4–14 July 1999. Montana State University, Bozeman, Montana, USA, pp. 931–40.

Kremer, R. J. and Kennedy, A. C. 1996. Rhizobacteria as biocontrol agents of weeds. *Weed Technology* 10: 601–9.

Li, J. and Kremer, R. J. 2000. Rhizobacteria associated with weed seedlings in different cropping systems. *Weed Science* 48: 734–41.

Li, J. and Kremer, R. J. 2006. Growth response of weed and crop seedlings to deleterious rhizobacteria. *Biological Control* 39: 58–65.

Littlefield, L. J., Bruckart, W. L., Luster, D. G., Pratt, P. W. and Scogin, V. L. 1998. First report of musk thistle rust (*Puccinia carduorum*) in Oklahoma. *Plant Disease* 82: 832.

Littlefield, J. L., Markin, G., Kashefi, J., de Meij, A., Runyon, J. 2013. The release and recovery of *Bradyrrhoa gilveolella* on rush skeletonweed in southern Idaho. In: Wu, Y., Johnson, T., Sing, S., et al., (Eds.) *Proceedings of the XIII International Symposium on Biological Control of Weeds*. Fort Collins (CO): USDA Forest Service, Forest Health Technology Enterprise Team

Louda, S. M., Rand, T. A., Arnett, A. E., McClay, A. S., Shea, K. and McEachern, A. K. 2005. Evaluation of ecological risk to populations of a threatened plant from an invasive biocontrol insect. *Ecological Applications* 15: 234–49.

Marrone Bio Innovations, 2015. *OPPORTUNE*™ *label*. Online: http://www.kellysolutions.com/erenewals/documentsubmit/KellyData%5CAK%5Cpesticide%5CProduct%20Label%5C84059%5C84059-12%5C84059-12_OPPORTUNE_10_2_2013_4_49_02_PM.pdf. Accessed 23 July 2017.

Mazzola, M., 2004. Assessment and management of soil microbial community structure for disease suppression. *Annual Review of Phytopathology* 42: 35–59.

McCarty, D. G., 2012. *Anaerobic Soil Disinfestation: Evaluation of Anaerobic Soil Disinfestation (ASD) for Warm-Season Vegetable Production in Tennessee.* Thesis. 28 July 2017, http://trace.tennessee.edu/utk_gradthes/1393/

McFadyen, R. 2008. Return on investment: Determining the economic impact of biological control programmes, pp. 67–74. In: M. Julien (Ed.) *XII International Symposium on Biological Control of Weeds*, 22–27 April 2007, La Grande-Motte, France.

McFadyen, R. E. C. 2000. Success in biological control of weeds, pp. 3–14. In: N. R. Spencer (Ed.) *Proceedings of the X International Symposium on Biological Control of Weeds*, 4–14 July 1999. Montana State University, Bozeman, Montana, USA.

Mitchell, J. K., Yerkes, C. N., Racine, S. R. and Lewis, E. H. 2008. The interaction of two potential fungal bioherbicides and a sub-lethal rate of glyphosate for the control of shattercane. *Biological Control* 46: 391–9.

Morin, L. 2015. Using pathogens to biologically control environmental weeds – updates. *Plant Protection Quarterly* 30: 82–5.

Morin, L., Reid, A. M., Sims-Chilton, N. M., Bucklely, Y. M., Dhileepan, K., Hastwell, G. T., Nordblom, T. L. and Raghu, S. 2009. Review of approaches to evaluate the effectiveness of weed biological control agents. *Biological Control* 51: 1–15.

Morris, M. J. 1997. Impact of the gall-forming rust fungus *Uromycladium tepperianum* on the invasive tree *Acacia saligna* in South Africa. *Biological Control* 10: 75–82.

Mowlick, S., Inoue, T., Takehara, T., Kaku, N., Ueki, K. and Ueki, A., 2013. Changes and recovery of soil bacterial communities influenced by biological soil disinfestation as compared with chloropicrin-treatment. *AMB Express* 3: 46.

Muramoto, J., Shennan, C., Fitzgerald, A., Koike, S., Bolda, M., Daugovish, O., Rosskopf, E., Kokalis-Burelle, N. and Butler, D. 2008. Effect of anaerobic soil disinfestation on weed seed germination. In: *Proceedings of the Annual International Research Conference on Methyl Bromide Alternatives and Emissions Reductions.* Orlando, FL, pp.11–14.

Organic Trade Association, 2016. *State of the Industry.* Online:28 July 2017, https://ota.com/sites/default/files/indexed_files/OTA_StateofIndustry_2016.pdf.

Pettersen, M. S., Charudattan, R., Hiebert, E., Zettler, F. W. and Elliott, M. S. 2000. Tobacco mild green mosaic tobamovirus strain U2 causes a lethal hypersensitive response in *Solanum viarum* Dunal (tropical soda apple). *WSSA Abstracts* 40: 84.

Politis, D. J., Watson, A. K. and Bruckart, W. L. 1984. Susceptibility of musk thistle and related composites to *Puccinia carduorum*. *Phytopathology* 74: 687–91.

Prevéy, J. S. and Seastedt, T. R. 2015. Increased winter precipitation benefits the native plant pathogen Ustilago bullata that infects an invasive grass. *Biological Invasions* 17: 3041–7.

Randles, J. W. 1986. Susceptibility of *Echium plantagineum* L. to tobacco mosaic, alfalfa mosaic, tobacco ringspot, and tobacco necrosis viruses. *Australasian Plant Pathology* 15: 74–7.

Ray, P. and Hill, M. P. 2012. Impact of feeding by *Neochetina* weevils on pathogenicity of fungi associated with waterhyacinth in South Africa. *Journal of Aquatic Plant Management* 50: 79–84.

Rayamajhi, M. B., Pratt, P. D., Center, T. D. and Van, T. K. 2010. Insects and a pathogen suppress *Melaleuca quinquenervia* cut-stump regrowth in Florida. *Biological Control* 53: 1–8.

Rayamajhi, M. B., Pratt, P. D., Center, T. D., Tipping, P. W. and Van, T. K. 2009. Decline in exotic tree density facilitates increased plant diversity: The experience from *Melaleuca quinquenervia* invaded wetlands. *Wetlands Ecological Management* 17: 455–67.

Rayamajhi, M. B., Van, T. K., Pratt, P. D., Center, T. D. and Tipping, P. W. 2007. *Melaleuca quinquenervia* dominated forests in Florida: Analyses of natural-enemy impacts on stand dynamics. *Plant Ecology* 192: 119–32.

Rosskopf, E. N., Charudattan, R. and Kadir, J. B. 1999. Use of plant pathogens in weed control In: Bellows, T. S. and Fischer, T. W. (Eds.) *Handbook of Biological Control*. Academic Press, San Diego, California, pp. 891–918.

Rosskopf, E. N., Burelle, N., Hong, J., Butler, D. M., Noling, J. W., He, Z., Booker, B. and Sances, F. 2014. Comparison of anaerobic soil disinfestation and drip-applied organic acids for raised-bed specialty crop production in Florida. *Acta Horticulture* 1044: 221–8.

Rosskopf, E. N., Serrano-Pérez, P., Hong, J., Shrestha, U., del Carmen Rodríguez-Molina, M., Martin, K., Kokalis-Burelle, N., Shennan, C., Muramoto, J. and Butler, D., 2015. Anaerobic soil disinfestation and soilborne pest management. In: Meghvansi, M. K. and Varma, A. (Eds.) *Organic Amendments and Soil Suppressiveness in Plant Disease Management.* Springer International, Switzerland, pp. 277–305.

Seastedt, T. R. 2015. Biological control of invasive plant species: A reassessment for the Anthropocene. *New Phytologist* 205: 490–502.

Seto, H., Sasaki, T., Imai, S., Tsuruoka, T., Ogawa, H., Satoh, A., Inouye, S., Niida, T. and Ōtake, N. 1983. Studies on the biosynthesis of Bialaphos (sf-1293). 2. Isolation of the first natural products with a C-P-H bond and their involvement in the C-P-C bond formation. *The Journal of Antibiotics* 36: 96–8. (Ogawa, Y., Tsuruoka, T., Inouye, S., and Niida, T. 1973. Studies on a new antibiotic SF-1293. II. Chemical structure of antibiotic SF-1293. Scientific Reports of Meiji Seika Kaisha 13: 42–8).

Shabbir, A., Dhileepan, K., Khan, N. and Adkins, S. W., 2014. Weed–pathogen interactions and elevated CO2: Growth changes in favour of the biological control agent. *Weed Research* 54: 217–22.

Shinmura, A. 2004. Principle and effect of soil sterilization method by reducing redox potential of soil (in Japanese) *The Phytopathological Society of Japan (PSJ) Soilborne Disease Workshop Report* 22: 2–12.

Shishkoff, N. and Bruckart, W. L. 1993. Evaluation of infection of target and nontarget hosts by isolates of the potential biocontrol agent *Puccinia jaceae* that infect *Centaurea* spp. *Phytopathology* 83: 894–8.

Simberloff, D. 2012. Risks of biological control for conservation purposes. *BioControl* 57: 263–76.

Sosnokie, L. M. and Culpepper, A. S. 2014. Glyphosate-resistant Palmer amaranth (*Amaranthus palmeri*) increases herbicide use, tillage, and hand-weeding in Georgia cotton. *Weed Science* 62: 393–402.

Stone, W. W., Gilliom, R. J. and Ryberg, K. R. 2014. Pesticides in U.S. streams and rivers: Occurrence and trends during 1992–2011. *Environmental Science and Technology* 48: 11025–30.

Stubbs, T. L. and Kennedy, A. C., 2012. Microbial weed control and microbial herbicides. In: Alvarez-Fernandez, R. (Ed.) *Herbicides-environmental Impact Studies and Management Approaches.* InTech-Open Access Publishers, Rijeka, Croatia, pp. 135-166, https://cdn.intechopen.com/pdfs-wm/25996.pdf.

Supkoff, D. M., Joley, D. B. and Marois, J. J. 1988. Effect of introduced biological control organisms on the density of *Chondrilla juncea* in California. *Journal of Applied Ecology* 25: 1089–95.

Trujillo, E. E. 2005. History and success of plant pathogens for biological control of introduced weeds in Hawaii. *Biological Control* 33: 113–22.

Tworkoski, T. 2002. Herbicide effects of essential oils. *Weed Science* 50: 425–31.

Van Driesche, R., Hoddle, M., and Center, T. 2008. Chapter 16: Non-target impacts of biological control agents, pp. 183–98. In: Gurr, G. and Wratten, S. (Eds) *Biological Control: Measures of Success.* Blackwell Publishing, Malden, MA, USA.

Van Driesche, R. G., Carruthers, R. I. and Center, T.2010. Classical biological control for protection of natural ecosystems. *Biological Control* 54: S2–S33.

Watson, A. K. 2017. Biocontrol and weed management in rice of Asian Pacific Region. In: *Proceedings of the 26th Asian Pacific Weed Science Society Conference,* 19–22 September, Kyoto, Japan.

Weaver, M. A., Lyn, M. E., Boyette, C. D. and Hoagland, R. E. 2007. *Bioherbicides for Weed Control. Non-chemical Weed Management.* CABI, International, Cambridge, UK, pp. 93–110.

Weaver M. A., Boyette, C. D. and Hoagland, R. E. 2016. Rapid kudzu eradication and switchgrass establishment through herbicide, bioherbicide and integrated programmes. *Biocontrol Science and Technology* 26: 640–50.

Winston, R. L., Schwarzländer, M., Hinz, H. L., Day, M. D., Cock, M. J. W. and Julien, M. H. (Eds). 2014. *Biological Control of Weeds: A World Catalogue of Agents and Their Target Weeds*, 5th Ed. USDA Forest Service, Forest Health Technology Enterprise Team, Morgantown, WV, 838p. FHTET-2014-04. Online: https://www.ibiocontrol.org/catalog/. Accessed17 July 2017.

Woods, D. M., Bruckart, W. L., Pitcairn, M, Popescu, V. and O'Brien, J. 2009. Susceptibility of yellow starthistle to *Puccinia jaceae* var. *solstitialis* and greenhouse production of inoculum for classical biological control programs. *Biological Control* 50: 275–80.

Woods, D. M., Fisher, A. J. and Villegas, B. 2010. Establishment of the yellow starthistle rust in California: Release, recovery, and spread. *Plant Disease* 94: 174–8.

Wood, A. R. and Morris, M. J. 2007. Impact of the gall-forming rust *Uromycladium terrerianum* on the invasive tree Acacia saligna in South Africa: 15 years of monitoring. *Biological Control* 41: 68–77.

Woods, D. M., Pitcairn, M. J., Luster, D. G. and Bruckart, W. L. 2002. First report of musk thistle rust (*Puccinia carduorum*) in California and Nevada. *Plant Disease* 86: 814.

Wu, F., Liu, Y. and Bhatnagar, D. 2008. Cost-effectiveness of Aflatoxin control methods: Economic incentives. *Toxin Reviews* 27: 203–25.

Wyss, G. S., Charudattan, R., Rosskopf, E. N. and Littell, R. C. 2004. Effects of selected pesticides and adjuvants on germination and vegetative growth of *Phomopsis amaranthicola*, a biocontrol agent for *Amaranthus* spp. *Weed Research* 44: 469–82.

Yandoc, C., Rosskopf, E. and Bull, C. T. 2014. Weed management in organic production systems. In: R. T. Lartey and A. J. Caesar (Eds) *Emerging Concepts in Plant Health Management*. Research Signpost, Kerala, India, pp. 213–54.

Yandoc, C., Rosskopf, E. N., Pitelli, R. L. C. M., and Charudattan, R. (2006) Effect of selected pesticides on conidial germination and mycelial growth of *Dactylaria higginsii*, a potential bioherbicide for purple nutsedge (*Cyperus rotundus*). *Weed Technology* 20: 255–60.

Young, S. L. 2004. Natural product herbicides for control of annual vegetation along roadsides. *Weed Technology* 18: 580–7.

Yu, J. and Morishita, D. W., 2014. Response of seven weed species to corn gluten meal and white mustard (*Sinapis alba*) seed meal rates. *Weed Technology* 28: 259–65.

Zelikova, T. J., Hufbauer, R. A., Reed, S. C., Wertin, T. Fettig, C. and Belnap, J. 2013. Eco-evolutionary responses of *Bromus tectorum* to climate change: Implications for biological invasions. *Ecology and Evolution* 3: 1374–87.

Ziska, L. H. 2014. Climate, CO_2, and invasive weed management. In: Ziska, L.H. and Dukes, J. S. (Eds) *Invasive Species and Global Climate Change* (Vol. 4). CAB, UK., pp. 293–305.

Chapter 19

The use of microorganisms in integrated weed management

Susan M. Boyetchko, Agriculture and Agri-Food Canada, Canada

1 Introduction

Weeds have been and remain the primary pest control target in agricultural, forestry and urban environments. More than $75 billion annual yield losses are attributed to them in the world's major food crops: rice, wheat, barley, corn, potato and soybean (Bhowmik 1999). Worldwide, 44% of all pesticides sold are herbicides. However, in North America, up to 80% of overall pesticide sales are herbicides (Bailey and Mupondwa 2006).

As crop production systems become more complex and diversified, weed communities evolve (Derksen et al. 1996). For example, changing a dominantly cereal-based system to a more diverse rotation will change the composition of the weed community, just as changing tillage from conventional to minimum or zero shifts the types of weeds. Synthetic herbicides have been the dominant weed control technique since the discovery of 2,4-D in the early 1940s (Bhowmik 1999). Although other weed control methods such as crop rotation, cultivation/zero tillage and increasing seeding rate are available, herbicides

http://dx.doi.org/10.19103/AS.2017.0025.21

are used predominantly because of their ease of handling, high efficacy, consistent field performance and relatively low cost (Kelly and Allen 2011). Herbicide use has led to a number of problems including residues in soil, water and food and spray drift, particularly to encroachment of urban communities and development of herbicide-resistant weeds. There are currently 26 known modes of action, but no new sites of action have been discovered in the herbicide market in the past 20 years (Kelly and Allen 2011). Banning the use of chemical pesticides in Canadian urban municipalities has created opportunities for bioherbicide discovery and development. However, it was assumed that there were alternative pest control products available to replace the chemicals that were being banned (Bailey et al. 2010). Nonetheless, these challenges have encouraged research to develop alternative biological control agents for weeds in urban and agricultural environments (Boyetchko and Rosskopf 2006; Bailey et al. 2010).

Biological control of weeds by fungal pathogens, bacteria and viruses has been studied for more than three decades. The aim is to suppress or reduce the weed population below an ecological or economic threshold (Charudattan 1991, 2001; Boyetchko and Rosskopf 2006; Bailey 2014). This is analogous to the application of synthetic chemical herbicides which are mass-produced, artificially formulated and applied annually in single or multiple sprays. Bioherbicides are applied in higher doses than normally found in nature, causing an epidemic amongst weed populations that may result in significant weed control – the inundative approach. However, comparison to synthetic chemicals often leads to unrealistically high expectations of biological weed control equivalent to that obtained from herbicides. This can result in scepticism about biocontrol without proper knowledge of its advantages and disadvantages. Production of natural products is often included in the definition of a bioherbicide because the activity may result from secondary metabolites produced by the microorganism. This chapter includes examples of microbes for weed control. The majority of research on bioherbicides has emphasized the development of fungal pathogens (Charudattan 2001; Rosskopf et al. 1999; Boyetchko 2005; Boyetchko and Rosskopf 2006; Bailey 2014). Examples of use of foliar and soil bacteria and a virus are included in the discussion.

2 The role of biopesticides

The lack of registered bioherbicides since the first registration of two fungal pathogens in the United States in the 1980s raises the question 'Where are the bioherbicides?' (Hallett 2005). There has been only a handful of subsequent registrations and the adoption of these bioherbicides has been rather fleeting. Hallett (2005) suggested that bioherbicides would fit best in niche markets such as parasitic weeds or narcotic plants. However, this may be a narrow view because bioherbicide research has progressed significantly due to advances in fermentation and formulation technology. The regulatory process in Canada and the United States has also matured (Ash 2010; Bailey et al. 2010; Bailey 2014) although important regulatory obstacles still exist.

In the United States and Canada, for example, use and registration of numerous synthetic pesticides in lawns and gardens have been phased out (Bailey et al. 2010). In addition, Canadian legislative authorities have banned the use of synthetic pesticides for cosmetic purposes in urban municipalities such as production of pristine weed-free lawns. Since Halifax became the first Canadian city to ban chemical pesticide use within the city limits in 2000, more than 100 municipalities have followed with similar bans (Bailey et al. 2010). Ontario followed with a province-wide ban of chemical pesticide use in urban environments in 2009. This has created a market for bioherbicides although registration of new products has still been modest (Ash 2010).

Another driver for research into bioherbicides is the growing number of reports of herbicide resistance to one or more modes of action of herbicides (Ash 2010; Bailey et al. 2010; Heap 2014). In order to mitigate this problem, researchers have reported the use of biological control agents in combination with reduced rates of herbicides (Boyette et al. 2007; Hoagland 1996; Peng and Byer 2005; Schnick et al. 2002).

Other social and economic drivers influencing the development and adoption of bioherbicides include the demand by consumers for environmentally safe products in agriculture (Ott et al. 1991; Bailey et al. 2010). The demand for organic and other pesticide-free products in grocery stores has further put pressure on the industry to provide products that do not contain pesticide residues. The organic movement has increased the importance of research into biological control as 'green' alternatives for production of field crops, vegetables, fruits and berries sold to the general public. The changing attitude of the producer and consumer and societal acceptance of the technology have led to a resurgence of interest in biological control. The development of initiatives such as farmers' markets in cities has helped develop a growing consumer awareness of the organic movement. This increasing pressure can no longer be ignored as agricultural practices have come under greater scrutiny by the public. It is becoming clear that citizens of a democratic society will not entrust their water, diets or natural resources blindly into the hands of farmers, agribusiness firms and agricultural scientists.

3 Historical accomplishments

The first two bioherbicides to be registered by the US EPA were Devine®, a composite of the soil-borne fungus *Phytophthora palmivora* Butler for control of strangler vine (*Morrenia odorata* (Hook. and Arn.) Lindl) in citrus crops (Kenney 1986) and Collego®, the fungal pathogen *Colletotrichum gloeosporioides* f.sp. *aeschynomene*, a post-emergent application for control of northern jointvetch [*Aeschynomene virginica* (L.) B.S.P.] in rice and soybean (TeBeest and Templeton 1985). Devine® is a stem and root rot pathogen of strangler vine, while Collego® causes anthracnose symptoms (i.e. leaf spots, stem canker and death). Both bioherbicides were considered niche-market products and were eventually dropped from registration due to limited markets. However, Collego® was re-registered under the name LockDown® and has been sold to farmers (Bailey 2014). In Canada, the fungus *Colletotrichum gloeosporioides* f.sp. *malvae* was registered under the name BioMal® for control of round-leaved mallow (*Malva* pusilla Sm.) in field crops by Philom Bios (Boyetchko 2005; Boyetchko et al. 2007; Bailey 2014). The pathogen caused leaf spots, stem lesions and stem dieback and was target specific to plants belonging to the Malvaceae family. Although the product was commercially available from 1992 to 1994, production ceased due to limited market size and cost. Efforts were made to pursue another industry partnership with Encore Technologies with a new registration as Mallet WP, but difficulties in manufacturing and commercialization led to the registration lapse and it is currently unavailable commercially.

Following these early discoveries, other potential bioherbicides followed. The very nature of their host specificity limited their market. Broad-spectrum activity became acceptable, while narrow specificity was rejected for economic reasons. The foliar-applied bacterial agent *Xanthomonas campestris* pv. *poae* was registered for use to control annual bluegrass (*Poa annual.*L.) as Camperico® on golf courses in Japan (Imaizumi et al. 1997; Boyetchko 2005). While being host-specific in nature, the market size of golf courses justified its production. The bacteria affect annual bluegrass, while golf course turf remains unharmed. Its registration and availability are currently uncertain. In addition, Dr. BioSedge®, an endemic rust pathogen (*Puccinia canaliculata* Schwein.) causing leaf infection on yellow nutsedge (*Cyperus esculentus* L.) was registered but is not commercially available. Similar events occurred with Woad Warrior®, a bioherbicide to control dyer's woad (*Isatis tinctoria* L.) (Bailey 2014). It is an obligate rust fungus and only reproduces on the weed host when applied as finely ground powder of leaf and stem pieces infected with the rust. The problem, as with Dr. BioSedge®, is that mass production is expensive and labour-intensive. It is not easily mass-produced through conventional means because it requires a living host to complete its life cycle. Therefore, there was little, if any, interest from industry in marketing and commercializing these pathogens.

Despite these early developments, the lack of commercialization raised scepticism about bioherbicides as viable weed control agents (Hallett 2005). Criteria used to assess a bioherbicide included host specificity. Initially, it was considered to be an asset, if not essential. Other reviewers have cited ease of use and formulation and stability as desirable characteristics (Auld and Morin 1995; Boyetchko 2005; Hallet 2005; Bailey 2014). Consistent field performance with weed control greater than 80% has led to comparisons with synthetic herbicides. Variable efficacy is viewed as a weakness. Unfortunately, inconsistencies observed with bioherbicides have been viewed more harshly than similar variability with herbicides. These are often dismissed due to weather conditions such as temperature that affect application and use. Often, the efficacy of bioherbicides has been considered before formulation and before full product development, unlike chemical herbicides.

4 Recent registrations

The fungus *Chondrostereum purpureum* (Pers.) Pouzar is a wound pathogen infecting the cambium of woody trees and shrubs on utility rights-of-way, and in forests and riparian lands (DeJong et al. 1990; Dumas et al. 1997; Shamoun et al. 1996; Wall 1994; Hintz 2007). It has a broad host range, prevents re-sprouting and promotes wood decay. It was initially studied in the Netherlands. Subsequent research was conducted at two institutes in Quebec and British Columbia. Although it is a ubiquitous wound pathogen, researchers discovered that the fungus shows weak pathogenicity and its release does not introduce novel virulence patterns in natural populations outside its range of introduction (Hintz 2007). A spin-off company, Mycologic Inc., resulted from the collaboration between the Pacific Forestry Centre and University of Victoria and the fungus was registered jointly between the Pest Management Regulatory Centre (PMRA) in Canada and the EPA in the United States. It is formulated as a paste and has a long shelf life at a titre of 10^5 to 10^7 cfu/kg (de la Bastide and Hintz 2007) It is registered under the trade name Chontrol® Paste.

In 2007, *Sclerotinia minor* Jagger was registered as Sarritor® for the control of dandelion (*Taraxacum officinale* Weber ex F. H. Wigg.) and other broadleaved weeds in turfgrass (Abu-Dieyeh and Watson 2007; Bailey 2010). The work was conducted at McGill University and a spin-off company resulted. The pathogen has broad-spectrum activity but is safe for use with cereals and grasses. The fungus is mass-produced on grain, dried and ground into granules that are broadcast with a fertilizer spreader or applied as a spot treatment by home owners and professional lawn care operators. This particular isolate of the pathogen does not produce sclerotia and therefore does not overwinter and persist in the environment. It is safe for all mammals, birds and fish. Full registration with PMRA in Canada was obtained in 2010, and it was considered a suitable alternative bioherbicide because of the ban on synthetic herbicides for domestic use in urban municipalities and the lack of alternatives in the marketplace. Its effectiveness appeared to be similar to that of the industry standard, 2,4-D. It now faces competition from a new bioherbicide (iron chelated with hydroxyethylenediamine triacetic acid) introduced into the marketplace and the company is restructuring its business plan to meet demand while addressing increased competition (Watson and Bailey 2013).

A group of lactic acid bacteria from fermented milk, *Lactobacillus rhamnosus*, *Lactococcus lactis* ssp. *lactis* strains LL64/CSL and LL102/CSL and *Lactococcus lactis* ssp. *cremoris* strain M11/CSL have been registered and commercialized by Lacto Pro-Tech Inc. under the name Organo-Sol® in Canada to suppress the growth of white and red clovers (*Trifolium* spp.), bird's foot trefoil (*Lotus corniculatus* L.) and black medic (*Medicago lupulina* L.) (Bailey 2014). These weeds are controlled in established lawns where the bacteria cause necrosis and suppress growth. Damage may occur if the bacteria are applied to newly seeded turf. The product, when diluted with water, can be broadcast or spot sprayed on foliage.

Phoma macrostoma Montagne, a soil-applied fungus, was isolated from Canada thistle (*Cirsium arvense* L.) but is used to selectively control broadleaved weeds, primarily dandelion, in turfgrass. The fungus causes photobleaching (severe chlorosis), inhibiting the development of roots and shoots (Graupner et al. 2003; Zhou et al. 2004; Bailey and Falk 2011; Watson and Bailey 2013; Bailey 2014). The mode of action is via macrocidins produced during fermentation, which inhibit carotenoid biosynthesis (Bailey 2014; Hubbard et al. 2015). The fungus is mass-produced through solid-state fermentation on grains and then ground and processed into granules which are spread by broadcast application. The target weeds belong to the family Asteraceae, and some effects are observed on other Brassicaceae and Fabaceae plants, while grasses remain unaffected (Zhou et al. 2004). No toxicological effects were observed on mammals, birds, fish, insects or other wild animals, and no residual effects were observed 12 months after application. While the primary use was intended and registered for lawn care, researchers are continuing their efforts to extend its use into agroforestry. The industry partner (The Scotts Company) obtained conditional registration in Canada (2011) for use in turf and full registration in the United States (2012), but they eventually pulled out of the project. Research is ongoing and a new industry partner has negotiated a licence for the technology.

Marrone Bio Innovations has developed and registered *Streptomyces acidiscabies*, an actinomycete bacterium, into a pre- and post-emergent bioherbicide product to control annual grasses and broadleaved sedge weeds in turf and several crops (Bailey 2014; Marrone Bio Innovations 2016). The product is registered under the name Opportune® and manufactured from spent fermented media containing killed bacterial cells. The natural product is thaxtomin A, a phytotoxin that affects cell biosynthesis and reduces seedling growth when present at low levels (less than 4 mg per ml). It has broad-spectrum activity but is not toxic to non-target organisms such as birds, fish and honeybees.

Table 1 Examples of microbial biological control (registered and commercialized)

Target weed	Pathogen (registered or trade mark name)	Country	Status[a]
Annual bluegrass	*Xanthomonas campestris* pv. *poae* (Camperico®)	Japan	4
Dandelion	*Sclerotinia minor*	Canada	5
Dandelion, Canada thistle, other broadleaved weeds	*Phoma macrostoma*	Canada	5*
Northern jointvetch	*Colletotrichum gloeosporioides* f.sp.*aeschynomene* (Lockdown®)	USA, Arkansas	5
Nutsedges	*Puccinia canaliculata* (Dr. BioSedge®)	USA, Georgia	5*
Round-leaved mallow	*Colletotrichum gloeosporioides* f.sp.*malvae* (Mallet WP)	Canada/USA	5*
Red and white clovers, bird's foot trefoil, black medic	*Lactobacillus rhamnosus,lactis*	Canada	5
Strangler vine	*Phytophthora palmivora* (DeVine®)	USA, Florida	5
Tropical Soda Apple	Tobacco mild green mosaic tobamovirus	USA	5*
Weedy hardwood	*Chondrostereum purpureum* (Chontrol® Paste)	Canada	5

Note: Compiled from published and unpublished reports.
[a] Status: 1 = exploratory phase; 2 = laboratory and/or greenhouse testing underway; 3 = field trials in progress; 4 = under early commercial or practical development; 4* = commercial development tried but registration uncertain; 5 = available for commercial or practical use and 5* = available registered as a microbial herbicide but currently unavailable commercially.

A bioherbicide containing a virus was developed by Dr. Raghavan Charudattan at the University of Florida (now retired) and was registered under the name SolviNix LC (Charudattan and Hiebert 2015a,b). It is comprised of the ubiquitous Tobacco mild green mosaic tobamovirus from tobacco (*Nicotiana tabacum* L.) and tree tobacco (*N. glauca* Glauca) and adapted for use in the Solanaceae. It is manufactured by the spin-off company, BioProdex Inc. in the United States, and provides 85–100% control of the noxious weed, tropical soda apple (*Solanum viarum* Dunal), causing localized necrotic lesions. Tropical soda apple is considered a large market because it is found in the United States, and Africa, Europe, Asian-Pacific countries, Central America, North and South America. It is the first plant virus to be registered under US EPA as a bioherbicide (Charudattan and Hiebert 2015b).

5 New discoveries under development

The biological control potential of soil-borne bacteria has great potential for weed control but is poorly understood. Rhizobacteria (root-colonizing bacteria) that have an inhibitory effect on plant growth are often referred to as deleterious rhizobacteria (DRB) (Kremer and Kennedy, 1996; Suslow and Schroth, 1982). They are nonparasitic (exopathogenic) and despite their ubiquity in soil, they have been ignored because their symptoms are subtle. A number of projects in the United States and Canada using DRB for biological control have been summarized (Boyetchko, 1999; Kremer and Kennedy 1996; Rosskopf et al. 1999). The potential of DRB as biological weed control agents has been explored for 18 weed species including velvetleaf, lamb's-quarters, pigweed, smartweed, cocklebur, downy brome, wild oat, green foxtail and jointed goatgrass (Begonia et al. 1990; Boyetchko 1999; Boyetchko and Mortensen 1993; Cherrington and Elliott 1987; Kennedy et al. 1991; Kremer et al. 1990; Kremer and Kennedy 1996). The majority of the projects targeted annual weeds in cereal and row crops, and several of these projects focused on preliminary screening of bacteria in laboratory experiments.

Kremer (1987) found specific bacterial strains that reduced seedling vigour, while seedling emergence was decreased 55% with the soil application of two bacterial isolates (Begonia 1989; Kremer 1993). Screening of several thousand bacterial strains against annual grass weeds such as downy brome, green foxtail and wild oat has been conducted in Canada and the United States (Boyetchko 1997; Boyetchko and Mortensen 1993; Kennedy et al. 1991). In many instances, the weed-suppressive activity of DRB has been attributed to the production of secondary metabolites and phytotoxins (Boyetchko 1999; Kremer and Kennedy 1996). Kennedy et al. (1991) discovered a weed-suppressive bacterium *Pseudomonas fluorescens* D7 strain that controls the winter annual grass, downy brome, in winter wheat. Partial purification of a phytotoxin from the *Pseudomonas fluorescens* D7 strain caused root discoloration and reduced root elongation (Tranel et al. 1993b). The toxin was composed of at least two polypeptides, fatty acid esters and lipopolysaccharides (Gurusiddaiah et al. 1992). Further investigation into the mode of action revealed that the active fraction from the D7 strain affected lipid synthesis and membrane integrity, with no effect on cell division, respiration or protein synthesis (Tranel et al. 1993a). Other researchers have also demonstrated that the overproduction of auxin (indole-3-acetic acid) in some bacterial strains, when L-tryptophan is added to the media, results in greater root length inhibition of field bindweed (Kremer and Sarwar 1995; Sarwar and Kremer 1995a, b).

In Canada, *Pseudomonas fluorescens* was reported as a pre-emergent bioherbicide to control green foxtail (Boyetchko and Mortensen 1993; Boyetchko 1997; Boyetchko et al. 2013). It also controls herbicide-resistant green foxtail: Group 1 (ACCase inhibitors) and Group 3 (trifluralin resistant) (S. Boyetchko, unpublished results). Chemical analysis of the secondary metabolites produced by *Pseudomonas fluorescens* strains BRG100 and 189 revealed the production of a complex mixture of cyclic peptides and X-ray crystallography allowed the determination of the chemical structure of two herbicidal and fungicidal compounds: pseudophomin A and pseudophomin B (Quail et al. 2002; Pedras et al. 2003). They are complex cyclic peptides with nine amino acid residues and a lipophilic portion with water-soluble peptides. The green fluorescent protein was introduced into *Pseudomonas fluorescens* BRG100 to visualize root colonization and the fate of the bacterium in the environment (Hanson 2008; Caldwell et al. 2012). The bacterium is an effective root colonizer and disperses from a granule formulation based on an oat flour matrix (called 'pesta') that is delivered as a pre-emergent bioherbicide to soil (Daigle et al. 2002). Soil moisture, temperature and soil type have an effect on the dispersal of the bacteria (Daigle et al. 2002; Hanson 2008; Hynes and Boyetchko, 2011). Improvements to the pesta formulation were made, and a shelf life of 16 months at 8.5 $\log_{10}$ colony-forming units (cfu) per gram of pesta was reported (Hynes and Boyetchko 2012). The pesta had better shelf life when stored at a water activity (moisture content) of 0.3 a_w compared to 0.8 a_w where the bacterial population decreased to 7.3 $\log_{10}$ cfu/g over only six months. Dispersal of the bacteria from the granule is critical to bioherbicide performance of *Pseudomonas fluorescens* BRG100.

The addition of various starches to pesta improved the dispersal of the bacterium from the granule. Pea, potato, corn and rice starches all increased the disintegration characteristics of the pesta granule, where peas starch performed better than potato, corn and rice starches. However, it could not be demonstrated that the bioherbicidal properties of the pesta-formulated *Pseudomonas fluorescens* were also improved. The formulation issue still needs to be resolved before these bacteria will become an effective bioherbicide. However, its potential to control green foxtail and herbicide-resistant weeds shows promise.

6 Target weed selection

In the past, a fungal pathogen was evaluated as a bioherbicide following the serendipitous or accidental discovery of a diseased weed plant, thus prompting the researcher to pursue further investigation (Boyetchko 2005; Bailey et al. 2009). Chemical companies invest in discovery and screening programmes with a particular target weed in mind and search for tens and hundreds of thousands of new active ingredients (Kelly and Allen 2011). Researchers should select weed targets that justify the substantial investment. Careful selection will make the research more attractive for investment and commercialization. The researcher may possess an excellent pathogen possessing weed-suppressive properties and high virulence or pathogenicity, but market forces may prevent commercialization (Cross and Polonenko 1996; Boyetchko et al. 2007).

Lack of industrial investment in bioherbicide research is reflective of low or not anticipated profit from the product. All the examples of registered bioherbicides are those where commercialization efforts were conducted by spin-off companies from universities or small-to-medium-sized enterprises or companies, not multinationals (Bailey et al. 2010). Recent acquisitions of biopesticide companies by multinationals have not resulted in increased investment in microbial-based technology. These large entities have pursued novel molecules from microorganisms as leads for new reduced-risk chemistries (Bailey et al. 2010). Examples include avermectin at Novartis, spinosyn at Dow AgroSciences and strobilurin compounds at BASF (Marrone 1999; Saxena and Pandey 2001). There is less investment in living organisms because it is difficult to modify existing chemical screening programmes and develop new technology in fermentation and formulation processes compatible with living organisms. Formulation technology has been developed for synthetic chemicals. Developing and screening formulation additives and adjuvants for bioherbicides is a greater, more expensive challenge because of the necessity of compatibility with living organisms (Prasad 1994). More importantly, the size of the bioherbicide market remains small compared to existing chemical sales potential. The target weeds for discovery are often related to major crops such as rice, soybean, corn, cotton, wheat and canola (Ash 2010). An example is grass weed control in cereal crops or Brassicaceae weed control in a crop of canola. The greatest bioherbicide potential may be amongst crops where weed control is required, but there are no or limited products available.

7 Early discovery and screening

The primary contribution of those who study bioherbicides has been discovery: searching and finding microbes (Charudattan 2001; Boyetchko 2005; Hallett 2005; Boyetchko and

Rosskopf 2007; Ash 2010). Finding potential microorganisms for weed control is the easier part; developing them is more challenging. The earlier studies in the development phase included efficacy testing, dose–response studies, host-range evaluation and rudimentary efforts in formulation testing. The utility of the microorganism for biological weed control becomes apparent when laboratory, greenhouse and field testing have been conducted, including environmental factors (e.g. temperature and moisture) affecting the efficacy and mode of action (Bailey et al. 2009). Earlier bioherbicides were not complex. They had a very simple formulation and a single spectrum of activity with a preference for a narrow host range to reduce the risk of non-target effects. The lack of host specificity frequently limited commercial success. Agroecosystems are commonly multispecies communities and producers would not invest in a product to control one weed species when a broader spectrum product would be more appropriate and economical. The broader have spectrum bioherbicides led to a greater chance of registration for Chontrol® Paste, Organo-Sol®, Phoma and Opportune®.

Instead of relying on serendipitous discovery of a diseased weed with a promising pathogen as a bioherbicide product, the target weed may be selected in a business-like manner by considering the effects and economics of the target weed, crop loss figures and even weed surveys that document their geographic distribution, abundance, density and cost of control in various crop production systems (Bailey et al. 2009). In addition, herbicide resistance, availability of other products, methods to control alien and invasive species and legislative regulations limiting the use of synthetic chemicals in certain environments all exert influence on success (Bailey and Mupondwa 2006; Bailey et al. 2010).

Once a target weed has been identified, screening bioassays need to be conducted in a screening programme, similar to screening candidate herbicides. Screening must be done on plant parts (leaves, stems and roots) to identify the plant part most susceptible to attack by the pathogen (Boyetchko 2005; Bailey et al. 2009). The earlier discoveries of bioherbicides relied on discovery of a pathogen that fulfilled all the criteria for Koch's postulates where the pathogenic agent was the same one isolated, purified and identified on the same plant and thus caused the same symptoms on the inoculated plant as those observed on the original diseased plant.

Of utmost importance is taxonomic identification of the microbial agent to ensure that it is not pathogenic to crops, especially those into which the pathogen will be introduced, and to ensure that the organism, particularly bacteria, is not a human or mammalian pathogen. The screening programme at Agriculture and Agri-Food Canada in Saskatoon also evaluated the culture media necessary to grow the microbe and, whether it could grow in liquid or on solid substrate agar or grains. Further, the taxonomy provides the researcher with clues into its biology and what propagules it produces to infect the plant and complete its life cycle (Bailey et al. 2009). Taxonomic designation to genus and species is a requirement for product registration.

Very quickly, the evaluation process must develop a product profile suggesting what market uses the bioherbicide might have. Potential uses range from home gardens, turf, field crops, vegetable or fruit crops and forestry? Will the bioherbicide be foliar-applied post-emergence or soil-applied pre-emergence? Potential markets define the tests that need to be conducted, under what environmental conditions and cropping systems, including the various crops that need to be selected to ensure tolerance and safety of the bioherbicide. The research continues to evaluate the efficacy of the bioherbicide under various field and environmental conditions, including optimum temperature required for weed control, whether the microbe is active and efficacious at cool spring temperatures or if it needs warmer temperatures to be effective at a later growth stage. Ultimately, the bioherbicide should effectively control the weeds with minimal yield loss.

8 Formulation and fermentation technologies

There is an abundance of research describing the screening of microbes for their potential as a bioherbicide, but the importance of formulation for commercial development is less prominent in the literature and poorly understood. The role of an effective formulation, when advancing a bioherbicide as a commercial product, is for it to maintain a stable population of the microbe when applied to the target weed (Hynes and Boyetchko 2006, Hynes et al. 2011). The formulation must also assist with the bioherbicide application and delivery, protect the microorganism from unfavourable environmental conditions but still result in sufficient bioherbicidal activity when it reaches the target weed. Nutrient supplements may be used in formulations to stimulate the infection process and protect the germinating propagule. These include simple sugars, amino acids, pectins, salts and plant extracts (Womack and Burge, 1993).

The formulation may be liquid for foliar application or a solid for soil application, either as a granule or seed treatment. It will depend on the nature of the weed and mode of action of the bioherbicide. Foliar-applied agents may be exposed to rainwash, UV irradiation and desiccation which may negatively affect the bioherbicide. Various adjuvants with sticking or adhesive, sun-blocking, or moisture-retention properties have been suggested to alleviate the negative environmental effects (Womack and Burge 1993; Green et al. 1998). Many of the environmental challenges such as the requirement for leaf wetness or period of dew may be overcome with formulation technology. Formulations can increase the moisture-retaining properties of the spray droplet, reduce the rate of evaporation and/or enhance the rate of infection. For example, humectants may be used to attract moisture from the air and thus keep the droplet from drying out. Since the active ingredient is a living organism which must penetrate and infect the target weed, exogenous ingredients may be added to stimulate germination and growth of the microbe.

Different types of formulation ingredients that may be used include surfactants and adjuvants, emulsions and vegetable oils, hydrophilic polymers and solid matrices encapsulating the active ingredient (Green et al. 1998; Boyetchko et al. 1999). The surfactant can facilitate dispersal and uniform distribution of the bacteria or fungus on the leaf surface or serve as a sticker, emulsifier or stabilizer and modify the surface tension of the spray droplet. For example, humectants and polyvinyl alcohol may be used in a formulation to reduce the rate of evaporation of the spray (Greaves et al. 2000). Silwet L-77 and Silwet 408 are organosilicone surfactants that reduce the surface tension of the droplet, resulting in bacterial or fungal penetration and entry to the plant stomata and hydathodes (Womack and Burge 1993; Zidack and Backman 1996).

Various oil emulsions have been used to alleviate moisture constraints, thereby enhancing field performance of the bioherbicide. Invert emulsions consist of water droplets suspended in oil, and evaporation of the trapped water is significantly reduced, thus protecting the microbial propagule from the harsh environment. This has been demonstrated by Boyette et al. (1993) when using the fungus *Colletotrichum truncatum* for control of hemp sesbania suspended in an invert emulsion. However, using higher oil content in formulation may lead to increased phytotoxicity and difficulty in application due to increased viscosity. Oil emulsions are considered more practical because the oil content is lower and they can be applied with existing spray equipment (Green et al. 1998). A water-in-oil-in-water formulation where the fungal propagule was contained in water entrapped in oil and surrounded by a continuous phase of water (Auld 1993) reduced the dew dependency of the fungus to cause infection on the weed.

Soil application of bioherbicides has been developed, particularly for those fungi or bacteria that affect the weed at or below the soil surface. Granular formulations protect the microorganism from the environment and other indigenous microorganisms (Walker and Connick 1983; Connick et al. 1991; Quimby et al. 1999; Daigle et al. 2002; Hynes and Boyetchko 2001). The 'pesta' formulation was developed as a solid matrix comprised of durum wheat encapsulating the microbe (Connick et al. 1996). This was then adapted to encapsulate bacteria in oat flour for biological control of grass weeds (Daigle et al. 2002, Hynes and Boyetchko 2011). Stabileze was used to formulate pseudomonad bacteria as a foliar spray by creating a starch/oil/sucrose matrix to stabilize the bacteria and resuspending them into a sprayable liquid formulation for the application of post-emergent bioherbicides (Zidack and Quimby 2002).The addition of sucrose to the formulation improved the survival rate of the bacteria.

9 Future trends and conclusion

The use of bioherbicides as a weed management tool began with a lot of scepticism but the future shows great promise. The application of microorganisms in weed control started with individual researchers registering fungal pathogens, Collego® and DeVine® for use in soybean or citrus for control of northern jointvetch and strangler vine, respectively, but their commercial success was short-lived. At the time when BioMal® was registered, there were no regulations for registering and commercializing bioherbicides in Canada. The lack of bioherbicides in the marketplace led to questions about their success and efficacy as a weed control option.

The early prototype bioherbicides had very simple formulations that could be suspended in liquid and foliar-applied. It soon became apparent that the complexity of developing a bioherbicide into a commercial product had been underestimated, and thorough investigation into the appropriate formulation, spray application methods and technologies required for mass production and stabilization was needed. Researchers had overlooked the regulatory climate for registering and commercializing the technology at the time, while ensuring that there were no detrimental effects on birds, fish, and so on and that the bioherbicides were safe to humans and mammals and not persistent in the environment. Our current weed management practices have led to the growing view of producers and consumers that traditional application of synthetic pesticides has created herbicide resistance, spray drift and the incidence of chemical residues in our soil and water. Increasing consumer awareness about chemical pesticide residues in food, the phasing out of registered synthetic herbicides and banning of chemicals for cosmetic use in lawns, gardens and urban municipalities have raised awareness of the development of 'green' technologies. The lack of new discoveries of herbicide modes of action over the last 20 years has underscored our need to investigate new technologies and sources of active ingredients for weed control.

If integrated weed management is to become a viable weed control option rather than a poorly defined system, application of bioherbicides needs to be considered. Teams of researchers are required to investigate how bioherbicides can be implemented into crop production systems in rural and urban landscapes. The teams must include plant pathologists, weed ecologists, agronomists, chemists, microbiologists and molecular biologists. Bioherbicides are not stand-alone products, just as synthetic herbicides should not be the lone weed control preference.

Researchers need to start thinking with an entrepreneurial spirit and consider the business side of science, while industry must also realize that scientists do not always have the inclination, expertise or business resources to provide all the market data to support a full business case justifying investment in their bioherbicide technology. On the other hand, industry has access to market and economic assessments by economists within their company. There needs to be a true partnership amongst all parties. What may be a small market to a multinational company may be a great opportunity to a smaller enterprise. The simplicity and ease of use of synthetic herbicides have encouraged their use as a quick fix, while we need to become better environmental stewards of the world. The use of bioherbicides in crop production as a complement to present pest control products should be actively encouraged.

10 Where to look for further information

Butt, T. M., C. W. Jackson, and N. Magan (Eds.), 2001. *Fungi as Biocontrol Agents: Progress, Problems and Potential*, CABI Publishing, Wallingford, UK.

Hall, F. R. and J. J. Menn (Eds.), 1998. *Biopesticides: Use and Delivery*, Humana Press, Totowa, New Jersey.

Singh, H. P., D. R. Batish, and R. K. Kohli (Eds.), 2006. *Handbook of Sustainable Weed Management*, The Haworth Press, Inc., New York.

Vincent, C., Goettel, M. S., and Lazarovits, G. (Eds.), 2007. *Biological Control: A Global Perspective*, CABI Publishing, Wallingford, UK.

11 References

Abu-Dieyeh, M. H. and Watson, A. K. 2007. Efficacy of Sclerotinia minor for dandelion control: Effect of dandelion accession, age, and grass competition. *Weed Research* 47:63–72

Ash, G. 2010 The science, art, and business of successful bioherbicides. *Biological Control* 52:230–40.

Auld, B. A. 1993. Vegetable oil suspension emulsions reduce dew dependence of a mycoherbicide. *Crop Protection* 12:477–9.

Auld, B. A. and Morin, L. 1995. Constraints in the development of bioherbicides. *Weed Technology* 9:638–52.

Bailey, K. L. 2010. Canadian innovations in microbial biopesticides. *Canadian Journal of Plant Pathology* 32:113–21.

Bailey, K. L. 2014. The bioherbicide approach to weed control using plant pathogens. Pages 245–65 In: Abrol, D. P. (Ed.), *Integrated Pest Management*, Elsevier, Inc.

Bailey, K. L. and Falk, S. 2011. Turning research on microbial bioherbicides into commencial products – A Phoma story. *Pest Technology* (Special Issue 1): 73–9.

Bailey K. L. and Mupondwa, E. K. 2006. Developing microbial weed control products: Commercial, biological, and technological considerations. Pages 431–73 In: H. P Singh , D. R. Batish and R. K. Kohli (Eds.), *Handbook of Sustainable Weed Management*, The Haworth Press, New York.

Bailey, K. L., Boyetchko, S. M., Peng, G., Hynes, R. K., Taylor, W. G., and Pitt, W. M. 2009. Developing weed control technologies with fungi. Pages 1–44 In: Rai, M. (Ed.), *Advances in Fungal Biotechnology*, I. K. International Pvt.Ltd., New Delhi.

Bailey, K. L., Boyetchko, S. M. and Längle, T. 2010. Social and economic drivers shaping the future of biological control: A Canadian perspective on the factors affecting the development and use of microbial biopesticides. *Biologilcal Control* 52:221–9. doi:10.1016/j.biocontrol.2009.05.

Begonia, M. F. T., Kremer, R. J., Stanley, L. and Jamshedi, A.. 1990. Association of bacteria with velvetleaf roots. *Transactions of Missouri Academy Science* 24:17–26.

Bhowmik, P. C. 1999. Herbicides in relation to food security and environment: a global perspective. *Indian Journal of Weed Science* 31:111–23.

Boyetchko, S. M. 1997. Efficacy of rhizobacteria as biological control agents of grassy weeds. Pages 460–5 In: Proc. Soils and Crops Workshop '97, Saskatoon, Saskatchewan, 20–21 February 1997.

Boyetchko, S. M. 1999. Innovative applications of microbial agents for biological weed control. Pages 73–97. In: K. G. Mukerji, B. P. Chamola and K. Upadhyay (Eds.), *Biotechnological Approaches in Biocontrol of Plant Pathogens*, Kluwer Academic/Plenum Publishers, London.

Boyetchko, S. M. 2005. Biological herbicides in the future. Pages 29–47 In: J. A. Ivany (Ed.), *Weed Management in Transition. Topics in Canadian Weed Science*, Volume 2, Canadian Weed Science Society – Societe canadienne de malherbologie,Sainte-Anne-de-Bellevue, Quebec.

Boyetchko, S. M. and Mortensen, K. 1993. Use of rhizobacteria as biological control agents of downy brome. Pages 443–8. In: Proc. Soils and Crops Workshop '93, Saskatoon, Saskatchewan, 25–26 February 1993.

Boyetchko, S. M. and Peng, G.. 2004. Challenges and strategies for development of mycoherbicides. Pages 111–21. In: D. K. Arora, P. Bridge and D. Bhatnagar (Eds.), *Fungal Biotechnology in Agricultural, Food, and Environmental Applications*, Volume 21, Marcel Dekker Inc.

Boyetchko, S. M. and Rosskopf, E. N. 2006. Strategies for developing bioherbicides for sustainable weed management. Pages 393–430. In: H. P. Singh, D. R. Batish and R. K. Kohli (Eds.), *Handbook of Sustainable Weed Management*, The Haworth Press, Inc., New York

Boyetchko , S. M., Bailey, R. K. and Peng, G. 2007. Development of an inundative mycoherbicide: BioMal®. Pages 274–83. In: C.Vincent, M. S.Goettel and G. Lazarovits (Eds), *Biological Control: A Global Perspective*, CABI Publishing, Wallingford, UK.

Boyette, C. D., Quimby, P. C. Jr., Bryson, C. T., Egley, G. H. and Fulgham, F. E. 1993. Biological control of hemp sesbania (Sesbania exalta) under field conditions with Colletotrichum truncatum formulated in an invert emulsion. *Weed Science* 41:497–500.

Boyette, C. D., Hoagland, R. E. and Weaver, M. A.. 2007. Interaction of a bioherbicide and glyphosate for controlling hemp sesbania in glyphosate-resistant soybean. *Weed Biology and Management* 8:18–24.

Caldwell, C. J. G., Hynes, R. K., Boyetchko, S. M. and Korber, D. 2012. Colonization and bioherbicidal activity on green foxtail by Pseudomonas fluorescens BRG100 in a pesta formulation. *Canadian Journal of Microbiology Canadian Journal of Microbiology* 58:1–9

Charudattan, R. 1991. The mycoherbicide approach with plant pathogens. Pages 24–57. In: D. O. TeBeest, (Ed.), *Microbial Control of Weeds*, Chapman and Hall, Inc., New York.

Charudattan R. 2001. Biological control of weeds by means of plant pathogens: significance for integrated weed management in modern agro-ecology. *BioControl* 46:229–60.

Cherrington, C. A. and Elliott, L. F. 1987. Incidence of inhibitory pseudomonads in the Pacific Northwest. *Plant Soil*. 101:159–65.

Connick, W. J. Jr., Boyette, C. D. and McAlpine, J. R.. 1991. Formulation of mycoherbicides using a pasta-like process. *Biological Control* 1:281–7.

Connick, W. J. Jr., Daigle, D. J., Boyette, C. D. and McAlpine, J. R.. 1996. Water activity and other factors that affect the viability of Colletotrichum truncatum conidia iln what flour-kaolin granules ('Pesta'). *Biocontrol Science and Technology* 1:277–84.

Cross, J. V. and Polonenko, D. R., 1996. An industry perspective on registration and commercialization of biocontrol agents in Canada. *Canadian Journal of Plant Pathology* 18: 446–54.

Daigle, D. J., Connick, Jr., W. J. and Boyetchko, S. M. 2002. Formulating a weed-suppressive bacterium in 'pesta'. *Weed Technology*. 16:407–13.

De la Bastide, P. Y. and Hintz, W. 2007 Developing the production system for Chondrostereum purpureum. Pages 291–9. In: C.Vincent, M. S.Goettel and G. Lazarovits (Eds.), *Biological Control: A Global Perspective*, CAB International, Wallingford, UK.

De Jong, M. D., Scheepens, P. C. and Zadoks, J. C. 1990. Risk analysis for biological control: A Dutch case study in biocontrol of Prunus serotina by the fungus Chondrostereum purpureum. *Plant Disease* 74:189–94.

Derksen, D. A., Blackshaw, R. E. and Boyetchko, S. M. 1996. Sustainability, conservation tillage and weeds in Canada. *Canadian Journal of Plant Science* 76:651–9.

Dumas, M. T., Wood, J. E., Mitchell, E. G. and Boyonoski, N. W. 1997. Control of stump sprouting of Populus tremuloides and P. grandidentata by inoculation with Chondrostereum purpureum. *Biological Control* 10:37–41.

Graupner P. R., Carr, A., Clancy, E., Gilbert, J., Bailey, K. L., Derby, J. and Gerwick, B. C. 2003. The macrocidins: Novel cyclic tetramic acids with herbicidal activity produced by Phoma macrostoma. *Journal of Natural Products* 66(12):1558–61.

Green, S., Stewart-Wade, S. M., Boland, G. J., Teshler, M. P. and Liu, S. H. 1998. Formulating microorganisms for biological control of weeds. Pages 249–81 In: G. J.Boland and L. D.Kuykendall (Eds.), *Plant-Microbe Interactions and Biological Control*, Marcel Dekker, Inc., New York.

Gurusiddaiah, S., Gealy, D. R., Kennedy, A. C. and Ogg, A. G. Jr. 1992. Production, isolation, and characterization of phytotoxic and fungistatic compounds for biocontrol of downy brome (Bromus tectorum L.) and plant pathogenic fungi. *Abstracts: Meeting of the Weed Science Society of America* 32:84.

Hallett S. G. 2005. Where are the bioherbicides? *Weed Science* 53: 404–15.

Hanson, C. 2008. Root colonization and environmental fate of the bioerbicide Pseudomonas fluorescens BRG100. M.Sc. thesis, University of Saskatchewan, Saskatoon, Saskatchewan.

Heap, I. Herbicide resistant weeds. 2014. Pages 281–301 In: D. Pimental and R. Peshin (Eds.), *Integrated Pest Management, Pesticide Problems*, Vol. 3, Springer, Netherlands.

Hintz, W. 2007 Development of Chondrostereum purpureum as a mycoherbicide for deciduous brush control. Pages 284–90 In: C.Vincent, M. S.Goettel and G. Lazarovits (Eds.), *Biological Control: A Global Perspective*, CAB International, Wallingford, UK.

Hoagland, R.E . 1996. Chemical interactions with bioherbicides to improve efficacy. *Weed Technology*. 10:651–74.

Hynes, R. K. and Boyetchko, S. M. 2006. Research initiatives in the art and science of biopesticide formulations. *Soil Biology and Biochemistry*. 38:845–9

Hynes, R. K. and Boyetchko, S. M. 2011. Improvements in the pesta formulation to promote survival of Pseudomonas fluorescens BRG100, green foxtail bioherbicide. *Pest Technology* (Special Issue 1):80–7

Hubbard, M. R. K., Hynes, M. Erlandson and Bailey, K. L. 2015. The biochemistry behind efficacy. *Sustainable Chemical Processes* 2:18–25.

Imaizumi, S., Nishino, T., Miyabe, K., Fujimori, T. and Yamada, M. 1997. Biological control of annual bluegrass (Poa annua L.) with a Japanese isolate of Xanthomonas campestris pv. poae (JT-P482). Biol. Control 8:7–14.

Kelly, I. D. and Allen, R. 2011. An industry perspective on challenges and hurdles faced in the development of agrochemicals. Pages 3–12 In: K. N. Harker (Ed.), *The Politics of Weeds. Topics in Canadian Weed Science*, Volume 7, Canadian Weed Science Society – Societe canadienne de malherbologie,Pinawa, Manitoba.

Kennedy, A. C., Elliott, L. F., Young, F. L. and Douglas, C. L. 1991. Rhizobacteria suppressive to the weed downy brome. *Soil Science Society of America Journal* 55:722–7.

Kenney D. S. 1986. DeVine® –The way it was developed – an industrialist's view. *Weed Science* 34 (Suppl. 1):15–16.

Kremer, R. J. and Kennedy, A. C. 1996. Rhizobacteria as biocontrol agents of weeds. *Weed Technology.* 10:601–9.

Kremer, R. J., Begonia, M. F. T., Stanley, L. and Lanham, E. T. 1990. Characterization of rhizobacteria associated with weed seedlings. *Applied and Environmental Microbiology* 56:1649–55.

Kremer, R. J. and Sarwar, M. 1995. Microbial metabolites with potential applications in weed management. *Plant Growth Regulation Society of America.* 21:48–51.

Marrone, P. G. 1999. Microbial pesticides and natural products as alternatives. *OutlookonAgriculture* 28:149–54.

Marrone Bio Innovations 2016, https://marronebioinnovations.com/marrone-bio-innovations-reports-fourth-quarter-and-record-full-year-2016-results/

Ott S. L., Huang, C. L. and Misra, S. K. 1991. Consumer's perceptions of risks from pesticide residues and demand for certification of residue-free produce. Pages 175–87 In: J. A.Caswell (Ed.),*Economics of Food Safety*, Elsevier Science Publishing Co., The Netherlands.

Pedras, M. S. C., Ismail, N., Quail, J. W. and Boyetchko, S. M. 2003. Structure, chemistry, and biological activity of pseudophomins A and B, new cyclic lipodepsipeptides isolated from the biocontrol bacterium Pseudomonas fluorescens. *Phytochemistry* 62:1105–14.

Peng, G. and Byer, K. N. 2005. Interacations of Pyricularia setariae with herbicides for control of green foxtail (Setaria viridis). *Weed Technology* 19:589–98.

Prasad, R. 1994. Influence of several pesticides and adjuvants on Chondrostereum purpureum – a bioherbicide agent for control of forest weeds. *Weed Technology* 8:445–9.

Quail, J. W., Ismail, N., Pedras, M. S. C. and Boyetchko, S. M. 2002. Pseudophomins A and B, a class of cyclic lipodepsipeptides isolated from a Pseudomonas species. *Acta Crystallographica* C58:o268–o271.

Quimby, P. C. Jr., Zidack, N. K., Bohette, C. D. and Grey, W. E. 1999. A simple method for stabilizing and granulating fungi. *Biocontrol Science and Technology* 9:5–8.

Rosskopf, E. N., Charudattan, R. and Kadir, J. B. 1999. Use of plant pathogens in weed control. Pages 891–918 In: T. S. Bellows and T. W. Fisher (Eds.), *Handbook of Biological Control*, Academic Press, Cambridge.

Sarwar, M. and Kremer, R. J. 1995a. Enhanced suppression of plant growth through production of L-tryptophan-derived compounds by deleterious rhizobacteria. *Plant Soil* 172:261–9.

Sarwar, M. and Kremer, R. J. 1995b. Determination of bacterially derived auxins using a microplate method. *Letters in Applied Microbiology* 20:282–5.

Saxena, S. and Pandey, A. K. 2001. Microbial metabolites as eco-friendly agrochemicals for the next millennium. *Applied Microbiology and Biotechnology* 55: 395–403.

Schnick, P. J., Stewart-Wade, S. M. and Boland, G. J. 2002. 2,4-D and Sclerotinia minor to control common dandelion. *Weed Science* 50:173–8.

Shamoun, S. F., Ramsfield, T. D., Shrimpton, G. and Hintz, W. E. 1996. Development of Chondrostereum purpureum as a mycoherbicide for red alder (Alnus rubra) in utility rights-of-way (Abstr). Page 199 In: P. Comeau and G. Harper (Eds.), *Proceedings of the National Meeting of the Expert Committee on Weeds*, 9–12 December 1996, Expert Committee on Weeds, Victoria, B.C.

Suslow, T. V. and Schroth, M. N. 1982. Role of deleterious rhizobacteria as minor pathogens in reducing crop growth. *Phytopathology* 72:111–15.

TeBeest, D. O. and Templeton, G. E. 1985. Mycoherbicides: Progress in the biological control of weeds. *Plant Disease* 69:6–10.

Tranel, P. J., Gealy, D. R. and Irzyk, G. P. 1993a. Physiological responses of downy brome (Bromus tectorum) roots to Pseudomonas fluorescens strain D7 phytotoxin. *Weed Science.* 41:483–9.

Tranel, P. J., Gealy, D. R. and Kennedy, A. C. 1993b. Inhibition of downy brome (Bromus tectorum) root growth by a phytotoxin from Pseudomonas fluorescens strain D7. *Weed Technology.* 7:134–9.

Walker, H. L. and Connick, W. J. Jr. 1983. Sodium alginate for production and formulation of mycoherbicides. *Weed Science* 31:333–8.

Wall, R. E. 1994. Biological control of red alder using stem treatments with the fungus Chondrostereum purpureum. *Canadian Journal of Forest Research* 24:1527–30.

Watson, A. K. and Bailey, K. L. 2013. Tarxacum officinale (Weber), Dandelion (Asteraceae) Pages 383–91 In: P. G. Mason, D. R. Gillespie (Eds.), *Biological Control Programmes in Canada, 2001-2012*, CAB International, Wallingford, UK.

Womack, J. G. and Burge, M. N. 1993. Mycoherbicide formulation and the potential for bracken control. *Pesticide Science* 37:337–41.

Zhou, L., Bailey, K. L., Derby, J. 2004. Plant colonization and environmental fate of the biocontrol fungus Phoma macrostoma. *Biological Control* 30:634–44.

Zidack, N. K. and Backman, P. A. 1996. Biological control of kudzu (Puerari lobata) with the plant pathogen Pseudomonas syringae pv. phaseolicola. *Weed Science* 44:645–9.

Zidack, N. K. and Quimby, P. C. Jr. 2002. Formulation of bacteria for biological weed control using the stabileze method. *Biocontrol Science and Technology* 12:67–74.

Chapter 20

The use of bacteria in integrated weed management

Ann C. Kennedy, USDA-ARS and Washington State University, USA

1 Introduction

Invasive grass weeds continue to gain ground against native plant species, especially in the arid crop and rangelands of the western United States, where an elegantly adaptive invasive grass weed, downy brome (cheatgrass, *Bromus tectorum* L.), dominates the landscape (Balch, 2013; Haubensak et al., 2009).

Understandably, the prospect of finding a successful biocontrol agent to manage weeds has its sceptics, perhaps because researchers have prematurely spoken or published on finding a pathogen of a problematic weed before conducting extensive, non-target studies, only to find later that the microorganism inhibits more than just the target weed (Cordeau et al., 2016; Ghosheh, 2005). It is not difficult to find bacteria or fungi that can inhibit weed growth. However, it is a completely different matter to find microorganisms that specifically suppress a weed and do not harm any other flora or fauna. A defining characteristic of a good herbicide is its selectivity. The same criterion must be paramount in the search for biological control organisms. Deleterious rhizobacteria were identified in the 1980s, and strains that could inhibit various crops and weeds were found (Kremer et al., 2006; de Luna et al., 2005; Nehl et al., 1997; Kremer, 1986, 1987; Frederick and Elliott, 1985; Suslow and Schroth, 1982), but few bacteria were screened across many plant families or studied extensively in the field over multiple years. Without performing

http://dx.doi.org/10.19103/AS.2017.0025.22

a comprehensive, methodical study of the effect of a biocontrol candidate in both small- and large-scale studies over many years in the field, accurate conclusions cannot be drawn.

Naturally occurring soil bacteria have been identified that specifically inhibit the growth of various invasive weeds by targeting the seed bank (Kennedy, 2016; Kennedy et al., 1991, 2007). This is the story of how these bacteria were isolated, identified and comprehensively evaluated as viable biocontrol agents.

2 The case of downy brome (cheatgrass, *Bromus tectorum* L.)

Downy brome has a number of unique traits that make it extremely competitive. Downy brome roots grow late into the fall, well after the growth of native plants has stopped. Downy brome roots begin growing in the spring long before other plant species awaken from winter dormancy. These two growth characteristics alone allow downy brome access to water and soil nutrients before other plants (Rice, 2005). In addition, downy brome flowers and matures early, well before most plant species, and it completes its life cycle before summer drought. Downy brome, which is extremely hardy and flexible, can flourish in the poor soils of scabland or in highly productive cropland and can produce seed in full sunlight or shade. Downy brome dramatically reduces yields in croplands (Stahlman and Miller, 1990; Thill et al., 1984) and replaces natives in rangelands (Rice, 2005; Duncan et al., 2004). Monocultures of downy brome smother native habitat and forage necessary for wildlife, birds such as the sage-grouse, and livestock (Maher et al., 2013; Epanchin-Niell et al., 2009; Brooks et al., 2004). Recreational lands (Rice, 2005; Duncan et al., 2004) and sacred Native American lands that produce medicinal plants (Borins, 1995) are also negatively affected by downy brome. The mat of highly flammable residue left by downy brome and other annual grass weeds provides fuel for wildfires (Balch, 2013; Epanchin-Niell et al., 2009; Brooks et al., 2004; USGS, 2002) and increases fire frequency and intensity in the western United States (Balch, 2013; Haubensak et al., 2009). After a fire, downy brome is the first plant to germinate and it quickly occupies the newly burned area before desirable species return (USGS, 2002). Downy brome residue has been held liable for wildfires that not only have a negative ecological effect on plant diversity, air, water and soil quality, but also result in staggering economic losses associated with fire suppression, structure loss, and cost of rangeland restoration (Wisdom and Chambers, 2009; Duncan et al., 2004). Wildfires fuelled by downy brome not only destroy native range, standing crops and property, but can result in the tragic loss of human life (US Fire Administration, 2012).

Traditional weed management practices involve annual application of synthetic herbicides. While the quick, visual disappearance of the above-ground weed is pleasing to the eye, what is hiding below the soil surface is of real concern. Many weeds, including downy brome, have copious seed banks that are untouched by synthetic herbicides and can remain viable in the soil for years, resulting in the annual appearance of weeds, even when herbicides are applied (Rice, 2005). Routine, annual herbicide application has failed to control downy brome and exacerbated the growing herbicide resistance problem. In the battle against downy brome and other similar weeds, the seed bank, rather than the standing plant, should be the target of suppression.

3 Finding a biocontrol agent to manage downy brome

3.1 Personal note

My home is nestled in the rolling hills of the Palouse in arid Eastern Washington. My neighbours grow cereal crops and legumes and raise livestock on rangelands along the nearby Snake River canyon. For three decades, I have observed the march of downy brome and the build-up of highly flammable residue, which increases the frequency and severity of wildfires across the West (Balch, 2013). My research has focused on developing weed-suppressive bacteria to stop invasive grass weeds (Kennedy, 2016; Kennedy et al., 1991). On 1 August 2016, my battle with downy brome became personal. A wildfire, fuelled by downy brome residue and pushed by high winds, raced along the breaks of the Snake River and towards my home. After a smoky, tense night, the fire was contained, but not before it consumed 4400 hectares (11 000 acres) of quality grazing land. Downy brome will be the first plant to return and over the next few years, it will eliminate most natives and further increase the frequency and severity of fires in the burned area (Epanchin-Niell et al., 2009). Clear, defensible agricultural, economic and personal arguments demand that downy brome must be stopped. Weed-suppressive bacteria coupled with other weed management tools can be utilized to delay the unbridled spread of downy brome and other invasive weeds. Native bunchgrasses can then re-establish, cereals can grow unhindered, and productivity can be restored.

3.2 The bacteria

In the early spring, soil and roots from patches of poorly growing grass or cereals (Fig. 1) in an otherwise healthy field harbour bacteria that suppress cereal growth and selectively inhibit certain cultivars and accessions of cereal crops (Fredrickson and Elliott, 1985). After years of studying freeze-thaw events, unique bacterial populations were found to

Figure 1 Patch of stunted wheat and other grasses (lighter colour at bottom slope) used as a sample site to screen for weed-suppressive bacteria.

be at their highest levels in soil during thaws that occurred after a hard freeze. The soil associated with these stressed plants was enriched in bacteria that suppress plant growth. In the Palouse region of Eastern Washington, soil and roots from areas of stunted grasses were collected in the early spring in hope of 'mining' for weed-suppressive bacteria (Kennedy, 2016; Kennedy et al., 1991). More than 25 000 bacterial isolates were collected over several years and subjected to a comprehensive laboratory screening protocol using root agar bioassays to identify those isolates that were selective in suppressing the growth of downy brome and other grass weeds, without harming crops, cereals, native grasses and forbs (Fig. 2). Weed-suppressive candidates were then tested in greenhouse assays and, finally, in field trials. Several bacterial isolates were found to inhibit downy brome in the field, without harming crops or native plants (Kennedy et al., 1991). One of these bacterial isolates recovered from the soil in 2001 was found to be a strain of *Pseudomonas fluorescens* biovar B and was designated ACK55. ACK55 exists in very low numbers or is dormant when soil temperatures are above 25°C (77°F), but unlike most soil bacteria, ACK55 thrives when temperatures cool below 10°C (50°F). *Pseudomonas fluorescens* strain ACK55 also inhibits medusahead (*Taeniatherum caput-medusae* (L.) Nevski) and jointed

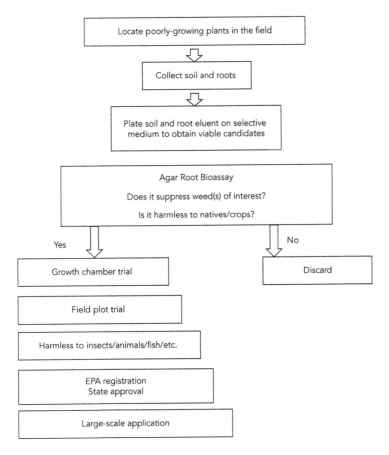

Figure 2 Flow chart of screening to obtain the weed-suppressive bacteria.

goatgrass (*Aegilops cylindrica* Host), but, here, the focus is on downy brome, also called cheatgrass.

Downy brome has a distinct, competitive advantage over native plants and crops (Rice, 2005; Knapp, 1996). Downy brome roots grow later in the fall and earlier in the spring than native plants and crops, thereby monopolizing soil water and nutrients (Thill et al., 1984). The bacteria also thrive late in the fall and early in the spring. This similarity of life cycles makes ACK55 an effective inhibitor of downy brome. ACK55 suppresses downy brome by inhibiting root cell elongation, and reducing root growth and tiller formation as do many of the weed-suppressive bacteria in our library (Kennedy et al., 1991). The reduction in root growth caused by ACK55 removes the competitive edge of the downy brome, allowing native plants and crops to utilize the soil moisture and nutrients that the downy brome would otherwise use.

Natives, having evolved over time in a specific area, adapt to the environmental conditions of their area. They are successful in growing within the specifics of the site, the particular nutrient balance, physical conditions, moisture patterns, and biotic community. An invasive weed, like downy brome, can alter the moisture regime of an area, which negatively impacts native plants. If the root elongation of the invasive weed is limited, the natives will quickly regain their position in their place of origin. Natives have a site-specific edge that invasive weeds and cultivars of natives from other geographic locations lack (Jerry Benson, BFI Native Seeds, Inc., Moses Lake WA, personal communication). Over time, the native plants flourish and compete with the remaining downy brome.

3.3 Screening

Host-range studies investigating more than 250 select grass species, various broadleaf plants in representative families, and 40 crop species confirmed that downy brome accessions were significantly inhibited by ACK55, without harming other plant species in root length bioassays in agar and growth chamber pot studies (Fig. 3 and 4). Wapshere's concept (1974) of concentric patterns chooses more plant species in related tribes and families closest to downy brome and fewer, but key representatives, of plant species in distantly related families. In addition, many other characteristics were evaluated (Table 1).

Figure 3 Downy brome plants grown in non-sterile soil in growth chamber. Left, control; middle and right, treated with two different weed-suppressive bacteria at 10^8 bacteria per pot. Pots were 15 cm in diameter and plants were grown for 21 days. Downy brome accession was from Lind, WA.

Figure 4 Winter wheat plants grown in non-sterile soil in growth chamber. Left, control; middle and right, treated with two different weed-suppressive bacteria at 10^8 bacteria per pot. Pots were 15 cm in diameter and plants were grown for 21 days. Winter wheat cultivar was 'Hill81'.

Table 1 Traits of the weed-suppressive bacterium *Pseudomonas fluorescens* strain ACK55

- ACK55 suppresses growth of downy brome, medusahead and jointed goatgrass.
- ACK55 does not suppress other weed species.
- Does not injure desirable plant species such as crops or native and near-native plant species in the field (over 250 grass species, 40 crop species and 20 broadleaf species tested).
- Increases in cell number in the soil only in the late fall, winter and early spring (<50°F). Weed-suppressive bacteria populations are low during the warmer temperatures of the plant growth season.
- Does not enter the plant cell.
- Produces a weed-suppressive compound in the root intercellular spaces. The compound breaks down very easily, has no residual and is not active in the soil.
- Is not competitive and survival in field soil is less than six years.
- Does not grow in natural waters including ditches, rivers, lakes and oceans.
- Has no anti-microbial activity.
- Does not produce enzymes that lyse cell membranes and enter plant cells.
- Has no protein secretions (Type I, II, III) that could harm non-target plants.
- Has low to no allergenic protein secretions.
- Does not elicit hypersensitivity reactions to humans or domestic animals.
- Has the weed-suppression gene located at multiple positions on the chromosome. This reduces risk of gene transfer to negligible levels.
- Has no plasmids that could be transferred among organisms.
- Does not alter soil microbial communities after application.
- Does not harm Daphnia, wireworm, Lemna, lady bugs, honey bees, birds, rabbits or rodents.

ACK55 is unable to enter the plant cell. It was selected as a weed-suppressive bacterium, because it lacks the enzymes necessary to break down the plant cell wall and membrane. The active compound produced by ACK55 is a complex of proteins and lipids that breaks down readily and is only active in nature when the bacterium inhabits the first few outside rows of root cells. This compound is similar to, but different from, the active compounds produced by the first group of ARS weed-suppressive bacteria from the late 1980s (Gurusiddaiah et al., 1994). There are no extrachromosomal plasmids in ACK55 and the genes responsible for the inhibitory compound are found at multiple locations on the chromosome, making transmission of genetic material between organisms highly unlikely.

4 Application and results

4.1 Field application

The bacteria can be applied to the soil as a spray, or coated onto crop or native seed and planted beneath the soil surface at levels of 4×10^{11} to 10^{12} colony forming units (CFU) per hectare (8×10^{11} to 10^{12} CFU acre^{-1}, Kennedy et al., 1991). The timing of ACK55 application is critical to the success of reduction and removal of downy brome. The harsh environment of the soil surface (ultraviolet light, low moisture, high temperatures) limits the survival of ACK55 (Fig. 5, #7, #8). However, if ACK55 is moved into the soil by rain or even melting snow, they survive well. Therefore, application of ACK55 should coincide with cool air temperatures (below 10°C (50°F)) and fall rains to ensure adequate survival. These weed-suppressive bacteria inhibit at the seedling stage and will not significantly inhibit the growth of standing plants. Weed growth in late summer from early fall rains requires treatment with a synthetic herbicide (low rates of imazapic or glyphosate) to reduce already standing plants. ACK55 can

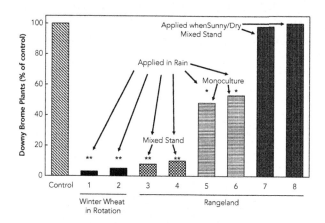

Figure 5 Downy brome plants (per cent of control) 5 years after different application treatments. ACK55 was applied to the soil as a spray at 4×10^{11} cells hectare^{-1} (8×1011 cells A^{-1}) in the fall of the year. Each 2 bar represents four sites with five replications of 0.4 hectare (1 acre) plots. Control for each site was set at 100%. Bars 1 through 6 represent plots in which ACK55 was applied in rain. Bars 7 and 8 are plots in 4 which ACK55 was applied when it was sunny and dry with no rain for at least two days after application. Wheat from a winter wheat/spring rotation in Washington is represented in bar1 and in Idaho is 6 represented in bar2. Bars 3, 4, 7, and 8 represent rangeland having a mixed stand of natives and downy brome. Bars 5 and 6 represented monoculture downy brome. Bars were significantly different from 8 control at P ≤ 0.10 = * and P ≤ 0.05 = **.

be tank-mixed with most herbicides; however, surfactants that are soapy or oily and adjuvants that are added to kill microbial growth may also kill the weed-suppressive bacteria.

4.2 Field results

It may take ACK55 several years to suppress weeds as the bacteria inhibit germinating seeds in the seed bank. In the first few years after field application, ACK55 inhibits downy brome growth by 20–50%. The inhibition increases with time, reaching maximum weed suppression (% of control) three to six years after application. In winter wheat fields, suppression of weed growth by ACK55 allowed the wheat to be more competitive, which in turn reduced weed populations further (Fig. 5, #1, #2, Fig. 6). In long-term rangeland field trials in WA, spray application of the ACK55 resulted in almost complete suppression of downy brome five to seven years after a single bacterial application when a crop or perennial species were present (Fig. 5, #1–4). Monoculture downy brome plots only reached 50% inhibition with ACK55 when no native seed was present (Fig. 5, #5, #6). A desirable plant is needed to further compete with the downy brome. The bacterium must survive to reduce the downy brome and if the bacteria is applied before or during sunny weather, little to no downy brome reduction is seen (Fig. 5, #7, #8). At each treated site where the

Figure 6 Winter wheat seeded with a no-till paired row drill in Grant County Washington. a) Control; b) ACK55 applied at 2×10^{11} cells per hectare (4×10^{11} cells A^{-1}). Plots were 33 hectares (350 ft^2) and replicated five times.

Figure 7 Burned land treated with ACK5. The control plot is to the left of the fencing. To the right of the fencing, ACK55 was applied as a spray at 4×10^{11} cells per hectare (8×10^{11} cells A^{-1}). The whole site was burned by wildfire in spring 2015; weed-suppressive bacteria were sprayed on December 2015; picture was taken on September 2016. The bacteria were sprayed along the fence row and to the right of the fence. Photo from Jerry Benson.

bacteria reduced downy brome, the populations of more desirable plant species increased as the invasive weed became less competitive. The bacteria, synthetic herbicide, plus the native plant, turf or wheat interact to reduce the downy brome. When applied to a wildfire site, these weed-suppressive bacteria can reduce the downy brome populations and allow desirable plant species to take hold (Fig. 7). As ACK55 reduces downy brome, voids are created that other weeds, especially broadleaf weeds fill, and broadleaf herbicides are necessary in the spring of each year. The coupling of ACK55 application and seeding of desirable/native plant species yields optimum results. Restoration and ridding an area of the invasive weed, downy brome, is a long-term undertaking that involves bacteria-weed-plant-herbicide interactions and an in-depth adaptive planning.

5 Summary

Sustainable crop production and pasture and rangeland improvements will benefit from research directed towards the discovery, characterization and utilization of soil bacteria that selectively suppress grass weeds. The soil contains weed-suppressive bacteria that are extremely selective, are well matched to the weed(s) of interest, and do not harm other members of the agroecosystem. These bacteria easily and successfully fit into invasive weed management plans in cropland and in rangeland restoration programmes. Weed-suppressive bacteria can reduce and complement synthetic herbicides, expand options in weed management, reduce farm and ranch costs, and encourage the use of ecologically based systems. Because weed-suppressive bacteria work at the seed bank and seedling level, they pair well with herbicides that work at the plant level. In addition, synthetic herbicides will also be used to manage the other weeds that ultimately germinate in the void created by bacterial suppression and loss of the target weed. These agents should reduce weeds effectively and economically, increase producer net profit, and lead to greater sustainability.

Weed-suppressive bacteria provide a novel way to think about weed management. When bacteria are applied in concert with native seed and the correct herbicide at the optimum rate and timing, the seed bank of the target weed is depleted and weed populations decrease to negligible levels. Weed-suppressive bacteria have an essential place in agriculture and rangeland weed management. They represent a novel approach to reduce weed populations and herbicide use. They increase yields by limiting invasive weeds and can lead to sustainable crop production. As invasive weeds disappear, rangeland plant diversity will increase, and forage quality will improve as native bunchgrasses and forbs return. Bacteria are an underdeveloped and underutilized tool for effective weed management.

6 Future trends in research

From a teaspoon of soil, tens of thousands of bacterial isolates can be vetted against nearly any plant at the seedling stage, including grasses, broadleaf plants, shrubs and trees. Research on weed-suppressive bacteria is not limited by the amount of material available for study, nor does it require sophisticated instrumentation or cutting-edge technology. It does, however, require insight – and methodical, comprehensive screening – to ensure the bacterial agent is specific to the weed target(s) and harmless to other life. New bacterial isolates that selectively suppress other weed species continue to emerge and validate the importance of pursuing additional weed-suppressive bacteria (Kennedy, 2016; Mejri et al., 2010; Kennedy and Stubbs, 2007; Zermane et al., 2007; Tichich and Doll, 2006).

The search for naturally occurring bacteria that reduce the growth of other annual grass weeds must continue. This chapter points the way. One of the next research challenges will be to find bacteria that specifically suppress the many broadleaf weeds for weed control not only in agricultural lands, rangeland and rural settings, but also in urban areas such as city parks, golf courses, lawns and gardens.

Discovery of bacterial strains that suppress a variety of weeds can be applied together as a standard weed management practice that is effective, sustainable and environmentally acceptable. Tank mixes consisting of synthetic herbicides and bacterial mixtures can be further developed to utilize every available means to combat weeds. Better plant establishment practices including drill seeding bacteria-coated seed will improve the establishment of crop and rangeland plants. Conceivably, crop seed could be coated with fungicides and weed-suppressive bacteria. Aircraft can be used to deliver the bacteria in wide swathes to pre-emptively establish firebreaks in rangeland to slow the spread of wildfires attributed to highly flammable weed residue. In the future, producers may select herbicides to be used in concert with various combinations of bacteria to diminish both the weed seed bank and weed growth.

Weed-suppressive bacteria can also be applied in urban areas, including residential lawns, parks and golf courses. Much to the chagrin of herbicide companies, there is every reason to believe a bacterial biocontrol agent can be found that inhibits dandelions, unwanted clover, or crabgrass and more. In the future, the seed that is bought to start a lawn or a garden may be coated with specific bacteria that provide a competitive edge to the desired plant. Perhaps tomorrow's bag of 'weed and feed' will contain fertilizer and weed-suppressive bacteria instead of synthetic herbicides. Using this concept, natural products that are herbicidal and environmentally friendly will be available not only to the large-scale grower, rancher and land manager, but also to the organic farmer, homeowner and gardener.

The use of bacteria in weed management is economical, effective and natural. Unlike viral or fungal agents, bacteria are relatively easy to propagate and produce in large quantities. Despite the long list of positive reasons to pursue these biological control agents, there are a number of hurdles to clear before they are accepted. Many users are reluctant to try bacterial products. Even though bacteria are common in food and drink, the public mistakenly perceives all bacteria as 'germs' and harmful causes of disease. Also, herbicide companies are not amenable to losing sales because of bacterial biocontrol agents. Perhaps the biggest obstacle preventing further discovery efforts and application of additional bacterial biocontrol products is the time, money and energy it takes to move these products from the lab to the field and, finally, through the federal bioherbicide registration process.

The US EPA Biopesticides and Pollution Prevention Division is interested in novel microbial products either live organisms or biochemical products that specifically inhibit weeds (Leahy et al., 2014; US EPA, 2016). In an EPA review of microbial herbicides, the first information needed is a clear identification of the microorganism and protocols for growth, and product analysis and manufacturing. The needed bioherbicide data also includes mammalian toxicity and health effects, including dermal and eye testing. The EPA uses a tiered approach in the bioherbicide just as in chemical herbicides. If the product shows no negative effects to the target in tier 1, then no further testing is needed in tier 2 and 3. Hypersensitivity to humans found in laboratory or field experiments is to be reported. Data on product performance and non-target impacts are included as well. Testing may not be required if ample information about the organism or related organisms is in the literature.

As research efforts continue, the number of safe, effective bacterial weed-management products will grow and social acceptance will increase. The majority of soil bacteria are beneficial and not harmful to humans and animals. Discovery and identification of selective weed-suppressive bacteria is time consuming, but further research will continue to yield extraordinary bacterial strains for the management of invasive weeds.

7 Where to look for further information

Although the first reports of deleterious rhizobacteria emerged over 30 years ago (Kremer, 1986; Fredrickson and Elliott, 1985; Suslow and Schroth, 1982), the discovery and application of effective bacterial weed biocontrol agents is still in its infancy. There are excellent review articles on the use of bacteria to suppress weeds (Harding and Raizada, 2015; Stubbs and Kennedy, 2012; Bailey, 2004). Additional information can be obtained from Cordeau et al. (2016), Bailey et al. (2010), Charudattan (2005) and Ghosheh (2005). Wapshere's (1974) concept of concentric spheres of related plant species is a starting point for investigations of bacterial effects on non-target plant species.

EPA regulations and requirements of the toxicological and pathological testing for EPA bioherbicide registration can be found online (US EPA, 2016). Various technologies have been used and will continue to be used to enhance biological weed control, including pairing bacteria with insects (Kremer et al., 2006; Denoth et al., 2002) and synthetic herbicides (Gressel, 2010). Host-range studies are needed to reduce potential risk and ensure that beneficial, non-target plant species are unaffected by the biocontrol agent. Rigorous testing against not only native flora and fauna but also laboratory animal models is required prior to the release of weed-suppressive bacteria to ensure the safety of humans, animals and the environment, while specifically targeting weed reduction.

Between 2000 and 2010, only five insect biocontrol agents were approved by the United States Department of Agriculture's Animal and Plant Health Inspection Services (APHIS) for use and not one insect biocontrol agent has been approved by APHIS between 2010 and the present (R. Zimdahl, personal communication). At a time when the public is calling for less herbicide use, the lack of biocontrol agents being released is of concern. More biological control agents are needed in weed management. The first bacterial bioherbicide to be registered by EPA was *Pseudomonas fluorescens* strain D7 (D7), one of the weed-suppressive bacteria discovered by the USDA-Agricultural Research Service (Kennedy et al., 1991). Unfortunately, the entity registering D7 did not realize that D7 was found to inhibit the growth of several native species when in stressed situation, which made it less specific than initially thought. Setbacks are part of the process of developing a novel system such as bacteria in weed management. The use of naturally occurring bacteria in weed management will have far-reaching benefits that include the removal of weed seed from the seed bank, the use of bioherbicides on a wider array of target weeds, and a reduction in the need for annual application of biological and synthetic herbicides.

8 References

Bailey, K. (2004), Microbial weed control: An off-beat application of plant pathology, *Can. J. Plant Pathol.*, 26(3), 239–44.

Bailey, K., Boyetchko, S. and Langle, T. (2010), Social and economic drivers shaping the future of biological control: A Canadian perspective on the factors affecting the development and use of microbial biopesticides, *Biol. Control*, 52, 221–9.

Balch, J. K. (2013), Introduced annual grass increase regional fire activity across the arid Western USA. (1980–2000), *Global Change Biol.*, 19, 173–83.

Borins, M. (1995), Native healing traditions must be protected and preserved for future generations, *Can. Med. Assoc. J.*, 153(9), 1356–7.

Brooks, M. L., D'Antonio, C. M., Richardson, D. M., Grace, J. B., Keeley, J. E., DiTomaso, J. M., Hobbs, R. J., Pellant, M. and Pyke, D. (2004), Effects of invasive alien plants on fire regimes, *BioScience*, 54, 677–88.

Charudattan, R. (2005), Ecological, practical, and political inputs into selection of weed targets: what makes a good biological control target?, *Biol. Control*, 35(3), 183–96.

Cordeau, S., Triolet, M., Wayman, S., Steinberg, C. and Guillemin, J.-P. (2016), Bioherbicides: Dead in the water? A review of the existing products for integrated weed management, *Crop Prot.*, 87, 44–9.

de Luna, L., Stubbs, T. L., Kennedy, A. C. and Kremer, R. (2005). Deleterious bacteria in the rhizosphere. In Zobel, R. and Wright, S. (Eds), *Roots and Soil Management: Interactions between Roots and the Soil*, Monograph no. 48. ASA: Madison, WI, pp. 233–61.

Denoth, M., Frid, L. and Myers, J. (2002), Multiple agents in biological control: Improving the odds?, *Biol. Control*, 24(1), 20–39.

Duncan, C. A., Jachetta, J. J., Brown, M. L., Carrithers, V. F., Clark, J. K., Ditomaso, J. M., Lym, R. G., McDaniel, K. C., Renz, M. J. and Rice, P. M. (2004), Assessing the economic, environmental, and societal losses from invasive plants on rangeland and wildlands, *Weed Tech.*, 18, 1411–16.

Epanchin-Niell, R., Englin, J. and Nalle, D. (2009), Investing in rangeland restoration in the Arid West, USA: Countering the effects of an invasive weed on the long-term fire cycle, *J. Environ. Manag.*, 91, 370–9.

Fredrickson, J. K. and Elliott, L. F. (1985), Effects on winter wheat seedling growth by toxin-producing rhizobacteria, *Plant Soil*, 83, 399–409.

Ghosheh, H. (2005), Constraints in implementing biological weed control: A review, *Weed Biol. Manag.*, 5(3), 83–92.

Gressel, J. (2010), Herbicides as synergists for mycoherbicides, and vice versa, *Weed Sci.*, 58(3), 324–8.

Gurusiddaiah, S., Gealy, D., Kennedy, A. C. and Ogg Jr., A. (1994), Isolation and characterization of metabolites from *Pseudomonas fluorescens* strain D7 for control of downy brome (*Bromus tectorum* L.), *Weed Sci.*, 42(3), 492–501.

Harding, D. P. and Raizada, M. N. (2015), Controlling weeds with fungi, bacteria and viruses: A review, *Front. Plant Sci.*, 6, 659. DOI:10.3389/fpls.2015.00659.

Haubensak, K., D'Antonio, C. and Wixon, D. (2009), Effects of fires and environmental variables on plant structure and composition in grazed salt desert shrublands of the Great Basin (USA), *J. Arid Environ.*, 73, 643–50.

Imaizumi, S., Nishino, T., Miyabe, K., Fujimori, T. and Yamada, M., 1997, Biological control of annual bluegrass (*Poa annua* L.) with a Japanese isolate of *Xanthomonas campestris* pv. *Poae* (JT-P482), *Biol. Control*, 8, 7–14.

Kennedy, A., Elliott, L., Young, F. and Douglas, C. (1991), Rhizobacteria suppressive to the weed downy brome, *Soil Sci. Soc. Am. J.*, 55(3), 722–7.

Kennedy, A., Johnson, B. and Stubbs, T. (2001), Host range of a deleterious rhizobacterium for biological control of downy brome, *Weed Sci.*, 49(6), 792–7.

Kennedy, A. C. and Stubbs, T. L. (2007), Management effects on the incidence of jointed goatgrass inhibitory rhizobacteria, *Biol. Control*, 40(2), 213–21.

Knapp, P. A. (1996), Cheatgrass (*Bromus tectorum* L.) dominance in the Great Basin Desert- history, persistence, and influences to human activities, *Glob. Environ. Change*, 6, 37–52.

Kremer, R. J. (1986), Bacteria can battle weed growth, *Am. Nurseryman*, 164, 162–3.

Kremer, R. J. (1987), Identity and properties of bacteria inhabiting seeds of selected broadleaf weed species, *Microb. Ecol.*, 14, 29–37.

Kremer, R., Caesar, A. and Souissi, T. (2006), Soilborne microorganisms of Euphorbia are potential biological control agents of the invasive weed leafy spurge, *Appl. Soil Ecol.*, 32(1), 27–37.

Leahy, J., Mendelsohn, M., Kough, J., Jones, R. and Berckes, N. (2014). Biopesticide Oversight and Registration at the U.S. Environmental Protection Agency. In Gross, A. D., Coats, J. R., Duke, S. O. and et al. (Eds), *Biopesticides: State of the Art and Future Opportunities*. ACS Symposium Series; American Chemical Society: Washington, DC, 2014. ACS Symposium Series; American Chemical Society: Washington, DC, 2014, pp. 4–18. DOI:10.1021/bk-2014-1172.ch001.

Maher, A. T., Tanaka, J. A. and Rimbey, N. (2013), Economic risks of cheatgrass invasion on a simulated Eastern Oregon ranch, *Rangeland Ecol. Manage.*, 66, 356–63.

Mejri, D., Gamalero, E., Tombolini, R., Musso, C., Massa, N., Berta, G. and Souissi, T. (2010), Biological control of great brome (*Bromus diandrus*) in durum wheat (*Triticum durum*): Specificity, physiological traits and impact on plant growth and root architecture of the fluorescent pseudomonad strain X33d, *Biocontrol*, 55(4), 561–72.

Mitkowski, N. (2005), First report of bacterial wilt of annual bluegrass caused by *Xanthomonas translucens* pv. *poae* in Montana, *Plant Dis.*, 89, 1016.

Nehl, D. B., Allen, S. J. and Brown, J. F. (1997), Deleterious rhizosphere bacteria: An integrating perspective, *Appl. Soil Ecol.*, 5(1), 1–20.

Rice, P. (2005). Downy brome *Bromus tectorum* L. In Duncan, C. A. and Clark, J. K. (Eds), *Invasive Plants of Range and Wildlands and Their Environmental, Economic, and Societal Impact*. Weed Science Society of America: Lawrence, KS, pp. 147–70.

Sands, D. C. and Rovira, A. D. (1970). Isolation of fluorescent pseudomonads with a selective medium, *Appl. Microbiol.*, 20, 513–14.

Stahlman, P. W. and S. D. Miller. (1990), Downy brome (*Bromus tectorum*) interference and economic thresholds in winter wheat (*Triticum aestivum*), *Weed Sci.*, 38, 224–8.

Stubbs, T. L. and Kennedy, A. C. (2012). Microbial weed control and microbial herbicides. In Alvarez-Fernandez, R. (Ed.), *Herbicides - Environmental Impact Studies and Management Approaches*. Chapter 8, pp. 135–66, DOI:10.5772/32705, 20 January 2012, InTech-Open Access Publisher: Rijeka, Croatia. http://www.intechopen.com/books/herbicides-environmental-impact-studies-and-management-approaches/microbial-weed-control-and-microbial-herbicides. (Accessed 25 October 2016).

Suslow, T. V. and Schroth, M. N. (1982), Role of deleterious rhizobacteria as minor pathogens in reducing crop growth, *Phytopathology*, 72, 111–15.

Thill, D. C., Beck, K. G. and Callihan, R. H. (1984), The biology of downy brome (*Bromus tectorum*), *Weed Sci.*, 32, 7–12.

Tichich, R. and Doll, J. (2006), Field-based evaluation of a novel approach for infecting Canada thistle (*Cirsium arvense*) with *Pseudomonas syringae* pv. *Tagetis*, *Weed Sci.*, 54(1), 166–71.

U.S. Fire Administration (2012), Firefighter fatalities in the United States in 2011, FEMA, 70.

US EPA (2016), 'Biopesticides', *EPA Pesticides*, https://www.epa.gov/pesticides/biopesticides. (Accessed 1 August 2016).

USGS (2002), Born of fire- restoring sagebrush steppe, USGS FS-126-02.

Wapshere, A. (1974), A strategy for evaluating the safety of organisms for biological weed control, *Ann. Appl. Biol.*, 77(2), 201–11.

Wisdom, M. J. and Chambers, J. C. (2009), A landscape approach for ecologically based management of great basin shrublands, *Restor. Ecol.*, 17(5), 740–9.

Zermane, N., Souissi, T., Kroschel, J. and Sikora, R. (2007), Biocontrol of broomrape (*Orobanche crenata* Forsk. and *Orobanche foetida* Poir.) by *Pseudomonas fluorescens* isolate Bf7-9 from the faba bean rhizosphere, *Biocontrol Sci. Technol.*, 17(5–6), 483–97.

Chapter 21

The use of insects in integrated weed management

Sandrine Petit and David A. Bohan, UMR Agroécologie, AgroSup Dijon, INRA, Université de Bourgogne Franche-Comté, France

1 Introduction

Many types of herbivorous insects feed on arable weeds, among which are leaf- and stem eaters, stem gallers, stem and root collar miners, capitula miners and seed feeders. The use of insects as biocontrol agents has mostly been investigated with a view of targeting exotic invasive weeds, and only to a limited extent has weed biocontrol been investigated in agricultural habitats. Numerous release programmes have been launched in Australia, New Zealand and North America over the last 100 years. Their success essentially hinges on agent establishment, effectiveness of control of the target weed and the risks to non-target plants. However, while there has been some success in controlling agricultural weeds, including control of tansy ragwort, *Senecio jacobaea* L., in New Zealand following the introduction of the flea beetle *Longitarsus flavicornis* (Suckling, 2013), the general consensus has been that these programmes have yielded highly variable results. In addition, reports of expansion of the range of plant hosts used by introduced biocontrol agents call into question the veracity of pre-release studies to evaluate the risks of an introduction for non-target organisms (Shaffner, 2001).

An alternative option to introducing alien biocontrol agents is to rely on insects that occur naturally at high abundance in arable agricultural fields, in what has been termed conservation biological control. Here the goal is to conserve and/or enhance populations

http://dx.doi.org/10.19103/AS.2017.0025.23

of biocontrol agents, through appropriate management, to promote the control of pests. Conservation biological control is often not targeted directly at the pest, as the naturally occurring insects tend to be less selective in their prey. Rather, the conservation management supports communities of insects that could exert effects on weeds. The hypothesis, which is rather similar to the *portfolio effect*, is that amongst the community there will be at least one insect species capable of exerting control of weeds. To date, possibly the best documented case is the interaction between the community of seed-eating carabid beetles and arable weeds. Seventy-four papers on this topic were identified from a literature review of the *Web of Science* (using 'weed seed predation' and 'carabid*' as search terms), with a steady increase since the first paper published in 1997; 27 papers were published between 2013 and 2016.

Carabid beetles are generalist predators that are very abundant in arable fields; many species contribute substantially to post-dispersal weed seed predation and could therefore represent valuable agents of weed control in agro-ecosystems (Honěk et al., 2003; Kulkarni et al., 2015). In contrast to herbicides, weed seed predators cannot robustly control standing weed numbers down to agreed sub-economic thresholds of damage to the crop. Rather, because seed predators affect weed seeds before they can germinate to become pestiferous, weed seed predation research aims to understand the relationship between carabid numbers and weed seed regulation, which can be defined as levels of predation that reduce the growth rate of the weed seedbank to below the replacement value (unity). The predation occurs in brief and intense periods of seed consumption (pulses) that follow recurrent weed seed rain episodes, and before seeds are protected by burial by the meteorological action of rain and wind (Westerman et al., 2009; Davis and Raghu, 2010). Carabid species exhibit preferences in the seeds they consume (Honěk et al., 2003; Petit et al., 2014), determined by the size of seeds and the body size of the predators. Larger carabids tend to consume larger weed seeds (Honěk et al., 2007). There is clear, but anecdotal, evidence of a role of other factors in mechanisms driving carabid preferences, such as the nature of resources stored in the seed, the thickness of the seed coat and the presence of carabid-predator and -competitor species.

The regulation of pests that results from the activity of naturally present predators is frequently cited as an important ecosystem service in arable agriculture (Losey and Vaughan, 2006). To date, however, few natural enemy functions have been demonstrated to elicit regulation or have been applied with robustness and generality in real agro-ecosystems. For weed regulation and control, three key issues need to be addressed to improve our ability to predict the intensity and resilience of weed seed predation and to foster wider adoption of conservation biocontrol management options in commercial fields. Specifically, we need to:

- Improve understanding of the complexity and variability in the structure of weed seed/seed predator trophic networks. Who eats whom?
- Determine the combination of in-field and landscape management options that promote weed seed predators and predation;
- Predict the effect of weed seed predation on the demography of weeds – how much control can be expected.

2 Deciphering complex interactions with generalist predator communities

The body of research currently available suggests that complex multi-trophic interactions influence the fate of weed seeds. The hypothesis that increasing seed-eating predator abundance may lead to enhanced weed suppression in many agricultural systems is widely accepted (Gallandt et al., 2005; Westerman et al., 2005). While in some instances seed-eating carabid abundance has been positively related to seed predation level (Menalled et al., 2007) or change in the weed seedbank (Bohan et al., 2011), other studies have failed to demonstrate such links (Davis and Raghu, 2010; Gaines and Claudio Gratton, 2010; Mauchline et al., 2005; Saska et al., 2008). The lack of a clear numerical response of weed predation to seed-eating carabid abundance possibly reflects differences in the species present in the experiment, and insufficient consideration may have been given to the diversity of trophic and functional roles within this group. The relative importance of 'granivorous' species, over omnivorous and carnivorous carabids, for the delivery of weed seed predation has been well studied (Trichard et al., 2013), even though granivorous carabids may represent only a small proportion of carabid community and many granivores may also feed on animal prey (Haschek et al., 2012).

Beyond the abundance of seed-eating carabids, carabid diversity may also drive the amount of seeds consumed in arable fields (Gaines and Claudio Gratton, 2010; Trichard et al., 2013). Increased species richness leads to an increase in the diversity of carabid body sizes and therefore an increase in the range of seeds consumed (Honěk et al., 2007). Such carabid diversity effects are also consistent with reinforcement of preferential associations between some granivorous functional groups and key weed functional groups, as established at a national scale by Brooks et al. (2012). For granivorous carabids, there appears to be a positive effect of seed predator diversity whereby resource use differences among species could lead to increased prey suppression by diverse communities (Finke and Snyder, 2010).

Despite a growing understanding that the variation in the interactions between carabids and weeds is important for explaining weed regulation, our ability to predict the intensity and resilience of weed seed predation delivered by carabid communities is limited by additional factors that can modulate these interactions. First, it is likely that density-dependent feeding effects are at play (Cardina et al., 1996; Cromar et al., 1999). Carabid satiation during periods of peak weed seed rain could have a large effect on the potential of seed-eating carabids to control weeds at certain times (Davis and Raghu, 2010). Overall seed consumption may therefore be affected by the density (population) of weed seeds and carabids (Frank et al., 2011). Second, there is evidence of intra-guild predation in carabid communities (Currie et al., 1996). Omnivorous carabids feed on one another and there are likely to be additional indirect (non-trophic) effects. Carabids might alter their behaviour, in response to the risk of intra-guild predation, by lowering their level of activity or by leaving prime foraging locations, in order to reduce their individual risk (Prasad and Snyder, 2004; Guy et al., 2008). Finally, several studies suggest the existence of shifts in diet, in seed-eating carabids, which depend upon the prevailing conditions (Marino et al., 2005; Mauchline et al., 2005; Brooks et al., 2012); indeed field surveys indicate that specific spatial associations vary in time (Trichard et al., 2014). Such prey switching has been found for carabids that can, for example, feed on weed seeds and animal pests such as slugs (Fig. 1). Perhaps more surprisingly, species that are often described as mostly 'granivorous' can feed nearly exclusively on animal prey depending upon the context. The nutritional status of *Amara similata* in oilseed rape fields, for example, has been shown to be positively related to the abundance of pollen beetle in the field (Haschek et al., 2012) and to decrease as insecticide use increases (Labruyere et al., 2016a). Alternative prey, and their availability, may therefore be key to the delivery of the weed seed predation service.

Recently, researchers have begun to examine, explicitly, the apparent complexity of interaction using trophic network approaches. Building food webs has proceeded using bibliographic approaches, to recover published data for the trophic interactions of carabids

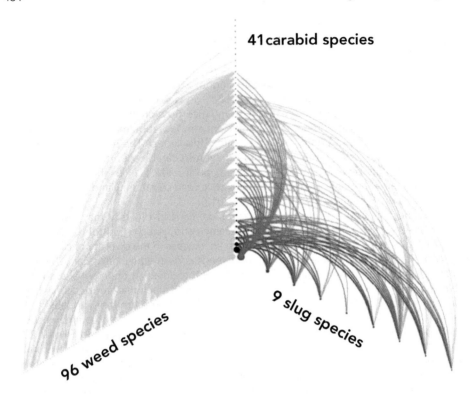

41 carabid species

9 slug species

96 weed species

Figure 1 Hive plot of a composite food web constructed from the literature, aiding the data from the farm-scale evaluations (Section 5).

and their prey (e.g. Fig. 1). Food webs constructed this way are subject to the vagaries of the quality of the literature. In any food web, there are 'recovered' links that will prove to be false, reflecting either an apparent but incorrectly evaluated trophic link in the literature or an inappropriate attribution of a feeding link in the bibliographic approach itself. The approach also cannot recover unpublished or unobserved links and so these food webs are likely to be incomplete.

Scrutiny of food webs constructed in this manner, which integrate the entirety of the carabid literature, supports many of the findings made to date from individual studies and underscores the complexity of interaction between ecological regulation functions that are commonly treated separately. In a network of carabids, slugs and weeds from arable agriculture, it is clear that there are a great many putative interactions already present in the literature (Fig. 1). The carabid species do not all share the same patterns of linkage. Some species are relatively specialist, concentrating on a few prey species. The majority of carabids appear quite generalist, being linked to a great many prey species. Among these generalists, some species are granivores, being linked only to plants, although most generalists appear to use a mixture of plant and animal prey species. There is also some evidence that predation of weeds by carabids is partly determined by carabid predation of slugs; there are interactions between these two ecological functions. The links to the weeds appear to be contingent upon the number of links to the slugs, and as the number of trophic links to the slugs increases, the number of links to the weeds declines. This suggests that prey switching depends upon the prevailing context of alternative prey.

The aim of this food web-based approach to carabid-weed interactions is to identify coherent groups of carabid species that deliver predictable trophic functions and may be managed together. While this work has begun using bibliographic approaches, the goal is to expand this research to include molecular trophic methods that much more precisely and robustly determine feeding interactions, exclude false-positive links and greatly increase the completeness of description of the trophic links in agriculture. Molecular approaches to carabid predation of prey commenced during the 1990s with monoclonal antibody-based approaches. The monoclonal antibodies were raised to particular proteins of a prey species, such as certain slug proteins (Symondson et al., 1997; Bohan et al., 2000), and used to screen the gut contents of sampled carabids for the presence of those proteins. While a great advance, the monoclonal approach suffered from both being limited to a species-by-species search for already-known prey and being destructive to the carabid predator. More recently, DNA-based approaches have begun to address these issues. Multiplex PCR primers allow many prey species to be screened simultaneously (King et al., 2011), and sampling carabid gut contents can now be done non-destructively using a method whereby individual carabids are provoked into regurgitating their stomach contents (Wallinger et al., 2015). The use of next-generation sequencing approaches, a catch-all name for a number of high-throughput sequencing approaches, also offers the potential for widening the assessment of prey species beyond those already known or suspected to affect plant species (Traugott et al., 2013). The reconstruction of ecological networks from DNA data is a growing topic of research (Vacher et al., 2016; Kamenova et al., 2017; Bohan et al., in press).

3 Managing fields and landscapes to enhance weed seed predation

A number of recent studies have attempted to identify in-field and landscape management options that might support predation of weed seeds. Managements that have been shown to substantially affect carabid richness, abundance or activity include in-field options and the compositional and structural aspects of the landscape surrounding fields, given that carabids are mobile organisms that respond at spatial scales much larger than the field (Kromp, 1999; Kulkarni et al., 2015).

Field management options can markedly affect weed seed fate. In general, weed seed predation has been found to be higher where in-field crop management intensity is lower, such as in no-till fields (Cromar, 1999; Menalled et al., 2007) and in fields with significant vegetation cover (Gallandt et al., 2005; Meiss et al., 2010; Sanguankeo and Leon, 2011). However, several studies have failed to detect any effect of in-field management on carabid abundance or weed seed predation, such as in organic farming (Diekötter et al., 2010; Jonason et al., 2013). In other situations, such as conservation agriculture, increased carabid abundance did not translate into enhanced weed seed predation, possibly because of a concomitant increase in the availability of alternative prey in these systems (Trichard et al., 2014).

At landscape scales, the amount of particular habitats and the complexity in spatial arrangement of those habitats could have important effects on weed seed predation.

The proportion of organic farming at the landscape scale was shown to enhance the activity and increase the body size of granivorous carabid species, thus improving their potential to control arable weeds (Diekötter et al., 2016). The effect of landscape properties on weed seed predation per se is, however, poorly documented to date, with often equivocal, probably context-dependent results (Menalled et al., 2000; Trichard et al., 2013; Jonason et al., 2013). Moreover, in-field management effects appear conditional on the landscape and *vice versa*. For example, seed predation in organic fields is enhanced in complex landscapes, whereas predation in conventional fields is enhanced only in simple, relatively homogeneous farm landscapes (Fisher et al., 2011). A recent large-scale study also demonstrated that weed seed predation co-varied with both landscape composition and in-field management, that is, the duration since conversion to conservation agriculture (Petit et al., 2017). Levels of weed seed predation in fields that were managed in conservation agriculture for less than four years were dependent on the properties of the surrounding landscape. Conversely, fields in conservation agriculture for four years, or more, tended to have higher weed seed predation with only a very limited effect due to the landscape context (Petit et al., 2017).

From an applied perspective, conservation management options that would promote weed seed predation remain unclear. In addition, management strategies aiming at supporting carabids across a farmland landscape will likely also benefit other organisms (Gonthier et al., 2014). Given that most seed-eating carabids are generalist predators (Sunderland, 2002; Tooley and Brust, 2002), any increase in the range of alternative prey available for carabids due to landscape management will permit prey switching and thus threaten the resilient delivery of weed seed regulation services.

This complexity of interaction between the biotic and abiotic elements of the agro-ecosystem makes the controlled study of carabid weed seed predation and weed regulation both difficult and imperative. In addition to the classical field manipulation experiments that have been done to date, we imagine the future use of long-term monitoring approaches and more finely controlled mesocosm experiments as a possible solution to elucidating the mechanisms that determine weed seed predation. Long-term monitoring might use a standard suite of methods to sample carabid abundance and estimate seed predation. Critically, this would be done in replicated landscapes across a gradient of landscape and management diversities that reflect the current farming situation. In France, the national monitoring network *SEBIOPAG* might provide a model for this kind of work (http://sebiopag.inra.fr). Mesocosms, which are mid-scale experimental systems that are used to evaluate the field environment under controlled conditions (e.g. Stewart et al., 2013), could be used to much more finely manipulate biotic conditions to estimate their effect on weed seed predation. In particular, both the diversity of cover crop plants, which are increasingly used in European agriculture, and alternative prey abundance and diversity could be precisely manipulated to understand their interaction, the lack of which currently renders the prediction of conservation management for seed predation so difficult.

4 Extent of regulation

The episodic nature of weed seed predation over the course of a year makes the estimation of annual seed loss due to seed predation in the field highly variable and subject to error without careful repeated measurement (Westerman et al., 2003). From ten published datasets, Davis et al. (2011) estimated that annual seed losses due to invertebrates averaged 40% and ranged from 8% to 70% depending on the weed species and the agronomic context. Subsequent studies on weed seed predation by invertebrates have yielded estimates within this range. The annual rate of seed depletion by invertebrate predators can therefore be substantial, but there has been little documentation of the effect of seed predation on weed regulation and control. Evidence that rates of predation reported in the literature can affect the demography of particular weed species is still scarce, and few modelling studies have addressed this question. Results suggest that an annual seed loss of 25–50% may be enough to slow down weed population growth substantially (Firbank and Watkinson, 1985; Westerman et al., 2005). Empirical evidence of effective regulation of the weed seedbank by carabid beetles is even scarcer; one of the few examples is the case study in Section 5. This lack of empirical evidence is the key information gap that remains to be filled. Demonstrating that ecological processes could be employed to replace herbicides and quantifying how much can be expected from these ecological processes represents the primary obstacle to the wider use of weed seed predation in agriculture.

5 Case study: the UK national survey farm-scale evaluation

One major limitation for a wider adoption of conservation biological control in arable agriculture is the lack of evidence and of quantification of the regulatory effect that can be expected. Establishing that carabids can elicit regulatory effects on the weed seedbank over a number of years and from fields undergoing the full range of management used in real-world farming would represent a major leap forward. Here, we present a demonstration case study, built on data collected during the farm-scale evaluation (FSE) – a UK national-scale field experiment done between 2000 and 2004.

5.1 The FSE data

The data on the abundance of the weed seedbanks, seed rain and carabids were collected from 66 spring-sown beet (both sugar and fodder), 59 spring maize, 67 spring oilseed rape and 65 winter oilseed rape fields (Bohan et al. 2005; Champion et al., 2003; Fig. 2). The fields were spread across the regions of the United Kingdom and sampled for one cropping year (Firbank et al., 2003) between 2000 and 2004. The pitfall trapping of carabids was conducted in the spring, summer and the late summer for the spring-sown crops and in the autumn, spring and summer for winter oilseed rape (Brooks et al., 2003; Bohan et al., 2005). Abundance of each carabid species, total carabids and the carabid functional groups of granivore and omnivore was available for each field. A total of 374 638 individuals of 126 species were identified. Seedbank samples were taken just prior to sowing in two consecutive years (Heard et al., 2003). The seedbank counts were pooled to give an estimate of the seedbank in each field (total weeds). A total of 38 402 seeds were sampled in the initial seedbank and 52 662 seeds were sampled in the follow-up seedbank, representing some 201 taxa. The return of weed seed to the seedbank (seed rain) was also measured using seed rain traps (Heard et al., 2003). In total, 508 777 seeds were shed into seed rain traps, and 211 taxa were identified.

Figure 2 Distribution of the 66 spring-sown beet, 59 spring maize, 67 spring oilseed rape and 65 winter oilseed rape fields sampled as part of the FSE.

5.2 Detecting signals of weed regulation in arable crops

The analysis tested for the existence of a negative relationship between the change in the weed seedbank over a year and the abundance of carabids in that particular year, all other factors being equal (Bohan et al., 2011). Here, the seedbank is expected to change as weed seeds are shed (as seed rain from seeding plants) and return to the soil. If some of this seed rain is intercepted at the soil surface and eaten by seed predator carabids, it will reduce the amount returned to the seedbank. Where the interception rate is high and enough seeds are eaten, there will be a net decline in the seedbank over the year.

Analysis of the FSE dataset revealed a significant negative coefficient between the abundance of omnivores and the change in the monocotyledon weed seedbank from one year to the next, across all crops (Fig. 3a). A similar relationship could be established between the abundance of granivorous carabids and the change in the monocotyledon weed seedbank change, across all crops (Fig. 3b). That seedbank change was also found to be positively related to seed rain abundance and the abundance of carabids was positively related to total seed rain abundance. This suggests that the predation model of carabid weed seed regulation that was tested for, namely that carabid-intercepted seed rain caused a change in the seedbank, was consistent with the data. The carabid beetles appeared to regulate the weed seedbank.

5.3 Identifying multiple-scale drivers of seed-eating carabid abundance in arable crops

The relative contribution of in-field and landscape-scale management factors on the abundance of nine individual seed-eating carabid species was then assessed on a subset of 161 fields from the FSE dataset (Labruyere et al., 2016b). In-field management was described with a composite indicator accounting for crop yield, quantity of inputs, field size and the presence of beetle banks or conservation headlands. The effect of land use outside the focal field focused on two specific land use types that were hypothesized

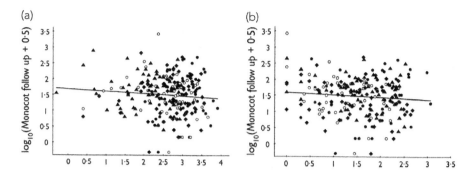

Figure 3 Multiple linear regression model fits between (a) the monocotyledon weed seedbank change and the abundance of omnivorous, and (b) the monocotyledon weed seedbank change and the abundance of granivorous (right plot) carabids in spring-sown beet (●), spring maize (○), spring oilseed rape (◆) and winter-sown oilseed rape (▲). Reproduced from Bohan et al. (2011).

to provide seed resources: oilseed rape crops and grassland. For each field, the study quantified (i) the presence of oilseed rape or grassland adjacent to the crop (neighbouring effect) and (ii) the proportion cover of these land use types in a 5 km² grid centred on the focal field (wider landscape effects).

The analysis revealed that the in-field carabid abundance responds to the spatial distribution of agricultural land use at three spatial scales. Locally, crop type strongly affected the abundance of individual species, in a species-specific manner, and the intensity of field management had a negative effect on the abundance of several species, irrespective of their trophic guild. In the neighbourhood, the occurrence of oilseed rape and grasslands was found to decrease the abundance of generalist seed-eating carabids, but it was the main factor positively affecting the abundance of the mainly granivorous *Amara aenea*, which occurred almost exclusively in our sampled oilseed rape fields. At the landscape scale, a generally positive effect of the cover of grassland and oilseed rape on in-field carabid abundance was detected, a finding which suggests that these habitats may provide alternative trophic resources to carabids, either seeds for granivorous carabids or alternative prey items for polyphagous species. These findings suggest that conservation management options would best be implemented at multiple spatial scales if used to promote weed seed predation in arable fields.

6 Conclusion

There is a critical need to assure future food security. Increasing emphasis will be placed on greater crop productivity while reducing environmental effects and reliance on chemical use in modern agriculture. Herbicides remain the predominant means of weed control, for instance, herbicides accounted for 42.3% of all pesticides used in Europe in 2010 (FAO). Policy-driven changes in herbicide use may lead to increases in weed plant densities in arable fields and reductions in crop productivity (Kim et al., 2002) and, more generally, the economic performance of agriculture. The move away from chemical weed control will thus only be possible if either there is an acceptance of a decline in yield or ecological functions and ecosystem services are available, which can be managed and function well enough to substitute for chemical inputs. For farmers to adopt these alternatives to herbicides, it will be necessary to show that ecological functions and services can replace herbicides with little or no additional risk to crop yield, farm productivity or profit. There is, however, a gap between the perceived need and the empirical evidence. This is mostly because regulation effects are neither robust/resilient nor is their management predictable under field conditions. Ecosystem services and conservation management currently represent a risk that is readily avoided by the continued use of herbicides.

Research can contribute to the overall goal of reducing reliance on herbicide use in several ways. The primary aim should be to produce a functional understanding of the conservation managements necessary to deliver the ecosystem service of carabid weed seed predation in place of herbicides. It is expected that local and landscape-scale management of arable fields could be used to increase the abundance of carabids, but that concomitant modification of the abundance and diversity of alternative prey may interfere with the delivery of high and resilient weed biocontrol. Quantifying and understanding the underlying mechanisms that link conservation managements to the delivery of the weed seed predation service will require the use of a combination of approaches that encompass large-scale surveys, mesocosms and manipulative experiments in which trophic feeding links between carabids, weed and alternative prey species are precisely determined by molecular analysis of carabid regurgitates (Wallinger et al., 2015). Such studies would test whether a high biodiversity of weed seed predators will assure natural weed regulation in agriculture and should elucidate the trophic consequences of contrasted agricultural management options. It will also generate the knowledge necessary to guide farming management strategies and will have a direct effect on all stakeholders, policy makers, farmers and the public by demonstrating a general, robust delivery of ecosystem services that are i) cost-effective, ii) credible for farmers to adopt and iii) able to maintain productivity services in the face of policy changes to pesticide inputs (e.g. Ecophyto 2018).

7 Future trends

Currently, we do not know whether, or how effectively, carabids can regulate the weed seedbank. Evidence such as the national-scale study by Bohan et al. (2011) would suggest that carabids can intercept weed seed rain and thereby cause changes in the weed seedbank that are consistent with regulation. However, the stomach contents of the carabids were never investigated for weed seed remains and, consequently, it is possible that this pattern was merely an artefact. It is also true that measuring weed seedbank change in the field is extremely difficult and costly. This means that sampling the weed seedbank is rarely done and most studies use seed predation estimates from seed predation cards, to estimate the effect of carabids. There are clear gaps in our current understanding and estimation of seed predation and regulation. In the future, it will be necessary to demonstrate clearly, using molecular approaches, that carabid feeding changes the weed seedbank and that seed predation cards are valid proxies for weed seedbank regulation.

Weeds and carabids exist within a wider ecological network of agricultural species (Pocock et al., 2012). This network is actually a meta-network of a wide variety of ecological interactions. It includes direct interactions, such as competition and trophic interactions studied by ecologists. It also includes other rather less well-studied indirect interactions, such as the provision of shelter. Weeds play a key role in this meta-network providing both food and shelter resources to a wide variety of animal species, including carabids. For the future, an important line of research will be to fully describe this network and place all species in their appropriate context. The importance of this is that these links (interactions) lie at the core of ecological functions. Trophic and competition interactions between the species of the network become the regulatory functions and ecosystem services we wish to harness in biological control. Understanding how the diversity and structure of these agricultural meta-networks vary with local and landscape management, particularly directed at weeds, will lead to better prediction of the resilience of regulation and the management that can better support biocontrol.

From a study of one of the most complete agricultural meta-networks yet produced, Pocock et al. (2012) demonstrated that the different ecological functions varied in their robustness, with pollinators being particularly fragile to loss of weed species, but there was no strong co-variation in function because the different interaction types were often in conflict. This network-based approach revealed there was no 'optimist's scenario' or 'win-win' management that benefited both biodiversity and multiple ecosystem functions. In essence, this suggests that only a subset of ecological functions can be managed, which places a great emphasis on identifying those interactions that are in conflict. As already discussed in Section 2.1, carabid predation of weed seeds and carabid predation of slugs may be antagonistic and, by extension, all alternative prey could have such interference effects. There is a need, therefore, to evaluate the interactions and antagonisms between ecological interactions across a framework of a meta-network, and it appears to us that the developing technology of next-generation sequencing offers great promise for the identification of both direct and indirect interactions (Vacher et al., 2016; Kamenova et al., 2013; Bohan et al., 2017).

8 Acknowledgements

S. Petit and D. A. Bohan acknowledge the support of the FACCE ERA-NET C-IPM project BioAWARE.

9 Where to look for further information

A great introduction to weed seed predation by carabids is given in the review by Kulkarni et al. (2015), which covers many of the critical issues. Readers will find the paper an easy-to-read introduction to the subject and a great source of further reading. A more general introduction to seed predation and plant population dynamics can be found in Crawley (2000).

Beyond the large ecological organizations, such as the BES and ESA, discussion of weed seed predation by insects can be found across a variety of international societies and their meetings. Primary amongst these are the European Weed Research Society – *Weeds and biodiversity working group* (http://www.ewrs.org) and a number of working groups of the International Organisation for Biological Control (IOBC – http://www.iobc-global.org). Specific sessions on weed seed predation by carabids can be found at the meetings of the European Carabidologists (e.g. https://colloque.inra.fr/18ecm/The-18th-EuropeanCarbidologist-Meeting/18th-ECM-2017), which have occurred approximately every two years since 1969. More recent developments include international meetings on molecular approaches and ecological networks including the joint *Symposium on the Molecular Analysis of Trophic Interactions* and *Symposium on Ecological Networks* (e.g. http://www.slu.se/ecology-symposium), which also occur every second year.

There are a considerable number of research groups active or previously active in research into carabid predation, including carabid predation of weed seeds. A rapid tour of the World Wide Web using Google will highlight many of these groups, including those from the United States. Here, as a guide for such a tour of the web, we list five groups that are currently active and doing important research. The newly formed group of Klaus Birkhofer, at Brandenburgische Technische Universität (BTU) Cottbus – Senftenberg in Germany, works on understanding how farm management affects predation and predation-derived services in agriculture. At Wageningen, the Netherlands, Wopke van der Werf and colleagues study the statistical ecology of regulation services, and in particular weed regulation. Paula Westerman, at the University of Rostock, seeks to understand seed predation ecology using a variety of invertebrate and vertebrate weed seed predators, including carabids. Pavel Saska and colleagues, at the Czech Crop Research Institute, use classical and molecular ecological approaches to understand weed seed predation. At the University of Innsbruck, Austria, the world-renowned team of Michael Traugott develop molecular approaches to understand the trophic ecology of predation, including carabid trophic interactions with weeds.

10 References

Bohan, D. A., Bohan, A., Glen, D. M., Symondson, W. O. C. and Wiltshire, C. (2000). Spatial dynamics of predation by carabid beetles on slugs. *Journal of Animal Ecology*. 69, 367–79.

Bohan, D. A., Boffey, C. W. H., Brooks, D. R., Clark, S. J., Dewar, A. M., Firbank, L. G., Haughton, A. J., Hawes, C., Heard, M. S., May, M. J., Osborne, J. L., Perry, J. N., Rothery, P., Roy, D. B., Scott, R. J., Squire, G. R., Woiwod, I. P. and Champion, G. T. (2005). Effects on weed and invertebrate abundance and diversity of herbicide management in genetically modified herbicide-tolerant winter-sown oilseed rape. *Proceedings of the Royal Society of London Series B, Biological Sciences*, 272, 463–74.

Bohan, D. A., Boursault, A., Brooks, D. R., Petit, S. (2011). National-scale regulation of the weed seedbank by carabid predators. *Journal of Applied Ecology*, 48, 888–98.

Bohan, D. A., Vacher, C., Tamaddoni-Nezhad, A., Raybould, A., Dumbrell, A. J. and Woodward, G. (2017). Next-generation Global Biomonitoring – large-scale, automated reconstruction of ecological networks. *Trends in Ecology and Evolution*, 32, 477–87.

Brooks, D. R., Bohan, D. A., Champion, G. T., Haughton, A. J., Hawes, C., Heard, M. S., Clark, S. J., Dewar, A. M., Firbank, L. G., Perry, J. N., Rothery, P., Scott, R. J., Woiwod, I. P., Birchall, C., Skellern, M. P., Walker, J. H., Baker, P., Bell, D., Browne, E. L., Dewar, A. J. G., Fairfax, C. M., Garner, B. H., Haylock, L. A., Horn, S. L., Hulmes, S. E., Mason, N. S., Norton, L. R., Nuttall, P., Randle, Z., Rossall, M. J., Sands, R. J. N., Singer, E. J. and Walker, M. J. (2003). Invertebrate responses to the management of genetically modified herbicide-tolerant and conventional crops. I. Soil-surface-active invertebrates. *Proceedings of the Royal Society of London Series B-Biological Sciences*, 358, 1847–62.

Brooks, D. R., Storkey, J., Clark, S. J., Firbank, L. G., Petit, S. and Woiwod, I. P. (2012). Trophic links between functional groups of arable plants and beetles are stable at a national scale. *Journal of Animal Ecology*, 81, 4–13.

Cardina, J., Norquay, H. M.; Stinner, B. R. and McCartney, D. A. (1996). Postdispersal predation of velvetleaf (*Abutilon theophrasti*) seeds. *Weed Science*, 44, 534–9.

Champion, G. T., May, M. J., Bennett, S., Brooks, D. R., Clark, S. J., Daniels, R. E., Firbank, L. G., Haughton, A. J., Hawes, C., Heard, M. S., Perry, J. N., Randle, Z., Rossall, M. J., Rothery, P., Skellern, M. P., Scott, R. J., Squire, G. R. and Thomas, M. R. (2003). Crop management and agronomic context of the Farm Scale Evaluations of genetically modified herbicide-tolerant crops. *Philosophical Transactions of The Royal Society of London, Series B*, 358, 1801–18.

Crawley, M. J. (2000). Seed predators and plant population dynamics. In M. Fenner (ed.), *Seeds: The Ecology of Regeneration in Plant Communities*. CABI, New York, USA, pp. 167–82.

Cromar, H. E., Murphy, S. D. and Swanton, C. J. (1999). Influence of tillage and crop residue on post dispersal predation of weed seeds. *Weed Science*, 47, 184–94.

Currie, C. R., Spence, J. R. and Niemelä, J. (1996). Competition, cannibalism and intra guild predation among ground beetles (Coleoptera: Carabidae): a laboratory study. *The Coleopterists Bulletin*, 50, 135–48.

Davis, A. S. and Raghu, S. (2010). Weighing abiotic and biotic influences on weed seed predation. *Weed Research*, 50, 402–12.

Davis, A. S., Daedlow, D., Schutte, B. J. and Westerman, P. R. (2011). Temporal scaling of episodic point estimates of seed predation to long-term predation rates. *Methods in Ecology and Evolution*, 2, 682–890.

Diekötter, T., Wamser, S., Wolters, V. and Birkhofer, K. (2010). Landscape and management effects on structure and function of soil arthropod communities in winter wheat. *Agriculture, Ecosystems and Environment*, 137, 108–12.

Diekötter, T., Wamser, S., Dörner, T. Wolters, V. and Birkhofer, K (2016). Organic farming affects the potential of a granivorous carabid beetle to control arable weeds at local and landscape scales. *Agricultural and Forest Entomology*, 18, 167–73.

Finke, D. L. and Snyder, W. E (2010). Conserving the benefits of predator biodiversity. *Biological conservation*, 143, 2260–9.

Firbank L. G. and Watkinson A. R. (1985). On the analysis of competition within two-species mixtures of plants. *Journal of Applied Ecology*. 22, 503–17.

Firbank, L. G., Heard, M. S., Woiwod, I. P., Hawes, C., Haughton, A. J., Champion, G. T., Scott, R .J., Hill, M. O., Dewar, A. M., Squire, G. R., May, M. J., Brooks, D. R., Bohan, D. A., Daniels, R. E., Osborne, J. L., Roy, D. B., Black, H. I. J., Rothery, P, and Perry, J. N. (2003). An introduction to the Farm-Scale Evaluations of genetically modified herbicide-tolerant crops. *Journal of Applied Ecology*, 40, 2–16.

Fischer, C., Thies, C. and Tscharntke, T. (2011). Mixed effects of landscape complexity and farming practice on weed seed removal. *Perspectives in Plant Ecology, Evolution and Systematics*, 13, 297–303.

Frank, S. D., Shrewsbury P. M. and Denno, R. F. (2011). Plant versus prey resources: Influence on omnivore behaviour and herbivore suppression. *Biological Control*, 57, 229–35.

Gaines, H. R. and Claudio Gratton, C. (2010). Seed predation increases with ground beetle diversity in a Wisconsin (USA) potato agroecosystem. *Agriculture, Ecosystems and Environment*, 137, 329–36.

Gallandt, E. R., Molloy, T., Lynch, R. P. and Drummond, F. A. (2005). Effect of cover-cropping systems on invertebrate seed predation. *Weed Science*, 53, 69–76.

Gonthier, D., Ennis, K., Farinas, S., Hsieh, H.-Y., Iverson, A., Batáry, P., Rudolphi, J., Tscharntke, T., Cardinale, B. and Perfecto, I. (2014). Biodiversity conservation in agriculture requires a multi-scale approach. *Proceedings of the Royal Society B: Biological Sciences*, 281 (1791).

Guy, A., Bohan, D. A., Powers, S. J. and Reynolds, A. M. (2008). Avoidance of conspecific odour by carabid beetles: a mechanism for the emergence of scale-free searching patterns. *Animal Behaviour*, 76, 585–91.

Haschek, C., Drapela, T., Schuller, N., Fiedler, K., Frank, T. (2012). Carabid beetle condition, reproduction and density in winter oilseed rape affected by field and landscape parameters. *Journal Applied Entomology*. 136, 665–74.

Heard, M. S., Hawes, C., Champion, G. T., Clark, S. J., Firbank, L. G., Haughton, A. J., Parish, A. M., Perry, J. N., Rothery, P., Scott, R. J., Skellern, M. P., Squire, G. R. and Hill, M. O. (2003). Weeds in fields with contrasting conventional and genetically modified herbicide-tolerant crops I. Effects on abundance and diversity. *Philosophical Transactions of The Royal Society of London, Series B*, 358, 1819–33.

Honěk, A., Martinkova, Z. and Jarošík, V. (2003). Ground beetles (Carabidae) as seed predators. *European Journal of Entomology*. 100, 531–44.

Honěk, A., Martinkova, Z., Saska, P. and Pekar, S. (2007). Size and taxonomic constraints determine seed preferences of Carabidae (Coleoptera). *Basic and Applied Ecology*, 8, 343–53.

Jonason, D., Smith, H. G., Bengtsson, J. and Birkhofer, K. (2013). Landscape simplification promotes weed seed predation by carabid beetles (Coleoptera: Carabidae). *Landscape Ecology*, 28, 487–94.

Kamenova, S., Bartley, T. J., Bohan, D., Boutain, J. R., Colautti, R. I., Domaizon, I., Fontaine, C., Lemainque, A., Le Viol, I., Mollot, G., Perga, M. E., Ravigné, V. and Massol, F. (2017). Invasions toolkit: current methods for tracking the spread and impact of invasive species. *Advances in Ecological Research*, 56, 85–182.

Kim, D. S., Brain, P., Marshall, E. J. P. and Caseley, J. C. (2002). Modelling herbicide dose and weed density effects on crop:weed competition. *Weed Research*, 42, 1–13.

King, R. A., Moreno-Ripoll, R., Agusti, N., Shayler, S. P., Bell, J. R., Bohan, D. A. and Symondson, W. O. C. (2011). Multiplex reactions for the molecular detection of predation on pest and nonpest invertebrates in agroecosystems. *Molecular Ecology Resources*, 11, 370–3.

Kromp, B. (1999). Carabid beetles in sustainable agriculture: a review on pest control efficacy, cultivation impacts and enhancement. *Agriculture, Ecosystems and Environment*. 74, 187–228.

Kulkarni, S. S., Dosdall, L. M. and Willenborg, C. J. (2015). The role of ground beetles (Coleoptera: Carabidae) in weed seed consumption: a review. *Weed Science*, 63, 355–76.

Labruyere S., Ricci B., Lubac A. and Petit S. (2016a). Crop type, crop management and grass margins affect the abundance and the nutritional state of seed eating carabid species in arable landscapes. *Agriculture Ecosystems and Environment*, 231, 183–92.

Labruyere, S., Bohan, D. A., Biju-Duval, L., Ricci, B. and Petit, S. (2016b). Local, neighbor and landscape effects on the abundance of weed seed-eating carabids in arable fields: a nationwide analysis. *Basic and Applied Ecology*. 17, 230–9.

Losey, J. E. and Vaughan, M. (2006). The economic value of ecological services provided by insects. *BioScience*, 56, 311–23.

Marino, P. C., Westerman, P. R., Pinkert, C. and van der Werf, W. (2005). Influence of seed density and aggregation on post-dispersal weed seed predation in cereal fields. *Agriculture, Ecosystems and Environment*, 106, 17–25.

Mauchline, A. L., Watson, S. J., Brown, V. K. and Froud-Williams, R. J. (2005). Post-dispersal seed predation of non-target weeds in arable crops. *Weed Research*, 45, 157–64.

Meiss, H., Le Lagadec, L., Munier-Jolain, N., Waldhardt, R. and Petit, S. (2010). Weed seed predation increases with vegetation cover in perennial forage crops. *Agriculture, Ecosystems and Environment*, 138, 10–16.

Menalled, F. D., Marino, P. C., Renner, K. A. and Landis, D. A. (2000). Post-dispersal weed seed predation in Michigan crop fields as a function of agricultural landscape structure. *Agriculture, Ecosystems and Environment*, 77, 193–202.

Menalled, F. D., Smith, R. G., Dauer, J. T. and Fox, T. B. (2007). Impact of agricultural management on carabid communities and weed seed predation. *Agriculture, Ecosystems and Environment*, 118, 49–54.

Petit, S., Boursault, A. and Bohan, D. A. (2014). Weed seed choice by carabid beetles (Coleoptera: Carabidae): Linking field measurements with laboratory diet assessments. *European Journal of Entomology*, 111, 1–6.

Petit, S., Trichard, A., Biju-Duval L., McLaughlin, Ó. B. and Bohan, D. A. (2017). Interactions between conservation agricultural practice and landscape composition promote weed seed predation by invertebrates. *Agriculture, Ecosystems and Environment*, 240, 45–53.

Pocock, M., Evans, D. and Memmott, J. (2012). The robustness and restoration of a network of ecological networks. *Science*, 335, 973–7.

Prasad, R. P. and Snyder, W. E. (2004). Predator interference limits fly egg biological control by a guild of ground-active beetles. *Biological Control*, 31, 428–37.

Sanguankeo, P. P. and Leon, R. G. (2011). Weed management practices determine plant and arthropod diversity and seed predation in vineyards. *Weed Research*, 51, 404–12.

Saska, P., Van der Werf, W., de Vries, E. and Westerman, P. R. (2008). Spatial and temporal patterns of carabid activity-density in cereals do not explain levels of predation on weed seeds. *Bulletin of Entomological Research*, 98, 169–81.

Shaffner, U. (2001). Host range testing of insects for biological weed control: How can tt be better interpreted? *BioScience* 51, 951–9.

Stewart, R. I. A., Dossena, M., Bohan, D. A., Jeppesen, E., Kordas, R. L., Ledger, M. E., Meerhoff, M., Moss, B., Mulder, C., Shurin, J. B., Suttle, B., Thompson, R., Trimmer, M. and Woodward, G. (2013). Mesocosm experiments as a tool for ecological climate-change research. In G. Woodward and E. J. O'Gorman (eds), *Advances in Ecological Research*, 48, 71–181.

Suckling, D. M. (2013). Benefits from biological control of weeds in New Zealand range from negligible to massive: a retrospective analysis. *Biological Control*, 66, 27–32.

Sunderland, K. (2002). *Invertebrate Pest Control by Carabids. The Agroecology of Carabid Beetles*. J. Holland. Andover, Intercept Publishers, pp. 165–214.

Symondson, W. O. C., Erickson, M. L. and Liddell, J. E. (1997). Species-specific detection of predation by Coleoptera on the milacid slug *Tandonia budapestensis* (Mollusca: Pulmonata). *Biocontrol Science and Technology*, 7, 457–465.

Tooley, J. and Brust, G. E. (2002). Weed seed predation by carabid beetles. In J. M. Holland (ed.), *The Agroecology of Carabid Beetles*, Intercept, Andover, pp. 215–29.

Traugott, M., Kamenova, S., Ruess, L., Seeber, J. and Plantegenest, M. (2013). Empirically characterising trophic networks: what emerging DNA-based methods, stable isotope and fatty acid analyses can offer. *Advances in Ecological Research*, 49, 177–224.

Trichard, A., Alignier, A., Biju-Duval, L. and Petit, S. (2013). The relative effects of local management and landscape context on weed seed predation and carabid functional groups. *Basic and Applied Ecology*, 14, 235–54.

Trichard, A., Ricci, B., Ducourtieux, C. and Petit, S. (2014). The spatio-temporal distribution of weed seed predation differs between conservation agriculture and conventional tillage. *Agriculture, Ecosystems and Environment*. 188, 40–7.

Vacher, C., Tamaddoni-Nezhad, A., Kamenova, S., Peyrard, N., Moalic, Y., Sabbadin, R., Schwaller, L., Chiquet, J., Alex Smith, M., Vallance, J., Fievet, V., Jakuschkin, B. and Bohan, D. A. (2016). Learning ecological networks from next-generation sequencing data. *Advances in Ecological Research*, 54, 1–39.

Wallinger, C., Sint, D., Baier, F., Schmid, C., Mayer, R. and Traugott, M. (2015). Detection of seed DNA in regurgitates of granivorous carabid beetles. *Bulletin of Entomological Research*, 105: 728–35.

Westerman, P. R., Wes, J. S., Kropff, M. J. and Van der Werf, W. (2003). Annual losses of weed seeds due to predation in organic cereal fields. *Journal of Applied Ecology*, 40, 824–36.

Westerman, P. R., Liebman, M., Menalled, F. D., Heggenstaller, A. H., Hartzler, R. G. and Dixon, P. M. (2005). Are many little hammers effective? Velvetleaf (*Abutilon theophrasti*) population dynamics in two- and four-year crop rotation systems. *Weed Science*, 53, 382–92.

Westerman, P. R., Dixon, P. M. and Liebman, M. (2009). Burial rates of surrogate seeds in arable fields. *Weed Research*, 49, 142–52.

Index

CPSIA information can be obtained
at www.ICGtesting.com
Printed in the USA
BVOW07*0750201217
503204BV00014B/4/P